CAMBRIDGE STUDIES IN ADVANCED MATHEMATICS 180

HIGHER CATEGORIES AND HOMOTOPICAL ALGEBRA

This book provides an introduction to modern homotopy theory through the lens of higher categories after Joyal and Lurie, giving access to methods used at the forefront of research in algebraic topology and algebraic geometry in the twenty-first century. The text starts from scratch – revisiting results from classical homotopy theory such as Serre's long exact sequence, Quillen's theorems A and B, Grothendieck's smooth/proper base change formulas and the construction of the Kan–Quillen model structure on simplicial sets – and develops an alternative to a significant part of Lurie's definitive reference *Higher topos theory*, with new constructions and proofs, in particular, the Yoneda lemma and Kan extensions. The strong emphasis on homotopical algebra provides clear insights into classical constructions such as calculus of fractions, homotopy limits and derived functors, which are revisited in this enhanced context.

For graduate students and researchers from neighbouring fields, this book is a user-friendly guide to the advanced tools that the theory provides for applications in such areas as algebraic geometry, representation theory, algebra and logic.

Denis-Charles Cisinski is Professor of Mathematics at the University of Regensburg, Germany. His research focuses on homotopical algebra, category theory, K-theory and the cohomology of schemes. He is also the author of a monograph entitled *Les préfaisceaux comme modèles des types d'homotopie*.

CAMBRIDGE STUDIES IN ADVANCED MATHEMATICS

All the titles listed below can be obtained from good booksellers or from Cambridge University Press. For a complete series listing, visit www.cambridge.org/mathematics.

Already Published
140 R. Pemantle & M. C. Wilson *Analytic Combinatorics in Several Variables*
141 B. Branner & N. Fagella *Quasiconformal Surgery in Holomorphic Dynamics*
142 R. M. Dudley *Uniform Central Limit Theorems (2nd Edition)*
143 T. Leinster *Basic Category Theory*
144 I. Arzhantsev, U. Derenthal, J. Hausen & A. Laface *Cox Rings*
145 M. Viana *Lectures on Lyapunov Exponents*
146 J.-H. Evertse & K. Győry *Unit Equations in Diophantine Number Theory*
147 A. Prasad *Representation Theory*
148 S. R. Garcia, J. Mashreghi & W. T. Ross *Introduction to Model Spaces and Their Operators*
149 C. Godsil & K. Meagher *Erdős–Ko–Rado Theorems: Algebraic Approaches*
150 P. Mattila *Fourier Analysis and Hausdorff Dimension*
151 M. Viana & K. Oliveira *Foundations of Ergodic Theory*
152 V. I. Paulsen & M. Raghupathi *An Introduction to the Theory of Reproducing Kernel Hilbert Spaces*
153 R. Beals & R. Wong *Special Functions and Orthogonal Polynomials*
154 V. Jurdjevic *Optimal Control and Geometry: Integrable Systems*
155 G. Pisier *Martingales in Banach Spaces*
156 C. T. C. Wall *Differential Topology*
157 J. C. Robinson, J. L. Rodrigo & W. Sadowski *The Three-Dimensional Navier–Stokes Equations*
158 D. Huybrechts *Lectures on K3 Surfaces*
159 H. Matsumoto & S. Taniguchi *Stochastic Analysis*
160 A. Borodin & G. Olshanski *Representations of the Infinite Symmetric Group*
161 P. Webb *Finite Group Representations for the Pure Mathematician*
162 C. J. Bishop & Y. Peres *Fractals in Probability and Analysis*
163 A. Bovier *Gaussian Processes on Trees*
164 P. Schneider *Galois Representations and* (φ, Γ)*-Modules*
165 P. Gille & T. Szamuely *Central Simple Algebras and Galois Cohomology (2nd Edition)*
166 D. Li & H. Queffelec *Introduction to Banach Spaces, I*
167 D. Li & H. Queffelec *Introduction to Banach Spaces, II*
168 J. Carlson, S. Müller-Stach & C. Peters *Period Mappings and Period Domains (2nd Edition)*
169 J. M. Landsberg *Geometry and Complexity Theory*
170 J. S. Milne *Algebraic Groups*
171 J. Gough & J. Kupsch *Quantum Fields and Processes*
172 T. Ceccherini-Silberstein, F. Scarabotti & F. Tolli *Discrete Harmonic Analysis*
173 P. Garrett *Modern Analysis of Automorphic Forms by Example, I*
174 P. Garrett *Modern Analysis of Automorphic Forms by Example, II*
175 G. Navarro *Character Theory and the McKay Conjecture*
176 P. Fleig, H. P. A. Gustafsson, A. Kleinschmidt & D. Persson *Eisenstein Series and Automorphic Representations*
177 E. Peterson *Formal Geometry and Bordism Operators*
178 A. Ogus *Lectures on Logarithmic Algebraic Geometry*
179 N. Nikolski *Hardy Spaces*

Higher Categories and Homotopical Algebra

DENIS-CHARLES CISINSKI
Universität Regensburg, Germany

CAMBRIDGE
UNIVERSITY PRESS

CAMBRIDGE
UNIVERSITY PRESS

University Printing House, Cambridge CB2 8BS, United Kingdom

One Liberty Plaza, 20th Floor, New York, NY 10006, USA

477 Williamstown Road, Port Melbourne, VIC 3207, Australia

314-321, 3rd Floor, Plot 3, Splendor Forum, Jasola District Centre, New Delhi - 110025, India

79 Anson Road, #06-04/06, Singapore 079906

Cambridge University Press is part of the University of Cambridge.

It furthers the University's mission by disseminating knowledge in the pursuit of education, learning and research at the highest international levels of excellence.

www.cambridge.org
Information on this title: www.cambridge.org/9781108473200
DOI: 10.1017/9781108588737

First published 2019

A catalogue record for this publication is available from the British Library

Library of Congress Cataloging in Publication data
Names: Cisinski, Denis-Charles, author.
Title: Higher categories and homotopical algebra / Denis-Charles Cisinski (Universitat Regensburg, Germany).
Description: Cambridge ; New York, NY : Cambridge University Press, 2019. | Series: Cambridge studies in advanced mathematics | Includes bibliographical references and index.
Identifiers: LCCN 2018048001 | ISBN 9781108473200 (hardback : alk. paper)
Subjects: LCSH: Homotopy theory. | Algebra, Homological. | Categories (Mathematics) | Presheaves.
Classification: LCC QA612.7 .C5645 2019 | DDC 514/.24–dc23
LC record available at https://lccn.loc.gov/2018048001

ISBN 978-1-108-47320-0 Hardback

For Isaac and Noé

Contents

Preface

A Couple of Perspectives and a Tribute

The aim of this book is to introduce the basic aspects of the theory of ∞-categories: a homotopy-theoretic variation on category theory, designed to implement the methods of algebraic topology in broader contexts, such as algebraic geometry [TV05, TV08, Lur09, Lur17] or logic [Uni13, KL16, Kap17]. The theory of ∞-categories is not only a new approach to the foundations of mathematics: it appears in many spectacular advances, such as the proof of Weil's conjecture on Tamagawa numbers over function fields by Lurie and Gaitsgory, or the modern approach to p-adic Hodge theory by Bhatt, Morrow and Scholze, for instance.

For pedagogical reasons, but also for conceptual reasons, a strong emphasis is placed on the following fact: the theory of ∞-categories is a semantic interpretation of the formal language of category theory.[1] This means that one can *systematically* make sense of any statement formulated in the language of category theory in the setting of ∞-categories.[2]

We also would like to emphasise that the presence of homotopical algebra in this book is not as an illustration, nor as a source of technical devices: it is

[1] To be precise, the language of category theory is the one provided by a Cartesian closed category endowed with an involution $X \mapsto X^{\mathrm{op}}$, called the 'opposite category functor', a monoidal structure defined by a 'join operation' $*$, whose unit is the initial object, and which is symmetric up to the opposite operation: $X * Y = (Y^{\mathrm{op}} * X^{\mathrm{op}})^{\mathrm{op}}$. Furthermore, for each object Y, we have the slice functor, obtained as a right adjoint of the functor $X \mapsto (Y \to X * Y)$. Finally, there is a final object Δ^0, and we get simplices by iterating the join operation with it: $\Delta^n = \Delta^0 * \Delta^{n-1}$. Category theory is obtained by requiring properties expressed in this kind of language.

[2] There is, more generally, a theory of (∞, n)-categories: a semantic interpretation of the language of (strict) n-categories (for various ordinals n). The theory of ∞-categories as above is thus the theory of $(\infty, 1)$-categories. Although we shall not say more on these higher versions here, the interested reader might enjoy a look at Baez's lectures [BS10] on these topics.

at the core of basic category theory. In classical category theory, homotopical algebra seems peculiar, because classical homotopy categories do not have (co)limits and are not concrete (i.e. cannot be embedded in the category of sets in a nice way), as the fundamental case of the homotopy category of CW-complexes shows [Fre70]. This is partly why some traditions seem to put classical category theory and classical homotopy theory apart. The story that we want to tell here is that the theory of ∞-categories involves a reunion: with this new semantic interpretation, homotopy theories define ∞-categories with (co)limits, and the classical methods of category theory do apply to them (and the problem of concreteness disappears because ∞-groupoids take the role of sets, not by choice, but under the rule of universal properties). In particular, in this book, model categories will eventually be allowed to be ∞-categories themselves, and we shall observe that the localisation of a model category is also a model category, where the weak equivalences are the invertible maps and the fibrations are all maps (for the reader who might not be familiar with such a language, the present text aims at explaining what such a sentence is about). This means that homotopy theories and their models do live in the same world, which changes dramatically our perspective on them. Finally, one may see homotopical algebra as the study of the compatibility of localisations with (co)limits. And the semantics of ∞-categories makes this a little more savoury because it provides much more powerful and flexible statements. Moreover, the fact that the free completion of a small category by small colimits can be described as the homotopy theory of presheaves of spaces on this category puts homotopical algebra at the very heart of the theory of Kan extensions, and thus of category theory itself. This enlightens many classical results of the heroic days of algebraic topology, such as Eilenberg and Steenrod's characterisation of singular homology, for instance. In some sense, this is the natural outcome of a historical process. Indeed category theory was born as a convenient language to express the constructions of algebraic topology, and the fact that these two fields were separated is a kind of historical accident whose effects only started to fade in the late 1990s, with the rise of ∞-categories as we know them today, after the contributions of André Joyal, Carlos Simpson, Charles Rezk, Bertrand Toën and Gabriele Vezzosi, and of course Jacob Lurie. A pioneer of higher category theory such as Daniel M. Kan was aware of the very fact that category theory should extend to homotopy theory already in the 1950s, and his contributions, all through his mathematical life, through the theory of simplicial categories, with William Dwyer, and, more recently, through the theory of relative categories, with Clark Barwick, for instance, are there to testify to this. The title of this book is less about putting higher category theory and homotopy theory side by side, than observing that higher category theory and homotopical algebra are essentially the same thing. However, a better tribute to Daniel M. Kan might have been to call it *Category theory*, plain and simple.

A Glimpse at the Narrative

As we already wrote above, this text emphasises the fact that the theory of ∞-categories is a semantic interpretation of the language of category theory. But, when it comes to language, there is syntax. And, if category theory is full of identifications which are not strict, such as isomorphisms, equivalences of categories, or even wider notions of weak equivalences, this does not get better with the theory of ∞-categories, which has an even greater homotopy-theoretic flavour. However, the only identification known by syntax is the identity. In practice, this means that we have to introduce various rectification tools, in order to bring back categorical constructions into our favourite language. In Lurie's book [Lur09], which is the standard reference on the subject, by its quality and its scope, this rectification appears early in the text, in several disguises, in the form of Quillen equivalences relating various model structures (e.g. to compare Joyal's model category structure, which encodes the homotopy theory of ∞-categories, with Bergner's model category structure, which expresses Dwyer and Kan's homotopy theory of simplicial categories). These Quillen equivalences consist in introducing several languages together with tools to translate statements from one language to another (for instance, the language provided by the category of simplicial sets, which is used to describe the Joyal model structure, and the language of simplicial categories). This is all good, since we can then extract the most convenient part of each language to express ourselves. But these Quillen equivalences are highly non-trivial: they are complex and non-canonical. And since they introduce new languages, they make unclear which aspects of a statement are independent of the theory we chose to express ourselves.

There are many models to describe ∞-categories, in the same way that there are many ways to describe homotopy types of CW-complexes (such as Kan complexes, or sheaves of sets on the category of smooth manifolds). All these models can be shown to be equivalent. For instance, as already mentioned above, in Lurie's book [Lur09], the equivalence between Kan's simplicial categories and Joyal's quasi-categories is proved and used all through the text, but there are plenty of other possibilities, such as Simpson's Segal categories [Sim12], or Rezk's complete Segal spaces [Rez01]. A reference where one may find all these comparison results is Bergner's monograph [Ber18], to which we should add the beautiful description of ∞-categories in terms of sheaves on an appropriate category of stratified manifolds by Ayala, Francis and Rozenblyum [AFR17]. Riehl and Verity's ongoing series of articles [RV16, RV17a, RV17b] aim at expressing what part of this theory is model independent.

In the present book, we choose to work with Joyal's model category structure only. This means that our basic language is the one of simplicial sets. In fact, the first half of the book consists in following Joyal's journey [Joy08a, Joy08b], step by step: we *literally* interpret the language of category theory in the category of simplicial sets, and observe, with care and wonder, that, although it might look naive at first glance, this defines canonically a homotopy theory such that all the constructions of interest are homotopy invariant in a suitable sense. After some work, it makes perfect sense to speak of the ∞-category of functors between two ∞-categories, to see homotopy types (under the form of Kan complexes) as ∞-groupoids, or to see that fully faithful and essentially surjective functors are exactly equivalences of ∞-categories, for instance. Still in the same vein, one then starts to speak of right fibrations and of left fibrations (i.e. discrete fibrations and discrete op-fibrations, respectively). This is an approach to the theory of presheaves which is interesting by itself, since it involves (generalisations of) Quillen's theorems A and B, revisited with Grothendieck's insights on homotopy Kan extensions (in terms of smooth base change formulas and proper base change formulas). This is where the elementary part ends, in the precise sense that, to go further, some forms of rectification procedure are necessary.

In classical category theory, rectification procedures are most of the time provided by (a variation on) the Yoneda lemma. In Lurie's work as well: the rectification (straightening) of Cartesian fibrations into simplicial contravariant functors is widely used, and this is strongly related to a homotopy-theoretic version of the Yoneda lemma for 2-categories.[3] Rectification is a kind of internalisation: we want to go from ∞-groupoids (or ∞-categories), seen as objects of the theory of ∞-categories, to objects of a suitable '∞-category of ∞-groupoids' (or 'of ∞-categories'). This step is non-trivial, but it is the only way we can see how objects defined up to homotopy are uniquely (and thus coherently) determined in a suitable sense. For instance, externally, the composition of two maps in an ∞-category C, is only well defined up to homotopy (i.e. there is a contractible space of choices) in the sense that, given

[3] There is no need to understand this to go through this book, but for the sake of completeness, let us explain what we mean here. From a Grothendieck fibration $p\colon X \to A$, we can produce a presheaf of categories F on A by defining $F(a)$ as the category of Cartesian functors from the slice category A/a to X (over A) for all a. The fact that p and F determine each other is strongly related to the 2-categorical Yoneda lemma, which identifies $F(a)$ with the category of natural transformation from the presheaf represented by a to F, and to its fibred counterpart: there is a canonical equivalence of categories from $F(a)$ to the fibre X_a of p at a.

three objects x, y and z in C, there is a canonical homotopy equivalence

$$C(x, y) \times C(y, z) \leftarrow C(x, y, z)$$

relating the ∞-groupoid $C(x, y, z)$ of pairs of maps of the form $x \to y \to z$, equipped with a choice of composition $x \to z$, with the product of the ∞-groupoid $C(x, y)$ of maps of the form $x \to y$ with the ∞-groupoid $C(y, z)$ of maps of the form $y \to z$, and there is a tautological composition law

$$C(x, y, z) \to C(x, z) .$$

Composing maps in C consists in *choosing* an inverse of the homotopy equivalence above and then applying the tautological composition law. In the case where C is an ordinary category, i.e. when $C(x, y)$ is a set of maps, the composition law is well defined because there really is a unique inverse of a bijective map. The fact that the composition law is well defined and associative in such an ordinary category C implies that the assignment $(x, y) \mapsto C(x, y)$ is actually a functor from $C^{\mathrm{op}} \times C$ to the category of sets. But, when C is a genuine ∞-category, such an assignment is not a functor any more. This is due to the fact that the above is expressed in the language of the category of ∞-categories (as opposed to the ∞-category of ∞-categories), so that the assignment $(x, y) \mapsto C(x, y)$ remains a functional from the set of pairs of objects of C to the collection of ∞-groupoids, seen as objects of the category of ∞-categories. Asking for functoriality is then essentially meaningless. However, internally, such compositions all are perfectly well defined in the sense that there is a genuine Hom functor with values in the ∞-category of ∞-groupoids: there is an appropriately defined ∞-category $\mathcal{S}$ of ∞-groupoids and a functor

$$\mathrm{Hom}_C \colon C^{\mathrm{op}} \times C \to \mathcal{S} .$$

Of course, for the latter construction to be useful, we need to make a precise link between ∞-groupoids, and the objects of $\mathcal{S}$, so that $C(x, y)$ corresponds to $\mathrm{Hom}_C(x, y)$ in a suitable way. And there is no easy way to do this.

Another example: the (homotopy) pull-back of Kan fibrations becomes a strictly associative operation once interpreted as composition with functors with values in the ∞-category of ∞-groupoids. And using the Yoneda lemma (expressed with the functor Hom_C above), this provides coherence results for pull-backs in general. More precisely, given a small ∞-groupoid X with corresponding object in $\mathcal{S}$ denoted by x, there is a canonical equivalence of ∞-groupoids between the ∞-category of functors $\underline{\mathrm{Hom}}(X, \mathcal{S})$ and the slice ∞-category $\mathcal{S}/x$ (this extends the well-known fact that the slice category of sets over a given small set X is equivalent to the category of X-indexed families of sets). Given a functor between small ∞-groupoids $F\colon X \to Y$ corresponding to a map $f\colon x \to y$ in $\mathcal{S}$, the pull-back functor

$$\mathcal{S}/y \to \mathcal{S}/x\,, \quad (t \to y) \mapsto (x \times_y t \to x)$$

corresponds to the functor

$$\underline{\mathrm{Hom}}(Y, \mathcal{S}) \to \underline{\mathrm{Hom}}(X, \mathcal{S})\,, \quad \Phi \mapsto \Phi F\,.$$

The associativity of composition of functors in the very ordinary category of ∞-categories thus explains how the correspondence

$$\underline{\mathrm{Hom}}(X, \mathcal{S}) \simeq \mathcal{S}/x$$

is a way to rectify the associativity of pull-backs of ∞-groupoids which only holds up to a canonical invertible map.

Rectification thus involves a procedure to construct and compute functors with values in the ∞-category of ∞-groupoids, together with the construction of a Hom functor (i.e. of the Yoneda embedding). In this book, we avoid non-trivial straightening/unstraightening correspondences which consist in describing ∞-categories through more rigid models. Instead, we observe that there is a purely syntactic version of this correspondence, quite tautological by nature, which can be interpreted homotopy-theoretically. Indeed, inspired by Voevodsky's construction of a semantic interpretation of homotopy type theory with a univalent universe within the homotopy theory of Kan complexes [KL16], we consider the universal left fibration. The codomain of this universal left fibration, denoted by $\mathcal{S}$, has the property that there is an essentially tautological correspondence between maps $X \to \mathcal{S}$ and left fibrations with small fibres $Y \to X$. In particular, the objects of $\mathcal{S}$ are nothing other than small ∞-groupoids (or, equivalently, small Kan complexes). In the context of ordinary category theory, such a category $\mathcal{S}$ would be the category of sets. In this book, we prove that, as conjectured by Nichols-Barrer [NB07], $\mathcal{S}$ is an ∞-category which is canonically equivalent to the localisation of the category of simplicial sets by the class of weak homotopy equivalences (hence encodes the homotopy theory of CW-complexes). Furthermore, the tautological correspondence alluded to above can be promoted to an equivalence of ∞-categories, functorially in any ∞-category X: an equivalence between an appropriate ∞-category of left fibrations of codomain X and the ∞-category of functors from X to $\mathcal{S}$. Even better, the ∞-category of functors from (the nerve of) a small category I to $\mathcal{S}$ is the localisation of the category of functors from I to simplicial sets by the class of levelwise weak homotopy equivalences. This description of the ∞-category of ∞-groupoids is highly non-trivial, and subsumes the result of Voevodsky alluded to above, about the construction of univalent universes within the homotopy theory of Kan complexes. But it has the advantage that the rectification of left fibrations is done without using the introduction of an extra language, and thus may be used at a rather early stage of the development of the theory of ∞-categories, while keeping an elementary level of expression.

In order to promote the correspondence between left fibrations $Y \to X$ and functors $X \to \mathcal{S}$ to an equivalence of ∞-categories, we need several tools. First, we extend this correspondence to a homotopy-theoretic level: we prove an equivalence of moduli spaces, i.e. we prove that equivalent left fibrations correspond to equivalent functors with values in $\mathcal{S}$ in a coherent way.[4] Subsequently, to reach an equivalence of ∞-categories, we need a series of results which are of interest themselves. We provide an *ad hoc* construction of the Yoneda embedding; this can be done quite explicitly, but the proof that it satisfies the very minimal properties we expect involves non-obvious computations, which we could only explain to ourselves by introducing a bivariant version of left fibrations. Then we develop, in the context of ∞-categories, all of classical category theory (the Yoneda lemma, the theory of adjoint functors, extensions of functors by colimits, the theory of Kan extensions) as well as all of classical homotopical algebra (localisations, calculus of fractions, ∞-categories with weak equivalences and fibrations, Reedy model structures, derived functors, homotopy limits). All these aspects are carried over essentially in the same way as in ordinary category theory (this is what internalisation is good for). The only difference is that inverting weak equivalences in complete ∞-categories gives, under suitable assumptions (e.g. axioms for complete model categories) ∞-categories with small limits. Furthermore, we have the following coherence property: the process of localisation for these commutes with the formation of functor categories (indexed by small 1-categories). This means that in the context of ∞-categories, the notions of homotopy limit and of limit are not only analogous concepts: they do coincide (in particular, homotopy limits, as usually considered in algebraic topology, really are limits in an appropriate ∞-category). Similarly, there are coherence results for finite diagrams. For instance, inverting maps appropriately in ∞-categories with finite limits commutes with the formation of slices. From all this knowledge comes easily the ∞-categorical correspondence between left fibrations $Y \to X$ and functors $X \to \mathcal{S}$. Furthermore, in the case where X is the nerve of a small category A, we observe immediately that the ∞-category of functors $X \to \mathcal{S}$ is the localisation of the category of simplicial presheaves on A by the fibrewise weak homotopy equivalences, which puts classical homotopy theory in perspective within ∞-category theory.

[4] Another way to put it, for type theorists, is that we prove Voevodsky's univalence axiom for the universal left fibration.

A Few Words on the Ways We May Read This Book

Although it presents an alternative approach to the basics of the theory of ∞-categories, and even contains a few new results which might make them of interest to some readers already familiar with higher categories after Joyal and Lurie, this book is really meant to be an introduction to the subject. It is written linearly, that is, following the logical order, which also corresponds to what was actually taught in a two-semester lecture series, at least for most of it. We have aimed at providing complete constructions and proofs, starting from scratch. However, a solid background in algebraic topology or in category theory would certainly help the reader: the few examples only appear at the very end, and, when we introduce a concept, we usually do not give any historical background nor pedestrian justification. We have tried to make clear why such concepts are natural generalisations of siblings from category theory, though. Despite this, apart from a few elementary facts from standard category theory, such as the contents of Leinster's book [Lei14] or parts of Riehl's [Rie17], there are no formal prerequisites for reading this text. A very few technical results, generally with an elementary set-theoretic flavour, are left as exercises, but always with a precise reference where to find a complete proof. In particular, we do not even require any previous knowledge of the classical homotopy theory of simplicial sets, nor of Quillen's model category structures. In fact, even the Kan–Quillen model category structure, corresponding to the homotopy theory of Kan complexes, is constructed in detail, as a warm-up to construct the Joyal model category structure, which corresponds to the homotopy theory of ∞-categories. We also revisit several classical results of algebraic topology, such as Serre's long exact sequence of higher homotopy groups, as well as Quillen's famous theorem A and theorem B. These well-known results are proven in full because they appear in this book in a rather central way. For instance, in order to prove that a functor is an equivalence of ∞-categories if and only if it is fully faithful and essentially surjective, one may observe that the particular case of functors between higher groupoids (i.e. Kan complexes) is a corollary of Serre's long exact sequence. Interestingly enough, the general case follows from this groupoidal version. Similarly, the account we give of Quillen's theorem A is in fact a preparation of the theory of Kan extension, and Quillen's theorem B is a way to understand locally constant functors (which will be a technical but fundamental topic in the computation of localisations).

For the readers who already know the basics of ∞-category theory (e.g. the five first chapters of [Lur09]), parts of Chapters 4, 5 and 6 might still be of interest, since they give an account of the basics which differs from Lurie's treatment. But such readers may go directly to Chapter 7, which deals with the general interpretation of homotopical algebra within the theory of higher categories. The treatment we give of homotopical techniques in this last chapter gives robust and rather optimal tools to implement classical homotopy theories in higher categories. This is a nice example which shows that apparently abstract concepts, such as that of Kan extensions, can be used intrinsically (without apparently more explicit tools, such a homotopy coherent nerves) to organise a theory (e.g. the localisation of higher categories) both conceptually and effectively (i.e. producing computational tools).

One of the interests of using a single formalism which is a literal semantic interpretation of the language of category theory is that, although the proofs can be rather intricate, most of the statements made in this book are easy enough to understand, at least for any reader with some knowledge of category theory. This hopefully should help the reader, whether she or he wants to read only parts of the book, or to follow it step by step. Furthermore, each chapter starts with a detailed description about its purposes and contents. This is aimed at helping the reader to follow the narrative as well as to facilitate the use of the book for reference.

Finally, as all introductions, this book ends when everything begins. The reader is then encouraged to go right away to Lurie's realm. And beyond.

Acknowledgments

This book grew out of several lectures I gave on these topics: a one-semester course at Université Paris 13 in 2009, a mini-course in the Winter School 'Higher Structures in Algebraic Analysis', at the University of Padova in 2014, and a full-year course during the academic year 2016–2017, at the University of Regensburg. While writing this book, I was partially supported by the SFB 1085 'Higher Invariants' funded by the Deutsche Forschungsgemeinschaft (DFG).

I wish to thank particularly Dimitri Ara, Christoph Eibl, Kévin François, Andrea Gagna, Adeel Khan, Han-Ung Kufner, Markus Land, Hoang Kim Nguyen, Marc Nieper-Wißkirchen, Benedikt Preis and George Raptis, for their questions and comments. Many discussions with Uli Bunke certainly provided motivation and inspiration. My knowledge of, and taste for, homotopy theory owe a lot to Georges Maltsiniotis, and this text bears witness of that debt (in particular, my understanding of Grothendieck's views on homotopy theory, especially on derivators and smooth/proper base change formulas, had a great influence on the way I wrote this text). I remember great conversations with Mark Weber and Michael Batanin on cosmoi and Yoneda structures, and these certainly made their way here. Working with Ieke Moerdijk, a few years ago, on ∞-operads as a generalisation of ∞-categories, was an opportunity to think about the foundations of this theory, and this contributed to my understanding of the subject. Joseph Tapia renewed my interest in logic, in particular, in the dialectic between syntax and semantics. N. Bourbaki gave me a great opportunity to think again about these matters by asking me to give a colloquium talk about Lurie's book on higher topoi. Finally, the influence of many open and friendly discussions with André Joyal cannot be overestimated.

1

Prelude

This short chapter is meant to introduce the definition of ∞-categories. However, it starts with a recollection on presheaves of sets on a small category, on the Yoneda lemma, as well as on the ramifications of the latter through extensions of functors by colimits (a particular case of left Kan extensions). This recollection is important because the main language we will use in this book is the one of presheaves of sets, since ∞-categories will be defined as simplicial sets with certain properties, and since simplicial sets are presheaves. On the other hand, extending functors by colimits via presheaves in the setting of ∞-categories may be seen as one of our main goals. In fact, it is probably what underlies the narrative all through this book.

The rest of the chapter recounts the basic features that allow one to understand the cellular structure of simplicial sets, as well as Grothendieck's description of nerves of small categories within simplicial sets. Then come the definitions of ∞-categories and of ∞-groupoids. We see that all Kan complexes are ∞-groupoids (the converse is true but non-trivial and will only be proved in the next chapter), and therefore see that the algebra of paths in topological spaces define ∞-groupoids. The proof of the theorem of Boardmann and Vogt, which describes the category associated to an ∞-category rather explicitly, is quite enlightening, as it is also a first test which strongly indicates that interpreting the language of category theory within the category of simplicial sets is sound.

1.1 Presheaves

Presheaves will reappear in this book many times, and in many disguises. This is the way we express ourselves, at least whenever we use the language of category theory, because of the ubiquitous use of the Yoneda lemma (which will be recalled below). However, the more we go into homotopical algebra, the more we will see this apparently innocent and rather formal looking result, and the more we will see how the Yoneda lemma ramifies into many refinements. We will recall here the basic results needed about presheaves (of sets). These will be used as tools right away, but they also will be revisited with the lenses of homotopical algebra, over and over again. The historical references for this part are D. M. Kan's paper [Kan58] (in which the notion of adjoint functor is introduced for the first time), as well as Grothendieck's [SGA72, Exposé I] (the presentation we give here is rather close to the latter).

We write *Set* for the category of sets.

Definition 1.1.1 Let A be a category. A *presheaf* over A is a functor of the form

$$X : A^{\mathrm{op}} \to Set.$$

For an object a of A, we will denote by

$$X_a = X(a)$$

the evaluation of X at a. The set X_a will sometimes be called the *fibre* of the presheaf X at a, and the elements of X_a thus deserve the name of *sections* of X over a. For a morphism $u : a \to b$ in A, the induced map from X_b to X_a often will be written

$$u^* = X(u) : X_b \to X_a \,.$$

If X and Y are two presheaves over A, a *morphism* of presheaves $f : X \to Y$ simply is a natural transformation from X to Y. In other words, such a morphism f is determined by a collection of maps $f_a : X_a \to Y_a$, such that, for any morphism $u : a \to b$ in A, the following square commutes.

$$\begin{array}{ccc} X_a & \xrightarrow{f_a} & Y_a \\ {\scriptstyle u^*}\uparrow & & \uparrow{\scriptstyle u^*} \\ X_b & \xrightarrow{f_b} & Y_b \end{array} \qquad f_a\, u^* = u^*\, f_b \,.$$

Presheaves naturally form a category. This category will be written $\widehat{A}$.

Remark 1.1.2 One checks that a morphism of presheaves $f : X \to Y$ is an isomorphism (a monomorphism, an epimorphism) if and only if, for any object a of A, the induced map $f_a : X_a \to Y_a$ is bijective (injective, surjective, respectively). Moreover, the evaluation functors $X \mapsto X_a$ preserve both limits and colimits (exercise: deduce this latter property by exhibiting a left adjoint and a right adjoint). As a consequence, if $F : I \to \widehat{A}$ is a diagram of presheaves and if X is a presheaf, the property that a cone from X to F (a cocone from

F to X) exhibits X as a limit (colimit) of F is local in the sense that it can be tested fibrewise. In other words, X is a limit (a colimit) of F if and only if, for any object a of A, the set X_a is a limit (a colimit) of the induced diagram $F_a\colon I \to Set$, respectively.

Definition 1.1.3 The *Yoneda embedding* is the functor

$$h\colon A \to \widehat{A} \tag{1.1.3.1}$$

whose value at an object a of A is the presheaf

$$h_a = \mathrm{Hom}_A(-, a)\,. \tag{1.1.3.2}$$

In other words, the evaluation of the presheaf h_a at an object c of A is the set of maps from c to a.

Theorem 1.1.4 (Yoneda lemma) *For any presheaf X over A, there is a natural bijection of the form*

$$\begin{aligned}\mathrm{Hom}_{\widehat{A}}(h_a, X) &\xrightarrow{\sim} X_a\\ (h_a \xrightarrow{u} X) &\mapsto u_a(1_a).\end{aligned}$$

Proof We only define the map in the other direction. Given a section s of X over a, we define a collection of morphisms

$$f_c\colon \mathrm{Hom}_A(c, a) \to X_c$$

(indexed by objects of A) as follows: for each morphism $u\colon c \to a$, the section $f_c(u)$ is the element $f_c(u) = u^*(s)$. One then checks that this collection defines a morphism $f\colon h_a \to X$, and that the assignment $s \mapsto f$ is a two-sided inverse of the Yoneda embedding. □

Corollary 1.1.5 *The Yoneda embedding $h\colon A \to \widehat{A}$ is a fully faithful functor.*

Notation 1.1.6 The author of this book prefers to write the isomorphism of the Yoneda embedding as an equality; we will often make an abuse of notation by writing again $f\colon a \to X$ for the morphism of presheaves associated to a section $f \in X_a$ (via the Yoneda lemma).

Definition 1.1.7 Let X be a presheaf on a category A. The *category of elements* of X (we also call it the *Grothendieck construction* of X) is the category whose objects are couples (a, s), where a is an object of A, while s is a section of X over a, and whose morphisms $u\colon (a, s) \to (b, t)$ are morphisms $u\colon a \to b$ in A, such that $u^*(t) = s$. If we adopt the abuse of notation of paragraph 1.1.6, this latter condition corresponds, through the Yoneda lemma, to the commutativity of the triangle below.

$$\begin{array}{ccc} h_a & \xrightarrow{u} & h_b \\ & {}_s\searrow \quad \swarrow_t & \\ & X & \end{array}$$

The *category of elements* of X is denoted by A/X. It comes equipped with a faithful functor

$$\varphi_X : A/X \to \widehat{A} \qquad (1.1.7.1)$$

defined on objects by $\varphi_X(a, s) = h_a$, and on morphisms, by $\varphi_X(u) = u$. There is an obvious cocone from φ_X to X defined by the following collection of maps:

$$s : h_a \to X\,, \quad (a, s) \in \mathrm{Ob}(A/X)\,. \qquad (1.1.7.2)$$

A variation on the Yoneda lemma is the next statement.

Proposition 1.1.8 *The collection of maps* (1.1.7.2) *exhibits the presheaf X as the colimit of the functor* (1.1.7.1).

Proof Let Y be another presheaf on the category A. We have to show that the operation of composing maps from X to Y with the maps (1.1.7.2) defines a (natural) bijection between morphisms from X to Y and cocones from the functor φ_X to Y in the category of presheaves over A. By virtue of the Yoneda lemma, a cocone from φ_X to Y can be seen as a collection of sections

$$f_s \in Y_a\,, \quad (a, s) \in \mathrm{Ob}(A/X)$$

such that, for any morphism $u : (a, s) \to (b, t)$ in A/X, we have the relation $u^*(f_t) = f_s$. This precisely means that the collection of maps

$$\begin{aligned} X_a &\to Y_a\,, \quad a \in \mathrm{Ob}(A) \\ s &\mapsto f_s \end{aligned}$$

is a morphism of presheaves. One then checks that this operation is a two-sided inverse of the operation of composition with the family (1.1.7.2). □

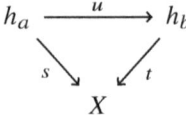

Remark 1.1.9 Until this very moment, we did not mention size (smallness) problems. Well, this is because there were not many. We will come back to size issues little by little. But, whenever we start to be careful with smallness, it is hard to stop. First, when we defined the Yoneda embedding (1.1.3.1), a first problem arose: for this construction to make sense, we need to work with *locally small categories.*[1] We might say: well, maybe we did not formulate things properly, since, for instance, even if their formulations seem to need the property that the category A is locally small, the proofs of the Yoneda lemma (1.1.4) and of its avatar (1.1.8) obviously are valid for possibly large categories. Or we could say: let us restrict ourselves to locally small categories, since, after all, most authors actually require the property of local smallness in the very definition of a category. But Definition 1.1.1 actually provides examples of categories which are not locally small: for a general locally small category A, the category of presheaves over A may not be locally small (exercise: find many examples). And there are other (less trivial but at least as fundamental) categorical constructions which do not preserve the property of being locally small (e.g. localisation). All this means that it might be wiser not to require that all categories are locally small, but, instead, to understand how and why, under appropriate assumptions, certain categorical constructions preserve properties of smallness, or of being locally small. For instance, we can see that, if ever the category A is small,[2] the category of presheaves $\widehat{A}$ is locally small. Moreover, the preceding theorem has the following consequence.

Theorem 1.1.10 (Kan) *Let A be a small category, together with a locally small category $\mathcal{C}$ which has small colimits. For any functor $u\colon A \to \mathcal{C}$, the functor of evaluation at u*

(1.1.10.1) $$u^* : \mathcal{C} \to \widehat{A}\,, \quad Y \mapsto u^*(Y) = (a \mapsto \mathrm{Hom}_{\mathcal{C}}(u(a), Y))$$

has a left adjoint

(1.1.10.2) $$u_! : \widehat{A} \to \mathcal{C}\,.$$

Moreover, there is a unique natural isomorphism

(1.1.10.3) $$u(a) \simeq u_!(h_a)\,, \quad a \in \mathrm{Ob}(A)\,,$$

such that, for any object Y of $\mathcal{C}$, the induced bijection

$$\mathrm{Hom}_{\mathcal{C}}(u_!(h_a), Y) \simeq \mathrm{Hom}_{\mathcal{C}}(u(a), Y)$$

[1] A category is locally small if, for any ordered pair of its objects a and b, morphisms from a to b do form a small set (depending on the set-theoretic foundations the reader would prefer, a small set must either be a set, as opposed to a proper class, or a set which is (in bijection with) an element of a fixed Grothendieck universe). Until we mention universes explicitly (which will happen in the second half of the book), we can be agnostic, at least as far as set theory is concerned. We refer to [Shu08] for an excellent account on the possible set-theoretic frameworks for category theory.

[2] We remind the reader that this means that it is locally small and that its objects also form a small set.

is the inverse of the composition of the Yoneda bijection

$$\mathrm{Hom}_{\mathcal{C}}(u(a), Y) = u^*(Y)_a \simeq \mathrm{Hom}_{\widehat{A}}(h_a, u^*(Y))$$

with the adjunction formula

$$\mathrm{Hom}_{\widehat{A}}(h_a, u^*(Y)) \simeq \mathrm{Hom}_{\mathcal{C}}(u_!(h_a), Y)\,.$$

Proof We shall prove that the functor u^* has a left adjoint (the second part of the statement is a direct consequence of the Yoneda lemma). For each presheaf X over A, we choose a colimit of the functor

$$A/X \to \mathcal{C}\,, \quad (a, s) \mapsto u(a)\,,$$

which we denote by $u_!(X)$. When $X = h_a$ for some object a of A, we have a canonical isomorphism $u(a) \simeq u_!(h_a)$ since $(a, 1_a)$ is a final object of A/h_a. Therefore, for any presheaf X over A, and any object Y of $\mathcal{C}$, we have the following identifications:

$$\begin{aligned}
\mathrm{Hom}_{\mathcal{C}}(u_!(X), Y) &\simeq \mathrm{Hom}_{\mathcal{C}}(\varinjlim_{(a,s)} u(a), Y) \\
&\simeq \varprojlim_{(a,s)} \mathrm{Hom}_{\mathcal{C}}(u(a), Y) \\
&\simeq \varprojlim_{(a,s)} \mathrm{Hom}_{\widehat{A}}(h_a, u^*(Y)) \quad \text{by the Yoneda lemma} \\
&\simeq \mathrm{Hom}_{\widehat{A}}(\varinjlim_{(a,s)} h_a, u^*(Y)) \\
&\simeq \mathrm{Hom}_{\widehat{A}}(X, u^*(Y)) \quad \text{by Proposition 1.1.8.}
\end{aligned}$$

In other words, the object $u_!(X)$ (co)represents the functor $\mathrm{Hom}_{\widehat{A}}(X, u^*(-))$. □

Remark 1.1.11 The functor $u_!$ will be called the *extension of u by colimits*. In fact, any colimit preserving functor $F\colon \widehat{A} \to \mathcal{C}$ is isomorphic to a functor of the form $u_!$ as above. More precisely, for any such colimit preserving functor F, if we put $u(a) = F(h_a)$, there is a unique natural isomorphism $u_!(X) = F(X)$ which is the identity whenever the presheaf X is representable (exercise). For instance, for $\mathcal{C} = \widehat{A}$, the identity of $\widehat{A}$ is (canonically isomorphic to) $h_!$, for h the Yoneda embedding.

Corollary 1.1.12 *Any colimit preserving functor $\widehat{A} \to \mathcal{C}$ has a right adjoint.*

Proof It is sufficient to consider functors of the form $u_!$, for a suitable functor $u\colon A \to \mathcal{C}$ (see the preceding remark). Therefore, by virtue of Theorem 1.1.10, it has a right adjoint, namely u^*. □

Notation 1.1.13 Let A be a small category. Then the category of presheaves over A is Cartesian closed: for any presheaves X and Y, there is an internal Hom, that is a presheaf $\underline{\mathrm{Hom}}(X, Y)$ together with natural bijections

$$\mathrm{Hom}_{\widehat{A}}(T, \underline{\mathrm{Hom}}(X, Y)) \simeq \mathrm{Hom}_{\widehat{A}}(T \times X, Y)\,.$$

As can be seen from Theorem 1.1.10 and Remark 1.1.11, this object is defined by the formula

$$\underline{\mathrm{Hom}}(X, Y)_a = \mathrm{Hom}_{\widehat{A}}(h_a \times X, Y)\,.$$

Remark 1.1.14 Given a presheaf X, it is equivalent to study maps of codomain X or to study presheaves on the category A/X. To be more precise, one checks that the extension by colimit of the composed functor $A/X \to A \xrightarrow{h} \widehat{A}$ sends the final object of $\widehat{A/X}$ to the presheaf X, and the induced functor

$$\widehat{A/X} \xrightarrow{\sim} \widehat{A}/X \tag{1.1.14.1}$$

is an equivalence of categories. For this reason, even though we will mainly focus on presheaves on a particular category (simplicial sets), it will be convenient to axiomatise our constructions in order to apply them to various categories of presheaves. Equivalence (1.1.14.1) will be at the heart of the construction of the ∞-category of small ∞-groupoids: this will appear in Section 5.2 below, and will be implicitly at the heart of much reasoning all through the second half of this book.

1.2 The Category of Simplicial Sets

We shall write $\mathbf{\Delta}$ for the category whose objects are the finite sets

$$[n] = \{i \in \mathbf{Z} \mid 0 \leq i \leq n\} = \{0, \ldots, n\}\,, \quad n \geq 0\,,$$

endowed with their natural order, and whose maps are the (non-strictly) order-preserving maps.

Definition 1.2.1 A *simplicial set* is a presheaf over the category $\mathbf{\Delta}$. We shall write $sSet = \widehat{\mathbf{\Delta}}$ for the category of simplicial sets.

Notation 1.2.2 For $n \geq 0$, we denote by $\Delta^n = h_{[n]}$ the *standard n-simplex* (i.e. the presheaf on $\mathbf{\Delta}$ represented by $[n]$).

For a simplicial set X and an integer $n \geq 0$, we write

$$X_n = X([n]) \simeq \mathrm{Hom}_{sSet}(\Delta^n, X) \tag{1.2.2.1}$$

for the set of *n-simplices of* X. A *simplex* of X is an element of X_n for some non-negative integer n. In agreement with the abuse of notation introduced in paragraph 1.1.6, an n-simplex x of X can also be seen as a morphism of simplicial sets $x\colon \Delta^n \to X$.

For integers $n \geq 1$ and $0 \leq i \leq n$, we let

$$\partial_i^n : \Delta^{n-1} \to \Delta^n \tag{1.2.2.2}$$

be the map corresponding to the unique strictly order preserving map from $[n-1]$ to $[n]$ which does not take the value i.

For integers $n \geq 0$ and $0 \leq i \leq n$, the map

$$\sigma_i^n : \Delta^{n+1} \to \Delta^n \tag{1.2.2.3}$$

corresponds to the unique surjective map from $[n+1]$ to $[n]$ which takes the value i twice.

Proposition 1.2.3 *The following identities hold:*

$$\partial_j^{n+1}\partial_i^n = \partial_i^{n+1}\partial_{j-1}^n \quad i < j, \tag{1.2.3.1}$$

$$\sigma_j^n\sigma_i^{n+1} = \sigma_i^n\sigma_{j+1}^{n+1} \quad i \leq j, \tag{1.2.3.2}$$

$$\sigma_j^{n-1}\partial_i^n = \begin{cases} \partial_i^{n-1}\sigma_{j-1}^{n-2} & i < j, \\ 1_{\Delta^{n-1}} & i \in \{j, j+1\}, \\ \partial_{i-1}^{n-1}\sigma_j^{n-2} & i > j+1. \end{cases} \tag{1.2.3.3}$$

The proof is straightforward.

Remark 1.2.4 One can prove that the category $\mathbf{\Delta}$ is completely determined by the relations above: more precisely, it is isomorphic to the quotient by these relations of the free category generated by the oriented graph which consists of the collection of maps ∂_i^n and σ_i^n (with the $[n]$ as vertices). In other words, a simplicial set can be described as a collection of sets X_n, $n \geq 0$, together with face operators $d_n^i = (\partial_i^n)^*\colon X_n \to X_{n-1}$ for $n \geq 1$, and degeneracy operators $s_n^i = (\sigma_i^n)^*\colon X_n \to X_{n+1}$ satisfying the dual version of the identities above. This pedestrian point of view is often the one taken in historical references.

Notation 1.2.5 For a simplicial set X, we shall write

$$d_n^i = (\partial_i^n)^*\colon X_n \to X_{n-1} \quad \text{and} \quad s_n^i = (\sigma_i^n)^*\colon X_n \to X_{n+1}$$

for the maps induced by the operators ∂_i^n and σ_i^n, respectively.

Although it follows right away from the notion of image of a map of sets, the following property is the source of many good combinatorial behaviours of the category $\mathbf{\Delta}$.

Proposition 1.2.6 *Any morphism $f\colon \Delta^m \to \Delta^n$ in $\mathbf{\Delta}$ admits a unique factorisation $f = i\pi$, into a split epimorphism $\pi\colon \Delta^m \to \Delta^p$ followed by a monomorphism (i.e. a strictly order preserving map) $i\colon \Delta^p \to \Delta^n$.*

Example 1.2.7 A good supply of simplicial sets comes from the category *Top* of topological spaces (with continuous maps as morphisms). For this, one defines, for each non-negative integer $n \geq 0$, the topological simplex

(1.2.7.1) $$|\Delta^n| = \left\{ (x_1, \ldots, x_n) \in \mathbf{R}^n_{\geq 0} \;\middle|\; \sum_{i=1}^{n} x_i \leq 1 \right\}.$$

Given a morphism $f\colon [m] \to [n]$ in $\mathbf{\Delta}$, we get an associated continuous (because affine) map

$$|f|\colon |\Delta^m| \to |\Delta^n|$$

defined by

$$|f|(x_0, \ldots, x_m) = (y_0, \ldots, y_n), \quad \text{where } y_j = \sum_{i \in f^{-1}(j)} x_i .$$

This defines a functor from Δ to *Top*. Therefore, by virtue of Theorem 1.1.10, we have the *singular complex functor*

(1.2.7.2) $$Top \to sSet\,, \quad Y \mapsto Sing(Y) = ([n] \to \mathrm{Hom}_{Top}(|\Delta^n|, Y))$$

and its left adjoint, the *realisation functor*

(1.2.7.3) $$sSet \to Top\,, \quad X \mapsto |X|\,.$$

This example already gives an indication of the possible semantics we can apply to simplicial sets. For instance, a 0-simplex $x\colon \Delta^0 \to X$ can be interpreted as a point of X, and a 1-simplex $f\colon \Delta^1 \to X$ as a path in X, from the point $x = d^1_1(f)$ to the point $y = d^0_1(f)$. This is already good, but we shall take into account that the orientation of paths can be remembered. And doing so literally, this will give semantics, in the category of simplicial sets, of the very language of category theory.

1.3 Cellular Filtrations

In this chapter, we shall review the combinatorial properties of simplicial sets which will be used many times to reduce general statements to the manipulation of finitely many operations on standard simplices. However, we shall present an axiomatised version (mainly to deal with simplicial sets over a given simplicial set X, or with bisimplicial sets, for instance). A standard source on this, in the case of simplicial sets themselves, is the appropriate chapter in the book of Gabriel and Zisman [GZ67]. What follows is to axiomatise the constructions and proofs of therein. For a nice axiomatic treatment of this kind of property, an excellent reference is Bergner and Rezk's paper [BR13].

Definition 1.3.1 An *Eilenberg–Zilber category* is a quadruple (A, A_+, A_-, d), where A is a small category, while A_+ and A_- are subcategories of A, and $d\colon \mathrm{Ob}(A) \to \mathbf{N}$ is a function with values in the set of non-negative integers, such that the following properties are verified:

EZ0. Any isomorphism of A is in both A_+ and A_-. Moreover, for any isomorphic objects a and b in A, we have $d(a) = d(b)$.

EZ1. If $a \to a'$ is a morphism in A_+ (in A_-) that is not an identity, then we have $d(a) < d(a')$ (we have $d(a) > d(a')$, respectively).

EZ2. Any morphism $u\colon a \to b$ in A has a unique factorisation of the form $u = ip$, with $p\colon a \to c$ in A_- and $i\colon c \to b$ in A_+.

EZ3. If a morphism $\pi\colon a \to b$ belongs to A_- there exists a morphism $\sigma\colon b \to a$ in A such that $\pi\sigma = 1_b$. Moreover, for any two morphisms in A_- of the form $\pi, \pi'\colon a \to b$, if π and π' have the same sets of sections, then they are equal.

We shall say that an object a of A is of *dimension* n if $d(a) = n$.

Example 1.3.2 The category $\mathbf{\Delta}$ is an Eilenberg–Zilber category, with $\mathbf{\Delta}_+$ the subcategory of monomorphisms, and $\mathbf{\Delta}_-$ the subcategory of epimorphisms, and $d(\Delta^n) = n$.

Example 1.3.3 If A is an Eilenberg–Zilber category, then, for any presheaf X, the category A/X is an Eilenberg–Zilber category: one defines the subcategory $(A/X)_+$ (the subcategory $(A/X)_-$) as the subcategory of maps whose image in A belongs to A_+ (to A_-, respectively), and one puts $d(a, s) = d(a)$.

Example 1.3.4 If A and B are two Eilenberg–Zilber categories, their product is one as well: one defines $(A \times B)_\varepsilon = A_\varepsilon \times B_\varepsilon$ for $\varepsilon \in \{+, -\}$, and one puts $d(a, b) = d(a) + d(b)$.

Let us fix an Eilenberg–Zilber category A.

Definition 1.3.5 Let X be a presheaf over A. A section x of X over some object a of A is *degenerate*, if there exists a map $\sigma\colon a \to b$ in A, with $d(b) < d(a)$, and a section y of X over b, such that $\sigma^*(y) = x$. Such a couple will be called a *decomposition of* x. A section of X is *non-degenerate* if it is not degenerate.

For any integer $n \geq 0$, we denote by $Sk_n(X)$ the maximal subpresheaf of X with the property that, for any integer $m > n$, any section of $Sk_n(X)$ over an object a of dimension m is degenerate. In other words, for any object a of A, the sections of $Sk_n(X)$ over a coincide with those of X for $d(a) \leq n$, and are those which are degenerations of sections of X over some b with $d(b) \leq n$ for $d(a) > n$. This construction is functorial: for any morphism of presheaves $f : X \to Y$, there is a unique morphism $Sk_n(f) : Sk_n(X) \to Sk_n(Y)$ such that the following square commutes.

$$\begin{array}{ccc} Sk_n(X) & \hookrightarrow & X \\ {\scriptstyle Sk_n(f)}\downarrow & & \downarrow{\scriptstyle f} \\ Sk_n(Y) & \hookrightarrow & Y \end{array}$$

Lemma 1.3.6 (Eilenberg–Zilber) *Let $x \in X_a$ be a section of a presheaf X over A. There exists a unique decomposition (σ, y) of x, such that σ is a morphism of A_-, while y is non-degenerate.*

Proof There are integers m such that there exists a decomposition (σ, y) of x, where $\sigma : a \to b$ is in A_- and $d(b) = m$ (e.g. $a = b$ and $\sigma = 1_a$). Therefore, there exists such a couple (σ, y) with $d(b) = m$ minimal. If y was degenerate, this would contradict the minimality of m, hence the section y must be non-degenerate. On the other hand, if we have another decomposition (σ', y') of x, with $\sigma' : a \to b'$ in A_-, and $d(b') = m$, since any morphism of A_- has a section in A, we can find a section ι of σ, and we get a map $u = \sigma'\iota : b \to b'$, such that $u^*(y') = y$. Moreover, by virtue of axiom EZ2, the morphism u has a unique factorisation $u = ip$ with i in A_+ and p in A_-. But axiom EZ1, together with the minimality of m, implies that p is an identity, and so i must be as well. Since axioms EZ0 and EZ1 also force the isomorphisms to be identities, we deduce that $b = b'$ and $y = y'$. We deduce from this that the two morphisms σ and σ' have the same set of sections, which implies, by virtue of axiom EZ3, that they are equal. □

Notation 1.3.7 For an object a of A, we put

$$\partial h_a = Sk_{d(a)-1}(h_a)\,.$$

This subobject is called the *boundary* of the representable presheaf h_a.

Theorem 1.3.8 *Let $X \subset Y$ be presheaves over A. For any non-negative integer n, there is a canonical push-out square*

$$\begin{array}{ccc} \coprod_{y \in \Sigma} \partial h_a & \longrightarrow & X \cup Sk_{n-1}(Y) \\ \downarrow & & \downarrow \\ \coprod_{y \in \Sigma} h_a & \longrightarrow & X \cup Sk_n(Y) \end{array}$$

where Σ denotes the set of non-degenerate sections of Y of the form $y: h_a \to Y$ which do not belong to X, and such that $d(a) = n$.

The proof is left as an exercise: the main ingredients are the preceding lemma and the excluded middle principle.

Definition 1.3.9 A class $\mathcal{C}$ of presheaves over A is *saturated by monomorphisms* if it has the following stability properties.

(a) For any small family of presheaves $(X_i)_i$, if each X_i belongs to $\mathcal{C}$, so does the coproduct $\coprod_i X_i$.

(b) For any push-out square of presheaves

$$\begin{array}{ccc} X & \longrightarrow & X' \\ \downarrow & & \downarrow \\ Y & \longrightarrow & Y' \end{array}$$

in which the vertical maps are monomorphisms, if X, X' and Y all are in $\mathcal{C}$, so is Y'.

(c) For any sequence of monomorphisms of presheaves

$$X_0 \hookrightarrow X_1 \hookrightarrow \cdots \hookrightarrow X_n \hookrightarrow X_{n+1} \hookrightarrow \cdots$$

in which each of the X_n is in $\mathcal{C}$, their reunion $\varinjlim_n X_n$ belongs to $\mathcal{C}$ as well.

Corollary 1.3.10 *If a class of presheaves over A is saturated by monomorphisms and contains all representable presheaves, then it contains all presheaves over A.*

Proof Let $\mathcal{C}$ be such a class. We apply Theorem 1.3.8 with X empty. We thus have push-out squares of the following form.

$$\begin{array}{ccc} \coprod_{y\in\Sigma} \partial h_a & \longrightarrow & Sk_{n-1}(Y) \\ \downarrow & & \downarrow \\ \coprod_{y\in\Sigma} h_a & \longrightarrow & Sk_n(Y) \end{array}$$

We prove by induction on n that each presheaf of the form $Sk_n(Y)$ is in $\mathcal{C}$. For $n = 0$, we see that $Sk_0(Y)$ is a small sum of representable presheaves, and therefore is in $\mathcal{C}$. For $n > 0$, the induction hypothesis means that both the domain and the codomain of the upper horizontal map of the commutative square above are in $\mathcal{C}$. Hence, using properties (a) and (b) of the definition of a saturated class by monomorphisms, we deduce that any presheaf of the form

$Sk_n(Y)$ is in $\mathcal{C}$. Since the union of the $Sk_n(Y)$ for $n \geq 0$ is Y itself, condition (c) above shows that any presheaf Y belongs to $\mathcal{C}$. □

Remark 1.3.11 Theorem 1.3.8 also implies that, for any simplicial set X, the realisation $|X|$ has a natural structure of CW-complex: this comes from the fact that the realisation functor preserves colimits (being a left adjoint) and that $|\partial\Delta^n| = S^{n-1}$ is the boundary of the topological simplex $|\Delta^n|$.

Corollary 1.3.12 *Let A be a small Eilenberg–Zilber category such that any representable presheaf only has finitely many non-degenerate sections. For any presheaf X on A with finitely many non-degenerate sections, the functor*

$$\mathrm{Hom}_{\widehat{A}}(X, -) \colon \widehat{A} \to \mathit{Set}$$

commutes with filtered colimits.

Proof Let $\mathcal{C}$ be the class of presheaves Y such that the functor $\mathrm{Hom}_{\widehat{A}}(Y, -)$ commutes with filtered colimits. Then the class $\mathcal{C}$ is stable under finite colimits. Indeed, since filtered colimits commute with finite limits (see for instance [Rie17, theorem 3.8.9]), if I is a finite category, and if $i \mapsto X_i$ is an I-indexed diagram of elements of $\mathcal{C}$, for any filtered diagram $j \mapsto Y_j$ of presheaves on A, the canonical map

$$\varinjlim_j \varprojlim_i \mathrm{Hom}_{\widehat{A}}(X_i, Y_j) \to \varprojlim_i \varinjlim_j \mathrm{Hom}_{\widehat{A}}(X_i, Y_j)$$

is bijective. Since we have a canonical bijection

$$\mathrm{Hom}_{\widehat{A}}(\varinjlim_i X_i, Y_j) \simeq \varprojlim_i \mathrm{Hom}_{\widehat{A}}(X_i, Y_j)$$

for all j, and since the canonical map

$$\varinjlim_j \mathrm{Hom}\,\widehat{A}(X_i, Y_j) \to \mathrm{Hom}_{\widehat{A}}(X_i, \varinjlim_j Y_j)$$

is invertible for all i (because X_i belongs to $\mathcal{C}$), this proves that the colimit of the X_i is an element of $\mathcal{C}$ as well. Let us prove that X is an element of $\mathcal{C}$. It follows right away from Theorem 1.3.8 that there is an integer n such that $X = Sk_n(X)$. We proceed by induction on n. If $n < 0$, then $X = \varnothing$, and the assertion follows from the fact that a filtered colimit of sets with one element is a set with one element. If $n \geq 0$, then, by virtue of Theorem 1.3.8, there is a coCartesian square of the form

$$\begin{array}{ccc} \coprod_{x \in \Sigma} \partial h_a & \longrightarrow & Sk_{n-1}(X) \\ \big\downarrow & & \big\downarrow \\ \coprod_{x \in \Sigma} h_a & \longrightarrow & X \end{array}$$

in which Σ is a finite set and all the a are of dimension n. It is clear that $\coprod_{x\in\Sigma} \partial h_a$ and $Sk_{n-1}(X)$ are in $\mathcal{C}$, by induction on n. Therefore, it is sufficient to prove that $\coprod_{x\in\Sigma} h_a$ belongs to $\mathcal{C}$. Since Σ is finite, it is sufficient to prove that each representable presheaf h_a belongs to $\mathcal{C}$. But the Yoneda lemma identifies $\mathrm{Hom}_{\widehat{A}}(h_a, -)$ with the evaluation functor at a, and the latter is known to commute with all colimits. □

1.4 Nerves

1.4.1 Any partially ordered set E gives rise to a category: the objects are the elements of E, and, for any couple of such elements (x, y), the set of morphisms from x to y has at most one element:

$$\mathrm{Hom}(x, y) = \begin{cases} \{\varnothing\} & \text{if } x \leq y, \\ \varnothing & \text{if } x > y. \end{cases}$$

The letter E can safely be used to represent this category, since, for two partially ordered sets E and F, there is a natural identification between the set of non-strict order preserving maps from E to F and the set of functors from the category associated to E to the one associated to F. In other words, it is a rather natural thing to see the category of partially ordered sets as a full subcategory of the category *Cat* of small categories (whose objects are small categories, while its morphisms are the functors between these).

In particular, we can restrict our attention to non-empty finite totally ordered sets, and we obtain a fully faithful inclusion functor

$$i\colon \mathbf{\Delta} \to Cat. \tag{1.4.1.1}$$

The *nerve* functor is defined as the evaluation at i:

$$N = i^*\colon Cat \to sSet\,, \quad C \mapsto N(C) = ([n] \mapsto \mathrm{Hom}_{Cat}([n], C))\,. \tag{1.4.1.2}$$

Hence an n-simplex of the nerve of a category C is a string of arrows of length n in C:

$$x_0 \xrightarrow{f_1} x_1 \xrightarrow{f_2} \cdots \xrightarrow{f_n} x_n\,. \tag{1.4.1.3}$$

By virtue of Theorem 1.1.10, the nerve functor has a left adjoint

$$\tau = i_!\colon sSet \to Cat\,. \tag{1.4.1.4}$$

The nerve of a category will serve as a paradigm to interpret the structure of a simplicial set.

Definition 1.4.2 Let X be a simplicial set.

An *object* of X is a 0-simplex of X, or equivalently a map $x\colon \Delta^0 \to X$.

An *arrow* of X (we shall also say a *morphism* of X, or even a *map* of X) is a 1-simplex of X, that is a map $f\colon \Delta^1 \to X$. Such an arrow has a source and a target, namely $f\partial_1^1$ and $f\partial_0^1$, respectively:

$$\Delta^0 \xrightarrow{\partial_1^1} \Delta^1 \xrightarrow{f} X, \qquad \Delta^0 \xrightarrow{\partial_0^1} \Delta^1 \xrightarrow{f} X.$$

Notation 1.4.3 A diagram of the form

$$x \xrightarrow{f} y$$

in a simplicial set X will mean that f is an arrow of X and that $x = f\partial_1^1$ is its source, while $y = f\partial_0^1$ is its target. In other words, this corresponds to a commutative diagram of simplicial sets of the following shape.

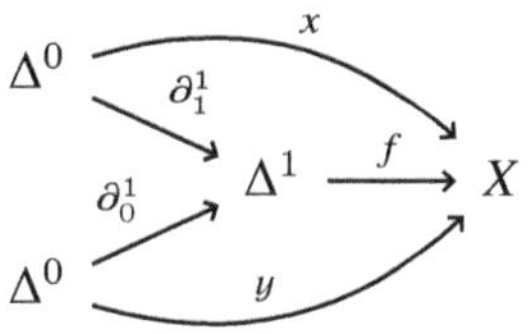

Example 1.4.4 Given an object x in a simplicial set X, the *identity* of x is the morphism of X corresponding to this composition of maps of simplicial sets:

$$1_x = x\sigma_0^0\colon \Delta^1 \xrightarrow{\sigma_0^0} \Delta^0 \xrightarrow{x} X\,. \tag{1.4.4.1}$$

1.4.5 Given a finite non-empty totally ordered set E, we will write

$$\Delta^E = N(E) \tag{1.4.5.1}$$

for its nerve. We have $\Delta^{[n]} = \Delta^{\{0,\dots,n\}} = \Delta^n$ for any integer $n \geq 0$. Any enumeration of E thus gives an isomorphism[3] of simplicial sets $\Delta^E \simeq \Delta^n$, and inclusions of finite non-empty totally ordered sets $E \subset F$ induce inclusions of simplicial sets $\Delta^E \subset \Delta^F$. For any integer $n \geq 0$, we define the *boundary* of the standard n-simplex to be

$$\partial\Delta^n = \partial h_{[n]} = \bigcup_{E \subsetneq [n]} \Delta^E \subset \Delta^n\,. \tag{1.4.5.2}$$

Similarly, for integers n, k, with $n \geq 1$ and $0 \leq k \leq n$, the kth *horn* of Δ^n is

$$\Lambda_k^n = \bigcup_{k \in E \subsetneq [n]} \Delta^E \subset \Delta^n\,. \tag{1.4.5.3}$$

[3] We remark that the simplicial sets of the form Δ^E do not have non-trivial automorphisms, so that the existence of an isomorphism between such objects (as opposed to the specification of such an isomorphism) is already meaningful.

Finally, for an integer $n \geq 1$, the *spine* of Δ^n is

(1.4.5.4)
$$Sp^n = \bigcup_{0 \leq i < n} \Delta^{\{i,i+1\}} \subset \Delta^n .$$

We will interpret these constructions in dimension $n = 2$, using the intuition provided by the nerve of small categories. First, note that $\Lambda^2_1 = Sp^2$.

1.4.6 Let X be a simplicial set.

A *triangle* in X is a map $t \colon \partial\Delta^2 \to X$. This can be seen as a triple (f, g, h) of morphisms in X, such that the target of f and the source of g coincide, while f and h have the same source x, and g and h have the same target z: since the boundary $\partial\Delta^2$ is the union of three copies of Δ^1, namely $\Delta^{\{i,j\}}$, for $i, j \in \{0, 1, 2\}$, with $i < j$, the morphism f corresponds to the map $\Delta^1 \simeq \Delta^{\{0,1\}} \subset \partial\Delta^2 \to X$, while g corresponds to the map $\Delta^1 \simeq \Delta^{\{1,2\}} \subset \partial\Delta^2 \to X$, and h to the map $\Delta^1 \simeq \Delta^{\{0,2\}} \subset \partial\Delta^2 \to X$. In other words, if we let y be the source of g, the triangle t can be faithfully represented as a diagram of the following shape.

(1.4.6.1)
$$\begin{array}{ccccc} & & y & & \\ & {}^{f}\nearrow & & \searrow^{g} & \\ x & & \xrightarrow{\;h\;} & & z \end{array}$$

Similarly, a map $\Lambda^2_1 = Sp^2 \to X$ can be seen as a couple (f, g) of morphisms in X, such that the target of f and the source of g coincide. Indeed, Sp^2 is the union of two copies of Δ^1, namely $\Delta^{\{0,1\}}$ and $\Delta^{\{1,2\}}$, which intersect at the point $\Delta^0 \simeq \Delta^{\{1\}}$, so that f corresponds to the map $\Delta^1 \simeq \Delta^{\{0,1\}} \subset Sp^2 \to X$, while g corresponds to the map $\Delta^1 \simeq \Delta^{\{1,2\}} \subset Sp^2 \to X$. Such a map (f, g) thus can be represented as a diagram of the form below.

(1.4.6.2)
$$\begin{array}{ccccc} & & y & & \\ & {}^{f}\nearrow & & \searrow^{g} & \\ x & & & & z \end{array}$$

Definition 1.4.7 A triangle (f, g, h) in X, as in diagram (1.4.6.1), is *commutative* (or simply *commutes*), if there exists a morphism of simplicial sets $c \colon \Delta^2 \to X$ whose restriction to the boundary coincides with (f, g, h):

$$\begin{array}{ccc} \partial\Delta^2 & \xrightarrow{(f,g,h)} & X \\ \downarrow & {}^{c}\nearrow & \\ \Delta^2 & & \end{array} \qquad\qquad c|_{\partial\Delta^2} = (f, g, h) .$$

Given a pair (f, g) of composable morphisms of X, as in diagram (1.4.6.2), a *composition* of f and g is a morphism h in X such that the triple (f, g, h) defines a commutative triangle in X.

Remark 1.4.8 If $X = N(C)$ is the nerve of a small category, the objects and arrows of X exactly are the objects and arrows of C, respectively. Furthermore, the commutative triangles of X precisely are the commutative triangles in the category C in the usual sense. A way to reformulate (and prove) this is to say that the operation of restriction along the inclusion $Sp^2 \subset \Delta^2$ induces a bijective map

(1.4.8.1) $$\mathrm{Hom}(\Delta^2, N(C)) \xrightarrow{\sim} \mathrm{Hom}(Sp^2, N(C))\,.$$

This bijection means that, in (the nerve of) a category, any couple of composable arrows has a *unique* composition. More generally, the description of the n-simplices of $N(C)$ as strings of arrows of the form (1.4.1.3) in C means that, for all integers $n \geq 2$, the restriction along the inclusion $Sp^n \subset \Delta^n$ induces a bijective map

(1.4.8.2) $$\mathrm{Hom}(\Delta^n, N(C)) \xrightarrow{\sim} \mathrm{Hom}(Sp^n, N(C))\,.$$

These bijections express the associativity of the composition law in a category (we shall be more explicit later). This implies that one can understand maps to nerves of small categories as follows. For any simplicial set X and any small category C, a map from X to $N(C)$ is completely determined by a map $u\colon X_1 \to \mathrm{Arr}(C)$, such that the two conditions below are satisfied:

(i) for all objects x of X, $u(1_x)$ is an identity;
(ii) for any commutative triangle (f, g, h) in X, we have $u(h) = u(g) \circ u(f)$.

This implies that, in particular, we have the following property.

Proposition 1.4.9 *For any simplicial set X, the inclusion $Sk_2(X) \subset X$ induces an isomorphism of categories $\tau(Sk_2(X)) \simeq \tau(X)$.*

1.4.10 The preceding remark is a description, in the language of simplicial sets, of (small) categories. To be more precise, we shall say that a simplicial set X satisfies the *Grothendieck–Segal condition* if the restriction along the inclusion map $Sp^n \subset \Delta^n$ induces a bijective map

(1.4.10.1) $$\mathrm{Hom}(\Delta^n, X) \xrightarrow{\sim} \mathrm{Hom}(Sp^n, X)\,, \quad \text{for all integers } n \geq 2.$$

Proposition 1.4.11 *The nerve functor is fully faithful: given two small categories C and D, the nerve functor defines a bijection*

$$\mathrm{Hom}_{\mathcal{C}at}(C, D) \simeq \mathrm{Hom}_{sSet}(N(C), N(D))\,.$$

Moreover, the essential image of the nerve functor precisely consists of the simplicial set which satisfies the Grothendieck–Segal condition. In other words, given a simplicial set X, the following conditions are equivalent.

(i) *There exists a small category C as well as an isomorphism of simplicial sets $X \simeq N(C)$.*
(ii) *The unit map $X \to N(\tau(X))$ is invertible.*
(iii) *The simplicial set X satisfies the Grothendieck–Segal conditions.*

Proof The fully faithfulness of the nerve functor is a corollary of bijection (1.4.8.2) (see the description of maps towards a nerve after that bijection). Condition (ii) is a reformulation of the property of fully faithfulness. Therefore, general facts about adjoint functors imply that conditions (i) and (ii) are equivalent. We already know that condition (i) implies condition (iii). It is thus sufficient to prove that the Grothendieck–Segal condition implies (ii). We thus have to check if the Grothendieck–Segal conditions ensure that, for any non-negative integer n, the maps

$$X_n \to N(\tau(X))_n$$

are bijective. But, for $n \leq 1$, one checks that this map is always the identity, from which one deduces, using the Grothendieck–Segal condition for $\tau(X)$, that, for $n \geq 2$, these maps coincide with (1.4.8.2). □

Remark 1.4.12 Let X be a topological space. A morphism in the simplicial set $Sing(X)$ is then a path in the space X, i.e. a continuous map from $|\Delta^1| = (0, 1)$ to X. Given two composable paths γ and γ', a composition of these in $Sing(X)$ precisely is a path λ whose starting point coincides with that of γ, and with the same end-point as γ', such that there exists a homotopy deformation of paths between λ and the concatenation of the paths γ and γ'. Therefore, the notion of composition of morphisms in a simplicial set encompasses both the notion of composition of morphisms in a category and the notion of composition of paths in a topological space.

A variation on the preceding proposition is the following.

Proposition 1.4.13 *A simplicial set X satisfies the Grothendieck–Segal conditions if and only if, for any integers $n \geq 2$ and $0 < k < n$, the restriction along the inclusion $\Lambda^n_k \subset \Delta^n$ induces a bijection*

$$\text{Hom}(\Delta^n, X) \xrightarrow{\sim} \text{Hom}(\Lambda^n_k, X). \tag{1.4.13.1}$$

We will only need to know that any simplicial set satisfying the Grothendieck–Segal condition has this property. Therefore, we will prove this fact, and leave the other direction as an exercise for the reader (although Proposition 3.7.4

below could be seen as a pedantic formulation of the solution). If X satisfies the Grothendieck–Segal condition, it is isomorphic to $N(C)$ for some small category C. Hence morphism $Y \to X$ always factors through $N(\tau(Y))$, and thus, by virtue of Proposition 1.4.9, only depends on their restriction to $Sk_2(Y)$. One checks that $Sk_2(\Lambda^n_k) = Sk_2(\Delta^n)$ whenever $n \geq 4$ because of dimension reasons. Since $\Lambda^2_1 = Sp^2$, it remains to check that $\tau(\Lambda^3_k) = \tau(\Delta^3)$ for $k = 1, 2$. This is done directly, using the explicit description of τ at the end of Remark 1.4.8 (for a hint, see the proof of Lemma 1.6.2 below, in the case where $X = N(C)$ is the nerve of a small category).

1.5 Definition of ∞-Categories

Definition 1.5.1 An ∞-*category* is a simplicial set X such that, for any integers $n \geq 2$ and $0 < k < n$, any morphism of the form $\Lambda^n_k \to X$ extends to Δ^n. In other words, the operation of restriction along the inclusion $\Lambda^n_k \subset \Delta^n$ induces a surjection

(1.5.1.1) $$\mathrm{Hom}(\Delta^n, X) \to \mathrm{Hom}(\Lambda^n_k, X)\,.$$

A morphism $f\colon x \to y$ in an ∞-category X is *invertible* if there exist two morphisms $g\colon y \to x$ and $h\colon y \to x$ such that 1_x is a composition of f and g, and that 1_y is a composition of h and f. In other words, both triangles

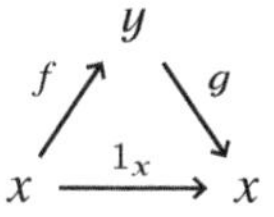

$$\begin{array}{ccc} & y & \\ {}^{f}\nearrow & & \searrow^{g} \\ x & \xrightarrow{\;1_x\;} & x \end{array} \qquad \text{and} \qquad \begin{array}{ccc} & x & \\ {}^{h}\nearrow & & \searrow^{f} \\ y & \xrightarrow{\;1_y\;} & y \end{array}$$

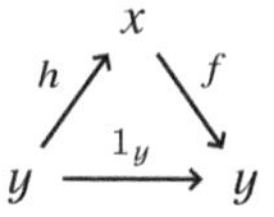

commute in X.

An ∞-*groupoid* is an ∞-category in which any morphism is invertible.

A *Kan complex* is a simplicial set X such that, for any integers $n \geq 1$ and $0 \leq k \leq n$, any morphism of the form $\Lambda^n_k \to X$ extends to Δ^n.

Remark 1.5.2 The surjectivity of the map (1.5.1.1) for $n = 2$ means that any composable pair of maps has a composition in any ∞-category. These compositions are not strictly unique, but we shall see in many ways that the surjectivity of the map (1.5.1.1) for $n > 2$ gives enough coherence to circumvent this apparent flaw, so that we shall have uniqueness of compositions up to homotopy, in a suitable sense.

Example 1.5.3 For any small category C, the nerve $N(C)$ is an ∞-category.

Proposition 1.5.4 *Any Kan complex is an ∞-groupoid.*

Proof It is clear that any Kan complex is an ∞-category. If $f : x \to y$ is a morphism in a Kan complex X, there is a unique morphism

$$\Lambda_0^2 \to X$$

which sends the non-degenerate 1-simplex of $\Delta^{\{0,1\}}$ to f and the non-degenerate 1-simplex of $\Delta^{\{0,2\}}$ to 1_x. Its extension to Δ^2 shows the existence of a commutative triangle

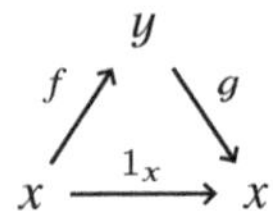

in X. The existence of the other commutative triangle is proved similarly. $\square$

Example 1.5.5 For any topological space X, the singular complex $Sing(X)$ is both a Kan complex (whence an ∞-category) and an ∞-groupoid. To prove this, one checks that, if we identify $|\Delta^n|$ with the hypercube $[0,1]^n$, the topological realisation of Λ_k^n corresponds to $[0,1]^{n-1} \times \{0\}$ (up to an affine automorphism of $[0,1]^n$). Using the adjunction formula

$$\mathrm{Hom}(|K|, X) \simeq \mathrm{Hom}(K, Sing(X))$$

this implies that the restriction along the inclusion $\Lambda_k^n \subset \Delta^n$ induces a surjection

$$\mathrm{Hom}(\Delta^n, Sing(X)) \to \mathrm{Hom}(\Lambda_k^n, Sing(X))$$

for any integers $n \geq 1$ and $0 \leq k \leq n$. Explicit inverses of paths come from the re-parametrisation of the interval given by the function $t \mapsto 1 - t$.

It was proved by Milnor that the homotopy theory of CW-complexes and the homotopy theory of Kan complexes essentially are the same. We shall see later that the converse of the preceding proposition is true: any ∞-groupoid is a Kan complex (see Theorem 3.5.1 below). Therefore, homotopy types (of CW-complexes) will play a central role in higher category theory. See Remark 7.8.11.

Remark 1.5.6 The notion of ∞-category as in Definition 1.5.1 was discovered and introduced by Boardman and Vogt in order to understand the theory of algebraic structures up to (coherent) homotopies, under the name of *weak Kan complexes*. They were developed by Joyal under the name of *quasi-categories*, and then by Lurie under the name of ∞-*categories*. Milnor's theorem alluded to above suggests that there are many different presentations (models) of homotopy types (of CW-complexes), and this is indeed the case, with no way to consider one of them as better than the others (see Grothendieck's theory of test categories [Mal05a, Cis06], for instance). The same thing happens with ∞-categories (see [Ber18] and [AFR17]).

1.5.7 Let $\rho\colon \mathbf{\Delta} \to \mathbf{\Delta}$ be the functor defined as the identity on objects and by the formula

$$\rho(f)(i) = n - f(m - i)$$

for any map $f\colon [m] \to [n]$, with $0 \leq i \leq m$. One has $\rho^2 = 1_{\mathbf{\Delta}}$ (one can check that ρ is the only non-trivial automorphism of the category $\mathbf{\Delta}$). Composing with ρ defines an automorphism of the category of simplicial sets

$$\rho^* : sSet \to sSet\,.$$

For a simplicial set X, one defines its *opposite* simplicial set as $X^{\mathrm{op}} = \rho^*(X)$. Given a morphism of simplicial sets $f\colon X \to Y$, we also get, by functoriality, a morphism $f^{\mathrm{op}} = \rho^*(f)\colon X^{\mathrm{op}} \to Y^{\mathrm{op}}$.

Proposition 1.5.8 *For any small category C, we have a canonical identification*

$$N(C)^{\mathrm{op}} = N(C^{\mathrm{op}})\,.$$

Moreover, for any ∞-category X, its opposite X^{op} also is an ∞-category.

(The proof is left as an exercise.)

1.6 The Boardman–Vogt Construction

1.6.1 Let X be an ∞-category. The end of this chapter will be devoted to giving an explicit description of the associated category $\tau(X)$.

For this purpose, we will have to study maps of the form

$$x\colon Sk_1(\Delta^3) \to X\,.$$

Since, by definition, the non-degenerate simplices of $Sk_1(\Delta^3)$ are of dimension ≤ 1, such a map precisely corresponds to a diagram in X of the form

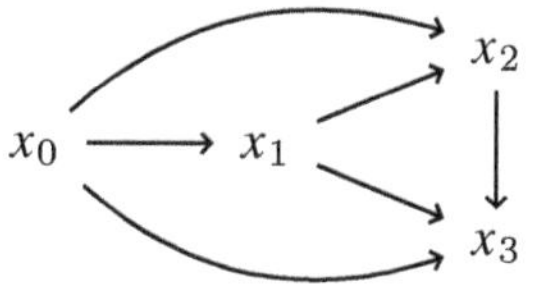

(in which none of the triangles is required to commute). There are four triangles

$d^i x\colon \partial\Delta^2 \to X$, corresponding to the restrictions of x to each of the subcomplexes $Sk_1(\Delta^{E_i})$, for each of the sets with three elements $E_i = \{0, 1, 2, 3\} - \{i\}$, for $i = 0, 1, 2, 3$.

$$d^0x\colon \quad \begin{array}{ccccc} & & x_2 & & \\ & \nearrow & & \searrow & \\ x_1 & & \longrightarrow & & x_3 \end{array} \qquad\qquad d^1x\colon \quad \begin{array}{ccccc} & & x_2 & & \\ & \nearrow & & \searrow & \\ x_0 & & \longrightarrow & & x_3 \end{array}$$

$$d^2x\colon \quad \begin{array}{ccccc} & & x_1 & & \\ & \nearrow & & \searrow & \\ x_0 & & \longrightarrow & & x_3 \end{array} \qquad\qquad d^3x\colon \quad \begin{array}{ccccc} & & x_1 & & \\ & \nearrow & & \searrow & \\ x_0 & & \longrightarrow & & x_2 \end{array}$$

Lemma 1.6.2 (Joyal's coherence lemma) *Assume that in a diagram $x\colon Sk_1(\Delta^3) \to X$ as above, the two triangles d^0x and d^3x commute. Then the triangle d^1x commutes if and only if the triangle d^2x commutes.*

Proof Since both triangles d^0x and d^3x commute, we may assume that two commutative triangles of the following form are given in the category of simplicial sets.

$$\begin{array}{ccc} \partial\Delta^2 & \xrightarrow{d^0x} & X \\ \downarrow & \nearrow_{y_0} & \\ \Delta^2 & & \end{array} \qquad \text{and} \qquad \begin{array}{ccc} \partial\Delta^2 & \xrightarrow{d^3x} & X \\ \downarrow & \nearrow_{y_3} & \\ \Delta^2 & & \end{array} .$$

Let us assume that d^1x commutes as well. There exists a commutative diagram

$$\begin{array}{ccc} \partial\Delta^2 & \xrightarrow{d^1x} & X \\ \downarrow & \nearrow_{y_1} & \\ \Delta^2 & & \end{array}$$

and the data of y_0, y_1, y_3 define a morphism $(y_0, y_1, y_3)\colon \Lambda^3_2 \to X$. The latter extends to a 3-simplex $y\colon \Delta^3 \to X$. If we put $y_2 = y\partial^3_2$, we see that the triangle of simplicial sets

$$\begin{array}{ccc} \partial\Delta^2 & \xrightarrow{d^2x} & X \\ \downarrow & \nearrow_{y_2} & \\ \Delta^2 & & \end{array}$$

commutes, and, therefore, that the triangle d^2x commutes in X.

If d^2x commutes, then applying what precedes to the opposite ∞-category X^{op} gives the commutativity of d^1x. □

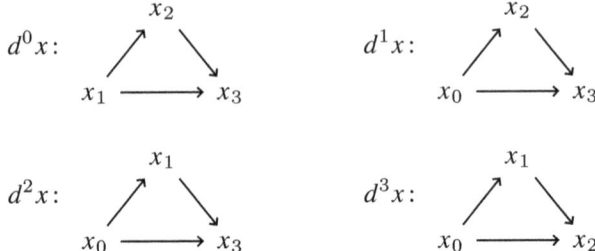

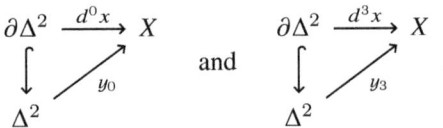

1.6.3 Given three morphisms f, g and h in X, we shall write $gf \sim h$ to mean that the triple (f, g, h) is a commutative diagram in X (i.e. a morphism $\partial\Delta^2 \to X$ which can be extended to Δ^2).

Let us fix two objects x and y of X. One defines four relations on the set of morphisms of X with source x and target y:

- $f \sim_1 g$ if $f1_x \sim g$;
- $f \sim_2 g$ if $1_y f \sim g$;
- $f \sim_3 g$ if $g1_x \sim f$;
- $f \sim_4 g$ if $1_y g \sim f$.

Lemma 1.6.4 *The four relations $\sim_i$, $i = 1, 2, 3, 4$, are equal. Moreover, they are equivalence relations on the set of morphisms from x to y in X.*

Proof For any two morphisms $f, g \colon x \to y$ in X, we have these two diagrams in X below.

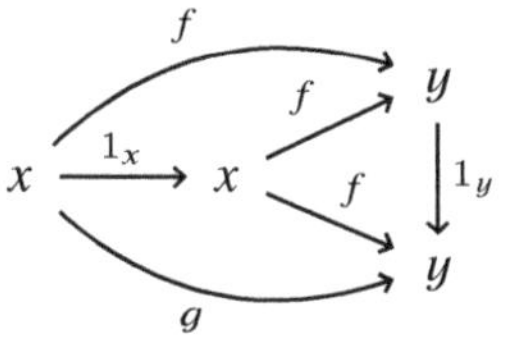

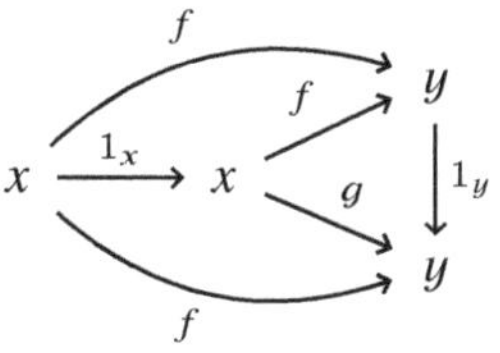

Since any triangle of the form

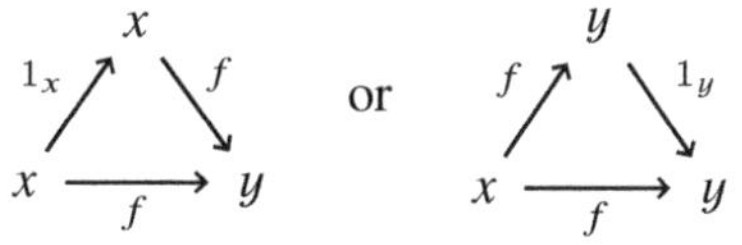

commutes (they are restrictions of 2-simplices of the form $\sigma^*(f)$, for appropriate surjective maps $\sigma \colon \Delta^2 \to \Delta^1$), we may apply the coherence lemma (1.6.2) and conclude that $1_y f \sim g \Leftrightarrow f1_x \sim g$ and that $1_y f \sim g \Rightarrow g1_x \sim f$. The same argument in the opposite ∞-category thus implies that these four relations are equal. Let us put $\simeq$ for this relation. It remains to prove that this is an equivalence relation. It is clear that $f \sim_1 f$ and that $f \sim_2 g \Leftrightarrow g \sim_4 f$. Therefore, it remains to prove the property of transitivity. Assume that $f \simeq g$ and $g \simeq h$. Then we have $g1_x \sim f$, $1_y g \sim h$ and $h1_x \sim h$. Applying the coherence lemma to the diagram

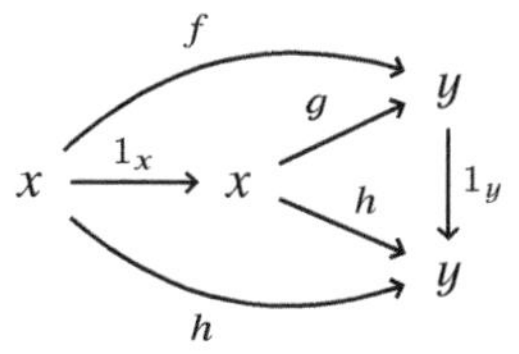

thus shows that $1_y f \sim h$. □

1.6.5 For two objects x and y, we define the set $\mathrm{Hom}_{ho(X)}(x, y)$ as the quotient of the set of morphisms from x to y in X by the equivalence relation $\sim_1$ given by Lemma 1.6.4 above. Given a morphism $f: x \to y$ in X, we write $[f]$ for its class in $\mathrm{Hom}_{ho(X)}(x, y)$. Finally, we define a composition law

$$\mathrm{Hom}_{ho(X)}(x, y) \times \mathrm{Hom}_{ho(X)}(y, z) \to \mathrm{Hom}_{ho(X)}(x, z)\,, \quad ([f], [g]) \mapsto [g] \circ [f]$$

by putting $[g] \circ [f] = [h]$ whenever h is a composition of f and g.

Theorem 1.6.6 (Boardman and Vogt) *The composition law constructed above is well defined, and this produces a category $ho(X)$. There is a unique morphism of simplicial sets $X \to N(ho(X))$ which is the identity on objects and which sends a morphism $f: x \to y$ in X to its class $[f]$ in $\mathrm{Hom}_{ho(X)}(x, y)$. Moreover, this morphism induces an isomorphism of categories*

$$\tau(X) \simeq ho(X)\,.$$

Proof If ever we have the relations $gf \sim h$ and $gf \sim h'$, applying the coherence lemma to the diagram

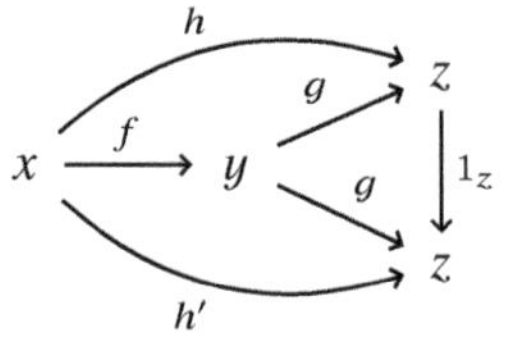

we see that we must have the relation $1_z h \sim h'$. Similarly, if $g \simeq g'$ and $gf \sim h$, applying the coherence lemma to the diagram

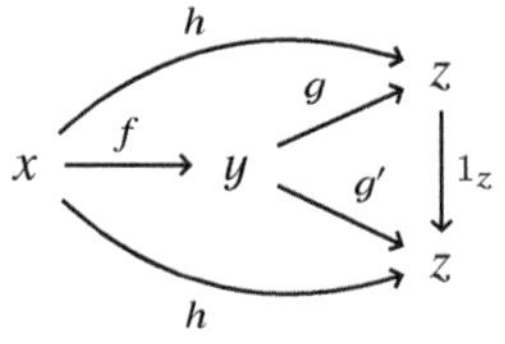

gives that $g'f \sim h$. The same argument in the opposite ∞-category shows that, if $f \simeq f'$ and if $gf \sim h$, then $gf' \sim h$. Therefore, the relation $gf \sim h$ only depends on the classes $[f]$, $[g]$ and $[h]$. This does prove, in particular, that the composition law is well defined. To check the associativity, we see that, for a triple of composable arrows (f, g, h), since compositions always exist in an ∞-category, one can complete the following diagram as follows: one chooses a composition a of (f, g), a composition b of (g, h), and, finally, a composition

c of (f, b).

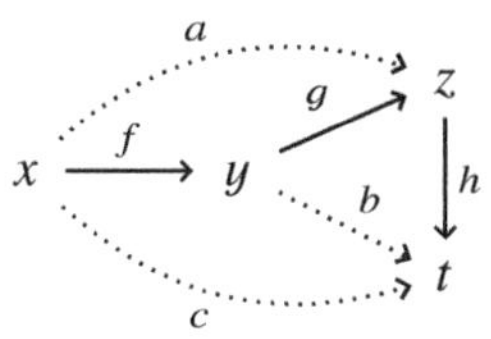

One then deduces from the coherence lemma that c is also a composition of (a, h). In other words, this composition law is associative:

$$([h] \circ [g]) \circ [f] = [b] \circ [f] = [c] = [h] \circ [a] = [h] \circ ([g] \circ [f]) .$$

The reflexivity of the equivalence relation $\sim_1 = \sim_2$ (1.6.4) means in particular that it is unital as well, with units the 1_x. The last assertions follows right away from the end of Remark 1.4.8: there is a unique map $X \to N(ho(X))$ as described in the statement of the theorem, and any map from X to the nerve of a small category factors uniquely through it. Hence $ho(X)$ and $\tau(X)$ are canonically isomorphic, because they represent the same functor. □

Corollary 1.6.7 *An ∞-category X is an ∞-groupoid if and only if the associated category $\tau(X) \simeq ho(X)$ is a groupoid.*

Corollary 1.6.8 *A morphism $f: x \to y$ in an ∞-category X is invertible if and only if there exists a morphism $g: y \to x$ such that 1_x is a composition of f and g, and that 1_y is a composition of g and f.*

Example 1.6.9 For any topological space X, the ∞-groupoid $Sing(X)$ thus defines a groupoid $\pi_1(X) = ho(Sing(X))$, which is nothing other than the *Poincaré groupoid*: the objects are the points of X, and the morphisms are the homotopy classes of continuous paths in X. In particular, for each point x of X, we have the *fundamental group* $\pi_1(X, x) = \mathrm{Hom}_{ho(Sing(X))}(x, x)$.

Definition 1.6.10 A *functor* between ∞-categories simply is a morphism of simplicial sets.

If X and Y are ∞-categories, and if $f, g: X \to Y$ are two functors, a *natural transformation from f to g* is a morphism $h: X \times \Delta^1 \to Y$ such that

$$h(x, 0) = f(x) \quad \text{and} \quad h(x, 1) = g(x)$$

or, more formally, if the following diagram commutes.

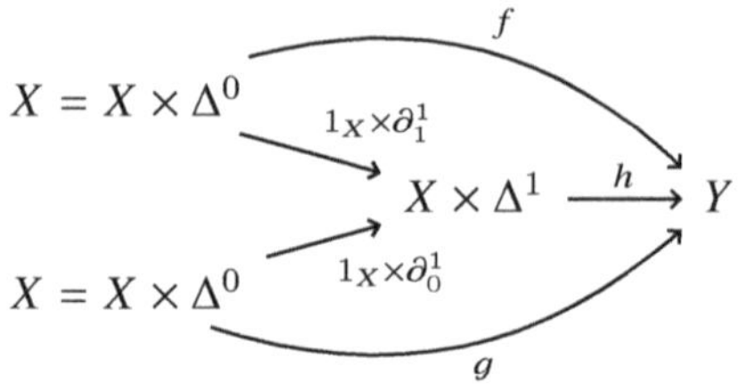

Such a natural transformation is *invertible* if, for any object x of X, the induced morphism $f(x) \to g(x)$ (corresponding to the restriction of h to $\Delta^1 = \{x\} \times \Delta^1$) is invertible in Y.

A functor $f\colon X \to Y$ is an *equivalence of* ∞*-categories* if there exist a functor $g\colon Y \to X$ as well as invertible natural transformations from fg to 1_Y as well as from 1_X to gf.

Remark 1.6.11 All these notions extend the usual ones for ordinary categories: the ordered set $[1]$ represents the set of morphisms in the category of small categories, and the nerve of the associated category $N([1])$ is canonically isomorphic to Δ^1, whence a natural transformation between ordinary functors is essentially the same thing as a natural transformation between their nerves. Since the nerve functor is fully faithful, we see that a functor between small categories is an equivalence of categories if and only if its nerve has the same property.

In order to work with ∞-categories, as above, we would like to have the usual categorical constructions: Cartesian products, limits and colimits, the ∞-category of functors between two ∞-categories, etc. If the latter exists, we would like natural transformations to be its morphisms, and the invertible natural transformations to be its invertible morphisms. This would imply that invertible natural transformations have inverses (whence their name). This is true, but non-trivial. Furthermore, we would like these categorical constructions to be compatible with equivalences of ∞-categories. As we will see in detail later, all these properties (and many more) will hold, but it comes with a cost: we must develop the homotopy theory of ∞-categories, in Quillen's setting of model categories (and we really must do it, since many of the properties we seek are equivalent to the verification of Quillen's axioms, at least in this particular context). The next chapter will be about the general theory of model categories, and about their constructions in the particular context of categories of presheaves.

2

Basic Homotopical Algebra

This chapter introduces Quillen's theory of model category structures. It starts with a recollection on factorisation systems: this level of generality will certainly help to understand much reasoning within Quillen's theory, but also to understand specific features of the theory of ∞-categories itself. We then give an exposition of all the basic constructions, such as the homotopy category, derived functors, or homotopy pull-backs. We give precise statements, often with the greatest level of generality we are aware of, but we sometimes give proofs that assume mild extra assumptions (which are always verified in the examples we will consider in this book), in which case we give a precise reference in the literature, pointing at a fully general proof (most of the time, such a reference is provided by Quillen's original monograph on *Homotopical algebra*, the reading of which we highly recommend). This part contains no original contribution, and only aims at introducing and producing the concepts needed to manipulate ∞-categories. However, the last chapter of this book will include homotopical algebra within the theory of ∞-category theories, hopefully enlightening and generalising this classical theory.

The second half of the chapter introduces a method to construct model category structures from scratch on categories of presheaves, when we choose to define the cofibrations to be the monomorphisms. This latter part is extracted right away from a little portion of [Cis06]; we only give an account of the constructions and proofs that are relevant for the present text, though. The model category structures defining the homotopy theory of ∞-categories will be constructed as a particular case in the next chapter. The last section of this chapter consists in observing that the data used to construct a given model category structure of that type can be used in fact to produce non-trivial *families* of model category structures. The class of absolute weak equivalences, i.e. of maps which can be interpreted as weak equivalences in the entire family, is an interesting subject of study. This apparently technical observation will not be used at first, when we will use these constructions to study ∞-groupoids and ∞-categories, but will be at the heart of our understanding of the homotopy theory of presheaves on an ∞-category in Chapter 4: the class of final functors is an instance of a class of absolute weak equivalences.

2.1 Factorisation Systems

Let us begin with some definitions.

Definition 2.1.1 Let $i\colon A \to B$ and $p\colon X \to Y$ be two morphisms in a category $\mathcal{C}$.

We say that i has the *left lifting property* with respect to p, or, equivalently, that p has the *right lifting property* with respect to i, if any commutative square of the form

$$\begin{array}{ccc} A & \xrightarrow{a} & X \\ {\scriptstyle i}\downarrow & & \downarrow{\scriptstyle p} \\ B & \xrightarrow[b]{} & Y \end{array}$$

has a diagonal filler

$$\begin{array}{ccc} A & \xrightarrow{a} & X \\ {\scriptstyle i}\downarrow & \nearrow^{h} & \downarrow{\scriptstyle p} \\ B & \xrightarrow[b]{} & Y \end{array}$$

(i.e. a morphism h such that $hi = a$ and $ph = b$).

Let F be a class of morphisms in $\mathcal{C}$. A morphism has the left (right) lifting property with respect to F if it has the left (right) lifting property with respect to any element of F.

One denotes by $l(F)$ (by $r(F)$) the class of morphisms which has the left (right) lifting property with respect to F.

Definition 2.1.2 An object X is a *retract* of another object U if there exists a commutative diagram of the following form.

$$X \overbrace{\xrightarrow{i} U \xrightarrow{p}}^{1_X} X$$

We say that a morphism $f\colon X \to Y$ is a retract of a morphism $g\colon U \to V$ if it is so in the category of morphisms: in other words, if there exists a commutative diagram of the form below.

$$
\begin{array}{ccccc}
 & & \overset{1_X}{\frown} & & \\
X & \xrightarrow{i} & U & \xrightarrow{p} & X \\
{\scriptstyle f}\downarrow & & \downarrow{\scriptstyle g} & & \downarrow{\scriptstyle f} \\
Y & \xrightarrow{j} & V & \xrightarrow{q} & Y \\
 & & \underset{1_Y}{\smile} & &
\end{array}
$$

A class of morphisms F is stable under *retracts* if any morphism which is a retract of an element of F belongs to F.

A class of morphisms F is stable under *push-outs* if, for any push-out square of the form

$$
\begin{array}{ccc}
X & \xrightarrow{a} & X' \\
{\scriptstyle f}\downarrow & & \downarrow{\scriptstyle f'} \\
Y & \xrightarrow{b} & Y'
\end{array}
$$

if f is in F, so is f'.

A class of morphisms F, in a category $\mathcal{C}$, is stable under *transfinite compositions* if, for any well-ordered set I, with initial element 0, for any functor $X\colon I \to \mathcal{C}$ such that, for any element $i \in I$, $i \neq 0$, the colimit $\varinjlim_{j<i} X(j)$ is representable and the induced map

$$\varinjlim_{j<i} X(j) \to X(i)$$

belongs to the class F, the colimit $\varinjlim_{i\in I} X(i)$ exists and the canonical morphism $X(0) \to \varinjlim_{i\in I} X(i)$ belongs to F as well.

A class of morphisms is *saturated* if it is stable under retracts, under push-outs and under transfinite compositions.

Remark 2.1.3 If a class $\mathcal{F}$ of morphisms contains all the identities and is stable under push-outs as well as by transfinite compositions, then it is stable under small sums: for any small set I and any family of maps $u_i\colon X_i \to Y_i$ in $\mathcal{F}$, indexed by $i \in I$, the induced map

$$\coprod_{i\in I} X_i \to \coprod_{i\in I} Y_i$$

is in $\mathcal{F}$.

The following proposition follows straight away from the definitions.

Proposition 2.1.4 *Let $\mathcal{C}$ be a category, together with two classes of morphisms F and F'. We have the following properties.*

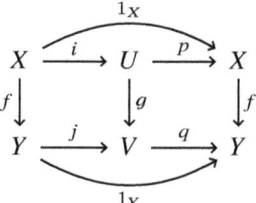

(a) $F \subset r(F') \Leftrightarrow F' \subset l(F)$.

(b) $F \subset F' \Rightarrow l(F') \subset l(F)$.

(c) $F \subset F' \Rightarrow r(F') \subset r(F)$.

(d) $r(F) = r(l(r(F)))$.

(e) $l(F) = l(r(l(F)))$.

(f) *The class $l(F)$ is saturated.*

(g) *The class $r(F)$ is cosaturated, i.e. is saturated as a class of morphisms of $\mathcal{C}^{\mathrm{op}}$.*

And here is a useful trick.

Proposition 2.1.5 (Retract lemma) *Assume that a morphism $f : X \to Y$ can be factored into $f = pi$.*

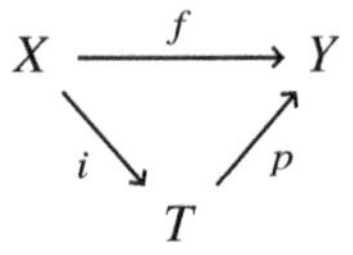

If f has the right (left) lifting property with respect to i (to p), then f is a retract of p (of i, respectively).

Proof If $f \in r(i)$, then the diagonal of the solid commutative square

$$
\begin{array}{ccc}
X & === & X \\
{\scriptstyle i}\downarrow & \overset{h}{\nearrow} & \downarrow{\scriptstyle f} \\
T & \xrightarrow{\ p\ } & Y
\end{array}
$$

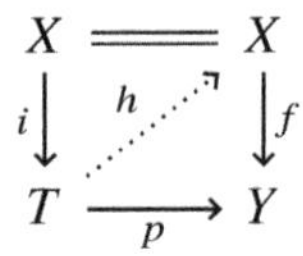

gives rise to a commutative diagram of the form

$$
\begin{array}{ccccc}
& & \overset{1_X}{\overparen{\qquad\qquad}} & & \\
X & \xrightarrow{\ i\ } & T & \xrightarrow{\ h\ } & X \\
{\scriptstyle f}\downarrow & & \downarrow{\scriptstyle p} & & \downarrow{\scriptstyle f} \\
Y & === & Y & === & Y
\end{array}
$$

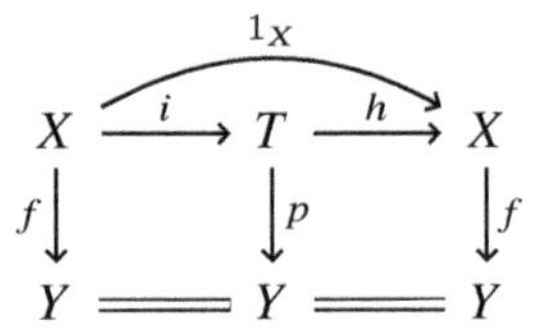

and thus turns f into a retract of p. The respective case follows by considering what precedes in the opposite category of $\mathcal{C}$. $\qquad\square$

Example 2.1.6   For $\mathcal{C} = Set$, and $i : \varnothing \to \{\text{point}\}$, the class $r(i)$ is the class of surjective maps, while a reformulation of the axiom of choice is the assertion that the class $l(r(i))$ is the class of injective maps. Since any small set is a small sum of sets with one element, the excluded middle principle means that the smallest saturated class of maps in *Set* which contains i is the class $l(r(i))$.

Definition 2.1.7 A *weak factorisation system* in a category $\mathcal{C}$ is a couple (A, B) of classes of morphisms satisfying the following properties:

(a) both A and B are stable under retracts;
(b) $A \subset l(B)$ ($\Leftrightarrow B \subset r(A)$);
(c) any morphism $f: X \to Y$ of $\mathcal{C}$ admits a factorisation of the form $f = pi$, with $i \in A$ and $p \in B$.

Remark 2.1.8 It follows from the retract lemma that we must have $A = l(B)$ as well as $B = r(A)$.

We recall the following general facts (for proofs, see [Hov99, theorem 2.1.14] or [Rie14, theorem 12.2.2], where more general assumptions will also be found).

Given a cardinal κ, a non-empty partially ordered set E is *κ-filtered* if, for any family of its elements x_j, $j \in J$, indexed through a set J of cardinal $\leq \kappa$, there exists an element x in E such that $x_j \leq x$ holds for each $j \in J$.

Proposition 2.1.9 (Small object argument) *Let $\mathcal{C}$ be a locally small category with small colimits, endowed with a small set of morphisms I. Assume that there exists a cardinal κ such that, for any element $i: K \to L$ of I, the functor*

$$\mathrm{Hom}_{\mathcal{C}}(K, -): \mathcal{C} \to Set$$

commutes with colimits indexed by κ-filtered well-ordered sets. Then the couple $(l(r(I)), r(I))$ is a weak factorisation system. Furthermore, $l(r(I))$ is the smallest saturated class containing I.

Corollary 2.1.10 *Let A be a small category, and I a small set of morphisms of presheaves over A. Then the couple $(l(r(I)), r(I))$ is a weak factorisation system in $\widehat{A}$.*

Proof We only need to check that the small object argument may be used here. Let κ be a regular cardinal. We want to prove that, for any presheaf of sets X on A, if κ is big enough, then the functor $\mathrm{Hom}(X, -)$ commutes with colimits indexed by κ-filtered well-ordered sets. If there is an object a of A such that X is isomorphic to the presheaf represented by a, then, by the Yoneda lemma, the functor $\mathrm{Hom}(X, -)$ is isomorphic to the functor of evaluation at a, which obviously commutes with all limits. If J is a set of cardinal $< \kappa$, the functor $(X_j)_{j \in J} \mapsto \prod_{j \in J} X_j$ commutes with colimits indexed by κ-filtered well-ordered sets (we leave the proof of this assertion to the reader). Therefore, if X is isomorphic to a J-indexed sum of representable presheaves, the functor $\mathrm{Hom}(X, -)$ is isomorphic to a J-indexed product of evaluation functors, and thus commutes with colimits indexed by κ-filtered well-ordered sets. In general, X is a small colimit of representable presheaves, whence a co-equaliser of maps between sums of representable presheaves. Since filtered colimits commute with finite limits [Rie17, theorem 3.8.9], we observe that the functor $\mathrm{Hom}(X, -)$ commutes with colimits indexed by κ-filtered well-ordered sets, for κ big enough. □

Example 2.1.11 In any category of presheaves over a small category A (or, more generally, in any Grothendieck topos), the class of monomorphisms is part of a weak factorisation system. Let us call *trivial fibrations* the morphisms of presheaves which have the right lifting property with respect to monomorphisms. We may prove that any morphism of presheaves over A can be factored as a monomorphism $i: X \to Y$ followed by a trivial fibration $p: Y \to Z$ in two ways.

First method. We have the subobject classifier Ω (also called the Lawvere object in Grothendieck's *Pursuing stacks*). In other words, for any object a in A, the set Ω_a is the set of subobjects of the representable presheaf h_a (the structure of a presheaf is given by pull-backs of subobjects). One checks that, for any presheaf X over A, there is a canonical bijection

$$\mathrm{Hom}(X, \Omega) \simeq \{\text{subobjects of } X\}.$$

We put $P(X) = \underline{\mathrm{Hom}}(X, \Omega)$. There is then a canonical embedding

$$\{.\}: X \to P(X),$$

which corresponds to the diagonal of $X \times X$ (seen as a subobject). We see that $P(X)$ is an injective object (i.e. that the map from $P(X)$ to the point is a trivial fibration): this is a reformulation of the fact that, for any monomorphism $K \to L$, any subobject of the Cartesian product $K \times X$ induces a subobject of $L \times X$. In other words, we have factored the map from X to the point into a monomorphism followed by a trivial fibration. The general case follows by applying the previous construction to the category of presheaves over A/Y, using the equivalence of categories $\widehat{A}/Y \simeq \widehat{A/Y}$, discussed in Remark 1.1.14.

Second method. One checks (using the axiom of choice) that the class of monomorphisms is equal to the class $l(r(I))$, for I the set of monomorphisms of presheaves of the form $K \to L$, with L a quotient of some representable presheaf over A. One can then apply the small object argument.

A basic, although extremely useful, recognition of lifting properties is the following one.

Proposition 2.1.12 *Let $F: \mathcal{C} \rightleftarrows \mathcal{C}': G$ be an adjunction. Assume that $\mathcal{C}$ and $\mathcal{C}'$ are endowed with weak factorisation systems (A, B) and (A', B'), respectively. Then we have $F(A) \subset A'$ if and only if $G(B') \subset B$.*

Proof We have, by adjunction, a natural correspondence of the following form.

$$\begin{array}{ccc} F(K) & \xrightarrow{a'} & X \\ {\scriptstyle F(i)}\big\downarrow & \overset{h'}{\nearrow} & \big\downarrow{\scriptstyle p} \\ F(L) & \xrightarrow[b']{} & Y \end{array} \quad \leftrightsquigarrow \quad \begin{array}{ccc} K & \xrightarrow{a} & G(X) \\ {\scriptstyle i}\big\downarrow & \overset{h}{\nearrow} & \big\downarrow{\scriptstyle G(p)} \\ L & \xrightarrow[b]{} & G(Y) \end{array}$$

□

2.2 Model Categories

The notion of model category was introduced by Quillen in [Qui67]. All the results of the next two sections are already in [Qui67].

Definition 2.2.1 A *model category* is a locally small category $\mathcal{C}$ endowed with three classes of morphisms W, Fib and Cof, such that the following three properties are verified.

(1) The category $\mathcal{C}$ has finite limits and finite colimits.
(2) The class W has the two-out-of-three property: for any commutative triangle of the form

$$\begin{array}{ccc} X & \xrightarrow{f} & Y \\ & {\scriptstyle h}\searrow \quad \swarrow{\scriptstyle g} & \\ & Z & \end{array}$$

if two, among f, g and h, are in W, so is the third.
(3) Both couples $(Cof, Fib \cap W)$ and $(Cof \cap W, Fib)$ are weak factorisation systems.

Here is the standard terminology in the presence of such a structure.

An element of W (of Fib, of Cof) is called a *weak equivalence* (a *fibration*, a *cofibration*, respectively). A morphism that is both a weak equivalence and a fibration (and a cofibration) is called a *trivial* fibration (a *trivial* cofibration, respectively).

We shall often write $\varnothing$ and e for the initial and final objects of $\mathcal{C}$, respectively. An object X of $\mathcal{C}$ is *fibrant* (*cofibrant*) if the unique morphism $X \to e$ ($\varnothing \to X$) is a fibration (a cofibration, respectively).

Remark 2.2.2 These axioms imply that any isomorphism is both a trivial fibration and a trivial cofibration (in particular, a weak equivalence).

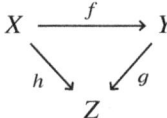

Remark 2.2.3 This definition is (equivalent to) the notion of *closed* model category introduced by Quillen in [Qui67]. Axiom 3 implies in particular that any morphism f has a factorisation of the form $f = pi$ where i is a cofibration, and p a trivial fibration, as well as a factorisation of the form $f = qj$, where j is a trivial cofibration, and q a fibration.

In some modern introductions to the theory of model categories, the underlying category is assumed to have small limits and small colimits (as opposed to finite ones), and the factorisations are required to exist functorially. All the examples we will consider here will be of this sort. There are many reasons not to rely on functorial factorisations, though. In a model category, the meaningful part is the class of weak equivalences. Fibrations and cofibrations really are intermediate tools (in particular, we should always feel free to replace these at will, as long as this makes sense, of course).

Example 2.2.4 For any category with finite limits and colimits, we have a model category structure for which the weak equivalences are the isomorphisms, any morphism being a cofibration as well as a fibration.

Example 2.2.5 It follows from Example 2.1.11 that, for any small category A, the category of presheaves over A has a model category structure whose class of weak equivalences is the class of all morphisms, while the cofibrations are the monomorphisms.

Proposition 2.2.6 *The notion of model category structure is stable under the following categorical constructions.*

(a) If $\mathcal{C}$ is a model category, so is $\mathcal{C}^{op}$: the weak equivalences are those of $\mathcal{C}$, while its fibrations (cofibrations) are the cofibrations (fibrations, respectively) of $\mathcal{C}$.

(b) For any object X in a model category $\mathcal{C}$, the slice category $\mathcal{C}/X$ has a natural structure of model category: the weak equivalences (fibrations, cofibrations) are the morphisms whose image in $\mathcal{C}$ are weak equivalences (the fibrations, the cofibrations, respectively).

(c) We can put the preceding two constructions together: the category of objects under X has a natural structure of model category.

Proposition 2.2.7 (Ken Brown's lemma) *Let $\mathcal{C}$ be a model category, together with a functor $F\colon \mathcal{C} \to \mathcal{D}$. Assume that the category $\mathcal{D}$ is endowed with a class of weak equivalences, by which we mean a class of morphisms which contains all isomorphisms and which has the two-out-of-three property. If ever the functor F sends trivial cofibrations between cofibrant objects to weak equivalences, then it sends weak equivalences between cofibrant objects to weak equivalences.*

Proof Let $f\colon X \to Y$ be a weak equivalence between cofibrant objects. We may form the push-out square

$$\begin{array}{ccc} \varnothing & \longrightarrow & Y \\ \downarrow & & \downarrow{\scriptstyle j} \\ X & \xrightarrow{i} & X \amalg Y \end{array}$$

and see that, since both X and Y are cofibrant, the canonical maps from X and Y to the coproduct $X \amalg Y$ are cofibrations. We may factor the map $(f, 1_Y) : X \amalg Y \to Y$ into a cofibration $k: X \amalg Y \to T$ followed by a weak equivalence (a trivial fibration) $p: T \to Y$. We thus have the following two commutative triangles.

$$\begin{array}{ccc} X & \xrightarrow{ki} & T \\ & {\scriptstyle f}\searrow \quad \swarrow{\scriptstyle p} & \\ & Y & \end{array} \qquad \begin{array}{ccc} Y & \xrightarrow{kj} & T \\ & {\scriptstyle 1_Y}\searrow \quad \swarrow{\scriptstyle p} & \\ & Y & \end{array}$$

Since the map $F(p)$ has a section which is the image of a trivial cofibration between cofibrant objects, it is a weak equivalence. On the other hand, the map ki is a trivial cofibration between cofibrant objects, and therefore its image $F(ki)$ is a weak equivalence. The two-out-of-three property for weak equivalences in $\mathcal{D}$ implies that $F(f)$ is a weak equivalence. □

Definition 2.2.8 Let $\mathcal{C}$ be a category endowed with a class of morphisms W. A *localisation* of $\mathcal{C}$ (by W) is a functor

(2.2.8.1) $$\gamma : \mathcal{C} \to ho(\mathcal{C})$$

that sends the elements of W to isomorphisms, and which is universal for this latter property. In other words, we ask that, for any category $\mathcal{D}$, if we denote by $\underline{\mathrm{Hom}}_W(\mathcal{C}, \mathcal{D})$ the full subcategory of the category of functors $\underline{\mathrm{Hom}}(\mathcal{C}, \mathcal{D})$ which consists of functors sending elements of W to isomorphisms, then the operation of composing with γ

(2.2.8.2) $$\gamma^* : \underline{\mathrm{Hom}}(ho(\mathcal{C}), \mathcal{D}) \to \underline{\mathrm{Hom}}_W(\mathcal{C}, \mathcal{D})$$

is an equivalence of categories. By abuse of terminology, we shall say that $ho(\mathcal{C})$ is the localisation of $\mathcal{C}$.

Proposition 2.2.9 *There always exists a localisation of $\mathcal{C}$ by W. Moreover, one can choose a localisation $\gamma : \mathcal{C} \to ho(\mathcal{C})$ in such a way that the map* (2.2.8.2) *is an isomorphism of categories. Under this more rigid constraint, the functor γ is a bijection on objects.*

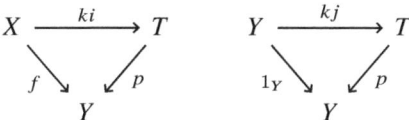

One defines the category $ho(\mathcal{C})$ as follows. The objects are those of $\mathcal{C}$, and the set of morphisms from X to Y is an appropriate quotient of the class of diagrams of the form

$$X = X_0 \longleftarrow X_1 \longrightarrow X_2 \longleftarrow \cdots \longrightarrow X_{n-1} \longleftarrow X_n \longrightarrow X_{n+1} = Y$$

where $n \geq 0$ is an integer, and where the maps of the form $\longleftarrow$ all are in W or are identities.[1] The equivalence relation we apply means that we can always replace such a diagram by a bigger one by adding identities anywhere we want (and in any direction we want). Any two such diagrams of the same length are equivalent whenever they can be connected through a commutative diagram in $\mathcal{C}$ of the form below.

$$\begin{array}{ccccccccccc} & & X_1 & \longrightarrow & X_2 & \longleftarrow & \cdots & \longrightarrow & X_{n-1} & \longleftarrow & X_n & & \\ & \swarrow & \downarrow & & \downarrow & & & & \downarrow & & \downarrow & \searrow & \\ X & & & & & & & & & & & & Y \\ & \nwarrow & X'_1 & \longrightarrow & X'_2 & \longleftarrow & \cdots & \longrightarrow & X'_{n-1} & \longleftarrow & X'_n & \nearrow & \end{array}$$

Remark 2.2.10 There is no reason, in general, why the sets $\mathrm{Hom}_{ho(\mathcal{C})}(X, Y)$ would be small. Therefore, for the moment, this construction only makes sense for non-necessarily locally small categories. We shall review first how to compute these sets of morphisms, and see in particular that, if W is the class of weak equivalences of a model category structure on $\mathcal{C}$, they are small sets.

From now on, let us consider a fixed model category $\mathcal{C}$.

Definition 2.2.11 A *cylinder* of an object A is a factorisation of the codiagonal of A into a cofibration followed by a weak equivalence, i.e. a commutative diagram of the form

(2.2.11.1)
$$A \amalg A \xrightarrow{(\partial_0, \partial_1)} IA \xrightarrow{\sigma} A \qquad \text{(with composite } (1_A, 1_A))$$

in which (∂_0, ∂_1) is a cofibration, and σ a weak equivalence. Dually, a *cocylinder* (we also say a *path object*) of an object X is a cylinder in the opposite category, i.e. a commutative diagram of the form

(2.2.11.2)
$$X \xrightarrow{s} X^I \xrightarrow{(d^0, d^1)} X \times X \qquad \text{(with composite } (1_X, 1_X))$$

in which the map s is a weak equivalence, and the map (d^0, d^1) is a fibration.

Let us consider two morphisms $f_0, f_1 \colon A \to X$.

[1] Or one can also assume that W contains all isomorphisms, since this does not affect the categories of the form $\underline{\mathrm{Hom}}_W(\mathcal{C}, \mathcal{D})$.

A *left homotopy* (a *right homotopy*) from f_0 to f_1 is given by a cylinder of A of the form (2.2.11.1) (a cocylinder of X of the form (2.2.11.2)), together with a morphism $h\colon IA \to X$ (a morphism $k\colon A \to X^I$), such that, for $i = 0, 1$, we have $h\partial_i = f_i$ (we have $d^i k = f_i$, respectively).

Lemma 2.2.12 *Let A and X be a cofibrant object and a fibrant object, respectively. For a pair of morphisms $f_i\colon A \to X$, $i = 0, 1$, the following conditions are equivalent.*

(a) *There exists a left homotopy from f_0 to f_1.*
(b) *There exists a right homotopy from f_0 to f_1.*
(c) *For any cylinder of A of the form* (2.2.11.1), *there exists a map $h\colon IA \to X$ such that $h\partial_i = f_i$ for $i = 0, 1$.*
(d) *For any cocylinder of X of the form* (2.2.11.2), *there exists a map $k\colon A \to X^I$ such that $d^i k = f_i$ for $i = 0, 1$.*

Proof Since we can replace the underlying category by its opposite $\mathcal{C}^{op}$, it is clearly sufficient to prove that (a) implies (d). Let us assume that there is a left homotopy from f_0 to f_1 given by a cylinder of A of the form (2.2.11.1), together with a map $h\colon IA \to X$, and let us consider an arbitrary cocylinder of X, of the form (2.2.11.2). Since the map $\partial_1\colon A \to IA$ is a trivial cofibration (because this is a weak equivalence, and since A is cofibrant, this is also a cofibration, for each canonical map $A \to A \amalg A$ is a push-out of the cofibration $\varnothing \to A$), and since the map (d^0, d^1) is a fibration, the following solid commutative square

$$\begin{array}{ccc} A & \xrightarrow{sf_1} & X^I \\ {\scriptstyle \partial_1}\downarrow & \overset{K}{\nearrow} & \downarrow{\scriptstyle (d^0,d^1)} \\ IA & \xrightarrow[(h, f_1\sigma)]{} & X \times X \end{array}$$

admits a filling K. Let us put $k = K\partial_0$. We then have:

$$d^1 k = d^1 K\partial_0 = f_1\sigma\partial_0 = f_1 1_A = f_1 \quad \text{and} \quad d^0 k = d^0 K\partial_0 = h\partial_0 = f_0\,.$$

In other words, condition (d) is verified. □

Lemma 2.2.13 *Let A and X be a cofibrant object and a fibrant object, respectively. We define an equivalence relation $\sim$ on the set of morphisms from A to X, by defining $f_0 \sim f_1$ whenever there exists a left (or right) homotopy from f_0 to f_1.*

Proof The reflexivity is clear: for any cylinder of the form (2.2.11.1), and for any map $f\colon A \to X$, the morphism $f\sigma$ defines a left homotopy from f to itself. The equivalence between conditions (a) and (c) of Lemma 2.2.12 implies that the relation $\sim$ is symmetric: indeed, for any cylinder of A of the form (2.2.11.1), the diagram

$$A \amalg A \xrightarrow{(\partial_1, \partial_0)} IA \xrightarrow{\sigma} A$$

is a cylinder of A as well. Assume that we have three maps u, v and w from A to X, as well as cylinders

$$A \amalg A \xrightarrow{(\partial_0, \partial_1)} IA \xrightarrow{\sigma} A, \qquad A \amalg A \xrightarrow{(\partial'_0, \partial'_1)} I'A \xrightarrow{\sigma'} A$$

and homotopies $h\colon IA \to X$ and $h'\colon I'A \to X$ satisfying $h\partial_0 = u$, $h\partial_1 = v = h'\partial'_0$ and $h'\partial'_1 = w$. We form the following push-out square.

$$\begin{array}{ccc} A & \xrightarrow{\partial_1} & IA \\ {\scriptstyle \partial'_0}\downarrow & & \downarrow{\scriptstyle e} \\ I'A & \xrightarrow{e'} & I''A \end{array}$$

There is a unique map $\sigma''\colon I''A \to A$ such that $\sigma''e = \sigma$ and $\sigma''e' = \sigma'$, and we get a cylinder of the form

$$A \amalg A \xrightarrow{(\partial''_0, \partial''_1)} I''A \xrightarrow{\sigma''} A$$

by defining $\partial''_0 = e\partial_0$ and $\partial''_1 = e'\partial'_1$. Indeed, it is clear that the map σ'' is a weak equivalence (because e is a trivial cofibration and σ a weak equivalence). One thus only has to show that the map $(\partial''_0, \partial''_1)$ is a cofibration. In the commutative diagram

$$\begin{array}{ccccc} A & \longrightarrow & A \amalg A & \xrightarrow{(\partial_0, \partial_1)} & IA \\ {\scriptstyle \partial'_0}\downarrow & & \downarrow{\scriptstyle 1_A \amalg \partial'_0} & & \downarrow{\scriptstyle e} \\ I'A & \longrightarrow & A \amalg I'A & \xrightarrow{(\partial''_0, e')} & I''A \end{array}$$

the left-hand square is the obvious push-out square, while the composed square is the previous push-out square, hence the right-hand square is coCartesian. In particular, the map (∂''_0, e') is a cofibration. Composing the latter with the cofibration $1_A \amalg \partial'_1$ thus gives a cofibration which is nothing other than $(\partial''_0, \partial''_1)$. Finally, we define a morphism $h''\colon I''A \to X$ as the unique one such that $h''e = h$ and $h''e' = h'$. It is clear that $h''\partial''_0 = u$ and $h''\partial''_1 = w$. □

Notation 2.2.14 Under the assumptions of the preceding lemma, we write

(2.2.14.1) $$[A, X] = \mathrm{Hom}_{\mathcal{C}}(A, X)/\sim$$

for the quotient of the set of morphisms from A to X by the relation of left (or right) homotopy $\sim$. It is clear that the relation of left homotopy is compatible with composition on the left, while the relation of right homotopy is compatible with composition on the right. In other words, we have defined a functor

$$[-,-]\colon \mathcal{C}_c^{\mathrm{op}} \times \mathcal{C}_f \to Set\,, \tag{2.2.14.2}$$

where $\mathcal{C}_c$ and $\mathcal{C}_f$ denote the full subcategories of $\mathcal{C}$ spanned by cofibrant objects and by fibrant objects, respectively.

Theorem 2.2.15 *The inclusion $\mathcal{C}_c \to \mathcal{C}$ induces an equivalence of categories of the form $ho(\mathcal{C}_c) \simeq ho(\mathcal{C})$ (where $ho(\mathcal{C}_c)$ denotes the localisation of $\mathcal{C}_c$ by the class of weak equivalences between cofibrant objects).*

Dually, the inclusion $\mathcal{C}_f \to \mathcal{C}$ induces an equivalence of categories of the form $ho(\mathcal{C}_f) \simeq ho(\mathcal{C})$.

This theorem is a triviality whenever the factorisations of the model structure can be obtained functorially (which will be the case in all the examples in this book). The general case is proven in [Qui67, chapter I, 1.13, theorem 1]; in Chapter 7, using an enhanced version of the theory of derived functors, we shall improve this statement a lot (Theorem 7.5.18).

Proposition 2.2.16 *The functor* (2.2.14.2) *is compatible with weak equivalences in $\mathcal{C}_c$ and in $\mathcal{C}_f$ (i.e. sends such weak equivalences to invertible natural transformations) and thus defines a functor*

$$[-,-]\colon ho(\mathcal{C}_c)^{\mathrm{op}} \times ho(\mathcal{C}_f) \to Set\,.$$

Proof Let $i\colon A \to B$ be a weak equivalence between cofibrant objects, and X a fibrant object. We shall prove that the induced map

$$i^*\colon [B,X] \to [A,X]$$

is bijective. By virtue of Ken Brown's lemma (Proposition 2.2.7), it is sufficient to consider the case where i is a trivial cofibration between cofibrant objects. The surjectivity is easy: for any morphism $f\colon A \to X$, the solid commutative square

$$\begin{array}{ccc} A & \xrightarrow{f} & X \\ {\scriptstyle i}\downarrow & \overset{g}{\nearrow} & \downarrow \\ B & \longrightarrow & e \end{array}$$

admits a filler g whose homotopy class is sent by i^* to the homotopy class of f. As for the injectivity, let us consider two morphisms $f, g\colon B \to X$ such that fi and gi are homotopic (i.e. their classes are equal in $[A,X]$). Then we may choose a right homotopy from fi to gi, defined by a cocylinder of the form

(2.2.11.2), together with a map $k\colon A \to X^I$ such that $d^0 k = fi$ and $d^1 k = gi$. The following solid commutative square

$$\begin{array}{ccc} A & \xrightarrow{\;k\;} & X^I \\ {\scriptstyle i}\downarrow & \overset{K}{\nearrow} & \downarrow{\scriptstyle (d^0,d^1)} \\ B & \xrightarrow[(f,g)]{} & X \times X \end{array}$$

admits a filler: the left-hand vertical map is a trivial cofibration, and the map (d^0, d^1) is a fibration. Therefore, f and g are homotopic.

Applying what precedes to $\mathcal{C}^{op}$, we see that, for any cofibrant object A and any weak equivalence between fibrant objects $p\colon X \to Y$, the induced map

$$p_*\colon [A, X] \to [A, Y]$$

is bijective. □

Theorem 2.2.17 *For any cofibrant object A and any fibrant object X, there is a canonical bijection*

$$[A, X] \simeq \mathrm{Hom}_{ho(\mathcal{C})}(A, X)$$

which is natural with respect to morphisms of $ho(\mathcal{C}_c)^{op} \times ho(\mathcal{C}_f)$.

Corollary 2.2.18 *Let $\mathcal{C}_{cf}$ be the full subcategory of objects which are both fibrant and cofibrant. The relation of homotopy of Lemma 2.2.13 defines an equivalence relation which is compatible with composition on $\mathcal{C}_{cf}$. The resulting quotient category $\pi(\mathcal{C}_{cf})$ thus has the fibrant-cofibrant objects of $\mathcal{C}$ as objects, while we have*

$$\mathrm{Hom}_{\pi(\mathcal{C}_{cf})}(A, X) = [A, X]\,.$$

The inclusion functor $\mathcal{C}_{cf} \subset \mathcal{C}$ induces a canonical equivalence of categories

$$\pi(\mathcal{C}_{cf}) \simeq ho(\mathcal{C})\,.$$

Theorem 2.2.17 is in fact a consequence of the corollary (using Theorem 2.2.15, we see that this is because any fibrant object X is isomorphic in $ho(\mathcal{C})$ to an object A which is both fibrant and cofibrant: simply consider a factorisation of the map $\varnothing \to X$ into a cofibration followed by a trivial fibration $A \to X$).

Let us assume that the inclusion functor $\mathcal{C}_{cf} \subset \mathcal{C}$ induces a canonical equivalence of categories $ho(\mathcal{C}_{cf}) \simeq ho(\mathcal{C})$, where $ho(\mathcal{C}_{cf})$ stands for the localisation of $\mathcal{C}_{cf}$ by the class of weak equivalences (this is easy to check when the factorisations can be obtained functorially). We can then directly prove Corollary 2.2.18 as follows.

First, one checks that if two maps $f, g\colon A \to X$ in $\mathcal{C}_{cf}$ are homotopic, then

$\gamma(f) = \gamma(g)$ in $ho(\mathcal{C})$. Let IA be a cylinder of A as in (2.2.11.1), and $h\colon A \to X$ be a map such that $h\partial_0 = f$ and $h\partial_1 = g$. Since $\sigma\partial_i = 1_X$ does not depend on i, and since σ becomes invertible in $ho(\mathcal{C})$, we must have $\gamma(\partial_0) = \gamma(\partial_1)$ in $ho(\mathcal{C})$. Therefore, the morphism $\gamma(g)$ must be equal to $\gamma(f)$.

As a consequence, any homotopy equivalence (i.e. any map of $\mathcal{C}_{cf}$ which becomes an isomorphism in $\pi(\mathcal{C}_{cf})$) induces an isomorphism in $ho(\mathcal{C})$. On the other hand, Proposition 2.2.16 implies that any weak equivalence $u\colon A \to B$ in $\mathcal{C}_{cf}$ must induce bijections $[B, X] \simeq [A, X]$ for any object X in $\mathcal{C}_{cf}$, and thus, by virtue of the Yoneda lemma for $\pi(\mathcal{C}_{cf})$, must become invertible in $\pi(\mathcal{C}_{cf})$. In other words, a morphism of $\mathcal{C}_{cf}$ is a weak equivalence if and only if it is a homotopy equivalence. Since the category $\pi(\mathcal{C}_{cf})$ is the localisation of $\mathcal{C}_{cf}$ by the class of homotopy equivalences, this shows that $\pi(\mathcal{C}_{cf}) = ho(\mathcal{C})$.

Corollary 2.2.19 *Let $\mathcal{C}$ be a model category endowed with a localisation functor $\gamma\colon \mathcal{C} \to ho(\mathcal{C})$. If $\mathcal{C}$ is locally small, then so is the homotopy category $ho(\mathcal{C})$. Furthermore, a morphism f of $\mathcal{C}$ is a weak equivalence if and only if its image $\gamma(f)$ is an isomorphism in $ho(\mathcal{C})$.*

2.3 Derived Functors

Definition 2.3.1 Let $\mathcal{C}$ be a model category, together with a localisation functor $\gamma\colon \mathcal{C} \to ho(\mathcal{C})$, as well as a functor $F\colon \mathcal{C} \to \mathcal{D}$.

A *left derived* functor of F is a functor $\mathbf{L}F\colon ho(\mathcal{C}) \to \mathcal{D}$ together with a functorial morphism $a_X\colon \mathbf{L}F(\gamma(X)) \to F(X)$ which turns $\mathbf{L}F$ into a right Kan extension of F along the localisation functor γ. In other words, for any functor $\Phi\colon ho(\mathcal{C}) \to \mathcal{D}$ and any natural morphism $\alpha_X\colon \Phi(\gamma(X)) \to F(X)$, there is a unique natural morphism $f_Y\colon \Phi(Y) \to \mathbf{L}F(Y)$ such that $\alpha_X = a_X f_{\gamma(X)}$ for all objects X of $\mathcal{C}$.

A *right derived* functor of F is a functor $\mathbf{R}F\colon ho(\mathcal{C}) \to \mathcal{D}$ together with a functorial morphism $b_X\colon F(X) \to \mathbf{R}F(\gamma(X))$ which turns $\mathbf{R}F$ into a left Kan extension of F along the localisation functor γ (i.e. $\mathbf{R}F^{\mathrm{op}}$ and b form a left derived functor of $F^{\mathrm{op}}\colon \mathcal{C}^{\mathrm{op}} \to \mathcal{D}^{\mathrm{op}}$).

2.3.2 Let $F\colon \mathcal{C} \to \mathcal{D}$ be a functor which sends trivial cofibrations between cofibrant objects to isomorphisms. By virtue of Ken Brown's lemma, the functor F sends weak equivalences between cofibrant objects to isomorphisms. Therefore, there is a unique functor $F_c\colon ho(\mathcal{C}_c) \to \mathcal{D}$ whose composition with the localisation functor $\mathcal{C}_c \to ho(\mathcal{C}_c)$ coincides with the restriction of F. Let us choose, for each object X of $\mathcal{C}$, a weak equivalence $a'_X\colon QX \to X$ with QX cofibrant. If we write $i\colon ho(\mathcal{C}_c) \to ho(\mathcal{C})$ for the canonical equivalence of categories induced by the inclusion $\mathcal{C}_c \subset \mathcal{C}$, there is a unique way to promote the collection of these choices to a functor

$$Q\colon ho(\mathcal{C}) \to ho(\mathcal{C}_c)$$

endowed with natural isomorphisms $i(Q(X)) \simeq X$ and $Q(i(X)) \simeq X$ defined as the images of the maps a'_X. One defines a functor $\mathbf{L}F$ by the formula $\mathbf{L}F(Y) = F_c(Q(Y))$. There is a unique natural morphism $a_X : \mathbf{L}F(\gamma(X)) \to F(X)$ which coincides with the image of a'_X.

Proposition 2.3.3 *For any functor $F\colon \mathcal{C} \to \mathcal{D}$ which sends trivial cofibrations between cofibrant objects to isomorphisms, the pair $(\mathbf{L}F, a)$ is a left derived functor of F.*

Proof Let $\Phi\colon ho(\mathcal{C}) \to \mathcal{D}$ be a functor, together with a natural morphism $\alpha_X\colon \Phi(\gamma(X)) \to F(X)$. By definition of localisations (applied to $\mathcal{C}_c$), and by virtue of the canonical equivalence of categories $ho(\mathcal{C}_c) \simeq ho(\mathcal{C})$, there is a unique natural morphism $f_Y\colon \Phi(Y) \to \mathbf{L}F(Y)$ such that $\alpha_X = a_X f_{\gamma(X)}$ for all objects X of $\mathcal{C}_c$. The latter property remains true for X running over the class of all objects of $\mathcal{C}$, simply because a, α and f are natural transformations, and because, for each object X, there exists a weak equivalence with cofibrant domain $Y \to X$. □

Corollary 2.3.4 *Under the hypothesis of the previous proposition, for any functor $G\colon \mathcal{D} \to \mathcal{E}$, the pair $(G\mathbf{L}F, Ga)$ is a left derived functor of GF.*

Definition 2.3.5 Let $\mathcal{C}$ and $\mathcal{C}'$ be two model categories endowed with two localisation functors $\gamma\colon \mathcal{C} \to ho(\mathcal{C})$ and $\gamma'\colon \mathcal{C}' \to ho(\mathcal{C}')$, respectively.

If a functor $F\colon \mathcal{C} \to \mathcal{C}'$ preserves trivial cofibrations, then the composed functor $\gamma' F$ sends trivial cofibrations between cofibrant objects to isomorphisms, and therefore, by virtue of the preceding proposition, admits a left derived functor. We also denote by

$$\mathbf{L}F\colon ho(\mathcal{C}) \to ho(\mathcal{C}')$$

the left derived functor of $\gamma' F$, which we call the *total left derived functor* of F.

Similarly, if a functor $F\colon \mathcal{C} \to \mathcal{C}'$ preserves trivial fibrations, we denote by

$$\mathbf{R}F\colon ho(\mathcal{C}) \to ho(\mathcal{C}')$$

the right derived functor of $\gamma' F$, which we call the *total right derived functor* of F.

Proposition 2.3.6 *If $F\colon \mathcal{C} \to \mathcal{C}'$ and $F'\colon \mathcal{C}' \to \mathcal{C}''$ both are functors between model categories which preserve cofibrant objects (fibrant objects) and trivial cofibrations (trivial fibrations), then at the level of total left derived functors (of total right derived functors), the canonical comparison map*

$$(\mathbf{L}F' \circ \mathbf{L}F)(X) \to \mathbf{L}(F' \circ F)(X),$$

$$(\mathbf{R}(F' \circ F)(X) \to (\mathbf{R}F' \circ \mathbf{R}F)(X), \textit{ respectively})$$

is an isomorphism for all objects X of $ho(\mathcal{C})$.

Proof It is sufficient to consider the case of a total left derived functor. One can prove this assertion using Corollary 2.3.4, but a very simple argument consists in remembering how we constructed the total left derived functors: for a given object X of $\mathcal{C}$, we choose a weak equivalence with cofibrant domain $p\colon X' \to X$ and $q\colon X'' \to F(X')$. We then have $\mathbf{L}F(X) = F(X')$, $\mathbf{L}(F' \circ F)(X) = F'(F(X'))$ and $(\mathbf{L}F' \circ \mathbf{L}F)(X) = F'(X'')$. The comparison map we want to understand is then the image in $ho(\mathcal{C}'')$ of the map $F'(q)$. Since q is a weak equivalence between cofibrant objects, by virtue of Ken Brown's lemma, such a map $F'(q)$ is a weak equivalence. □

Definition 2.3.7 Let $\mathcal{C}$ and $\mathcal{C}'$ be two model categories.

A *Quillen adjunction* is a pair of adjoint functors

$$F\colon \mathcal{C} \rightleftarrows \mathcal{C}' : G$$

such that F preserves cofibrations and G preserves fibrations. A *left Quillen functor* (a *right Quillen functor*) is a functor F (G, respectively) with a right adjoint G (a left adjoint F) such that (F, G) is a Quillen adjunction.

Remark 2.3.8 By virtue of Remark 2.1.8 and of Proposition 2.1.12, for a pair of adjoint functors F and G as above, the following conditions are equivalent.

(i) The pair (F, G) is a Quillen adjunction.
(ii) The functor F preserves cofibrations as well as trivial cofibrations.
(iii) The functor G preserves fibrations as well as trivial fibrations.

In particular, for any Quillen adjunction (F, G), the functor F admits a total left derived functor, and the functor G a total right derived functor.

Theorem 2.3.9 *Any Quillen adjunction $F\colon \mathcal{C} \rightleftarrows \mathcal{C}' : G$ naturally induces an adjunction*

$$\mathbf{L}F\colon ho(\mathcal{C}) \rightleftarrows ho(\mathcal{C}') : \mathbf{R}G.$$

Proof It follows from Ken Brown's lemma that, for any cofibrant object A of $\mathcal{C}$, the functor F sends cylinders of A to cylinders of $F(A)$. Since the functor

G preserves fibrant objects, it follows that, for any cofibrant object A of $\mathcal{C}$ and any fibrant object X of $\mathcal{C}'$, the bijection

$$\mathrm{Hom}_{\mathcal{C}}(A, G(X)) \simeq \mathrm{Hom}_{\mathcal{C}'}(F(A), X)$$

is compatible with the relation of homotopy, and thus, by virtue of Proposition 2.2.16, induces a natural bijection

$$[A, G(X)] \simeq [F(A), X]$$

as functors from $ho(\mathcal{C}_c)^{\mathrm{op}} \times ho(\mathcal{C}'_f)$ to the category of sets. Therefore, Theorems 2.2.15 and 2.2.17, together with the very construction of the functors $\mathbf{L}F$ and $\mathbf{R}G$, end the proof. □

Examples of useful derived functors consist in considering the total derived functor of basic categorical operations, such as limits and colimits. Although we will not consider the general case here, we will need special cases which can be understood at a rather elementary level.

2.3.10 Let $\mathcal{C}$ be a model category.

Given a small category I, let us consider the category $\mathcal{C}^I = \underline{\mathrm{Hom}}(I, \mathcal{C})$. We define the weak equivalences of $\mathcal{C}^I$ as the morphisms $F \to G$ such that, for any object i of I, the evaluation at i is a weak equivalence $F_i \to G_i$ in $\mathcal{C}$.

We want to define a model structure on $\mathcal{C}^I$ by defining the class of fibrations as the one which consists of morphisms $F \to G$ such that, for any object i of I, the evaluation at i is a fibration $F_i \to G_i$ in $\mathcal{C}$ (the cofibrations are defined by the condition of left lifting property with respect to trivial fibrations). The problem is that there is no known result asserting the existence of such a model structure, unless we make further assumptions on either I or $\mathcal{C}$. However, when such a model category structure exists, it is called the *projective model structure*. We will consider several basic examples of small categories I such that the projective model structure always exists (for any model category $\mathcal{C}$).

For instance, if I is a small discrete category, then the projective model structure exists: the cofibrations simply are the morphisms $F \to G$ such that, for any object i of I, the evaluation at i is a cofibration $F_i \to G_i$ in $\mathcal{C}$; all the axioms are simply verified levelwise. Here are slightly less trivial examples.

Proposition 2.3.11 *If I is the free category generated by the oriented graph $0 \to 1$ (so that the category $\mathcal{C}^I$ simply is the category of arrows $X_0 \to X_1$ in $\mathcal{C}$, with commutative squares as morphisms), then the projective model structure exists.*

Proof We shall adopt the convention that, when an object of $\mathcal{C}^I$ is denoted by an uppercase letter such as X, for instance, the corresponding arrow $X_0 \to$

X_1 will be denoted by the corresponding lowercase letter $x\colon X_0 \to X_1$. A morphism $f\colon X \to Y$ in $\mathcal{C}^I$ thus corresponds to a commutative square in $\mathcal{C}$ of the form below.

$$\begin{array}{ccc} X_0 & \xrightarrow{f_0} & Y_0 \\ \downarrow{\scriptstyle x} & & \downarrow{\scriptstyle y} \\ X_1 & \xrightarrow{f_1} & Y_1 \end{array}$$

One defines cofibrations (trivial cofibrations) as follows: a morphism $f : X \to Y$ is a (trivial) cofibration if the map f_0 is a (trivial) cofibration and if the canonical map $(f_1, y)\colon X_1 \amalg_{X_0} Y_0 \to Y_1$ is a (trivial) cofibration. One checks that trivial cofibrations precisely are the cofibrations which are weak equivalences. The verification of the lifting properties is straightforward. To obtain factorisation into a cofibration followed by a trivial fibration, we proceed as follows (the case of a trivial cofibration followed by a fibration is similar). Given a map $f : X \to Y$ in $\mathcal{C}^I$, we factor f_0 as a cofibration $i_0 : X_0 \to T_0$ followed by a trivial fibration $p_0 : T_0 \to Y_0$. We then form the push-out square

$$\begin{array}{ccc} X_0 & \xrightarrow{i_0} & T_0 \\ \downarrow{\scriptstyle x} & & \downarrow{\scriptstyle t'} \\ X_1 & \xrightarrow{i'_1} & T'_1 \end{array}$$

and we choose a factorisation of the map $(f_1, p_0)\colon T'_1 \to Y_1$ into a cofibration $j\colon T'_1 \to T_1$ followed by a trivial fibration $p_1 : T_1 \to Y_1$. We put $i_1 = ji'_1$. We thus have factored the map f into a cofibration $i : X \to T$ followed by a fibration $p\colon T \to Y$. □

Proposition 2.3.12 *In the case of the free category I generated by the oriented graph $(0, 1) \leftarrow (0, 0) \to (1, 0)$ (so that the category $\mathcal{C}^I$ simply is the category of diagrams of the form $X_{0,1} \leftarrow X_{0,0} \to X_{1,0}$ in $\mathcal{C}$), the projective model structure exists. The cofibrant objects are the diagrams of the form $X_{0,1} \leftarrow X_{0,0} \to X_{1,0}$ in which all the objects are cofibrant and all the maps are cofibrations.*

Proposition 2.3.13 *If $\mathcal{C}$ has small colimits, then for any small well-ordered set I with initial element 0, the projective model structure exists. The cofibrant objects are the functors $X : I \to \mathcal{C}$ such that, for any element $i \in I$, the map $\varinjlim_{j<i} X_j \to X_i$ is a cofibration (in particular, all the X_i must be cofibrant).*

The proofs of Propositions 2.3.12 and 2.3.13 are left as exercises.[2]

[2] These are special cases of a more general fact: if I is a direct category, the projective model structure exists. We refer to [Hov99, theorem 5.1.3] for the proof (which involves some knowledge of the special cases above anyway).

2.3.14 Whenever the projective model category structure exists, we then have a Quillen adjunction

$$\varinjlim_I : \mathcal{C}^I \rightleftarrows \mathcal{C} : \delta, \tag{2.3.14.1}$$

where $\delta(X) = X_I$ denotes the constant diagram indexed by I with value X. This means that, for any functor $F : I \to \mathcal{C}$ and any object X, there is an adjunction of the form

$$\mathrm{Hom}_{ho(\mathcal{C}^I)}(F, X_I) \simeq \mathrm{Hom}_{ho(\mathcal{C})}(\mathbf{L}\varinjlim_I F, X). \tag{2.3.14.2}$$

Proposition 2.3.15 *Assume that the projective model category structure exists on $\mathcal{C}^I$ and that $\mathcal{C}$ has I-indexed colimits. Any natural transformation $f : X \to Y$ between cofibrant functors from I to $\mathcal{C}$, which is a weak equivalence $f_i : X_i \to Y_i$ at each object i of I, induces a weak equivalence $\varinjlim_I X \to \varinjlim_I Y$.*

Proof Such a map $X \to Y$ is an isomorphism in $ho(\mathcal{C}^I)$ and therefore its image by the functor $\mathbf{L}\varinjlim_I$ is an isomorphism of $ho(\mathcal{C})$. But, by construction of the derived functor $\mathbf{L}\varinjlim_I$ (see 2.3.2), this isomorphism is the image of the map $\varinjlim_I X \to \varinjlim_I Y$. Since any map inducing an isomorphism in the homotopy category is a weak equivalence (2.2.19), this proves the proposition. □

Taking the case of a discrete category, as well as the examples provided by Propositions 2.3.12 and 2.3.13, the preceding proposition takes the following explicit forms, respectively.

Corollary 2.3.16 *Let I be a small set such that I-indexed sums exist, and $f_i : X_i \to Y_i$ a family of weak equivalences between cofibrant objects. Then the induced map*

$$\coprod_{i\in I} f_i : \coprod_{i\in I} X_i \to \coprod_{i\in I} Y_i$$

is a weak equivalence.

Corollary 2.3.17 *Consider the following commutative diagram in which all the horizontal maps are cofibrations between cofibrant objects.*

$$\begin{array}{ccccc} X' & \xleftarrow{x'} & X & \xrightarrow{x''} & X'' \\ \downarrow{\scriptstyle f'} & & \downarrow{\scriptstyle f} & & \downarrow{\scriptstyle f''} \\ Y' & \xleftarrow{y'} & Y & \xrightarrow{y''} & Y'' \end{array}$$

If the three vertical maps are weak equivalences, then the induced morphism

$$X' \amalg_X X'' \to Y' \amalg_Y Y''$$

is a weak equivalence.

Corollary 2.3.18 *Assume that $\mathcal{C}$ has small colimits. Let I be a small well-ordered set, and $X \to Y$ be a natural transformation between functors from I to $\mathcal{C}$. Assume that, for any element i of I, the maps*

$$\varinjlim_{j<i} X_j \to X_i \quad \textit{and} \quad \varinjlim_{j<i} Y_j \to Y_i$$

are cofibrations, and the map $X_i \to Y_i$ is a weak equivalence. Then the induced morphism

$$\varinjlim_{i\in I} X_i \to \varinjlim_{i\in I} Y_i$$

is a weak equivalence.

In the case where I is the smallest infinite ordinal, the preceding corollary takes an even more concrete form.

Corollary 2.3.19 *Consider the following commutative diagram, which consists of a morphism f of sequences of cofibrations between cofibrant objects indexed by non-negative integers.*

$$\begin{array}{ccccccccccccc}
X_0 & \longrightarrow & X_1 & \longrightarrow & X_2 & \longrightarrow & \cdots & \longrightarrow & X_n & \longrightarrow & X_{n+1} & \longrightarrow & \cdots \\
\downarrow{\scriptstyle f_0} & & \downarrow{\scriptstyle f_1} & & \downarrow{\scriptstyle f_2} & & & & \downarrow{\scriptstyle f_n} & & \downarrow{\scriptstyle f_{n+1}} & & \\
Y_0 & \longrightarrow & Y_1 & \longrightarrow & Y_2 & \longrightarrow & \cdots & \longrightarrow & Y_n & \longrightarrow & Y_{n+1} & \longrightarrow & \cdots
\end{array}$$

If the vertical maps all are weak equivalences, then the induced map

$$\varinjlim_{n} X_n \to \varinjlim_{n} Y_n$$

is a weak equivalence as well.

Definition 2.3.20 Let $F: I \to \mathcal{C}$ be a functor, and X an object of the model category $\mathcal{C}$. A cocone $F \to X_I$ (i.e. a map in $\mathcal{C}^I$) exhibits the object X as the *homotopy colimit* of F if the induced map $\mathbf{L}\varinjlim_I F \to X$ is an isomorphism in $ho(\mathcal{C})$.

Homotopy limits are defined similarly (as homotopy colimits in $\mathcal{C}^{\mathrm{op}}$).

Example 2.3.21 Assuming that $\mathcal{C}$ has I-indexed colimits, for any functor $F: I \to \mathcal{C}$ which is cofibrant in the projective model category structure on $\mathcal{C}^I$ (and provided that the latter exists), the colimit cocone exhibits $\varinjlim_I F$ as a homotopy colimit of F. In fact, provided that the projective model category structure exists on $\mathcal{C}^I$, we can characterise homotopy colimits as follows: given a functor F from I to $\mathcal{C}$, a cocone $F \to X_I$ exhibits X as a homotopy colimit of F if and only if there exists a weak equivalence $F' \to F$, with cofibrant domain in the projective model category structure, such that the induced map $\varinjlim_I F' \to X$ is a weak equivalence.

Definition 2.3.22 A commutative square

$$\begin{array}{ccc} X & \xrightarrow{x} & X' \\ \downarrow{\scriptstyle f} & & \downarrow{\scriptstyle f'} \\ Y & \xrightarrow{y} & Y' \end{array}$$

is *homotopy coCartesian* if it exhibits Y' as the homotopy push-out (i.e. the homotopy colimit) of the diagram $Y \leftarrow X \rightarrow X'$.

Dually, such a square is said to be *homotopy Cartesian* if it exhibits X as the homotopy pull-back of the diagram $Y \rightarrow Y' \leftarrow X'$ (or, equivalently, if it is a homotopy coCartesian square in the opposite category).

Example 2.3.23 Any coCartesian square in which all maps are cofibrations between cofibrant objects is homotopy coCartesian.

Remark 2.3.24 Let $\square = [1] \times [1]$, so that functors from $\square$ to $\mathcal{C}$ precisely are the commutative squares in $\mathcal{C}$. We then have a projective model category structure on $\mathcal{C}^\square$ (because $\mathcal{C}^\square \simeq (\mathcal{C}^{[1]})^{[1]}$, so that we may apply Proposition 2.3.11 twice). A commutative square of $\mathcal{C}$ is homotopy coCartesian if and only if it is isomorphic in $ho(\mathcal{C}^\square)$ to a commutative square of $\mathcal{C}$ which is coCartesian and in which all the maps are cofibrations between cofibrant objects. The next two statements are direct consequences of this characterisation.

Proposition 2.3.25 *One of the two squares*

$$\begin{array}{ccc} X & \xrightarrow{x} & X' \\ \downarrow{\scriptstyle f} & & \downarrow{\scriptstyle f'} \\ Y & \xrightarrow{y} & Y' \end{array} \qquad \text{and} \qquad \begin{array}{ccc} X & \xrightarrow{f} & Y \\ \downarrow{\scriptstyle x} & & \downarrow{\scriptstyle y} \\ X' & \xrightarrow{f'} & Y' \end{array}$$

is homotopy coCartesian if and only if the other one has the same property.

Proposition 2.3.26 *Any commutative square of the form*

$$\begin{array}{ccc} X & \xrightarrow{x} & X' \\ \downarrow{\scriptstyle f} & & \downarrow{\scriptstyle f'} \\ Y & \xrightarrow{y} & Y' \end{array}$$

in which both x and y are weak equivalences is homotopy coCartesian.

Proposition 2.3.27 *Consider a coCartesian square of the form below.*

$$\begin{array}{ccc} X & \xrightarrow{x} & X' \\ \downarrow{\scriptstyle f} & & \downarrow{\scriptstyle f'} \\ Y & \xrightarrow{y} & Y' \end{array} \tag{2.3.27.1}$$

Assume that the morphism f is a cofibration and that both X and X' are cofibrant. If moreover the map x is a weak equivalence, so is y.

Proof Given any map $x\colon X \to X'$, there is an adjunction of the form

$$x_!\colon X\backslash\mathcal{C} \rightleftarrows X'\backslash\mathcal{C} : x^!,$$

where $x^!$ denotes the functor induced by composition with x:

$$x^!(f'\colon X' \to Y') = (f'x\colon X \to Y')\,.$$

The left adjoint $x_!$ is the functor which associates to any map $f\colon X \to Y$ its push-out along the map x, so that we get a push-out of the form (2.3.27.1). The functor $x^!$ obviously preserves fibrations and cofibrations, so that we have a Quillen adjunction, and therefore a derived adjunction

$$\mathbf{L}x_!\colon ho(X\backslash\mathcal{C}) \rightleftarrows ho(X'\backslash\mathcal{C}) : \mathbf{R}x^!\,.$$

The cofibrant objects of $X\backslash\mathcal{C}$ precisely are the cofibrations $f\colon X \to Y$, so that the unit of the derived adjunction evaluated at such a cofibration f is the image of the map y in the coCartesian square (2.3.27.1). Therefore, if, for any cofibration $f\colon X \to Y$, the map y, in diagram (2.3.27.1), is a weak equivalence, then the functor $\mathbf{L}x_!$ is fully faithful. By virtue of Corollary 2.2.19, the reverse is true: if this functor is fully faithful, then the map y is a weak equivalence for any cofibration $f\colon X \to Y$. On the other hand, the functor $\mathbf{R}x^!$ is always conservative: it is sufficient to check this property for maps between objects which are both fibrant and cofibrant in $X'\backslash\mathcal{C}$, in which case this follows from the fact that, by definition, a map in $X'\backslash\mathcal{C}$ is a weak equivalence if and only if its image in $\mathcal{C}$ has the same property. Therefore, the functor $\mathbf{L}x_!$ is an equivalence of categories if and only if it is fully faithful.[3] Finally, we conclude that the functor $\mathbf{R}x^!$ is an equivalence of categories if and only if, for any cofibration $f\colon X \to Y$, the associated map $y\colon Y \to Y'$ is a weak equivalence. For instance, this is the case whenever the map x is a trivial cofibration (since the class of trivial cofibrations is stable under push-outs). It is time to remark that the functor $x^!$ preserves weak equivalence, from which one checks that the functors $\mathbf{R}x^!$ turn the map

$$X \mapsto ho(X\backslash\mathcal{C})$$

into a functor from $\mathcal{C}$ to the opposite of the category of locally small categories.

[3] Given an adjunction $u\colon A \rightleftarrows B : v$, if the left adjoint u is fully faithful (or, equivalently, if the unit map $a \to v(u(a))$ is invertible for any object a of A), then the functor v is a localisation of B by the class of maps whose image by v is invertible. Therefore, if u is fully faithful and if v is conservative, both u and v are equivalences of categories and are quasi-inverses to each other.

Since the class of equivalences of categories satisfies the 'two-out-of-three' property, by virtue of Ken Brown's lemma (Proposition 2.2.7), we conclude that any weak equivalence between cofibrant objects $x\colon X \to X'$ induces an equivalence of categories $\mathbf{R}x^!$, and this achieves the proof. □

Corollary 2.3.28 *Consider a coCartesian square of the form* (2.3.27.1). *If the morphism f is a cofibration and if both X and X' are cofibrant, then this is a homotopy coCartesian square.*

Proof We choose a factorisation of the map x into a cofibration $i\colon X \to X''$ followed by a weak equivalence $p\colon X'' \to X'$. The commutative square above now is the composition of two coCartesian squares of the form below.

$$\begin{array}{ccccc} X & \xrightarrow{i} & X'' & \xrightarrow{p} & X' \\ \downarrow{\scriptstyle f} & & \downarrow{\scriptstyle f''} & & \downarrow{\scriptstyle f'} \\ Y & \xrightarrow{j} & Y'' & \xrightarrow{q} & Y' \end{array}$$

The left-hand commutative square is homotopy coCartesian (2.3.23), and, therefore, as explained in Remark 2.3.24, it is sufficient to prove that the induced map q is a weak equivalence. In other words, we may assume that, furthermore, the map x is a weak equivalence, and it is sufficient to prove that, under this additional assumption, the map y is a weak equivalence, which follows from the preceding proposition. □

Corollary 2.3.29 *In Corollary 2.3.17, one may only assume that all objects are cofibrant and that x' and y' are cofibrations, and still get to the same conclusion.*

2.4 Model Structures *ex Nihilo*

2.4.1 As we have mentioned in Proposition 2.1.9, the small object argument is one of the possible tools to construct weak factorisation systems. When, in a model category structure, the weak factorisation systems $(Cof, W \cap Fib)$ and $(W \cap Cof, Fib)$ can be constructed out of the small object argument, we say that the model category is *cofibrantly generated*. In that case, we thus have the existence of a small set I of cofibrations, as well as of a small set J of trivial cofibrations, such that $l(r(I)) = Cof$ and $l(r(J)) = W \cap Cof$. We then say that I and J *generate* the class of cofibrations and the class of trivial cofibrations, respectively.

Example 2.4.2 Let A be a ring, and $\mathcal{C} = Comp(A)$ be the category of (possibly unbounded) cochain complexes of (left) A-modules. This category has a structure of cofibrantly generated model category, for which the weak equivalences are the quasi-isomorphisms and the fibrations are the epimorphisms (i.e. the maps which are surjective in each degree); see the second chapter of [Hov99], for instance. A generic set of cofibrations is given by all the shifts of the inclusion of the ring A (seen as a complex of A-modules concentrated in degree zero) into the mapping cone of the identity of A. A generating set of trivial cofibrations is given by all shifts of inclusion of 0 into the mapping cone of the identity of A. (This example will not be used in this book.)

2.4.3 The purpose of this chapter is to explain a general procedure to construct cofibrantly generated model category structures on categories of presheaves of sets over a fixed small category, following a previous work [Cis06] of the author of this book. The idea is simple: it consists of following step by step most of the book of Gabriel and Zisman [GZ67] on the homotopy theory of Kan complexes, and seeing that a significant part of it makes sense in a wide generality. The idea of promoting the book of Gabriel and Zisman into a general way to define homotopy theories *ex nihilo* comes from the early work of Fabien Morel on homotopy theory of schemes [Mor06].

From now on, we fix a small category A. In this chapter, all presheaves are presheaves of sets over A.

Definition 2.4.4 A *cellular model* is a small set M of monomorphisms of presheaves such that the class $l(r(M))$ is the class of all monomorphisms of presheaves.

Cellular models always exist: for instance, the set of monomorphisms of the form $K \to L$, where L runs over the quotients of representable presheaves, is a cellular model; see Example 2.1.11.

But some cellular models are nicer than others.

Example 2.4.5 If A is an Eilenberg–Zilber category (e.g. $A = \mathbf{\Delta}$), it follows right away from Theorem 1.3.8 that the boundary inclusions $\partial h_a \to h_a$, $a \in \mathrm{Ob}(A)$, do form a cellular model: indeed, this theorem shows that any monomorphism belongs to the smallest saturated class containing boundary inclusions, and conversely, since the class of monomorphisms or presheaves of sets is saturated, it contains the smallest saturated class containing boundary inclusions; we conclude with the last assertion of Proposition 2.1.9.

Definition 2.4.6 A *cylinder* of a presheaf X is a commutative diagram

(2.4.6.1) $$X \amalg X \xrightarrow{(\partial_0,\partial_1)} IX \xrightarrow{\sigma} X, \qquad \sigma \circ (\partial_0,\partial_1) = (1_X, 1_X)$$

in which the map (∂_0, ∂_1) is a monomorphism.[4]

Remark 2.4.7 Given a category $\mathcal{C}$, we have the category $\underline{End}(\mathcal{C})$ of endofunctors of $\mathcal{C}$. This is a monoidal category whose tensor product is defined by composition of functors. Of course, the category $\underline{End}(\mathcal{C})$ acts on the left on $\mathcal{C}$:

$$(F, X) \mapsto F \otimes X = F(X).$$

With this convention, any natural transformation $u\colon F \to G$ in $\underline{End}(\mathcal{C})$ and any morphism $p\colon X \to Y$ in $\mathcal{C}$ induce a morphism

$$u \otimes p\colon F \otimes X \to G \otimes Y\,.$$

If we write 1 for the identity of $\mathcal{C}$, we also have $1 \otimes X = X$.

Definition 2.4.8 A *functorial cylinder* is an endofunctor I of the category of presheaves endowed with a morphism $(\partial_0, \partial_1)\colon 1 \amalg 1 \to I$ as well as a morphism $\sigma\colon I \to 1$ such that, for any presheaf X, the diagram

$$X \amalg X \xrightarrow{(\partial_0 \otimes 1_X, \partial_1 \otimes 1_X)} I \otimes X \xrightarrow{\sigma \otimes 1_X} X \tag{2.4.8.1}$$

is a cylinder of X.

An *exact cylinder* is such a functorial cylinder satisfying the following properties.

DH1. The functor I commutes with small colimits and preserves monomorphisms.

DH2. For any monomorphism of presheaves $j\colon K \to L$, the commutative square

$$\begin{array}{ccc} K & \xrightarrow{j} & L \\ {\scriptstyle \partial_\varepsilon \otimes 1_K}\downarrow & & \downarrow{\scriptstyle \partial_\varepsilon \otimes 1_L} \\ I \otimes K & \xrightarrow{1_I \otimes j} & I \otimes L \end{array}$$

is Cartesian for $\varepsilon = 0, 1$.

Remark 2.4.9 Given a Cartesian square of presheaves

$$\begin{array}{ccc} X & \longrightarrow & Y \\ \downarrow & & \downarrow \\ Z & \longrightarrow & T \end{array}$$

[4] This means both that maps $\partial_i : X \to IX$ are monomorphisms and that the intersection of the image of ∂_0 and the image of ∂_1 is empty.

in which all maps are monomorphisms, the induced morphism $Y \amalg_X Z \to T$ is a monomorphism, whose image is denoted by $Y \cup Z$. We thus have an inclusion

$$Y \cup Z \subset T\,.$$

We will write $\{\varepsilon\}$ for the subobject of I determined by the monomorphism $\partial_\varepsilon : 1 \to I$. We thus have canonical inclusions

$$K \simeq \{\varepsilon\} \otimes K \subset I \otimes K$$

for $\varepsilon = 0, 1$. The axiom DH2 now means that we have inclusions

$$I \otimes K \cup \{\varepsilon\} \otimes L \subset I \otimes L$$

for $\varepsilon = 0, 1$.

We shall also write

$$\partial I = \{0\} \amalg \{1\}\,.$$

Since colimits are universal in any category of presheaves (i.e. pulling back along a morphism of presheaves defines a functor which preserves small colimits), the inclusion $\partial I \to I$, together with any monomorphism of presheaves $j : K \to L$, gives a Cartesian square made of monomorphisms

$$\begin{array}{ccc} \partial I \otimes K & \hookrightarrow & I \otimes K \\ \downarrow{\scriptstyle 1_{\partial I} \otimes j} & & \downarrow{\scriptstyle 1_I \otimes j} \\ \partial I \otimes L & \hookrightarrow & I \otimes L \end{array}$$

and, therefore, induces a canonical inclusion

$$I \otimes K \cup \partial I \otimes L \subset I \otimes L\,.$$

Example 2.4.10 If I is an interval (i.e. a cylinder object of the final presheaf), the functor $X \mapsto I \times X$ is an exact cylinder. For instance, one may always take the subobject classifier $I = \Omega$; see Example 2.1.11.

Definition 2.4.11 Given an exact cylinder I, a class of morphisms of presheaves An is a class of *I-anodyne extensions* if it satisfies the following properties.

An0. There exists a small set of monomorphisms of presheaves Λ such that $An = l(r(\Lambda))$ (in particular, the class An is saturated and is a subclass of the class of monomorphisms).

An1. For any monomorphism of presheaves $K \to L$, the induced morphism $I \otimes K \cup \{\varepsilon\} \otimes L \to I \otimes L$ is in An for $\varepsilon = 0, 1$.

An2. For any map $K \to L$ in An, the induced map $I \otimes K \cup \partial I \otimes L \to I \otimes L$ is in An.

A *homotopical structure* is the data of an exact cylinder I, together with a class of I-anodyne extensions.

Example 2.4.12 The class of monomorphisms is a class of I-anodyne extensions (for any I).

Example 2.4.13 Given an exact cylinder I, together with a small set of monomorphisms of presheaves S, one can construct the smallest class of I-anodyne extensions containing S as follows. First, we choose a cellular model M, and we define the set $\Lambda_I^0(S, M)$ as

$$\Lambda_I^0(S, M) = S \cup \{I \otimes K \cup \{\varepsilon\} \otimes L \to I \otimes L \mid K \to L \in M\,,\ \varepsilon = 0, 1\}\,. \tag{2.4.13.1}$$

For $n \geq 0$, we put

$$\Lambda_I^{n+1}(S, M) = \{I \otimes K \cup \partial I \otimes L \to I \otimes L \mid K \to L \in \Lambda_I^n(S, M)\}\,, \tag{2.4.13.2}$$

and finally

$$\Lambda_I(S, M) = \bigcup_{n \geq 0} \Lambda_I^n(S, M)\,. \tag{2.4.13.3}$$

Then the class $An_I(S) = l(r(\Lambda_I(S, M)))$ (i.e. the smallest saturated class of maps containing $\Lambda_I(S, M)$) is the smallest class of I-anodyne extensions containing S. Indeed, it is clear that any class of I-anodyne extensions containing S also contains $\Lambda_I(S, M)$, and thus the smallest saturated class of maps containing the latter, which is $An_I(S)$. Therefore, it is sufficient to check that the class $An_I(S)$ itself is a class of I-anodyne extensions. Axiom An0 is true by construction. For this we note that, since it preserves small colimits, the functor I has a right adjoint $X \mapsto X^I$. The inclusion $\{\varepsilon\} \subset I$ induces a natural map $X^I \to X^{\{\varepsilon\}} = X$. Henceforth, any morphism of presheaves $X \to Y$ defines a commutative square

$$\begin{array}{ccc} X^I & \longrightarrow & Y^I \\ \downarrow & & \downarrow \\ X & \longrightarrow & Y \end{array}$$

and thus a canonical map $X^I \to Y^I \times_Y X$. For any monomorphism $K \to L$ we get a correspondence of the following form

$$\begin{array}{ccc} I \otimes K \cup \{\varepsilon\} \otimes L & \xrightarrow{a'} & X \\ \downarrow & \overset{h'}{\nearrow} & \downarrow \\ I \otimes L & \xrightarrow[b']{} & Y \end{array} \quad \leftrightsquigarrow \quad \begin{array}{ccc} K & \xrightarrow{a} & X^I \\ \downarrow & \overset{h}{\nearrow} & \downarrow \\ L & \xrightarrow[b]{} & Y^I \times_Y X \end{array} \tag{2.4.13.4}$$

In the case where the map $X \to Y$ has the right lifting property with respect to elements of $\Lambda_I(S, M)$, we see from formula (2.4.13.1) and from correspondence (2.4.13.4) that the map $X^I \to Y^I \times_Y X$ has the right lifting property with respect to any element of the cellular model M, and thus with respect to any monomorphism (since M generates the class of monomorphisms). Using correspondence (2.4.13.4) again, this proves axiom An1. The proof of axiom An2 is similar (replacing (2.4.13.1) by (2.4.13.2), M by $\Lambda_I(S, M)$ and $\{\varepsilon\}$ by ∂I).

From now on, we assume that a homotopical structure is given: an exact cylinder I together with a class An of I-anodyne extensions.

Definition 2.4.14 Let $f_0, f_1 \colon K \to X$ be two morphisms of presheaves. An *I-homotopy* from f_0 to f_1 is a map $h \colon I \otimes K \to X$ such that $h(\partial_\varepsilon \otimes 1_K) = f_\varepsilon$ for $\varepsilon = 0, 1$.

We denote by $[K, X]$ the quotient of the set $\mathrm{Hom}(K, X)$ by the smallest equivalence relation generated by the relation of being connected by an I-homotopy. Two maps $f_0, f_1 \colon K \to X$ are said to be *I-homotopic* if they have the same class in $[K, X]$; equivalently, this means that there is a finite sequence of maps $h_i \colon I \otimes K \to X$, $1 \le i \le n$, such that $h_1(\partial_0 \otimes 1_K) = f_0$ and $h_n(\partial_1 \otimes 1_K) = f_1$, and such that, for each i, $1 < i \le n$, there exists ε and η in $\{0, 1\}$ such that $h_{i-1}(\partial_\eta \otimes 1_K) = h_i(\partial_\varepsilon \otimes 1_K)$.

The functoriality of the cylinder I ensures that this equivalence relation is compatible with composition. There is thus a well-defined category of presheaves up to I-homotopy (whose objects are presheaves, and whose sets of morphisms are the quotients of the form $[K, X]$).

A morphism of presheaves is an *I-homotopy equivalence* if it defines an isomorphism in the category of presheaves up to I-homotopy.

The Yoneda lemma applied to the category of presheaves up to I-homotopy gives the following characterisation of I-homotopy equivalences.

Proposition 2.4.15 *For a morphism of presheaves $f \colon X \to Y$, the following conditions are equivalent.*

(a) *The map f is an I-homotopy equivalence.*
(b) *There exists a morphism $g \colon Y \to X$ such that gf is I-homotopic to 1_X and fg is I-homotopic to 1_Y.*
(c) *For any presheaf K, the induced map $f_* \colon [K, X] \to [K, Y]$ is bijective.*
(d) *For any presheaf W, the induced map $f^* \colon [Y, W] \to [X, W]$ is bijective.*

Definition 2.4.16 A morphism of presheaves $f \colon X \to Y$ is a *strong deformation retract* (*the dual of a strong deformation retract*) if there exists a morphism

$g: Y \to X$ as well as a morphism $h: I \otimes Y \to Y$ (as well as a morphism $k: I \otimes X \to X$, respectively) such that:

(i) $gf = 1_X$ ($fg = 1_Y$, respectively);
(ii) $h(\partial_0 \otimes 1_Y) = 1_Y$ and $h(\partial_1 \otimes 1_Y) = fg$ ($k(\partial_0 \otimes 1_X) = 1_X$ and $k(\partial_1 \otimes 1_X) = gf$, respectively);
(iii) $h(1_I \otimes f) = \sigma \otimes f$ ($fk = \sigma \otimes f$, respectively).

Recall from Example 2.1.11 that we call trivial fibrations the morphisms of presheaves which have the right lifting property with respect to monomorphisms.

Proposition 2.4.17 *Any trivial fibration is the dual of a strong deformation retract, and any section of a trivial fibration is a strong deformation retract. In particular, any trivial fibration has a section, and is an I-homotopy equivalence.*

Proof Let $p: X \to Y$ be a trivial fibration. Since the map from the empty presheaf to any presheaf is a monomorphism, the following solid commutative square admits a filler.

$$\begin{array}{ccc} \varnothing & \longrightarrow & X \\ \downarrow & \overset{s}{\nearrow} & \downarrow{\scriptstyle p} \\ Y & = & Y \end{array}$$

Hence the existence of sections. Let us assume that such a section s is provided. Since s must be a monomorphism, the expression $I \otimes Y \cup \partial I \otimes X$ makes sense. We thus have a solid commutative diagram

$$\begin{array}{ccc} I \otimes Y \cup \partial I \otimes X & \xrightarrow{(s(\sigma \otimes 1_Y),(1_X, sp))} & X \\ \downarrow & \overset{k}{\nearrow} & \downarrow{\scriptstyle p} \\ I \otimes X & \xrightarrow{p(\sigma \otimes 1_X)} & Y \end{array}$$

which admits a filler k. □

Definition 2.4.18 A *naive fibration* is a morphism with the right lifting property with respect to the given class *An* of I-anodyne extensions. A presheaf X is *fibrant* if the map from X to the final presheaf is a naive fibration.

A morphism of presheaves is a *cofibration* if it is a monomorphism.

A morphism of presheaves $f: X \to Y$ is a *weak equivalence* if, for any fibrant presheaf W, the induced map

$$f^*: [Y, W] \to [X, W]$$

is bijective.

Theorem 2.4.19 *With this definition of weak equivalences and of cofibrations, we have a cofibrantly generated model category structure on the category of presheaves over A. Moreover, a presheaf is fibrant precisely when its map to the final presheaf is a naive fibration, and a morphism between fibrant objects is a fibration if and only if it is a naive fibration.*

Remark 2.4.20 Any cofibrantly generated model category structure on the category of presheaves over A in which the cofibrations are the monomorphisms can be obtained in this way: if J is a generating set of the class of trivial cofibrations (i.e. such that $l(r(J))$ is the class of trivial cofibrations), we can take as exact cylinder the one defined as the Cartesian product with the subobject classifier Ω (2.4.10), and consider the smallest class of Ω-anodyne maps generated by J (2.4.13). Using the fact that the projections $\Omega \times X \to X$ are trivial fibrations, we see that the model category structure obtained by the theorem above is the one we started from.

The proof of Theorem 2.4.19 will require quite a few steps, and is the goal of the remaining part of this chapter.

Remark 2.4.21 By virtue of the characterisation of I-homotopy equivalences given by condition (d) of Proposition 2.4.15, the class of weak equivalences contains the class of I-homotopy equivalences. Furthermore, it is clear that the class of weak equivalences has the two-out-of-three property and is stable under retracts. Proposition 2.4.17 tells us that any trivial fibration is a weak equivalence. As explained in Example 2.1.11, any morphism can be factored into a cofibration followed by a trivial fibration (possibly using the small object argument).

Remark 2.4.22 For any generating set Λ of the class of I-anodyne maps, one can apply the small object argument (2.1.9). In particular, any morphism of presheaves f admits a (functorial) factorisation of the form $f = pi$ with i an I-anodyne map and p a naive fibration.

Lemma 2.4.23 *A morphism of presheaves is a trivial fibration if and only if it has both properties of being a weak equivalence and of having the right lifting property with respect to trivial cofibrations.*

Proof Since any trivial cofibration is a monomorphism and since we already know that any trivial fibration is a weak equivalence, this is certainly a necessary condition. Conversely, let $p: X \to Y$ be a morphism which is a weak equivalence and has the right lifting property with respect to trivial cofibrations. We may choose a factorisation $p = qj$ where j is a cofibration and q is a trivial fibration. The morphism j must be a weak equivalence, and, therefore, by virtue of the retract lemma (2.1.5), the morphism p is a retract of q. Hence it is a trivial fibration. □

Lemma 2.4.24 *Let X and W be two presheaves, with W fibrant. Then the relation defined by the existence of an I-homotopy is an equivalence relation on the set* $\operatorname{Hom}(X, W)$.

In particular, for any pair of morphisms $u, v : X \to W$, if $u = v$ in the category of presheaves up to I-homotopy equivalence, then there exists an I-homotopy from u to v.

Proof It is clear that, for any map u from X to W, there is an I-homotopy from u to itself (e.g. $h = \sigma \otimes u$). Let us prove that, for three morphisms $u, v, w : X \to W$, if there is an I-homotopy h from u to v as well as an I-homotopy k from u to w, then there exists an I-homotopy from v to w. Taking into account the identifications

$$I \otimes \partial I \otimes X \simeq I \otimes X \amalg I \otimes X \quad \text{and} \quad \{0\} \otimes I \otimes X \simeq I \otimes X\,,$$

we have a map

$$((h, k), \sigma \otimes u) : I \otimes \partial I \otimes X \cup \{0\} \otimes I \otimes X \to W\,,$$

and the inclusion

$$I \otimes \partial I \otimes X \cup \{0\} \otimes I \otimes X \to I \otimes I \otimes X$$

is an I-anodyne extension. Therefore, since W is fibrant, there exists a morphism $H : I \otimes I \otimes X \to W$ whose restriction to $I \otimes X \simeq I \otimes \{0\} \otimes X$ coincides with h, whose restriction to $I \otimes X \simeq I \otimes \{1\} \otimes X$ coincides with k, and whose restriction to $I \otimes X \simeq \{0\} \otimes I \otimes X$ is the constant homotopy $\sigma \otimes u$. Let us define the morphism $\eta : I \otimes X \to W$ by the equality $\eta = H(\partial_1 \otimes 1_I \otimes 1_X)$. We then have

$$\begin{aligned} \eta(\partial_0 \otimes 1_X) &= H(\partial_1 \otimes 1_I \otimes 1_X)(\partial_0 \otimes 1_X) \\ &= H(1_I \otimes \partial_0 \otimes 1_X)(\partial_1 \otimes 1_X) \\ &= h(\partial_1 \otimes 1_X) = v\,. \end{aligned}$$

Similarly, one checks that $\eta(\partial_1 \otimes 1_X) = w$.

For $w = u$ and $k = \sigma \otimes u$ the constant I-homotopy from u to itself, this shows that the relation of I-homotopy is symmetric. The general case, together with the property of symmetry, proves the property of transitivity. □

Proposition 2.4.25 *Any I-anodyne map is a weak equivalence.*

Proof Let $j: X \to Y$ be an I-anodyne map (i.e. an element of the class An), and W a fibrant presheaf. We must prove that the induced map

$$j^*: [Y, W] \to [X, W]$$

is bijective. The surjectivity is clear (W being fibrant and j an I-anodyne extension, it is already true at the level of morphisms of presheaves). Let $y_0, y_1: Y \to W$ be two morphisms of presheaves such that $y_0 j$ is I-homotopic to $y_1 j$. Then, by virtue of the preceding lemma, there exists an I-homotopy $h: I \otimes X \to W$ from $y_0 j$ to $y_1 j$. This defines a morphism

$$(h, (y_0, y_1)): I \otimes X \cup \partial I \otimes Y \to W\,.$$

On the other hand, since j is an I-anodyne extension, the induced map

$$I \otimes X \cup \partial I \otimes Y \to I \otimes Y$$

is I-anodyne. Therefore, since W is fibrant, there exists a map $H: I \otimes Y \to W$ which is an I-homotopy from y_0 to y_1. □

Proposition 2.4.26 *A morphism between fibrant presheaves is a weak equivalence if and only if it is an I-homotopy equivalence.*

Proof This follows right away from the Yoneda lemma applied to the category of fibrant presheaves up to I-homotopy. □

Lemma 2.4.27 *A naive fibration is a trivial fibration if and only if it is the dual of a strong deformation retract.*

Proof We already know that this is a necessary condition (2.4.17). Let $p: X \to Y$ be a naive fibration which is the dual of a strong deformation retract. We may assume that we have morphisms $s: Y \to X$ as well as $k: I \otimes X \to X$ such that $ps = 1_Y$, $k(\partial_0 \otimes 1_X) = 1_X$, $k(\partial_1 \otimes 1_X) = sp$ and $pk = \sigma \otimes p$. We must prove that any commutative square of the form

$$\begin{array}{ccc} K & \xrightarrow{a} & X \\ {\scriptstyle i}\downarrow & & \downarrow{\scriptstyle p} \\ L & \xrightarrow{b} & Y \end{array}$$

in which i is a monomorphism has a filler $l: L \to X$. We have a morphism

$$u = (k(1_I \otimes a), sb): I \otimes K \cup \{1\} \otimes L \to X\,.$$

And the monomorphism i induces an I-anodyne extension

$$j: I \otimes K \cup \{1\} \otimes L \to I \otimes L\,.$$

Therefore, the commutative square

$$\begin{array}{ccc} I \otimes K \cup \{1\} \otimes L & \xrightarrow{u} & X \\ {\scriptstyle j}\downarrow & \overset{h}{\nearrow} & \downarrow{\scriptstyle p} \\ I \otimes L & \xrightarrow{\sigma \otimes b} & Y \end{array}$$

admits a filler h. We put $l = h(\partial_0 \otimes 1_L)$. We then have

$$\begin{aligned} pl &= ph(\partial_0 \otimes 1_L) = (\sigma \otimes b)(\partial_0 \otimes 1_L) = b \\ li &= h(\partial_0 \otimes 1_L)i = h(1_I \otimes i)(\partial_0 \otimes 1_K) \\ &= k(1_I \otimes a)(\partial_0 \otimes 1_K) = k(\partial_0 \otimes 1_X)a = a, \end{aligned}$$

and this achieves the proof. □

Lemma 2.4.28 *A naive fibration with fibrant codomain is a weak equivalence if and only if it is a trivial fibration.*

Proof Let $p: X \to Y$ be a naive fibration, with Y fibrant, and assume that p is a weak equivalence as well. By virtue of Proposition 2.4.26, the map p is an I-homotopy equivalence. Furthermore, Lemma 2.4.24 ensures that there exists a morphism $t: Y \to X$ as well as an I-homotopy $k: I \otimes Y \to Y$ from 1_Y to pt. In particular, we have the following commutative square.

$$\begin{array}{ccc} Y & \xrightarrow{t} & X \\ {\scriptstyle \partial_1 \otimes 1_Y}\downarrow & \overset{k'}{\nearrow} & \downarrow{\scriptstyle p} \\ I \otimes Y & \xrightarrow{k} & Y \end{array}$$

Let us put $s = k'(\partial_0 \otimes 1_Y)$. We then have

$$ps = pk'(\partial_0 \otimes 1_Y) = k(\partial_0 \otimes 1_Y) = 1_Y .$$

Since p is an I-homotopy equivalence, using Lemma 2.4.24 once more, there is an I-homotopy h from the identity of X to the map sp. In particular, under the identifications

$$I \otimes X \simeq \{1\} \otimes I \otimes X \quad \text{and} \quad I \otimes X \amalg I \otimes X \simeq I \otimes \partial I \otimes X,$$

this defines a morphism

$$(\sigma \otimes sp, (h, sph)): \{1\} \otimes I \otimes X \cup I \otimes \partial I \otimes X \to X .$$

Finally we have a commutative square

$$\begin{array}{ccccc}
\{1\}\otimes I\otimes X\cup I\otimes\partial I\otimes X & & \xrightarrow{(\sigma\otimes sp,(h,sph))} & & X \\
\big\downarrow & & \overset{H}{\nearrow} & & \big\downarrow{\scriptstyle p} \\
I\otimes I\otimes X & \xrightarrow[1_I\otimes\sigma\otimes 1_X]{} & I\otimes X & \xrightarrow[ph]{} & Y
\end{array}$$

which has a filler H. We define $K = H(\partial_0 \otimes 1_{I\otimes X})$. The identities

$$\begin{aligned}
K(\partial_0 \otimes 1_X) &= H(\partial_0 \otimes 1_{I\otimes X})(\partial_0 \otimes 1_X) = h(\partial_0 \otimes 1_X) = 1_X \\
K(\partial_1 \otimes 1_X) &= H(\partial_0 \otimes 1_{I\otimes X})(\partial_1 \otimes 1_X) = sph(\partial_0 \otimes 1_X) = sp \\
pK &= pH(\partial_0 \otimes 1_{I\otimes X}) = ph(\partial_0 \otimes \sigma \otimes 1_X) = p(\sigma \otimes 1_X)
\end{aligned}$$

show that p is the dual of a strong deformation retract. We conclude with Lemma 2.4.27. The converse follows from Proposition 2.4.17. □

Corollary 2.4.29 *A cofibration with fibrant codomain is a weak equivalence if and only if it is an I-anodyne extension.*

Proof We already know that any I-anodyne extension is a weak equivalence (see Proposition 2.4.25). Let i be a cofibration with fibrant codomain. There exists a factorisation of the form $i = qj$ with j an I-anodyne map and q a naive fibration. Since j is a weak equivalence, if ever i is a weak equivalence, so will be q, and therefore, by virtue of Lemma 2.4.28, the morphism q is a trivial fibration. It thus follows from the retract lemma (2.1.5) that i is a retract of j and thus is I-anodyne. □

Proposition 2.4.30 *A cofibration is a weak equivalence if and only if it has the left lifting property with respect to the class of naive fibrations with fibrant codomain.*

Proof Let $i\colon K \to L$ be a cofibration. Let us choose an I-anodyne extension with fibrant codomain $j\colon L \to L'$.

Assume that i is a trivial cofibration. For any commutative square

$$\begin{array}{ccc}
K & \xrightarrow{a} & X \\
{\scriptstyle i}\big\downarrow & & \big\downarrow{\scriptstyle p} \\
L & \xrightarrow{b} & Y
\end{array}$$

with p a naive fibration with fibrant codomain, we may assume that there is a map $b'\colon L' \to Y$ such that $b'j = b$. Since ji is a trivial cofibration with fibrant codomain, the previous corollary tells us that there exists a map $l'\colon L' \to X$ such that $pl' = b'$ and $l'ji = a$. Therefore, the map $l = l'j$ is a filler of the square above.

Conversely, if i has the left lifting property with respect to the class of naive fibrations with fibrant codomain, we may choose a factorisation of the map ji into an I-anodyne extension k followed by a naive fibration (with fibrant codomain) p. The retract lemma implies that ji is a retract of k, whence is I-anodyne. In particular, the map ji is a trivial cofibration. Since j has the same property, the map i itself must be a trivial cofibration as well. □

Corollary 2.4.31 *The class of trivial cofibrations is saturated.*

Proof Any class of maps defined by the left lifting property with respect to a given class of maps is saturated. □

The last part of the proof of Theorem 2.4.19 consists of exhibiting a small set of generating trivial cofibrations. This requires a few more steps, including a closer look at the proof of the small object argument in the case of monomorphisms of presheaves.

Lemma 2.4.32 *Any strong deformation retract is an I-anodyne extension.*

Proof Let $i \colon K \to L$ be a strong deformation retract. Then there is a morphism $r \colon L \to K$ such that $ri = 1_K$, as well as an I-homotopy h from 1_L to ir, such that $h(1_I \otimes i) = (\sigma \otimes 1_L)(1_I \otimes i) = \sigma \otimes i$. Let us consider a commutative square of the following form

$$\begin{array}{ccc} K & \xrightarrow{a} & X \\ {\scriptstyle i}\big\downarrow & & \big\downarrow{\scriptstyle p} \\ L & \xrightarrow{b} & Y \end{array}$$

with p a naive fibration. We want to prove the existence of a filler $l \colon L \to X$. The solid commutative square below has a filler k.

$$\begin{array}{ccc} I \otimes K \cup \{0\} \otimes L & \xrightarrow{(\sigma\otimes a,\, ar)} & X \\ \big\downarrow & \overset{k}{\nearrow} & \big\downarrow{\scriptstyle p} \\ I \otimes L & \xrightarrow{bh} & Y \end{array}$$

(In the diagram, k is a dotted arrow.)

We define $l = k(\partial_1 \otimes 1_L)$ and check that $li = a$ and $pl = b$. □

Lemma 2.4.33 *Any I-anodyne extension between fibrant presheaves is a strong deformation retract.*

Proof Let $i \colon K \to L$ be an I-anodyne extension, with both K and L fibrant.

There exists a retraction r of i. The filler h in the diagram

$$
\begin{array}{ccc}
I \otimes K \cup \partial I \otimes L & \xrightarrow{(\sigma\otimes i,(1_L, ir))} & L \\
\big\downarrow & \overset{h}{\nearrow} & \\
I \otimes L & &
\end{array}
$$

turns i into a strong deformation retract. □

Proposition 2.4.34 *For a monomorphism* $i \colon K \to L$*, with both* K *and* L *being fibrant presheaves, the following conditions are equivalent:*

(i) it is a weak equivalence;
(ii) it is an I*-anodyne extension;*
(iii) it is a strong deformation retract.

Proof The equivalence between conditions (i) and (ii) is a special case of Corollary 2.4.29. The equivalence between conditions (ii) and (iii) follows from Lemmas 2.4.32 and 2.4.33. □

We leave the proof of the following lemma as an exercise (it is sufficient to prove it in the category of sets to make it true of presheaves). A full proof can be found in [Cis06, lemme 1.2.32], though.

Lemma 2.4.35 *In any category of presheaves, we have the following properties.*

(a) For any commutative diagram

$$
\begin{array}{ccccc}
X_1 & \xleftarrow{x_1} & X_0 & \xrightarrow{x_2} & X_2 \\
\big\downarrow{\scriptstyle i_1} & & \big\downarrow{\scriptstyle i_0} & & \big\downarrow{\scriptstyle i_2} \\
S_1 & \xleftarrow{s_1} & S_0 & \xrightarrow{s_2} & S_2
\end{array}
$$

in which the map i_2 *as well as the canonical map* $X_1 \amalg_{X_0} S_0 \to S_1$ *are monomorphisms, the induced map*

$$X_1 \amalg_{X_0} X_2 \to S_1 \amalg_{S_0} S_2$$

is a monomorphism.

(b) *For any commutative diagram of the form*

$$\begin{array}{ccccc}
X_1 & \xleftarrow{x_1} & X_0 & \xrightarrow{x_2} & X_2 \\
\downarrow{\scriptstyle i_1} & & \downarrow{\scriptstyle i_0} & & \downarrow{\scriptstyle i_2} \\
S_1 & \xleftarrow{s_1} & S_0 & \xrightarrow{s_2} & S_2 \\
\uparrow{\scriptstyle j_1} & & \uparrow{\scriptstyle j_0} & & \uparrow{\scriptstyle j_2} \\
Y_1 & \xleftarrow{y_1} & Y_0 & \xrightarrow{y_2} & Y_2
\end{array}$$

if the vertical maps as well as the maps x_1, s_1 and y_1 all are monomorphisms, and if the commutative squares of the left column are Cartesian, then the induced map

$(X_1 \times_{S_1} Y_1) \amalg_{(X_0 \times_{S_0} Y_0)} (X_2 \times_{S_2} Y_2) \to (X_1 \amalg_{X_0} X_2) \times_{(S_1 \amalg_{S_0} S_2)} (Y_1 \amalg_{Y_0} Y_2)$

is an isomorphism.

(c) *For any small family of maps of the form $X_i \to S_i \leftarrow Y_i$, the canonical map*

$$\coprod_i (X_i \times_{S_i} Y_i) \to (\coprod_i X_i) \times_{(\amalg_i S_i)} (\coprod_i Y_i)$$

is invertible.

2.4.36 Let us recall a special case of the construction provided by the small object argument. Let us fix a set Λ of monomorphisms. We define two functors S and B as follows: for any presheaf X, we have

$$S(X) = \coprod_{i\colon K \to L \in \Lambda} \mathrm{Hom}(K, X) \times K \quad \text{and} \quad B(X) = \coprod_{i\colon K \to L \in \Lambda} \mathrm{Hom}(K, X) \times L.$$

There is a canonical map $s_X\colon S(X) \to X$ induced by the evaluation maps from $\mathrm{Hom}(K, X) \times K$ to X, and we define $L_1(X)$ by forming the following coCartesian square.

$$\begin{array}{ccc}
S(X) & \xrightarrow{s_X} & X \\
\downarrow & & \downarrow{\scriptstyle \lambda_X^{(1)}} \\
B(X) & \longrightarrow & L_1(X)
\end{array}$$

Given a well-ordered set E, we define, for each element $e \in E$, the functor L_e, together with the natural map $\lambda_X^{(e)}\colon X \to L_e(X)$, as follows. If e is the successor of an element e_0, we put $L_e(X) = L_1(L_{e_0}(X))$, and we define the map $\lambda_X^{(e)}$ as the composition of $\lambda_{L_{e_0}(X)}^{(1)}$ with the map $\lambda_X^{(e_0)}$. If e is a limit element, we define $L_e(X)$ as the colimit of the $L_i(X)$ for $i < e$. Finally, we define $L(X)$ as the colimit of the $L_e(X)$. There is a canonical embedding $X \to L(X)$ which is a transfinite composition of push-outs of elements of Λ, and, if E is large enough, the map from $L(X)$ to the final presheaf has the right lifting property with respect to Λ.

In the case where Λ generates the given class An of I-anodyne extensions, this is the way we get a functorial I-anodyne extension $X \to L(X)$ with fibrant codomain.[5]

We shall say that a presheaf X is of size $< \alpha$ if α is a cardinal which is greater than the set of morphisms of the category A/X.

Lemma 2.4.37 *The functors L and I have the following properties.*

(i) They preserve monomorphisms.
(ii) They preserve intersections of subobjects.
(iii) There exists an infinite cardinal α such that:

(a) for any presheaf X, the union of subobjects of size $< \alpha$ indexed by a set of cardinal $< \alpha$ is of size $< \alpha$;
(b) for any presheaf of size $< \alpha$, the presheaves $L(X)$ and $I \otimes X$ are of size $< \alpha$;
(c) for any presheaf X, the presheaf $L(X)$ is the union of its subobjects of the form $L(V)$, where V runs over the subobjects of size $< \alpha$ in X;
(d) for any well-ordered set E of cardinal α and any increasing family of subobjects $(V_e)_{e \in E}$, the union of the $L(V_e)$ is canonically isomorphic to the image by L of the union of the V_e.

It follows from the preceding lemma that the functor L_1 preserves monomorphisms as well as intersections of subobjects. Since filtered colimits are exact in any category of presheaves, this proves properties (i) and (ii). Property (iii) is then a lengthy but elementary exercise of set theory that we leave to the reader. For a full proof, we refer to [Cis06, propositions 1.2.16, 1.2.17 and 1.2.35], for instance.

Lemma 2.4.38 *There exists an infinite cardinal α such that, for any trivial cofibration $i: X \to Y$ and any subpresheaf U of size $< \alpha$ in Y, there exists a subpresheaf V of size $< \alpha$ in Y, containing U, such that the inclusion $V \cap X \to V$ is a trivial cofibration.*

Proof We choose a generating set Λ of the class of I-anodyne extensions An, so that $An = l(r(\Lambda))$. We construct a functorial I-anodyne map $l_X : X \to L(X)$ with fibrant codomain, as explained in paragraph 2.4.36. Finally, we choose a

[5] This construction provides a way to factorise any map: for a map $X \to Y$, seen as a presheaf over A/Y, we may apply this to the set Λ/Y of maps of presheaves over A/Y whose image in the category of presheaves over A belongs to Λ.

cardinal α such that the conclusions of Lemma 2.4.37 hold. Let $i\colon X \to Y$ be a monomorphism of presheaves. We have a commutative square

$$\begin{array}{ccc} X & \xrightarrow{l_X} & L(X) \\ {\scriptstyle i}\downarrow & & \downarrow{\scriptstyle L(i)} \\ Y & \xrightarrow{l_Y} & L(Y) \end{array}$$

in which both horizontal maps are I-anodyne, and therefore are weak equivalences. Since $L(i)$ is a monomorphism between fibrant presheaves, we conclude from Proposition 2.4.34 that i is a weak equivalence if and only if $L(i)$ is a strong deformation retract.

From now on, we assume that $L(i)$ is a strong deformation retract. So there is a map $r\colon L(Y) \to L(X)$ such that $rL(i) = 1_{L(X)}$, as well as an I-homotopy $h\colon I \otimes L(Y) \to L(Y)$ from $1_{L(Y)}$ to $L(i)r$ whose restriction to $I \otimes L(X)$ is the constant homotopy $\sigma \otimes 1_{L(i)}$.

Let U be a subpresheaf of size $< \alpha$ in Y. We shall choose a well-ordered set E of cardinal α and construct an increasing sequence $(V_e)_{e\in E}$ of subobjects of Y containing U, such that, for any $e \in E$, if $j_e\colon V_e \to Y$ denotes the inclusion, the map h sends $I \otimes L(V_e)$ to $L(V_{e+1})$ (where $e+1$ denotes the successor of e). For $e = 0$ the initial element of E, we simply put $V_0 = U$. Assume that V_i is already constructed for all $i < e$. Let us define V_e' as the union of the V_i for $i < e$. Since $I \otimes L(V_e')$ is of size $< \alpha$, so is its image by the map h, and since $L(Y)$ is the union of its subobjects of the form $L(T)$ with $T \subset Y$ of size α, we see that there exists $V_e'' \subset Y$ of size $< \alpha$ such that the image of $I \otimes L(V_e')$ by h is contained in $L(V_e'')$. We define $V_e = V_e' \cup V_e''$, and let V be the union of the V_e. Since the union of the $L(V_e)$ is canonically isomorphic to $L(V)$, we check that r maps $L(V)$ to $L(V \cap X)$ and that the restriction of the map h to $I \otimes L(V)$ turns the inclusion $L(V \cap X) \to L(V)$ into a strong deformation retract. Therefore, the inclusion $V \cap X \to V$ is a trivial cofibration. □

Lemma 2.4.39 *There exists a small set J of trivial cofibrations such that the class $l(r(J))$ coincides with the class of trivial cofibrations.*

Proof Let us choose a cardinal α such that the conclusion of the preceding lemma holds. We define J to be the set of trivial cofibrations $U \to V$ with V of size $< \alpha$.

Since the class of trivial cofibrations is saturated (Corollary 2.4.31), it is clear that any element of the smallest saturated class of maps containing J (i.e. any element of $l(r(J))$) is a trivial cofibration. It is thus sufficient to prove that any trivial cofibration belongs to $l(r(J))$.

Let $i\colon X \to Y$ be a trivial cofibration. Since any totally ordered set has a cofinal well-ordered subset, by Zorn's lemma, there exists a maximal subobject W of Y containing X such that the map $X \to W$ belongs to $l(r(J))$. Let a be a representable presheaf, and $y \in Y_a$. The image U of the map $y \colon h_a \to Y$ is of size $< \alpha$, and, therefore, there exists a subobject V of Y of size $< \alpha$, containing U, such that the inclusion $V \cap W \to V$ belongs to J. Its push-out along $V \cap W \to W$, namely the map $W \to V \cup W$, thus belongs to $l(r(J))$. Since W is maximal, this means that $V \cup W = W$, and thus that $y \in W_a$. In other words, we have $W = Y$. □

This achieves the proof of Theorem 2.4.19: once a cellular model has been chosen, the small object argument and Lemma 2.4.23 show that cofibrations and fibrations which are weak equivalences do form a weak factorisation system; the small object argument and Lemma 2.4.39 show that trivial cofibrations and fibrations form a weak factorisation system; Proposition 2.4.30 also asserts that fibrations between fibrant objects are nothing other than naive fibrations between fibrant presheaves.

Proposition 2.4.40 *Assume that $\mathcal{C}$ is a model category, and let $F \colon \widehat{A} \to \mathcal{C}$ be a colimit preserving functor which sends monomorphisms to cofibrations. Then F is a left Quillen functor (i.e. F and its right adjoint form a Quillen adjunction) if and only if F sends I-anodyne extensions to trivial cofibrations. If Λ is a generating set of the specified class An of I-anodyne extensions, then this property holds whenever F sends Λ to the class of trivial cofibrations.*

Proof Since the class of trivial cofibrations always is saturated, and since F preserves small colimits, the class of morphisms of presheaves which are sent to a trivial cofibration by F is saturated. Therefore, if F sends a generating set of the class of An to the class of trivial cofibrations, then it sends any element of An to a trivial cofibration. If this is the case, for any morphism of presheaves $f \colon X \to Y$, there is a commutative square

$$\begin{array}{ccc} X & \longrightarrow & L(X) \\ {\scriptstyle f}\downarrow & & \downarrow{\scriptstyle L(f)} \\ Y & \longrightarrow & L(Y) \end{array}$$

in which the horizontal maps are I-anodyne. Therefore, the map $F(f)$ is a weak equivalence if and only if $F(L(f))$ has the same property. Furthermore, if f is a monomorphism, we may assume that $L(f)$ is a monomorphism between fibrant presheaves. Therefore, $L(f)$ is a weak equivalence if and only if it is an I-anodyne extension. Thus, if f is a trivial cofibration, $L(f)$ is an I-anodyne extension and this implies that $F(f)$ must be a weak equivalence. □

Remark 2.4.41 It follows from the proof of the preceding proposition that inverting the elements of An inverts all weak equivalences. However, it might be the case that the class An is strictly smaller than the class of trivial cofibrations. We shall see natural examples in both cases: where these two notions coincide (Theorem 3.1.29), and where they do not (since there are categories whose nerve is weakly contractible but without any final object (e.g. the set of finite subsets of an infinite set), Proposition 4.1.7, Corollary 4.1.9 and Theorem 4.3.9 provide a counterexample).

Remark 2.4.42 Lemmas 2.4.37 and 2.4.38 simply are pedestrian ways to prove a particular case of Jeff Smith's theorem [Lur09, proposition A.2.6.13] for constructing combinatorial model categories.[6] Note however that Theorem 2.4.19 gives more than the mere existence of a model category structure: since it relies on the description of fibrations with fibrant codomain in terms of a class which might be smaller than the class of trivial cofibrations, it gives the possibility to describe such fibrations in several different ways. This will be useful when we start doing such constructions in families (i.e. working over a presheaf X which is allowed to vary); see the proof of Theorem 4.1.5.

2.5 Absolute Weak Equivalences

In this chapter, we still consider a fixed small category A, and we will work in the category of presheaves of sets over A. We also assume that an exact cylinder I is given, together with a given class An of I-anodyne maps.

2.5.1 For each presheaf S, the category of presheaves over A/S is canonically equivalent to the slice category $\widehat{A}/S$.

One defines an exact cylinder I_S on the category of presheaves over A/S by describing its effect on morphisms $f\colon X \to S$, seen as objects of $\widehat{A}/S$. The object $I_S \otimes (X, f)$ is the object corresponding to the map

$$\sigma \otimes f = f(\sigma \otimes 1_X)\colon I \otimes X \to S\,.$$

Similarly, one defines An_S as the class of morphisms in $\widehat{A}/S$ whose image in $\widehat{A}$ belongs to An. If Λ is a set of generators of An, then we have $An_S = l(r(\Lambda_S))$,

[6] For the reader who already knows what this means: it is easy to see that the class of I-homotopy equivalences is accessible and that the functor L is accessible, so that the class of weak equivalences is accessible, so that, by virtue of Corollary 2.4.31, we may apply Jeff Smith's theorem, and thus get a proof of Theorem 2.4.19.

where Λ_S stands for the morphisms of $\widehat{A}/S$ of the form

$$\begin{array}{ccc} K & \xrightarrow{u} & L \\ & {\scriptstyle a}\searrow \quad \swarrow{\scriptstyle b} & \\ & S & \end{array}$$

such that $u \in \Lambda$. The class An_S is a class of I_S-anodyne extensions in $\widehat{A}/S$. Therefore, applying Theorem 2.4.19 to the category of presheaves over A/S with the cylinder I_S and the class An_S gives a unique model category structure whose cofibrations are the monomorphisms and whose fibrant objects are the naive fibrations of codomain S.

Definition 2.5.2 A morphism of presheaves over S is an *S-weak equivalence* if it is a weak equivalence in the model category structure associated to I_S and An_S.

A morphism of presheaves $f: X \to Y$ (over A) is an *absolute weak equivalence* if, for any presheaf S and any morphism $b: Y \to S$, the map

$$\begin{array}{ccc} X & \xrightarrow{f} & Y \\ & {\scriptstyle a=bf}\searrow \quad \swarrow{\scriptstyle b} & \\ & S & \end{array}$$

is an S-equivalence.

Proposition 2.5.3 *A monomorphism is in An if and only if it is an absolute weak equivalence. Similarly, a morphism of presheaves is both a naive fibration and an absolute weak equivalence if and only if it is a trivial fibration.*

Proof It is clear that any I-anodyne extension is an absolute weak equivalence. For the converse, let $i: X \to Y$ be a monomorphism which is an absolute weak equivalence. Then, in particular, it is a trivial cofibration with fibrant codomain in $\widehat{A}/Y$. By virtue of Corollary 2.4.29, this proves that i is an I-anodyne extension.

It is obvious that any trivial fibration is both a naive fibration and an absolute weak equivalence. Conversely, let $p: X \to Y$ be a morphism which is both a naive fibration and an absolute weak equivalence. Then one can consider f as a naive fibration with fibrant codomain in $\widehat{A}/Y$. Therefore, the map f is a trivial fibration in $\widehat{A}/Y$, and thus in $\widehat{A}$. □

Proposition 2.5.4 *The class of absolute weak equivalences W^a is the smallest class C satisfying the following conditions:*

(a) the class C is stable under composition;

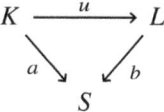

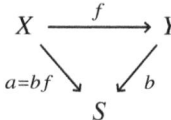

(b) for any pair of composable morphisms $f\colon X \to Y$ and $g\colon Y \to Z$, if both f and gf are in C, so is g;

(c) $An \subset \mathsf{C}$.

Moreover, a morphism of presheaves is an absolute weak equivalence if and only if it admits a factorisation into an I-anodyne extension followed by a trivial fibration.

Proof Let us prove that the class of absolute weak equivalences satisfies these three properties. Let $f\colon X \to Y$ and $g\colon Y \to Z$ be a pair of composable morphisms, and let us assume that f is an absolute weak equivalence. Then, for any morphism $Z \to S$, we may see the maps g and gf as maps in $\widehat{A}/S$, and it is clear that g is an S-equivalence if and only if gf is an S-equivalence. This proves that W^a satisfies properties (a) and (b). Property (c) is already known.

Let us prove the last assertion. Since trivial fibrations and I-anodyne extensions are in W^a, it follows from (a) that any composition of such morphisms is in W^a. Conversely, if f is an absolute weak equivalence, then it admits a factorisation of the form $f = pi$ with i an I-anodyne extension, and p a naive fibration. Condition (b) ensures that p is an absolute weak equivalence as well, and thus the second assertion of Proposition 2.5.3 implies that p is a trivial fibration.

To finish the proof, let C be a class of morphisms satisfying the conditions (a), (b) and (c). To prove that $W^a \subset \mathsf{C}$, it is sufficient to prove that any trivial fibration is in C. But this follows from the fact that any trivial fibration has a section and that any such section is I-anodyne (being a strong deformation retract). □

Corollary 2.5.5 *Let us consider a pair of composable monomorphisms $f\colon X \to Y$ and $g\colon Y \to Z$, and assume that f is an I-anodyne extension. Then the map g is an I-anodyne extension if and only if the map gf has the same property.*

Proof This follows right away from Propositions 2.5.3 and 2.5.4. □

Proposition 2.5.6 *Let $f\colon X \to Y$ be a map, and $p\colon Y \to S$ a naive fibration. Then f is an absolute weak equivalence if and only if the map p turns f into an S-weak equivalence.*

Proof We factor f into a cofibration $i\colon X \to X'$ followed by a trivial fibration $q\colon X' \to Y$. Then, by Corollary 2.4.29, i is an S-weak equivalence if and only if it is in An. Since q is a trivial fibration, i is an S-weak equivalence if and only if f is an S-weak equivalence. Therefore, if f is an S-weak equivalence, it is the composition of an I-anodyne map with a trivial fibration, and thus is an absolute weak equivalence. □

Recall that a supply of I-anodyne extensions comes from strong deformation retracts (Lemma 2.4.32). The following statement might thus be meaningful when it comes to prove that a map is an absolute weak equivalence.

Proposition 2.5.7 *For any Cartesian square*

$$\begin{array}{ccc} X' & \xrightarrow{i} & X \\ {\scriptstyle q}\downarrow & & \downarrow{\scriptstyle p} \\ Y' & \xrightarrow{j} & Y \end{array}$$

in which p is a naive fibration and j is a strong deformation retract, the map i is a strong deformation retract.

Proof Let us choose a retraction $r\colon Y \to Y'$ of j as well as a homotopy $h\colon I \otimes Y \to Y$ from 1_Y to jr, such that $h(1_I \otimes j) = \sigma \otimes j$. The commutative square below then has a filler.

$$\begin{array}{ccc} I \otimes X' \cup \{0\} \otimes X & \xrightarrow{(\sigma \otimes i, 1_X)} & X \\ \downarrow & \overset{k}{\nearrow} & \downarrow{\scriptstyle p} \\ I \otimes X & \xrightarrow{h(1_I \otimes p)} & Y \end{array}$$

The maps $u = k(\partial_1 \otimes 1_X)\colon X \to X$ and $v = rp\colon X \to Y'$ satisfy the relations $pu = jv$ (because $h(\partial_1 \otimes 1_X) = jr$) and thus define a unique map $s\colon X \to X'$ such that $qs = v$ and $is = u$. We have $si = 1_{X'}$ because we check that $qsi = q$ and $isi = i$. Hence i is a strong deformation retract. □

3

The Homotopy Theory of ∞-Categories

We will start here by constructing the classical Kan–Quillen model category structure on the category of simplicial sets, which encodes the homotopy theory of Kan complexes. We will do this using the general method of the previous chapter. However, even though this method gives that the fibrant objects are the Kan complexes, extra work is needed to prove that the fibrations precisely are the Kan fibrations. This goal is achieved using Kan's subdivision functor, through classical arguments on diagonals of bisimplicial sets. Since Kan complexes will be shown to be the ∞-groupoids (i.e. the ∞-categories in which all maps are invertible), and since ∞-groupoids are to ∞-categories what sets are to categories, this precise understanding of the homotopy theory of Kan complexes is not only a warm-up before defining the homotopy theory of ∞-categories: it will play a fundamental role all through the book (to understand many fundamental notions such as locally constant presheaves and localisations). Even the proofs we choose are meaningful in this respect, either because they express a categorical intuition (we strongly encourage the reader to look at Kan's Ex^{∞} functor with scrutiny, and to see how obviously it is related with the idea of inverting arrows), or because possibly generalised versions of these methods will apply in the homotopy theory of ∞-categories (generalising the trick of the diagonal of bisimplicial sets is the subject of the whole of Section 5.5 below).

The second section is technical, but fundamental: it is all about the compatibility of the homotopy theory of ∞-categories with finite Cartesian products. In particular, it gives sense to the ∞-category of functors between two given ∞-categories. The third section defines the Joyal model category structure. However, at that stage, it will not be obvious that the class of fibrant objects exactly is the class of ∞-categories. The third section, which we took entirely from Joyal's work, introduces fundamental constructions such as joins and slices, as well as a non-trivial lifting property expressing the fact that, although one cannot compose maps canonically, one can choose inverses coherently in ∞-categories. The latter will be used in the fourth section. First, to prove that the Kan complexes exactly are the ∞-groupoids. Second, to prove that a natural transformation is invertible if and only if it is fibrewise invertible. This will imply that ∞-categories are precisely the fibrant objects of the Joyal model category structure, and that the fibrations between fibrant objects are the isofibrations.

After revisiting features of the Joyal model category structures for a couple of sections, we will come back in Section 3.8 to classical homotopy theory: the Serre long exact sequence associated to a Kan fibration, from which we will prove the simplicial version of Whitehead's theorem: a morphism of Kan complexes is a homotopy equivalence if and only if it induces an isomorphism of higher homotopy groups (and a bijection on the sets of connected components). This will be used in Section 3.9 to prove a kind of generalisation to ∞-categories: a functor between ∞-categories is an equivalence of ∞-categories if and only if it is fully faithful and essentially surjective.

3.1 Kan Fibrations and the Kan–Quillen Model Structure

The simplicial set Δ^1 has an obvious structure of interval, so that the Cartesian product functor $\Delta^1 \times (-)$ defines an exact cylinder on the category of simplicial sets.

Definition 3.1.1 An *anodyne extension* is an element of the smallest class of $\Delta^1 \times (-)$-anodyne maps (see Example 2.4.13).

Proposition 3.1.2 (Gabriel and Zisman) *The following three classes of morphisms of simplicial sets are equal:*

(a) the class of anodyne extensions;
(b) the smallest saturated class of maps containing inclusions of the form

$$\Delta^1 \times \partial\Delta^n \cup \{\varepsilon\} \times \Delta^n \to \Delta^1 \times \Delta^n \text{ for } n \geq 0 \text{ and } \varepsilon = 0, 1;$$

(c) the smallest saturated class containing inclusions of the form

$$\Lambda^n_k \to \Delta^n \text{ for } n \geq 1 \text{ and } 0 \leq k \leq n.$$

Proof We know that the set M of boundary inclusions $\partial\Delta^n \to \Delta^n$, $n \geq 0$, is a cellular model. With the notation introduced in Example 2.4.13, the class of anodyne extensions is the class $l(r(\Lambda_I(\varnothing, M)))$ with $I = \Delta^1 \times (-)$. The class described in (b) above is the class $l(r(\Lambda^0_I(\varnothing, M)))$. To prove that these two classes coincide, since, by construction, $\Lambda^0_I(\varnothing, M) \subset \Lambda_I(\varnothing, M)$, it is sufficient to prove that any element of $\Lambda_I(\varnothing, M)$ belongs to $l(r(\Lambda^0_I(\varnothing, M)))$. The explicit description of $\Lambda_I(\varnothing, M)$ shows that it is sufficient to prove that, for any monomorphism of simplicial sets $K \to L$ and for $\varepsilon = 0, 1$, the inclusion

$$\Delta^1 \times K \cup \{\varepsilon\} \times L \to \Delta^1 \times L$$

belongs to the class $l(r(\Lambda^0_I(\varnothing, M)))$. Using the correspondence (2.4.13.4), we see that the class of maps $K \to L$ having this property is saturated. Therefore, it is sufficient to check this property for the inclusions $\partial\Delta^n \to \Delta^n$, in which case this holds by definition. The equality between the classes (b) and (c) follows from the lemma below (and the dual version obtained by applying the auto-equivalence $X \mapsto X^{\mathrm{op}}$), which is a slightly more precise version of the proposition. □

Lemma 3.1.3 *The following two classes of morphisms are equal:*

(a) the smallest saturated class of maps containing inclusions of the form

$$\Delta^1 \times \partial\Delta^n \cup \{1\} \times \Delta^n \to \Delta^1 \times \Delta^n \text{ for } n \geq 0;$$

(b) the smallest saturated class containing inclusions of the form

$$\Lambda^n_k \to \Delta^n \text{ for } n \geq 1 \text{ and } 0 < k \leq n.$$

Proof We define two maps

$$s\colon [n] \to [1] \times [n] \quad \text{and} \quad r\colon [1] \times [n] \to [n]$$

as follows. We define

$$s(i) = \begin{cases} (0, i) & \text{if } i < k, \\ (1, i) & \text{otherwise.} \end{cases}$$

We also put

$$r(0, i) = \begin{cases} i & \text{if } i \leq k, \\ k & \text{otherwise} \end{cases} \quad \text{and} \quad r(1, i) = \begin{cases} k & \text{if } i \leq k, \\ i & \text{otherwise.} \end{cases}$$

Since the nerve functor preserves products and $N([m]) = \Delta^m$, these maps define morphisms

$$s\colon \Delta^n \to \Delta^1 \times \Delta^n \quad \text{and} \quad r\colon \Delta^1 \times \Delta^n \to \Delta^n\,.$$

The image of Λ^n_k by s must be in the union $\Delta^1 \times \partial\Delta^n \cup \partial\Delta^1 \times \Delta^n$ (for cardinality reasons), and none of the faces of Λ^n_k can be sent to the component $\{0\} \times \Delta^n$ (because $k > 0$). Using that $(1, k)$ is reached by each face of Λ^n_k, we see that the image of s must be in $\Delta^1 \times \Lambda^n_k \cup \{1\} \times \Delta^n$. On the other hand, the image of the union $\Delta^1 \times \Lambda^n_k \cup \{1\} \times \Delta^n$ by r fits in Λ^n_k. To see this, recall that Λ^n_k is the union of the images of the face maps δ^n_i for $i \neq k$. This means that the non-degenerate simplices of Λ^n_k correspond to the maps $u\colon [n-1] \to [n]$ whose image contains k. Since $k > 0$, the image of $\{1\} \times \Delta^n$ is in the image of δ^n_0, and thus fits in Λ^n_k. It remains to prove that r sends $\Delta^1 \times \Lambda^n_k$ to Λ^n_k. This follows from the fact that, for any injective map $(u, v)\colon \Delta^n \to \Delta^1 \times \Delta^n$ such that u and v reach 0 and k, respectively, with v not an identity, the map $ru\colon \Delta^n \to \Delta^n$ reaches k and is not surjective. The composed map $r \circ s$ is the identity of Δ^n. Therefore, we have a commutative diagram of the following form.

$$
\begin{array}{ccccc}
\Lambda^n_k & \xrightarrow{s} & \Delta^1 \times \Lambda^n_k \cup \{1\} \times \Delta^n & \xrightarrow{r} & \Lambda^n_k \\
\downarrow & & \downarrow & & \downarrow \\
\Delta^n & \xrightarrow{s} & \Delta^1 \times \Delta^n & \xrightarrow{r} & \Delta^n
\end{array}
\qquad r \circ s = 1_{\Lambda^n_k}\ \text{(top)},\quad r \circ s = 1_{\Delta^n}\ \text{(bottom)}
\tag{3.1.3.1}
$$

This proves that the class (b) is contained in the class (a). For the converse, we define, for each integer i, $0 \leq i \leq n$, the map $c_i\colon [n+1] \to [1] \times [n]$ as the unique strictly increasing map which reaches both values $(0, i)$ and $(1, i)$. We write $c_i\colon \Delta^{n+1} \to \Delta^1 \times \Delta^n$ for the induced morphisms of simplicial sets, and C_i for its image. We define a finite filtration

$$\Delta^1 \times \partial\Delta^n \cup \{1\} \times \Delta^n = A_{-1} \subset A_0 \subset A_1 \subset \cdots \subset A_n = \Delta^1 \times \Delta^n \tag{3.1.3.2}$$

by the formula $A_i = A_{i-1} \cup C_{n-i}$ for $0 \leq i \leq n$. The isomorphism $\Delta^{n+1} \simeq C_{n-i}$ defined by c_{n-i} induces an isomorphism $\Lambda^{n+1}_{n-i+1} \simeq C_{n-i} \cap A_{i-1}$. Let us check that, for any face map $\delta^{n+1}_j\colon \Delta^n \to \Delta^{n+1}$, with $j \neq n-i+1$, the image of $c_i \circ \delta^{n+1}_j$ is in A_{i-1}. If $j \neq n-i$, the image is in $\Delta^1 \times \partial\Delta^n$, while, if $n-i = j$, we have two cases: for $n-i = j = 0$, the image precisely is $\{1\} \times \Delta^n$, while for $n > i$, it is in C_{n-i+1}. Conversely, since C_{n-i} is not contained in A_{i-1}, we must have $c^{-1}_{n-i}(C_{n-i} \cap A_{i-1}) \subset \partial\Delta^{n+1}$. This means that the simplices of $C_{n-i} \cap A_{i-1}$ must all factor through a simplex of dimension at most n. One checks that any injective map $\Delta^m \to C_{n-i} \cap A_{i-1}$ (with $m \leq n$) must factor through the image of Λ^{n+1}_{n-i+1}. In conclusion, we have a biCartesian square of the following form.

$$
\begin{array}{ccc}
\Lambda^{n+1}_{n-i+1} & \longrightarrow & A_{i-1} \\
\downarrow & & \downarrow \\
\Delta^{n+1} & \xrightarrow{c_{n-i}} & A_i
\end{array}
\tag{3.1.3.3}
$$

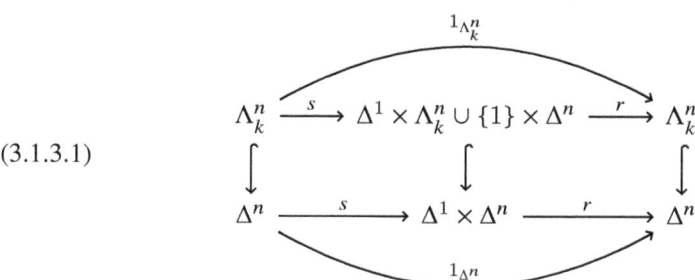

This shows that the generators of the class (b) all belong to the class (a) and thus achieves the proof of the lemma. □

Corollary 3.1.4 *For any anodyne extension $K \to L$ and any monomorphism $U \to V$, the induced inclusion $K \times V \cup L \times U \to L \times V$ is an anodyne extension.*

Definition 3.1.5 A *Kan fibration* is a morphism of simplicial sets with the right lifting property with respect to the inclusions of the form $\Lambda^n_k \to \Delta^n$, for $n \geq 1$ and $0 \leq k \leq n$.

A *Kan complex* is a simplicial set X such that the morphism from X to the final simplicial set is a Kan fibration.

Corollary 3.1.6 *For any monomorphism $i : U \to V$ and for any Kan fibration $p \colon X \to Y$, the canonical map*

$$(i^*, p_*) \colon \underline{\mathrm{Hom}}(V, X) \to \underline{\mathrm{Hom}}(U, X) \times_{\underline{\mathrm{Hom}}(U,Y)} \underline{\mathrm{Hom}}(V, Y)$$

is a Kan fibration.

Corollary 3.1.7 *For any anodyne extension $i \colon K \to L$ and for any Kan fibration $p \colon X \to Y$, the canonical map*

$$(i^*, p_*) \colon \underline{\mathrm{Hom}}(L, X) \to \underline{\mathrm{Hom}}(K, X) \times_{\underline{\mathrm{Hom}}(L,Y)} \underline{\mathrm{Hom}}(K, Y)$$

is a trivial fibration (i.e. has the right lifting property with respect to monomorphisms).

Theorem 3.1.8 *There is a unique model category structure on the category of simplicial sets whose class of cofibrations is the class of monomorphisms, and whose fibrant objects are the Kan complexes. Moreover, any anodyne extension is a trivial cofibration, and the fibrations between fibrant objects exactly are the Kan fibrations between Kan complexes.*

Proof By virtue of Proposition 3.1.2, this follows from Theorem 2.4.19 applied to the category of simplicial sets for the exact cylinder $\Delta^1 \times (-)$ and for the class *An* of anodyne extensions. □

Definition 3.1.9 The *Kan–Quillen model category structure* is the model category structure on the category of simplicial sets obtained by Theorem 3.1.8. The weak equivalences of this model category structure are called the *weak homotopy equivalences*.

Corollary 3.1.10 *The class of weak homotopy equivalences is closed under finite products.*

Proof This assertion is equivalent to the property that, for any simplicial set Y, the functor $X \mapsto X \times Y$ preserves weak equivalences. Since all simplicial sets are cofibrant, this can be reformulated by saying that it is part of a Quillen adjunction from the Kan–Quillen model category structure to itself. This functor obviously preserves cofibrations, so that it is sufficient to prove that it preserves trivial cofibrations. By virtue of Proposition 2.4.40, it is sufficient to prove that it preserves anodyne extensions, which is a particular case of Corollary 3.1.4. □

Lemma 3.1.11 *Let $\mathcal{C}$ be a category in which a commutative square*

$$\begin{array}{ccc} A & \xrightarrow{i} & B \\ {\scriptstyle j}\downarrow & & \downarrow{\scriptstyle l} \\ C & \xrightarrow{k} & D \end{array}$$

is given. Assume that the maps j, k and l have retractions r, q and p, respectively, and that $pk = ir$. Then this square is absolutely Cartesian (i.e. its image by any functor of domain $\mathcal{C}$ is Cartesian).

Proof It is sufficient to prove that such a square is Cartesian. Let $u\colon X \to C$ and $v\colon X \to B$ be two morphisms such that $ku = lv$. We must prove that there is a unique map $w\colon X \to A$ such that $jw = u$ and $iw = v$. Since $pj = 1_A$, we must have $w = rjw = ru$, so that the uniqueness is clear. For the existence, we put $w = ru$. Since both k and l are monomorphisms, it is sufficient to check that $kjw = ku$ and that $liw = lv$, which one sees right away. □

Lemma 3.1.12 *For $n \geq 2$ and $0 \leq i < j \leq n$, the following square is absolutely Cartesian.*

$$\begin{array}{ccc} \Delta^{n-2} & \xrightarrow{\delta_i^{n-1}} & \Delta^{n-1} \\ {\scriptstyle \delta_{j-1}^{n-1}}\downarrow & & \downarrow{\scriptstyle \delta_j^n} \\ \Delta^{n-1} & \xrightarrow{\delta_i^n} & \Delta^n \end{array}$$

Proof It is sufficient to check the assumptions of the previous lemma. We choose a retraction q of δ_n^i. In the case where $j < n$, we may take $r = \sigma_{n-2}^{j-1}$ and $p = \sigma_{n-1}^j$. If $j = n$, then we may assume that $i \neq n-1$, because otherwise we apply the functor $X \mapsto X^{\mathrm{op}}$ and get back to this case. Under this extra assumption, we put $r = \sigma_{n-2}^{j-2}$ and $p = \sigma_{n-1}^{j-1}$. □

Proposition 3.1.13 *Let C be a small category and $A\colon \mathbf{\Delta} \to \widehat{C}$ a functor. Then the induced colimit preserving functor $A_!\colon sSet \to \widehat{C}$ preserves* *monomorphisms if and only if the map*

$$(A(\delta^1_0), A(\delta^1_1)) \colon A^0 \amalg A^0 \to A^1 = A_!(\Delta^1)$$

is a monomorphism.

Proof The class of monomorphisms of simplicial sets is the smallest saturated class which contains the inclusions $\partial\Delta^n \to \Delta^n$. Therefore, the functor $A_!$ preserves monomorphisms if and only if, for any integer $n \geq 0$, it sends $\partial\Delta^n \to \Delta^n$ to a monomorphism. We shall see that the map

$$A_!(\partial\Delta^n) \to A_!(\Delta^n) = A^n$$

always is a monomorphism for $n \neq 1$, which will prove the proposition. The case where $n = 0$ comes from the fact that, since the functor $A_!$ preserves colimits, it sends the empty simplicial set to the empty presheaf over C. Let us assume that $n \geq 2$. The boundary $\partial\Delta^n$ is the union of representable subpresheaves of the form $F_i = Im(\delta^n_i)$, $0 \leq i \leq n$, and we have Cartesian squares of subobjects of A^n of the form

$$\begin{array}{ccc} A_!(F_i \cap F_j) & \hookrightarrow & A_!(F_i) \\ \downarrow & & \downarrow \\ A_!(F_j) & \hookrightarrow & A^n \end{array}$$

(because either $F_i \cap F_j = \varnothing$, or this square is the image by $A_!$ of an absolute Cartesian square provided by Lemma 3.1.12). In other words, $A_!(\partial\Delta^n)$ is canonically isomorphic to the union of the $A_!(F_i)$, and thus, in particular, is a subobject of A^n. □

Proposition 3.1.14 *Let $A, B \colon \mathbf{\Delta} \to \mathcal{C}$ be two functors with values in a model category. We denote by $A_!$ and $B_!$ their extensions by colimits, respectively, and we assume that both of them send monomorphisms to cofibrations. If a natural transformation $u \colon A \to B$ induces a weak equivalence $A^n \to B^n$ for all $n \geq 0$, then, for any simplicial set X, the map*

$$u_X \colon A_!(X) \to B_!(X)$$

is a weak equivalence.

Proof It follows from Corollaries 2.3.16, 2.3.18 and 2.3.29 that the class of simplicial sets X such that the map u_X is a weak equivalence is saturated by monomorphisms (see Definition 1.3.9). Since the category of simplices is an Eilenberg–Zilber category, we may apply Corollary 1.3.10 for $A = \mathbf{\Delta}$, and conclude that the smallest class of simplicial sets which is saturated by monomorphisms and which contains all representable simplicial sets is the class of all simplicial sets. Therefore, since the map u_X is a weak equivalence for $X = \Delta^n$, it must have the same property for any X. □

3.1.15 A *bisimplicial set* is a presheaf over the product $\mathbf{\Delta}\times\mathbf{\Delta}$. For a bisimplicial set X, its evaluation at $([m],[n])$ is denoted by $X_{m,n}$.

For two simplicial sets X and Y, we write $X \boxtimes Y$ for the bisimplicial set defined by

$$(X \boxtimes Y)_{m,n} = X_m \times Y_n \,.$$

Note that this operation preserves representable presheaves: for $m, n \geq 0$, we have

$$\Delta^m \boxtimes \Delta^n = h_{([m],[n])} \,.$$

For a bisimplicial set X and a simplicial set K, we write X^K for the simplicial set defined by

$$(X^K)_m = \varprojlim_{\Delta^n \to K} X_{m,n} \,.$$

In other words, the functor $K \mapsto X^K$ is the extension by colimits of the functor $\mathbf{\Delta} \to sSet^{\mathrm{op}}$ which sends $[n]$ to the simplicial set $X^{\Delta^n} = ([m] \mapsto X_{m,n})$.

Finally, for a bisimplicial set X, we write $\operatorname{diag}(X)$ for the simplicial set defined by

$$\operatorname{diag}(X)_n = X_{n,n} \,.$$

Theorem 3.1.16 *If a morphism of bisimplicial sets $X \to Y$ is a levelwise weak homotopy equivalence (i.e. induces a weak homotopy equivalence $X^{\Delta^m} \to Y^{\Delta^m}$ for all $m \geq 0$), then the diagonal map*

$$\operatorname{diag}(X) \to \operatorname{diag}(Y)$$

is a weak homotopy equivalence.

Proof It follows from Theorem 1.3.8, applied to $A = \mathbf{\Delta} \times \mathbf{\Delta}$ (see Example 1.3.4), that a cellular model for the category of bisimplicial sets is given by the maps of the form

$$\Delta^m \boxtimes \partial\Delta^n \cup \partial\Delta^m \boxtimes \Delta^n \to \Delta^m \boxtimes \Delta^n$$

for $m, n \geq 0$ (because the domain of this embedding is the boundary of the codomain in the sense of Notation 1.3.7). In other words, in the context of bisimplicial sets, the trivial fibrations are the maps $X \to Y$ such that the induced morphism of the simplicial set

$$X^{\Delta^m} \to Y^{\Delta^m} \times_{Y^{\partial\Delta^m}} X^{\partial\Delta^m} \tag{3.1.16.1}$$

is a trivial fibration for all $m \geq 0$.

There is an exact cylinder I defined as the Cartesian product with the interval $\Delta^0 \boxtimes \Delta^1$. Let An be the smallest class of I-anodyne extensions containing the maps of the form

$$\Delta^m \boxtimes \Lambda^n_k \to \Delta^m \boxtimes \Delta^n \tag{3.1.16.2}$$

for $m, n \geq 0$ and $0 \leq k \leq n$. Using the explicit construction of the class An given in Example 2.4.13 together with Proposition 3.1.2, we see that the corresponding naive fibrations precisely are the morphisms $X \to Y$ such that, for any $m \geq 0$, the map (3.1.16.1) is a Kan fibration. In the sequel of this proof, we will consider the category of bisimplicial sets as endowed with the model category structure associated to I and An by Theorem 2.4.19.

On the other hand, the class of morphisms of bisimplicial sets whose image by the diagonal functor diag is a trivial cofibration of the Kan–Quillen model structure is saturated. Since the diagonal of any map of the form (3.1.16.2) is a weak homotopy equivalence, we deduce that the functor diag sends any element of An to a weak homotopy equivalence. Therefore, by virtue of Proposition 2.4.40, the functor diag sends weak equivalences of the model category structure associated to I and An to weak homotopy equivalences.

Since levelwise trivial cofibrations also form a saturated class of morphisms of bisimplicial sets and since the maps of the form (3.1.16.2) have this property, we also know that any element of An is a levelwise weak homotopy equivalence. For a general morphism of bisimplicial sets $f \colon U \to V$, we can form a commutative diagram of shape

$$\begin{array}{ccc} U & \xrightarrow{i} & X \\ {\scriptstyle f}\downarrow & & \downarrow{\scriptstyle p} \\ V & \xrightarrow{j} & Y \end{array}$$

in which i and j are I-anodyne, while p is a naive fibration with fibrant codomain. If f is a levelwise weak equivalence, then so is p. Applying Proposition 3.1.14 for $\mathcal{C} = sSet^{\mathrm{op}}$, we deduce that, for any simplicial set K, the induced map $X^K \to Y^K$ is a weak homotopy equivalence between Kan complexes. Using the fact that pulling back weak equivalences between fibrant objects along fibrations gives weak equivalences, together with the two-out-of-three property for weak equivalences, we deduce that the map (3.1.16.1) is a trivial fibration for all $m \geq 0$. In particular, f is a weak equivalence, and thus $\mathrm{diag}(f)$ is a weak homotopy equivalence. □

3.1.17 Given a partially ordered set E, we write $s(E)$ for the set of finite non-empty totally ordered subsets of E, ordered by inclusion. This defines a functor from the category of partially ordered sets to the category of small categories: given a morphism $f: E \to F$, the induced map $s(f): s(E) \to s(F)$ simply is defined by

$$s(f)(U) = f(U) = \{f(x) \mid x \in U\}.$$

The (barycentric) *subdivision* functor

$$Sd: sSet \to sSet$$

is the extension by colimits of the functor $[n] \mapsto s([n])$. It has a right adjoint

$$Ex: sSet \to sSet$$

defined by $Ex(X)_n = \mathrm{Hom}(Sd(\Delta^n), X)$. Taking the maximal element of non-empty subsets of totally ordered finite sets, define a functorial order preserving map

$$a_n: s([n]) \to [n], \quad U \mapsto \max(U).$$

This extends to a natural transformation

$$a_X: Sd(X) \to X,$$

which, by transposition, defines an embedding

$$b_X: X \to Ex(X)$$

(obtained by composing the Yoneda isomorphism $X_n \simeq \mathrm{Hom}(\Delta^n, X)$ with the map $a_n^*: \mathrm{Hom}(\Delta^n, X) \to \mathrm{Hom}(Sd(\Delta^n), X)$).

Proposition 3.1.18 *The functor Sd preserves monomorphisms as well as anodyne extensions.*

Proof The fact that this functor preserves monomorphisms is a direct application of Proposition 3.1.13.

For each $n \geq 0$, the simplicial set $Sd(\Delta^n)$ is the nerve of a partially ordered set with a final element (namely $[n]$) and the inclusions $S \subset [n]$ define a Δ^1-homotopy from the identity of $Sd(\Delta^n)$ to the constant map with value $[n]$. In particular, the map $\omega: \Delta^0 \to Sd(\Delta^n)$ corresponding to the final element is a strong deformation retract, and thus is an anodyne extension (see Lemma 2.4.32).

Let us prove that any point $\eta: \Delta^0 \to Sd(\Delta^n)$ is an anodyne extension. There is a homotopy $h: \Delta^1 \to Sd(\Delta^n)$ such that $h(0) = \eta$ and $h(1) = \omega$. By virtue of Proposition 2.5.4, we see that h and η must be absolute weak equivalences and thus that η must be an anodyne extension.

Using Proposition 2.5.4 again, we deduce that, for any morphism $u: \Delta^m \to$

Δ^n the map $Sd(u)\colon Sd(\Delta^m) \to Sd(\Delta^n)$ is an absolute weak equivalence (because we may always choose a base point).

Let W be the class of morphisms whose image by Sd becomes an absolute weak equivalence. We want to prove that all inclusions of the form $\Lambda^n_k \to \Delta^n$ belong to W, knowing that all the maps $\Delta^m \to \Delta^n$ are in W. We proceed by induction on $n \geq 1$. The case $n = 1$ is easy: we have to check that the maps $\Delta^0 \to \Delta^1$ are in W, which we already know. For a subset $I \subset [n]$, we put

$$\Lambda^n_I = \bigcup_{i \notin I} Im(\delta^n_i)\,.$$

Hence $\Lambda^n_{\{k\}} = \Lambda^n_k$. We shall prove that, for any non-empty proper subset I, the map $\Lambda^n_I \to \Delta^n$ is in W. We proceed by induction on the number c of elements in the complement of I. If $c = 1$, this means that we have to prove that any face map $\Delta^{n-1} \to \Delta^n$ is in W, which is clear. If $c > 1$, we pick an element $k \notin I$, and, using Lemma 3.1.12, we see that there is a unique non-empty proper subset I' of $[n-1]$ such that we have biCartesian squares of the following form.

$$\begin{array}{ccc} \Lambda^{n-1}_{I'} & \longrightarrow & \Lambda^n_{I\cup\{k\}} \\ \downarrow & & \downarrow \\ \Delta^{n-1} & \xrightarrow{\delta^n_k} & \Lambda^n_I \end{array}$$

Therefore, by induction on n, the inclusion $\Lambda^n_{I\cup\{k\}} \to \Lambda^n_I$ is in W. Since the inclusion $\Lambda^n_{I\cup\{k\}} \to \Delta^n$ is in W by induction on c, this proves that the inclusion $\Lambda^n_k \to \Delta^n$ is in W. □

Proposition 3.1.19 *For any simplicial set X, the map $a_X\colon Sd(X) \to X$ is a weak homotopy equivalence.*

Proof By virtue of Proposition 3.1.14, we are reduced to proving that the map a_X is a weak homotopy equivalence for $X = \Delta^n$, $n \geq 0$. Therefore, using the two-out-of-three property for weak homotopy equivalences, we obtain the property we seek. □

Lemma 3.1.20 *Let $f, g\colon K \to L$ be two morphisms of simplicial sets, such that there exists a homotopy $h\colon \Delta^1 \times K \to L$ from f to g. Then, for any simplicial set X, there exists a homotopy from f^* to g^*, where $f^*, g^*\colon \underline{\mathrm{Hom}}(L, X) \to \underline{\mathrm{Hom}}(K, X)$ are the morphisms induced by f and g, respectively.*

Proof The map h defines a morphism

$$h^*\colon \underline{\mathrm{Hom}}(L, X) \to \underline{\mathrm{Hom}}(\Delta^1 \times K, X) \simeq \underline{\mathrm{Hom}}(\Delta^1, \underline{\mathrm{Hom}}(K, X))$$

which induces, by transposition, a morphism

$$\tilde{h}\colon \Delta^1 \times \underline{\mathrm{Hom}}(L, X) \to \underline{\mathrm{Hom}}(K, X)\,.$$

The latter is a homotopy from f^* to g^*. □

Proposition 3.1.21 (Kan) *For any simplicial set X, the map $b_X\colon X \to Ex(X)$ is a weak homotopy equivalence.*

Proof In the particular case where X is a Kan complex, the assertion follows from Proposition 3.1.19. Indeed, this proposition means that the pair Sd and Ex form a Quillen adjunction, which induces a derived adjunction $\mathbf{L}Sd$ and $\mathbf{R}Ex$, and that the total left derived functor $\mathbf{L}Sd$ is isomorphic to the identity via the natural maps a_X. By transposition, this means that the functor $\mathbf{R}Ex$ is isomorphic to the identity, which precisely means that b_X is a (weak) homotopy equivalence for any Kan complex X. For the general case, we consider an anodyne extension $i\colon X \to Y$ with Y a Kan complex, and we consider the commutative diagram below.

$$\begin{array}{ccc} X & \xrightarrow{\;i\;} & Y \\ {\scriptstyle b_X}\downarrow & & \downarrow{\scriptstyle b_Y} \\ Ex(X) & \xrightarrow{Ex(i)} & Ex(Y) \end{array}$$

The maps i and b_Y are weak homotopy equivalences. To finish the proof, it is thus sufficient to prove that the map $Ex(i)$ is a weak homotopy equivalence. We will now prove that the functor Ex preserves weak homotopy equivalences, which will allow us to conclude.

For each simplicial set X we define a bisimplicial set $E(X)$ by the formula

$$E(X)_{m,n} = \mathrm{Hom}(\Delta^m \times Sd(\Delta^n), X)\,.$$

The projections $\Delta^m \leftarrow \Delta^m \times Sd(\Delta^n) \to Sd(\Delta^n)$ induce functorial morphisms of bisimplicial sets

$$X \boxtimes \Delta^0 \to E(X) \leftarrow \Delta^0 \boxtimes Ex(X)\,.$$

For any simplicial set K such that the map $K \to \Delta^0$ is a homotopy equivalence, it follows from the preceding lemma that the induced map $X \to \underline{\mathrm{Hom}}(K, X)$ is a homotopy equivalence. In particular, the map $X \to \underline{\mathrm{Hom}}(Sd(\Delta^n), X)$ is a homotopy equivalence for all $n \geq 0$. Applying Theorem 3.1.16 (possibly after permuting the factors in $\mathbf{\Delta} \times \mathbf{\Delta}$), we conclude that the induced map

$$X \simeq \mathrm{diag}(X \boxtimes \Delta^0) \to \mathrm{diag}(E(X))$$

is a weak homotopy equivalence.

Similarly, the map $X \to \underline{\mathrm{Hom}}(\Delta^m, X)$ is a homotopy equivalence, and thus, since the functor Ex commutes with finite products, the map

$$Ex(X) \simeq (\Delta^0 \boxtimes Ex(X))^{\Delta^m} \to Ex(\underline{\mathrm{Hom}}(\Delta^m, X)) = E(X)^{\Delta^m}$$

is a homotopy equivalence with respect to the cylinder associated to the interval $Ex(\Delta^1)$. The map $b_{\Delta^1} : \Delta^1 \to Ex(\Delta^1)$ turns any $Ex(\Delta^1)$-homotopy into a Δ^1-homotopy. Applying Theorem 3.1.16 once more, we deduce that the functorial map

$$Ex(X) \simeq \mathrm{diag}(\Delta^0 \boxtimes Ex(X)) \to \mathrm{diag}(E(X))$$

is a weak homotopy equivalence. For any morphism $f: X \to Y$, we now have the commutative diagram below, in which all horizontal maps are weak homotopy equivalences.

$$\begin{array}{ccccc} X & \longrightarrow & \mathrm{diag}(E(X)) & \longleftarrow & Ex(X) \\ {\scriptstyle f}\downarrow & & {\scriptstyle \mathrm{diag}(E(f))}\downarrow & & \downarrow{\scriptstyle Ex(f)} \\ Y & \longrightarrow & \mathrm{diag}(E(Y)) & \longleftarrow & Ex(Y) \end{array}$$

Therefore, the map f is a weak homotopy equivalence if and only if the map $Ex(f)$ has this property. □

3.1.22 For any simplicial set X, and for any integer $n \geq 1$, we define Ex^n by induction:

(3.1.22.1) $$Ex^{n+1}(X) = Ex(Ex^n(X)) .$$

It follows right away from Proposition 3.1.18 that the functors Ex^n all preserve Kan fibrations as well as trivial fibrations. We have natural morphisms

(3.1.22.2) $$b_{Ex^n(X)} : Ex^n(X) \to Ex^{n+1}(X)$$

and thus a sequence of morphisms

(3.1.22.3) $$X \to Ex(X) \to Ex^2(X) \to \cdots \to Ex^n(X) \to Ex^{n+1}(X) \to \cdots .$$

One defines

(3.1.22.4) $$Ex^\infty(X) = \varinjlim_{n \geq 0} Ex^n(X) .$$

By virtue of Proposition 3.1.21, the canonical map

(3.1.22.5) $$\beta_X : X \to Ex^\infty(X)$$

is a trivial cofibration.

Proposition 3.1.23 *The functor Ex^∞ preserves Kan fibrations as well as trivial fibrations.*

Proof Since the analogous property is known for Ex^n with n finite, this follows from the next lemma. □

Lemma 3.1.24 *The classes of Kan fibrations and of trivial fibrations are closed under filtered colimits.*

Proof For any set J of morphisms between finite simplicial sets (i.e. simplicial sets which are finite colimits of representable presheaves, or equivalently, which have finitely many non-degenerate simplices), the class of morphisms with the right lifting property with respect to J is closed under filtered colimits. Indeed, for any finite simplicial set K, the functor $\operatorname{Hom}(K, -)$ commutes with filtered colimits (Corollary 1.3.12), and a morphism $p\colon X \to Y$ has the right lifting property with respect to a map $u\colon K \to L$ if and only if the canonical

$$(u^*, p_*)\colon \operatorname{Hom}(L, X) \to \operatorname{Hom}(K, X) \times_{\operatorname{Hom}(K,Y)} \operatorname{Hom}(L, Y)$$

is surjective. With u a fixed map between finite simplicial sets, the formation of such a map (u^*, p_*) commutes with filtered colimits of the p, and this proves our assertion. □

Lemma 3.1.25 *For any partially ordered set E, there is a canonical isomorphism*

$$Sd(N(E)) \simeq N(s(E)) .$$

Proof There is a canonical functorial map $Sd(N(E)) \to N(s(E))$ which is characterised by being the identity for $E = [n]$, $n \geq 0$. One checks that

$$\varinjlim_{P \in s(E)} N(P) \simeq N(E) .$$

In other words, the nerve of E is the union of the subobjects of the form $N(P)$ (one sees this thanks to the fact that, for two finite non-empty totally ordered subsets P and Q in E, the intersection $N(P) \cap N(Q)$ is either empty or the nerve of a finite non-empty totally ordered subset). Therefore, it is sufficient to check that the natural map

$$\varinjlim_{P \in s(E)} N(s(P)) \to N(s(E))$$

is an isomorphism. Since the functor s preserves monomorphisms, intersection of subobjects, and since $s(\varnothing) = \varnothing$, we see that the left-hand side is the union, on the right-hand side, of the subobjects of the form $N(s(P))$ for $P \in s(E)$.

Therefore, it is sufficient to check the surjectivity, which is obvious: any finite sequence $P_0 \subset P_1 \subset \cdots \subset P_n$ in $s(E)$ can be seen as a sequence in $s(P_n)$. □

Lemma 3.1.26 *There are canonical isomorphisms*

$$Sd(\Lambda^n_k) \simeq N(\Phi^n_k) \quad \textit{and} \quad Sd(\partial\Delta^n) \simeq N(\partial\Phi^n),$$

where Φ^n_k denotes the set of non-empty subsets of $[n]$ which do not contain the complement of $\{k\}$, and $\partial\Phi^n$ is the set of proper non-empty subsets of $[n]$ (ordered by inclusion).

Proof We already know that $Sd(\Lambda^n_k)$ is a subobject of $Sd(\Delta^n)$, and it is clear, by virtue of the functoriality of the isomorphism provided by the preceding lemma, that it is contained in $N(\Phi^n_k)$. Any m-simplex of $N(\Phi^n_k)$ is of the form $P_0 \subset P_1 \subset \cdots \subset P_m$ and thus defines an m-simplex of $N(s(P_m))$. Since the inclusion $N(P_m) \subset \Delta^n$ factors through Λ^n_k, this shows that $Sd(\Lambda^n_k) = N(\Phi^n_k)$. The case of $Sd(\partial\Delta^n)$ is proven similarly. □

Theorem 3.1.27 (Kan) *For any simplicial set X, we have a Kan complex $Ex^\infty(X)$.*

Proof We define a morphism $\psi^n_k \colon s(s([n])) \to \Phi^n_k$ by the formula

$$\psi^n_k(P) = \{c^n_k(S) \mid S \in P\},$$

where, for a non-empty subset $S \subset [n]$, we put

$$c^n_k(S) = \begin{cases} \max(S) & \text{if } S \in \Phi^n_k, \\ k & \text{otherwise.} \end{cases}$$

By virtue of the two preceding lemmas, the nerve of ψ^n_k defines a morphism

$$u^n_k = N(\psi^n_k) \colon Sd^2(\Delta^n) \to Sd(\Lambda^n_k),$$

and we check that the following triangle commutes.

$$\begin{array}{ccc} Sd^2(\Lambda^n_k) & \xrightarrow{Sd(a_{\Lambda^n_k})} & Sd(\Lambda^n_k) \\ \downarrow & \nearrow_{u^n_k} & \\ Sd^2(\Delta^n) & & \end{array}$$

Let $x \colon \Lambda^n_k \to Ex^\infty(X)$ be a morphism. The previous two lemmas also imply that the simplicial set $Sd^m(\Lambda^n_k)$ has finitely many non-degenerate simplices for any $m \geq 0$. Therefore, by virtue of Corollary 1.3.12, there exists $m \geq 1$ such that the map x factors through $Ex^m(X)$. The map x corresponds by adjunction to a map $\tilde{x}\colon Sd(\Lambda^n_k) \to Ex^{m-1}(X)$. The morphism $\tilde{x}u^n_k$ defines a morphism $y\colon \Lambda^n_k \to Ex^{m+1}(X)$ such that the following diagram commutes.

$$\begin{array}{ccccc} \Lambda^n_k & \xrightarrow{x} & Ex^m(X) & \longrightarrow & Ex^\infty(X) \\ \big\downarrow & & \big\downarrow{\scriptstyle b_{Ex^m(X)}} & \nearrow & \\ \Delta^n & \xrightarrow{y} & Ex^{m+1}(X) & & \end{array}$$

This proves that $Ex^\infty(X)$ is a Kan complex. □

Corollary 3.1.28 *For any Cartesian square of simplicial sets*

$$\begin{array}{ccc} X' & \xrightarrow{f} & X \\ {\scriptstyle p'}\big\downarrow & & \big\downarrow{\scriptstyle p} \\ Y' & \xrightarrow{g} & Y \end{array}$$

in which p is a Kan fibration, if g (or p) is a weak homotopy equivalence, so is the map f (or p', respectively).

Proof Applying the functor Ex^∞ to this square, we obtain a Cartesian square whose vertices all are Kan complexes, and in which the map $Ex^\infty(p)$ is a Kan fibration. If the map $Ex^\infty(g)$ is a weak equivalence, by virtue of the dual version of Proposition 2.3.27 (i.e. applying the latter for $\mathcal{C} = (sSet)^{\mathrm{op}}$), we see that the map $Ex^\infty(f)$ is a weak equivalence. If the map $Ex^\infty(p)$ is a weak equivalence, then it is a trivial fibration (because it is a naive fibration between fibrant objects, hence a fibration). The existence of the natural weak equivalence from the identity to Ex^∞ (3.1.22.5) thus implies that f (respectively p) is a weak homotopy equivalence. □

Theorem 3.1.29 (Quillen) *A morphism of simplicial sets is a fibration of the Kan–Quillen model category structure if and only if it is a Kan fibration. A morphism of simplicial sets is a trivial cofibration of the Kan–Quillen model category structure if and only if it is an anodyne extension.*

Proof It is sufficient to prove the second assertion. Factorising such a trivial cofibration into an anodyne extension followed by a Kan fibration, we see from the retract lemma that it is sufficient to prove that any Kan fibration which is a weak homotopy equivalence is a trivial fibration. Since there is a cellular model which consists of monomorphisms with representable codomain, and since, by virtue of Corollary 3.1.28, the class of Kan fibrations which are weak homotopy equivalences is closed under pull-backs, it is sufficient to prove the special case where the codomain is representable. Let $p\colon X \to \Delta^n$ be a Kan fibration which is a weak homotopy equivalence. Let F be the fibre of p at 0. Then, by virtue of Proposition 2.5.7, the inclusion $F \to X$ is a strong deformation retract, and, therefore, is an anodyne extension (see Lemma 2.4.32). The map $F \to \Delta^0$ is a weak homotopy equivalence and a fibration, which implies that it is a trivial fibration, hence an absolute weak equivalence. It follows from Proposition 2.5.4 that the map p is an absolute weak equivalence, from which we deduce, thanks to Proposition 2.5.3, that it is a trivial fibration. □

3.1.30 The inclusion functor $Set \to sSet$ has a left adjoint

$$\pi_0 \colon sSet \to Set\,.$$

In other words, $\pi_0(X)$ is the colimit of X in the category of sets.

Proposition 3.1.31 *The functor π_0 sends weak homotopy equivalences to bijections and commutes with finite products. Furthermore, for any Kan complex X, the set $\pi_0(X)$ may be identified with the set of Δ^1-homotopy classes of maps $\Delta^0 \to X$.*

Proof We first prove that π_0 is the left adjoint in a Quillen adjunction, where the category of small sets is equipped with the model category structure whose weak equivalences are the bijections and the cofibrations are all maps. By virtue of Propositions 2.4.40 and 3.1.2, it is sufficient to prove that $\pi_0(\Lambda^n_k)$ is the one-point set, which follows from an easy induction on n, using the push-out square at the end of the proof of Proposition 3.1.18. The last assertion of the proposition, about the description of $\pi_0(X)$ when X is a Kan complex, thus follows from Theorem 2.3.9. For any simplicial set X, there is a functorial bijection $\pi_0(X) \simeq \pi_0(Ex^\infty(X))$. Since the functor Ex^∞ commutes with finite products (as a filtered colimit of limit preserving functors), it is sufficient to prove that π_0 commutes with finite products when restricted to Kan complexes. But the homotopy category of Kan complexes has finite products which correspond to finite products of the underlying simplicial sets. Since, for a Kan complex X, the set $\pi_0(X)$ is the set of maps from Δ^0 to X in the homotopy category, this proves that the functor π_0 commutes with finite products. □

3.2 Inner Anodyne Extensions

Definition 3.2.1 An *inner anodyne extension* is an element of the smallest saturated class of morphisms of simplicial sets which contains the set of inclusions of the form $\Lambda^n_k \to \Delta^n$ with $n \geq 2$ and $0 < k < n$.

We already used the following lemma in the case where $m = 1$ (and it may be seen as a direct consequence of the computation of the colimit in the proof of Lemma 3.1.25 for $E = [m] \times [n]$).

Lemma 3.2.2 *For any non-negative integers m and n, the simplicial set $\Delta^m \times \Delta^n$ is the union of its subobjects of the form $N(P)$, where P runs over the set of totally ordered subsets of cardinal $m + n + 1$.*

Proposition 3.2.3 (Joyal) *The following three classes of morphisms of simplicial sets are equal:*

(a) the class of inner anodyne extensions;

(b) the smallest saturated class of maps containing inclusions of the form

$$\Delta^2 \times \partial\Delta^n \cup \Lambda^2_1 \times \Delta^n \to \Delta^2 \times \Delta^n \text{ for } n \geq 0;$$

(c) the smallest saturated class containing inclusions of the form

$$\Delta^2 \times K \cup \Lambda^2_1 \times L \to \Delta^2 \times L \text{ for any monomorphism } K \to L.$$

Proof The fact that the classes (b) and (c) coincide is a formal consequence of the fact that the smallest saturated class of maps containing the boundary inclusions $\partial\Delta^n \to \Delta^n$, for $n \geq 0$, is the class of all monomorphisms, and from the fact that the class of monomorphisms $K \to L$ such that the induced map $\Delta^2 \times K \cup \Lambda^2_1 \times L \to \Delta^2 \times L$ is inner anodyne is saturated (because it can be characterised in terms of left lifting property with respect to a certain class of morphisms).

Let us prove that the class (a) is in the class (b). For $n \geq 2$ and $0 < k < n$, we define morphisms

$$s\colon [n] \to [2] \times [n] \quad \text{and} \quad r\colon [2] \times [n] \to [n]$$

by the formulas

$$s(j) = \begin{cases} (0, j) & \text{if } j < k, \\ (1, j) & \text{if } j = k, \\ (2, j) & \text{if } j > k, \end{cases} \quad \text{and} \quad r(i, j) = \begin{cases} \min\{j, k\} & \text{if } i = 0, \\ k & \text{if } i = 1, \\ \max\{j, k\} & \text{if } i = 2. \end{cases}$$

We clearly have $rs = 1_{[n]}$, and we denote by the same letters

$$s\colon \Delta^n \to \Delta^2 \times \Delta^n \quad \text{and} \quad r\colon \Delta^2 \times \Delta^n \to \Delta^n$$

the corresponding morphisms of simplicial sets. We claim that these maps s

and r induce a commutative diagram of the form below.

$$\begin{array}{ccccc} \Lambda^n_k & \xrightarrow{s'} & \Delta^2 \times \Lambda^n_k \cup \Lambda^2_1 \times \Delta^n & \xrightarrow{r'} & \Lambda^n_k \\ \downarrow & & \downarrow & & \downarrow \\ \Delta^n & \xrightarrow{s} & \Delta^2 \times \Delta^n & \xrightarrow{r} & \Delta^n \end{array}$$

To prove that the map s' is well defined, we have to check that the image of any composition of the form $s\delta^n_i$, $i \neq k$, is in $\Delta^2 \times \Lambda^n_k \cup \Lambda^2_1 \times \Delta^n$. Such an image must land in $\Delta^2 \times \partial\Delta^n \cup \partial\Delta^2 \times \Delta^n$ for cardinality reasons. Therefore, it is sufficient to check that the image is not contained in $Im(\delta^2_1) \times \Delta^n$ nor in $\Delta^2 \times Im(\delta^n_k)$, which follows, in both cases, from the fact that the point $(1, k)$ must be reached. To prove that r' is well defined, we have several cases to consider. The image of $\Delta^{\{0,1\}} \times \Delta^n$ by r is contained in $\Delta^{\{0,\ldots,k\}} \subset \Lambda^n_k$, and, dually, the image of $\Delta^{\{1,2\}} \times \Delta^n$ is contained in $\Delta^{\{k,\ldots,n\}} \subset \Lambda^n_k$. It remains to prove that r sends $\Delta^2 \times \Lambda^n_k$ to Λ^n_k. But, for $0 \leq l \leq n$, $l \neq k$, we have $r(i, j) \neq l$ for any $j \neq l$. In other words, the image of $\Delta^2 \times Im(\delta^n_l)$ by r does not reach the value i, and is thus contained in $Im(\delta^n_i) \subset \Lambda^n_k$.

It remains to check that the class (b) is in the class (a). For $0 \leq i \leq j < n$, we let $U_{i,j}$ be the image of the map

$$u_{i,j} \colon \Delta^{n+1} \to \Delta^2 \times \Delta^n$$

defined as the nerve of the map

$$k \mapsto \begin{cases} (0, k) & \text{if } 0 \leq k \leq i, \\ (1, k-1) & \text{if } i < k \leq j+1, \\ (2, k-1) & \text{otherwise.} \end{cases}$$

For $0 \leq i \leq j \leq n$, we let $V_{i,j}$ be the image of the map

$$v_{i,j} \colon \Delta^{n+2} \to \Delta^2 \times \Delta^n$$

defined as the nerve of the map

$$k \mapsto \begin{cases} (0, k) & \text{if } 0 \leq k \leq i, \\ (1, k-1) & \text{if } i < k \leq j+1, \\ (2, k-2) & \text{otherwise.} \end{cases}$$

We define

$$X(-1, -1) = \Delta^2 \times \partial\Delta^n \cup \Lambda^2_1 \times \Delta^n$$

and, for $n > j \geq 0$ and $0 \leq i \leq j$, we put

$$X(i, j) = X(j-1, j-1) \cup \bigcup_{0 \leq l \leq i} U_{l,j} \,.$$

Using the fact that $u_{i,j}^{-1}(\Delta^{\{0,2\}} \times \Delta^n) = \Delta^{\{0,\ldots,n\}-\{i+1,\ldots,j+1\}}$, we see that we have $u_{0,j}^{-1}(\Delta^{\{0,2\}} \times \Delta^n) = Im(\delta_{j+1}^{n+1})$, and that $u_{i,j}^{-1}((\Delta^{\{0,2\}} \times \Delta^n) \cup X(i, j+1)) = Im(\delta_{i+1}^{n+1})$, from which we deduce biCartesian squares of the form below.

$$\begin{array}{ccc} \Lambda_{j+1}^{n+1} & \longrightarrow & X(j-1, j-1) \\ \downarrow & & \downarrow \\ \Delta^{n+1} & \xrightarrow{u_{0,j}} & X(0, j) \end{array} \qquad \begin{array}{ccc} \Lambda_{i+1}^{n+1} & \longrightarrow & X(i, j) \\ \downarrow & & \downarrow \\ \Delta^{n+1} & \xrightarrow{u_{i+1,j}} & X(i+1, j) \end{array}$$

In particular, the inclusion $X(-1, -1) \to X(n, n)$ is an inner anodyne extension. Similarly, we define

$$Y(-1, -1) = X(n, n)$$

and, for $n + 2 > j \geq 0$ and $0 \leq i \leq j$, we put

$$Y(i, j) = Y(j-1, j-1) \cup \bigcup_{0 \leq l \leq i} V_{l,j} \,.$$

One then checks as above that each inclusion of the form $Y(j-1, j-1) \to Y(0, j)$ as well as each inclusion of the form $Y(i, j) \to Y(i+1, j)$ is a push-out of a horn inclusion of type $\Lambda_k^{n+2} \to \Delta^{n+2}$ for $0 < k < n+2$, and, since $Y(n+2, n+2) = \Delta^2 \times \Delta^n$, this ends the proof. □

Corollary 3.2.4 *For any inner anodyne extension $K \to L$ and any monomorphism $X \to Y$, the induced inclusion*

$$K \times Y \cup L \times X \to L \times Y$$

is an inner anodyne extension.

Definition 3.2.5 An *inner fibration* is a morphism of simplicial sets which has the right lifting property with respect to the class of inner anodyne extensions.

Example 3.2.6 A simplicial set X is an ∞-category if and only if the morphism $X \to \Delta^0$ is an inner fibration.

Example 3.2.7 For any integers $n \geq 2$ and $0 < k < n$, it follows from Proposition 1.4.13 that the natural map $\tau(\Lambda_k^n) \to \tau(\Delta^n)$ is an isomorphism of categories. This implies right away that, for any functor between two small categories $C \to D$, the induced morphism of simplicial sets $N(C) \to N(D)$ is an inner fibration. This gives an even larger supply of examples: more generally, for any ∞-category X and any small category C, any morphism $X \to N(C)$ is an inner fibration.

Corollary 3.2.8 *For a morphism of simplicial sets $p\colon X \to Y$, the following conditions are equivalent.*

(*i*) *The morphism p is an inner fibration.*

(*ii*) *For any inner anodyne extension* $i\colon K \to L$*, the restriction along* i *induces a trivial fibration*

$$\underline{\mathrm{Hom}}(L, X) \to \underline{\mathrm{Hom}}(K, X) \times_{\underline{\mathrm{Hom}}(K,Y)} \underline{\mathrm{Hom}}(L, Y)\,.$$

(*iii*) *The restriction along the inclusion of* Λ^2_1 *into* Δ^2 *induces a trivial fibration*

$$\underline{\mathrm{Hom}}(\Delta^2, X) \to \underline{\mathrm{Hom}}(\Lambda^2_1, X) \times_{\underline{\mathrm{Hom}}(\Lambda^2_1,Y)} \underline{\mathrm{Hom}}(\Delta^2, Y)\,.$$

(*iv*) *For any monomorphism* $i\colon K \to L$*, the restriction along* i *induces an inner fibration*

$$\underline{\mathrm{Hom}}(L, X) \to \underline{\mathrm{Hom}}(K, X) \times_{\underline{\mathrm{Hom}}(K,Y)} \underline{\mathrm{Hom}}(L, Y)\,.$$

Corollary 3.2.9 *A simplicial set X is an ∞-category if and only if the canonical map*

$$\underline{\mathrm{Hom}}(\Delta^2, X) \to \underline{\mathrm{Hom}}(\Lambda^2_1, X)$$

is a trivial fibration.

Corollary 3.2.10 *For any ∞-category X and any simplicial set A, the simplicial set* $\underline{\mathrm{Hom}}(A, X)$ *of morphisms from A to X is an ∞-category. In particular, the functors from an ∞-category A to an ∞-category X do form an ∞-category.*

3.3 The Joyal Model Category Structure

3.3.1 Let X be a simplicial set and $f\colon x \to y$ a morphism in X. A *left inverse* of f is a morphism $g\colon y \to x$ such that the triangle

$$\begin{array}{ccc} & y & \\ {}^{f}\nearrow & & \searrow^{g} \\ x & \xrightarrow{\;1_x\;} & x \end{array}$$

commutes. One can freely add a left inverse as follows. One first adds a map $g\colon y \to x$ by forming the push-out below.

$$\begin{array}{ccc} \partial\Delta^1 & \xrightarrow{(y,x)} & X \\ \downarrow & & \downarrow \\ \Delta^1 & \xrightarrow{\;g\;} & X[g] \end{array}$$

Finally, we form the following push-out.

$$\begin{array}{ccc} \partial\Delta^2 & \xrightarrow{(f,g,1_x)} & X[g] \\ \big\downarrow & & \big\downarrow \\ \Delta^2 & \xrightarrow{G} & X[gf = 1] \end{array}$$

We still denote by f the image of f in $X[gf = 1]$. Applying this procedure to X^{op} gives a way to freely add a right inverse, which gives a simplicial set $X[fh = 1]$ together with a 2-simplex H witnessing that the new 1-simplex h is a right inverse of f, and we also denote by f the image of f in $X[fh = 1]$.

Therefore, we have defined a procedure to freely invert a morphism f: we define

$$X[f^{-1}] = (X[gf = 1])[fh = 1]\,.$$

The canonical inclusion $X \to X[f^{-1}]$ has the following universal property. For any morphism of simplicial sets $u \colon X \to Y$, and for any pair of 2-simplices of $G', H' \colon \Delta^2 \to Y$ whose boundaries respectively are triangles of the form

$$\begin{array}{ccccc} & & u(y) & & \\ & {\scriptstyle u(f)}\nearrow & & \searrow{\scriptstyle g'} & \\ u(x) & & \xrightarrow{1_{u(x)}} & & u(x) \end{array} \qquad \text{and} \qquad \begin{array}{ccccc} & & u(x) & & \\ & {\scriptstyle h'}\nearrow & & \searrow{\scriptstyle u(f)} & \\ u(y) & & \xrightarrow{1_{u(y)}} & & u(y) \end{array}$$

in Y, there exists a unique morphism $v \colon X[f^{-1}] \to Y$ whose restriction to X is u, and such that $v(G) = G'$ and $v(H) = H'$.

Proposition 3.3.2 *The inclusion map $X \to X[f^{-1}]$ is an anodyne extension.*

Proof Since the functor $X \mapsto X^{\mathrm{op}}$ preserves anodyne extensions, it is sufficient to prove that the inclusion $X \to X[gf = 1]$ is an anodyne extension. An alternative construction of $X[fg = 1]$ is the following. One first forms the push-out below.

$$\begin{array}{ccc} \Delta^1 & \xrightarrow{f} & X \\ {\scriptstyle \delta^2_2}\big\downarrow & & \big\downarrow \\ \Delta^2 & \xrightarrow{s} & X' \end{array}$$

Finally, we see that $X[gf = 1]$ naturally fits in the following coCartesian square.

$$\begin{array}{ccc} \Delta^1 & \xrightarrow{s\delta^2_1} & X' \\ \big\downarrow & & \big\downarrow \\ \Delta^0 & \xrightarrow{g} & X[gf = 1] \end{array}$$

The map $X \to X'$ is an anodyne extension because it is the push-out of such a thing. By virtue of Proposition 2.3.27, the map $X' \to X[gf = 1]$ is a weak homotopy equivalence because it is the push-out of the weak equivalence $\Delta^1 \to \Delta^0$ along the cofibration $s\delta^2_1$. Therefore, the map $X \to X[gf = 1]$ is both a monomorphism and a weak homotopy equivalence, whence an anodyne extension. □

Definition 3.3.3 We define the interval J as

$$J = \Delta^1[f^{-1}],$$

where $f: 0 \to 1$ is the unique non-trivial map in Δ^1 (corresponding to the identity of Δ^1). The class of *categorical anodyne extensions* is the smallest class of J-anodyne extensions containing the set of inner horn inclusions $\Lambda^n_k \to \Delta^n$, for $n \geq 2$ and $0 < k < n$ (see Example 2.4.13).

Remark 3.3.4 The class of categorical anodyne extensions is the smallest saturated class of morphisms of simplicial sets containing the inner horn inclusions as well as the inclusion maps of the form

$$J \times \partial\Delta^n \cup \{\varepsilon\} \times \Delta^n \to J \times \Delta^n$$

for $n \geq 0$ and $\varepsilon = 0, 1$. This follows from Proposition 3.2.3. One deduces the next proposition right away.

Proposition 3.3.5 *For any categorical anodyne extension $K \to L$ and any monomorphism $X \to Y$, the induced map*

$$K \times Y \cup L \times X \to L \times Y$$

is a categorical anodyne extension.

Proposition 3.3.6 *Any inner anodyne extension is a categorical anodyne extension. Any categorical anodyne extension is an anodyne extension.*

Proof The first assertion is obvious. The second one follows from Corollary 3.1.4 and Proposition 3.3.2. □

Definition 3.3.7 The *Joyal model category structure* is the model category structure obtained by applying Theorem 2.4.19 to the exact cylinder defined as the Cartesian product of J and to the class of categorical anodyne extensions. The weak equivalences of this model category structure are called the *weak categorical equivalences*.

Remark 3.3.8 A reformulation of the explicit description of a generating family of the saturated class of categorical anodyne extensions given in

Remark 3.3.4 is that a morphism of simplicial sets $p\colon X \to Y$ has the right lifting property with respect to the class of categorical anodyne extensions if and only if the operation of evaluation at ε (i.e. of restriction along $\{\varepsilon\} \to J$), $\varepsilon = 0, 1$, induces trivial fibrations

$$\underline{\mathrm{Hom}}(J, X) \to X \times_Y \underline{\mathrm{Hom}}(J, Y),$$

and the operation of restriction along $\Lambda^2_1 \to \Delta^2$ induces trivial fibrations

$$\underline{\mathrm{Hom}}(\Delta^2, X) \to \underline{\mathrm{Hom}}(\Lambda^2_1, X) \times_{\underline{\mathrm{Hom}}(\Lambda^2_1, Y)} \underline{\mathrm{Hom}}(\Delta^2, Y).$$

In particular, the fibrant objects of the Joyal model category structure are the simplicial sets X such that the two restriction maps

$$\underline{\mathrm{Hom}}(J, X) \to \underline{\mathrm{Hom}}(\{0\}, X) = X \quad \text{and} \quad \underline{\mathrm{Hom}}(\Delta^2, X) \to \underline{\mathrm{Hom}}(\Lambda^2_1, X)$$

are trivial fibrations.

It is clear that any fibrant object of the Joyal model category structure is an ∞-category. We will prove, among other things, that the converse is true: the fibrant objects of this model category precisely are the ∞-categories; see Theorem 3.6.1. The proof of this fact will require quite a few intermediate results which will be meaningful in their own right, for our understanding of the theory of ∞-categories as a semantic of the language of category theory. More generally, we will also characterise the fibrations between fibrant objects of the Joyal model category structure.

In order to do this, it is enlightening to compare the Joyal model category structure with the usual homotopy theory of categories (which consists in inverting the class of equivalences of categories).

Definition 3.3.9 A functor $p\colon X \to Y$ is an *isofibration* if, for any object x_0 in X and any invertible map of the form $g\colon y_0 \to y_1$ in Y with $p(x_0) = y_0$, there exists an invertible map $f\colon x_0 \to x_1$ in X such that $p(f) = g$.

Theorem 3.3.10 *The category Cat of small categories admits a cofibrantly generated model category structure whose weak equivalences are the equivalences of categories, whose cofibrations are the functors which induce an injective map at the level of objects, and whose fibrations are the isofibrations.*

Proof Here are several functors of interest. The inclusion functor $\eta\colon \{0\} \to \tau(J)$, which picks the object 0 in the unique category whose set of objects is $\{0, 1\}$ and which is equivalent to the final category; the functor $i\colon \varnothing \to e$, from the empty category to the final category; the functor $j\colon \{0, 1\} \to [1]$; and the unique functor $k\colon S \to [1]$, which is the identity on objects, where S is the free category whose set of objects is $\{0, 1\}$ with two parallel maps from 0 to 1. The class of isofibrations is the class of morphisms with the right lifting property with respect to $\{\eta\}$. The morphisms with the right lifting property with respect to $\{i, j, k\}$ precisely are the functors which induce a surjective map at the level of objects and which are fully faithful. Conversely, one checks that the morphisms with the left lifting property with respect to the class of equivalences of categories which are surjective on objects precisely is the class of functors which are injective on objects. It is clear that any equivalence of categories which is surjective on objects is an isofibration, and that any isofibration which is an equivalence of categories is surjective on objects. Using the small object argument for each set of maps $\{\eta\}$ and $\{i, j, k\}$ provides the existence of the expected factorisations. □

Definition 3.3.11 The model category structure provided by the preceding proposition is called the *canonical model category structure*.

Lemma 3.3.12 *The left adjoint of the nerve functor sends inner anodyne extensions to isomorphisms.*

Proof The class of morphisms of simplicial sets whose image by the functor τ is an isomorphism of categories is saturated. Therefore, it is sufficient to check that τ sends each inclusion $\Lambda^n_k \to \Delta^n$ to an invertible map of *Cat* whenever $n \geq 2$ and $0 < k < n$. But this latter property is a reformulation of Proposition 1.4.13. □

Lemma 3.3.13 *The functor* $\tau\colon sSet \to Cat$ *commutes with finite products.*

Proof Let X and Y be two simplicial sets. We choose two inner anodyne extensions $X \to X'$ and $Y \to Y'$ such that both X' and Y' are ∞-categories. Then the product map $X \times Y \to X' \times Y'$ is an inner anodyne extension (this is a consequence of Corollary 3.2.4). Therefore, by virtue of Lemma 3.3.12, we have canonical isomorphisms $\tau(X) \simeq \tau(X')$, $\tau(Y) \simeq \tau(Y')$ and $\tau(X \times Y) \simeq \tau(X' \times Y')$. The explicit description of $\tau(W)$, for any ∞-category W, provided by the theorem of Boardman and Vogt (Theorem 1.6.6), implies that the natural map from $\tau(X' \times Y')$ to $\tau(X') \times \tau(Y')$ is an isomorphism. □

Proposition 3.3.14 *The adjunction*

$$\tau\colon sSet \rightleftarrows Cat : N$$

is a Quillen adjunction from the Joyal model category structure to the canonical model category structure.

Proof Since, for any simplicial set X, the set of objects of the associated category $\tau(X)$ coincides with X_0, the functor τ sends monomorphisms to functors which are injective on objects. In other words, it preserves cofibrations. For any simplicial set X, the functor τ sends the inclusion $X = \{0\}\times X \to J\times X$ to the inclusion $\tau(X) = \{0\} \times \tau(X) \to \tau(J) \times \tau(X)$. Since $\tau(J)$ is equivalent to the final category, the explicit description of the class of categorical anodyne extensions given in Remark 3.3.4 and Lemma 3.3.12 implies that τ sends categorical anodyne extensions to trivial cofibrations. This proves that τ and N define a Quillen adjunction; see Proposition 2.4.40. □

Definition 3.3.15 A morphism of simplicial sets $p\colon X \to Y$ is an *isofibration* if it is an inner fibration, and if, for any object x_0 in X and any invertible map of the form $g\colon y_0 \to y_1$ in Y with $p(x_0) = y_0$, there exists an invertible map $f\colon x_0 \to x_1$ in X such that $p(f) = g$.

Remark 3.3.16 Lemma 3.3.12 means that the nerve of any functor between small categories is an inner fibration. One easily checks that a functor $p\colon X \to Y$ is an isofibration if and only if its nerve $N(p)\colon N(X) \to N(Y)$ is an isofibration. We will prove later that a functor between ∞-categories is an isofibration if and only if it is a fibration of the Joyal model category structure; see Theorem 3.6.1.

Remark 3.3.17 Any fibration of the Joyal model category structure is an isofibration. This is a consequence of the fact that, by construction, a map $J \to W$ is the same thing as a morphism $f\colon x \to y$ in W equipped with a proof that it is invertible.

Proposition 3.3.18 *A morphism of ∞-categories $p\colon X \to Y$ is an isofibration if and only if it is an inner fibration and if, for any object x_1 in X and any invertible map of the form $g\colon y_0 \to y_1$ in Y with $p(x_1) = y_1$, there exists an invertible map $f\colon x_0 \to x_1$ in X such that $p(f) = g$.*

Proof Assume that p has the property stated in the proposition, and let us prove that it is an isofibration. Let us choose an object x_0 in X and an invertible map $g\colon y_0 \to y_1$ such that $p(x_0) = y_0$. We choose a map $g'\colon y_1 \to y_0$ which is a left inverse of g. Then, by assumption on p, we can find an invertible morphism $f'\colon x_1 \to x_0$ in X such that $p(f') = g'$. Let us choose a map $\tilde{f}\colon x_0 \to x_1$ which is a right inverse of f'. Then, $p(\tilde{f})$ and g coincide in $\tau(X)$, and thus, by virtue of Theorem 1.6.6, there exists a map $c\colon \Delta^2 \to Y$ whose boundary is the following triangle.

$$c_{|\partial\Delta^2} = \begin{array}{ccccc} & & y_1 & & \\ & {\scriptstyle p(\tilde{f})}\nearrow & & \searrow{\scriptstyle 1_{y_1}} & \\ y_0 & & \xrightarrow{\quad g \quad} & & y_1 \end{array}$$

There is a unique morphism $b\colon \Lambda^2_1 \to X$ corresponding to the diagram below.

$$b = \begin{array}{ccc} & x_1 & \\ {\scriptstyle \tilde{f}}\nearrow & & \searrow{\scriptstyle 1_{x_1}} \\ x_0 & & x_1 \end{array}$$

Since p is an inner fibration, the solid commutative square

$$\begin{array}{ccc} \Lambda^2_1 & \xrightarrow{b} & X \\ \big\downarrow & \overset{a}{\nearrow} & \big\downarrow{\scriptstyle p} \\ \Delta^2 & \xrightarrow{c} & Y \end{array}$$

has a filler a. For $f = a\delta^2_1$, the image of the map $f\colon x_0 \to x_1$ by p is equal to g.

Conversely, applying what precedes to $X^{\mathrm{op}} \to Y^{\mathrm{op}}$ shows that any isofibration satisfies the assumption of the proposition. □

Corollary 3.3.19 *A morphism of ∞-categories $X \to Y$ is an isofibration if and only if the induced morphism $X^{\mathrm{op}} \to Y^{\mathrm{op}}$ is an isofibration.*

3.4 Left or Right Fibrations, Joins and Slices

Definition 3.4.1 A *left anodyne extension* (a *right anodyne extension*) is an element of the smallest saturated class of morphisms of simplicial sets containing the horn inclusions of the form $\Lambda^n_k \to \Delta^n$ for $n \geq 1$ and $0 \leq k < n$ (and $0 < k \leq n$, respectively).

A *left fibration* (a *right fibration*) is a morphism of simplicial sets with the right lifting property with respect to the class of left (right) anodyne extensions.

Remark 3.4.2 Right fibrations are a generalisation of the notion of Grothendieck fibration with discrete fibres; see paragraph 4.1.1 and Proposition 4.1.2 below. However, in this section, we shall focus on technical properties involving coherence issues related to providing inverses of invertible maps in ∞-categories.

The functor $X \mapsto X^{\mathrm{op}}$ maps the horn Λ^n_k to Λ^n_{n-k}. Therefore, it sends right fibrations to left fibrations and right anodyne maps to left anodyne maps, and vice versa. In particular, any statement about right fibrations and right anodyne extensions always has a counterpart in terms of left fibrations and left anodyne extensions. The following three propositions are interpretations of Lemma 3.1.3.

Proposition 3.4.3 *For any left (right) anodyne extension $K \to L$ and any monomorphism $X \to Y$, the induced inclusion*

$$L \times X \cup K \times Y \to L \times Y$$

is a left (right) anodyne extension.

Proposition 3.4.4 *Let $X \to Y$ be a left (right) fibration. If the map $K \to L$ is a monomorphism (a left anodyne extension), then the induced morphism*

$$\underline{\mathrm{Hom}}(L, X) \to \underline{\mathrm{Hom}}(K, X) \times_{\underline{\mathrm{Hom}}(K,Y)} \underline{\mathrm{Hom}}(L, Y)$$

is a left fibration (a trivial fibration, respectively).

Proposition 3.4.5 *A morphism of simplicial sets $X \to Y$ is a left fibration (a right fibration) if and only if the evaluation at 1 (the evaluation at 0, respectively) induces a trivial fibration of the form*

$$\underline{\mathrm{Hom}}(\Delta^1, X) \to X \times_Y \underline{\mathrm{Hom}}(\Delta^1, Y)\,.$$

Definition 3.4.6 A morphism of simplicial sets $p\colon X \to Y$ is *conservative* if, for any morphism $f\colon x \to x'$ in X, if $p(f)\colon p(x) \to p(x')$ is invertible in Y, then so is f in X.

Remark 3.4.7 If $p\colon X \to Y$ is a functor between ∞-categories, the explicit description of $\tau(X)$ and $\tau(Y)$ given by the theorem of Boardman and Vogt (Theorem 1.6.6) implies that p is conservative if and only if the functor $\tau(p)\colon \tau(X) \to \tau(Y)$ is conservative.

Proposition 3.4.8 *Let $p\colon X \to Y$ be a morphism of ∞-categories. If p is either a left fibration or a right fibration, then it is a conservative isofibration.*

Proof It is sufficient to prove the case of a left fibration. Let us prove first that p is conservative. Let $x\colon x_0 \to x_1$ be a morphism in X whose image $p(x) = y\colon y_0 \to y_1$ is invertible. This implies that y has a left inverse g. In other words, there is a map $c\colon \Delta^2 \to Y$ whose restriction to $\partial\Delta^2$ is a commutative triangle of the form below.

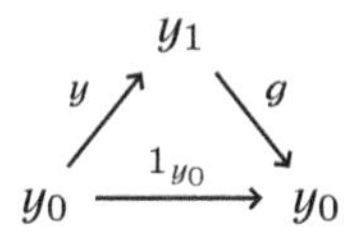

Let $b\colon \Lambda^2_0 \to X$ be the morphism corresponding to the diagram

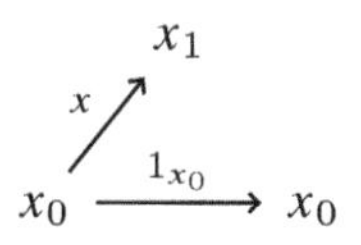

in X. The solid commutative square

$$\begin{array}{ccc} \Lambda^2_0 & \xrightarrow{b} & X \\ \big\downarrow & \nearrow & \big\downarrow{\scriptstyle p} \\ \Delta^2 & \xrightarrow{c} & Y \end{array}$$

has a filler whose restriction to $\partial\Delta^2$ provides a commutative triangle of the following form.

$$\begin{array}{ccccc} & & x_1 & & \\ & {\scriptstyle x}\nearrow & & \searrow{\scriptstyle f} & \\ x_0 & & \xrightarrow{1_{x_0}} & & x_0 \end{array}$$

Since $p(x)$ is invertible, so is $p(f)$, and we can repeat the procedure above to get a commutative triangle of the form below.

$$\begin{array}{ccccc} & & x_0 & & \\ & {\scriptstyle f}\nearrow & & \searrow{\scriptstyle f'} & \\ x_1 & & \xrightarrow{1_{x_1}} & & x_1 \end{array}$$

In other words, the morphism f is invertible. This implies that the morphism corresponding to x in $\tau(X)$ is invertible, and thus that x itself is invertible.

Since p is conservative and has the right lifting property with respect to the inclusion $\{0\} \to \Delta^1$, it must be an isofibration. □

3.4.9 Let $\mathbf{\Delta}_{aug}$ be the category whose objects are the ordered sets $[n] = \{0, \ldots, n\}$ for $n \geq -1$, and whose morphisms are the (non-strictly) increasing maps. In other words, the category $\mathbf{\Delta}_{aug}$ is the category obtained from the usual category of simplices $\mathbf{\Delta}$ by adjoining an initial object $[-1]$ (the empty ordered set). One can see a presheaf over the category $\mathbf{\Delta}_{aug}$ as a triple (X, a, E), where X is a simplicial set, E is a set, and a is an augmentation from X to E (i.e. a morphism from X to the constant simplicial set with value E). We denote by $sSet_{aug}$ the category of presheaves over $\mathbf{\Delta}_{aug}$, which we shall call the *augmented simplicial sets*. The operation of composition with the inclusion functor $i : \mathbf{\Delta} \to \mathbf{\Delta}_{aug}$ induces a functor

$$i^* : sSet_{aug} \to sSet$$

which can be described as $(X, E, a) \mapsto X$ in the language of augmented simplicial sets. This functor has a left adjoint $i_!$ and a right adjoint i_*; the left adjoint sends a simplicial set X to the triple $(X, p_X, \pi_0(X))$, where $\pi_0(X)$ is the set of connected components of X (see paragraph 3.1.30), and $p_X : X \to \pi_0(X)$ is the canonical map. The functor i_* sends a simplicial set X to the triple (X, a_X, Δ^0), with $a_X : X \to \Delta^0$ the obvious map.

3.4.10 The category $\mathbf{\Delta}_{aug}$ is equipped with a monoidal structure induced by the sum of ordinals

$$\begin{aligned} \mathbf{\Delta}_{aug} \times \mathbf{\Delta}_{aug} &\to \mathbf{\Delta}_{aug} \\ ([m],[n]) &\mapsto [m+1+n] \end{aligned}$$

(the unit object thus is $[-1]$). By a repeated use of Theorem 1.1.10, there is a unique monoidal structure on the category of augmented simplicial sets whose tensor product commutes with small colimits in each variable, which extends the sum of ordinals in $\mathbf{\Delta}_{aug}$. The tensor product of two augmented simplicial sets X and Y is denoted by $X * Y$, so that, in particular, $h_{[m]} * h_{[n]} = h_{[m+1+n]}$. The unit object is the representable presheaf $h_{[-1]} = i_*(\varnothing)$. However, we also have an explicit description of this tensor product.

Proposition 3.4.11 *For any augmented simplicial sets X and Y, and for any integer $n \geq -1$, there is a canonical identification*

$$(X * Y)_n = \coprod_{i+1+j=n} X_i \times Y_j \, .$$

Proof For $X = h_{[p]}$ and $Y = h_{[q]}$, since we restrict ourselves to consider maps which never decrease, one has

$$\operatorname{Hom}([n],[p+1+q]) = \coprod_{i+1+j=n} \operatorname{Hom}([i],[p]) \times \operatorname{Hom}([j],[q]) \, .$$

We then extend these identifications by colimits. □

3.4.12 For two simplicial sets X and Y, we define the *join* of X and Y as

$$X * Y = i^*(i_*(X) * i_*(Y)) \, .$$

In other words, for each $n \geq 0$, the simplicial set $X * Y$ evaluated at n is

$$(X * Y)_n = \coprod_{i+1+j=n} X_i \times Y_j \, .$$

In other words, a map $f : \Delta^n \to X * Y$ is uniquely determined by a decomposition of n into a sum $n = p + 1 + q$ together with maps $a : \Delta^p \to X$ and $b : \Delta^q \to Y$; one recovers f as $f = a * b$.

This defines a monoidal structure on the category of simplicial sets whose unit object is the empty simplicial set $\varnothing = i^*(h_{[-1]})$. Therefore, there is a functorial injective map

$$X \amalg Y \to X * Y$$

induced by functoriality from the inclusions $\varnothing \to W$ for $W = X, Y$.

Given a fixed simplicial set T, we thus get two functors:

$$(-) * T \colon sSet \to T\backslash sSet$$
$$X \mapsto (T = \varnothing \amalg T \to X * T),$$

$$T * (-) \colon sSet \to T\backslash sSet$$
$$X \mapsto (T = T \amalg \varnothing \to T * X).$$

Proposition 3.4.13 *Both functors $(-) * T$ and $T * (-)$ commute with small colimits.*

Proof These functors both preserve initial objects. It is thus sufficient to prove that the join operation preserves connected colimits in each variable. For this, since the tensor product preserves small colimits in each variable in the category of augmented simplicial sets, it is sufficient to prove that the functor i_* commutes with small connected colimit. Since it obviously preserves filtered colimits, we only need to check that it preserves coCartesian squares, which follows right away from its explicit description. □

3.4.14 Let $t \colon T \to X$ be a morphism of simplicial sets. We shall write X/t or X/T ($t\backslash X$ or $T\backslash X$, respectively) for the image of (X, t) by the right adjoint of the functor $(-) * T$ (of the functor $T * (-)$, respectively), which exists by virtue of the preceding proposition.

In the case where $T = \Delta^0$ and $X = N(C)$ is the nerve of a category, we observe that the identification $\Delta^{n+1} = \Delta^n * \Delta^0$ implies that $X/t = N(C/t)$. In other words, the construction X/t extends the usual construction of slice categories.

Remark 3.4.15 The join operation is compatible with the opposite operations as follows:

$$(T * S)^{op} = S^{op} * T^{op} .$$

Therefore, we have

$$(X/T)^{op} = T^{op}\backslash X^{op} .$$

Since the join operation is associative, we also have the following formulas:

$$X/(S * T) \simeq (X/T)/S \qquad \text{and} \qquad (S * T)\backslash X \simeq T\backslash(S\backslash X) .$$

Proposition 3.4.16 *For any monomorphisms of simplicial sets $U \to V$ and*

$K \to L$, the induced commutative square

$$\begin{array}{ccc} K * U & \hookrightarrow & K * V \\ \downarrow & & \downarrow \\ L * U & \hookrightarrow & L * V \end{array}$$

is Cartesian, and all its maps are monomorphisms. We thus have a canonical inclusion map

$$L * U \cup K * V \to L * V\,.$$

Proof This follows right away from Proposition 3.4.11 and the analogous property for the binary Cartesian product of sets. □

The following statement is obvious.

Proposition 3.4.17 *As subobjects of $\Delta^{m+1+n} = \Delta^m * \Delta^n$, we have the following identities:*

$$\partial\Delta^m * \Delta^n \cup \Delta^m * \partial\Delta^n = \partial\Delta^{m+1+n},$$
$$\Lambda^m_k * \Delta^n \cup \Delta^m * \partial\Delta^n = \Lambda^{m+1+n}_k,$$
$$\partial\Delta^m * \Delta^n \cup \Delta^m * \Lambda^n_k = \Lambda^{m+1+n}_{m+1+k}.$$

Theorem 3.4.18 (Joyal) *Let X and Y be ∞-categories, and $p\colon X \to Y$ an inner fibration. We consider that we are given a commutative square of the form*

$$\begin{array}{ccc} \Lambda^n_0 & \xrightarrow{a} & X \\ \downarrow & & \downarrow p \\ \Delta^n & \xrightarrow{b} & Y \end{array} \qquad n \geq 2,$$

and we assume that the morphism $\alpha\colon a_0 \to a_1$, defined by the restriction of a on $\Delta^{\{0,1\}}$, is invertible in X. Then there exists a morphism $h\colon \Delta^n \to X$ such that the restriction of h on Λ^n_0 is equal to a, and such that $ph = b$.

By contemplating the identification $\Lambda^1_0 * \Delta^{n-2} \cup \Delta^1 * \partial\Delta^{n-2} = \Lambda^n_0$, this theorem is a particular case of the following one.

Theorem 3.4.19 (Joyal) *Let $p\colon X \to Y$ be an inner fibration, and assume that Y is an ∞-category. We consider a monomorphism $S \to T$ as well as a commutative square of the form*

$$\begin{array}{ccc} \{0\} * T \cup \Delta^1 * S & \xrightarrow{a} & X \\ \downarrow & & \downarrow p \\ \Delta^1 * T & \xrightarrow{b} & Y \end{array}$$

and we assume that the morphism $\alpha\colon a_0 \to a_1$*, obtained as the restriction of* a *on* $\Delta^1 * \varnothing \subset \Delta^1 * S$*, is invertible in* X*. Then the square above has a filler* $c\colon \Delta^1 * T \to X$*.*

The proof of this theorem will require a few preliminary results. Observe that, for a morphism of simplicial sets $S \to T$ and a map $T \to X$, there is an induced map $X/T \to X/S$ which is functorial in X. Therefore, if $X \to Y$ is a morphism of simplicial sets, we obtain a canonical commutative square

$$\begin{array}{ccc} X/T & \longrightarrow & Y/T \\ \downarrow & & \downarrow \\ X/S & \longrightarrow & Y/S \end{array}$$

and thus a canonical map $X/T \to X/S \times_{Y/S} Y/T$. We leave the next lemma as an exercise for the reader.

Lemma 3.4.20 *Let* $i\colon A \to B$ *and* $j\colon S \to T$ *be two monomorphisms. For any morphism of simplicial sets* $p\colon X \to Y$*, we have the following correspondence of lifting properties.*

$$\begin{array}{ccc} A * T \cup B * S & \xrightarrow{a} & X \\ \big\downarrow & \overset{h}{\nearrow} & \big\downarrow p \\ B * T & \xrightarrow[b]{} & Y \end{array} \quad \leftrightsquigarrow \quad \begin{array}{ccc} A & \xrightarrow{\tilde{a}} & X/T \\ i\big\downarrow & \overset{\tilde{h}}{\nearrow} & \big\downarrow \\ B & \xrightarrow[\tilde{b}]{} & X/S \times_{Y/S} Y/T \end{array}$$

Lemma 3.4.21 *Let* $p\colon X \to Y$ *be an inner fibration. Given any integers* $n \geq 1$ *and* $0 \leq k < n$ *and any morphism* $t\colon \Delta^n \to X$*, the induced map*

$$X/\Delta^n \to X/\Lambda^n_k \times_{Y/\Lambda^n_k} Y/\Delta^n$$

is a trivial fibration.

Proof Since $0 < m + 1 + k < m + 1 + n$, the third equality in Proposition 3.4.17 and the preceding lemma explain everything we need to know. □

Theorem 3.4.22 *Let* $p\colon X \to Y$ *be an inner fibration. We consider a monomorphism* $S \to T$*, together with a morphism* $t\colon T \to X$*. Then the canonical projection*

$$X/T \to X/S \times_{Y/S} Y/T$$

is a right fibration. Furthermore, if ever Y *is an* ∞*-category, then so are* X/T *as well as the fibre product* $X/S \times_{Y/S} Y/T$*.*

Proof Let $n \geq 1$ and $0 < k \leq n$. We then have the following correspondence of lifting problems.

$$
\begin{array}{ccc}
\Lambda^n_k * T \cup \Delta^n * S & \xrightarrow{a} & X \\
\downarrow & \overset{h}{\nearrow} & \downarrow p \\
\Delta^n * T & \xrightarrow[b]{} & Y
\end{array}
\quad \leftrightsquigarrow \quad
\begin{array}{ccc}
\Lambda^n_k & \xrightarrow{\tilde{a}} & X/T \\
\downarrow & \overset{\tilde{h}}{\nearrow} & \downarrow \\
\Delta^n & \xrightarrow[\tilde{b}]{} & X/S \times_{Y/S} Y/T
\end{array}
$$

Similarly, we have a correspondence of the form below.

$$
\begin{array}{ccc}
\Lambda^n_k * T \cup \Delta^n * S & \xrightarrow{a} & X \\
\downarrow & \overset{h}{\nearrow} & \downarrow p \\
\Delta^n * T & \xrightarrow[b]{} & Y
\end{array}
\quad \leftrightsquigarrow \quad
\begin{array}{ccc}
S & \xrightarrow{a'} & \Delta^n\backslash X \\
\downarrow & \overset{h'}{\nearrow} & \downarrow \\
T & \xrightarrow[b']{} & \Lambda^n_k\backslash X \times_{\Lambda^n_k\backslash Y} \Delta^n\backslash Y
\end{array}
$$

Therefore, it is sufficient to prove that the map $\Delta^n\backslash X \to \Lambda^n_k\backslash X \times_{\Lambda^n_k\backslash Y} \Delta^n\backslash Y$ is a trivial fibration. We even may check this property after applying the functor $W \mapsto W^{op}$, so that the first assertion follows from Lemma 3.4.21. In the case where Y is a final object, we obtain that the map $X/T \to X/S$ is a right fibration, and the case where S is empty tells us that the map $X/T \to X$ is a right fibration. Since right fibrations are inner fibrations, we deduce that X/T and X/S always are ∞-categories. The projection of $X/S \times_{Y/S} Y/T$ on X/S is a right fibration as well (as the pull-back of the right fibration $Y/T \to Y/S$), from which we see that $X/S \times_{Y/S} Y/T$ is an ∞-category. □

Proof of Theorem 3.4.19 Let $p\colon X \to Y$ be an inner fibration between ∞-categories, $S \to Y$ a monomorphism, and a commutative square

$$
\begin{array}{ccc}
\{0\} * T \cup \Delta^1 * S & \xrightarrow{a} & X \\
\downarrow & & \downarrow p \\
\Delta^1 * T & \xrightarrow{b} & Y
\end{array}
$$

such that the morphism $\alpha\colon a_0 \to a_1$, given by the restriction of a on $\Delta^1 * \varnothing$, is invertible in X. By virtue of Lemma 3.4.20, it is sufficient to prove that the induced commutative square

$$
\begin{array}{ccc}
\{0\} & \longrightarrow & X/T \\
\downarrow & & \downarrow \\
\Delta^1 & \xrightarrow{\tilde{\alpha}} & X/S \times_{Y/S} Y/T
\end{array}
$$

has a filler, where $\tilde{\alpha}$ is the morphism induced by α by transposition. But Theorem 3.4.22 ensures that the right-hand vertical map is a right fibration between ∞-categories, and thus, by virtue of Proposition 3.4.8, is an isofibration. Therefore, we are reduced to prove that the morphism $\tilde{\alpha}$ is invertible in $X/S \times_{Y/S} Y/T$. On the other hand, by virtue of Theorem 3.4.22, the canonical morphism $X/S \times_{Y/S} Y/T \to X$ is a right fibration between ∞-categories, whence, applying Proposition 3.4.8 again, is conservative. In other words it is now sufficient to prove that this map sends the morphism $\tilde{\alpha}$ to an invertible morphism in X. But, by definition, it sends $\tilde{\alpha}$ to α, which is invertible by assumption. □

3.5 Invertible Natural Transformations

Theorem 3.5.1 (Joyal) *An ∞-category is a Kan complex if and only if it is an ∞-groupoid.*

Proof If X is an ∞-category, then the map $X \to \Delta^0$ has the right lifting property with respect to horns of the form $\Lambda^n_k \to \Delta^n$ with $n = 1$ or $0 < k < n$. This theorem thus follows from applying Theorem 3.4.18 to the ∞-category X as well as to its opposite X^{op}. □

3.5.2 Let *Gpd* be the full subcategory of the category of small categories whose objects are the groupoids. The inclusion functor *Gpd* → *Cat* has a left adjoint π_1 and a right adjoint k. For a small category C, the groupoid $\pi_1(C)$ is obtained as the localisation of C by all its morphisms, while $k(C)$ is the groupoid whose objects are those of C, and whose morphisms are the invertible morphisms of C. The latter construction can be extended to ∞-categories as follows. For an ∞-category X, we form the pull-back square below.

$$\begin{array}{ccc} k(X) & \longrightarrow & X \\ \downarrow & & \downarrow \\ N(k(\tau(X))) & \longrightarrow & N(\tau(X)) \end{array}$$

We remark that the inclusion $k(\tau(X)) \to \tau(X)$ is an isofibration, hence so is its nerve. Therefore, the morphism $k(X) \to X$ is an inner fibration and $k(X)$ is an ∞-category. It immediately follows from Theorem 1.6.6 that the map $k(X) \to N(k(\tau(X)))$ induces an isomorphism after we apply the functor τ. By virtue of the preceding theorem, the ∞-category $k(X)$ is a Kan complex.

Corollary 3.5.3 *The ∞-category $k(X)$ is the largest Kan complex contained in the ∞-category X.*

In other words, the functor k is a right adjoint of the inclusion functor from the category of Kan complexes to the category of ∞-categories.

Corollary 3.5.4 *For any* ∞*-category* X*, the canonical map* $k(X) \to X$ *is a conservative isofibration.*

Proposition 3.5.5 *Any left (right) fibration whose codomain is a Kan complex is a Kan fibration between Kan complexes.*

Proof Indeed, such a map is conservative (Proposition 3.4.8), so that it is a morphism of Kan complexes. By virtue of Theorem 3.4.18 (and of its dual version, replacing Λ^n_0 by $\Lambda^n_n = (\Lambda^n_0)^{\mathrm{op}}$), any inner fibration between Kan complexes has the right lifting property with respect to horns of the form $\Lambda^n_k \to \Delta^n$ with $n \geq 2$ and $0 \leq k \leq n$. Since left (right) fibrations between ∞-categories also are isofibrations (Proposition 3.4.8), it follows from Corollary 3.3.19 that any left (right) fibration between Kan complexes has the right lifting property with respect to $\Lambda^1_k \to \Delta^1$ for $k = 0, 1$. □

Corollary 3.5.6 *Let* $p\colon X \to Y$ *be a left (or right) fibration. For any object* y *of* Y*, the fibre of* p *at* y *is a Kan complex.*

Proof The fibre of p at y is the pull-back of p along the map $y : \Delta^0 \to Y$ and thus is a left (or right) fibration of the form $q\colon F \to \Delta^0$. Since Δ^0 obviously is a Kan complex, the previous proposition applied to q implies that F must have the same property. □

3.5.7 Let A be a simplicial set. We write $\mathrm{Ob}(A)$ for the constant simplicial set whose set of objects is A_0. There is a unique morphism

$$\mathrm{Ob}(A) \to A$$

which is the identity on sets of objects. Let X be an ∞-category. By virtue of Corollary 3.2.8, the restriction map

$$\begin{aligned} \underline{\mathrm{Hom}}(A, X) &\to \underline{\mathrm{Hom}}(\mathrm{Ob}(A), X) = X^{A_0} \\ F &\mapsto (F(a))_{a \in A_0} \end{aligned}$$

is an inner fibration between ∞-categories.

One defines the ∞-category $k(A, X)$ by forming the following Cartesian square.

$$\begin{array}{ccc} k(A,X) & \longrightarrow & \underline{\mathrm{Hom}}(A,X) \\ \downarrow & & \downarrow \\ k(\underline{\mathrm{Hom}}(\mathrm{Ob}(A),X)) & \longrightarrow & \underline{\mathrm{Hom}}(\mathrm{Ob}(A),X) \end{array} \tag{3.5.7.1}$$

Note that $k(\underline{\mathrm{Hom}}(\mathrm{Ob}(A), X)) = k(\mathrm{Ob}(A), X) = k(X)^{A_0}$ is the maximal Kan complex in X^{A_0}. The ∞-category $k(A, X)$ is the ∞-category of functors

$F: A \to X$, with morphisms $F \to G$ those natural transformations such that, for any object a of A, the induced map $F(a) \to G(a)$ is invertible in X. More precisely, a simplex $x: \Delta^n \to \underline{\mathrm{Hom}}(A, X)$ belongs to $k(A, X)$ if and only if, for any object a of A, the map $x(a): \Delta^n \to X$, obtained by composing x with the evaluation at a, factors through $k(X)$.

The main goal of this section is to prove that the inclusion $k(\underline{\mathrm{Hom}}(A, X)) \subset k(A, X)$ is in fact an equality. In words: we want to prove that any objectwise invertible natural transformation is indeed invertible as a map of the ∞-category of functors. This will be achieved in Corollary 3.5.12. This fundamental characterisation of invertible natural transformations is very strongly related to the characterisation of ∞-categories as fibrant objects in the Joyal model category structure; see the proof of Theorem 3.6.1 below.

We need a companion of the construction $k(A, X)$. For a simplicial set B, we introduce the simplicial set $h(B, X)$ as the subobject of $\underline{\mathrm{Hom}}(B, X)$ whose simplices correspond to maps $\Delta^n \to \underline{\mathrm{Hom}}(B, X)$ such that the associated morphism $B \to \underline{\mathrm{Hom}}(\Delta^n, X)$ factors through $k(\Delta^n, X)$.

One checks that the bijections $\mathrm{Hom}(A, \underline{\mathrm{Hom}}(B, X)) \simeq \mathrm{Hom}(B, \underline{\mathrm{Hom}}(A, X))$ induce canonical bijections

$$\mathrm{Hom}(A, h(B, X)) \simeq \mathrm{Hom}(B, k(A, X)) . \tag{3.5.7.2}$$

One may think of $h(B, X)$ as the full subcategory of $\underline{\mathrm{Hom}}(B, X)$ which consists of functors $B \to X$ sending all maps of B to invertible maps of X.

3.5.8 Let $p: X \to Y$ be an inner fibration between ∞-categories. Let us consider the morphism

$$ev_1: h(\Delta^1, X) \to X \times_Y h(\Delta^1, Y) \tag{3.5.8.1}$$

induced by the inclusion $\{1\} \subset \Delta^1$.

We remark that the morphism (3.5.8.1) has the right lifting property with respect to $\partial\Delta^0 \to \Delta^0$ (i.e. is surjective on objects) if and only if p is an isofibration. Indeed, we have a correspondence of the form

$$\begin{array}{ccc} \varnothing & \longrightarrow & h(\Delta^1, X) \\ \downarrow & \nearrow & \downarrow{\scriptstyle ev_1} \\ \Delta^0 & \longrightarrow & X \times_Y h(\Delta^1, Y) \end{array} \quad \leftrightsquigarrow \quad \begin{array}{ccc} \{1\} & \longrightarrow & X \\ \downarrow & \overset{x}{\nearrow} & \downarrow{\scriptstyle p} \\ \Delta^1 & \longrightarrow & Y \end{array} \tag{3.5.8.2}$$

with x invertible.

Theorem 3.5.9 *For any inner fibration between ∞-categories $p: X \to Y$, the evaluation at 1 map $ev_1: h(\Delta^1, X) \to X \times_Y h(\Delta^1, Y)$ has the right lifting property with respect to inclusions of the form $\partial\Delta^n \to \Delta^n$, $n > 0$.*

Proof For any map $A \to B$ inducing a bijection $A_0 \simeq B_0$ and any ∞-category W, the obvious commutative square

$$\begin{array}{ccc}
k(B,W) & \hookrightarrow & \underline{\mathrm{Hom}}(B,W) \\
\downarrow & & \downarrow \\
k(A,W) & \hookrightarrow & \underline{\mathrm{Hom}}(A,W)
\end{array}$$

is Cartesian. Since the inclusion $\partial\Delta^n \to \Delta^n$ induces a bijection on objects, we deduce from this property that there is a correspondence between lifting problems of shape

$$\begin{array}{ccc}
\partial\Delta^n & \longrightarrow & h(\Delta^1, X) \\
\downarrow & \nearrow & \downarrow{\scriptstyle ev_1} \\
\Delta^n & \longrightarrow & X \times_Y h(\Delta^1, Y)
\end{array}$$

and lifting problems of shape

$$\begin{array}{ccc}
\Delta^1 \times \partial\Delta^n \cup \{1\} \times \Delta^n & \xrightarrow{a} & X \\
\downarrow & \overset{c}{\nearrow} & \downarrow{\scriptstyle p} \\
\Delta^1 \times \Delta^n & \xrightarrow[b]{} & Y
\end{array}$$

in which b corresponds to an element of $k(\Delta^n, Y)_1$ and the restriction of a to $\Delta^1 \times \partial\Delta^n$ to an element of $k(\partial\Delta^n, X)_1$ (the lifting c, if ever it exists, will always correspond to an element of $k(\Delta^n, X)_1$ thanks to the Cartesian square above for $X = W$, $A = \partial\Delta^n$ and $B = \Delta^n$).

Let us prove the existence of a lifting c (in the presence of the extra hypothesis on a and b as above). We recall that there is a finite filtration of the form

$$\Delta^1 \times \partial\Delta^n \cup \{1\} \times \Delta^n = A_{-1} \subset A_0 \subset \cdots \subset A_n = \Delta^1 \times \Delta^n,$$

each step $A_{i-1} \to A_i$ being obtained through a biCartesian square of the form

$$\begin{array}{ccc}
\Lambda^{n+1}_{i+1} & \xrightarrow{c_i'} & A_{i_1} \\
\downarrow & & \downarrow \\
\Delta^{n+1} & \xrightarrow{c_i} & A_i
\end{array}$$

where c_i is induced by the unique order preserving map $[n+1] \to [1] \times [n]$ which reaches both points $(0, i)$ and $(1, i)$. In particular, the map $A_{-1} \to A_{n-1}$ is an inner anodyne extension, so that we may assume that the map a is the restriction of a map $a' \colon A_{n-1} \to X$ such that pa' equals the restriction of b to A_{n-1}. It is thus sufficient to prove that the commutative square

$$\begin{array}{ccc} \Lambda^{n+1}_{n+1} & \xrightarrow{a'c'_n} & X \\ \big\downarrow & & \big\downarrow{\scriptstyle p} \\ \Delta^{n+1} & \xrightarrow[b]{} & Y \end{array}$$

admits a filler. The map $\alpha \in X_1$, corresponding to the image of $\Delta^{\{n-1,n\}}$ by $c'_n a'$ is the image of the map $(0,n) \to (1,n)$ by a; in particular, it is invertible. Therefore, by virtue of Theorem 3.4.18 (applied to $X^{\mathrm{op}} \to Y^{\mathrm{op}}$), the expected filler $\Delta^{n+1} \to X$ exists. □

Corollary 3.5.10 *An inner fibration between ∞-categories $p \colon X \to Y$ is an isofibration if and only if the evaluation at 1 map $ev_1 \colon h(\Delta^1, X) \to X \times_Y h(\Delta^1, Y)$ is a trivial fibration.*

Proof This follows from (3.5.8.2) and from Theorem 3.5.9. □

Theorem 3.5.11 *Let $p \colon X \to Y$ be an isofibration between ∞-categories. For any monomorphism of simplicial sets $i \colon A \to B$, the induced map*

$$(i^*, p_*) : k(B, X) \to k(A, X) \times_{k(A,Y)} k(B, Y)$$

is a Kan fibration between Kan complexes.

Proof We will prove first that the map $q = (i^*, p_*)$ is a left Kan fibration. More precisely, we will prove that it has the right lifting property with respect to the maps

$$\Delta^1 \times \partial\Delta^n \cup \{1\} \times \Delta^n \to \Delta^1 \times \Delta^n\,, \quad n \geq 0\,.$$

Let us first observe that the correspondence (3.5.7.2) induces the following one.

$$\begin{array}{ccc} \{1\} & \longrightarrow & k(B,X) \\ \big\downarrow & \nearrow & \big\downarrow{\scriptstyle q} \\ \Delta^1 & \longrightarrow & k(A,X) \times_{k(A,Y)} k(B,Y) \end{array} \quad \leftrightsquigarrow \quad \begin{array}{ccc} A & \longrightarrow & h(\Delta^1, X) \\ \big\downarrow & \nearrow & \big\downarrow{\scriptstyle ev_1} \\ B & \longrightarrow & X \times_Y h(\Delta^1, Y) \end{array}$$

Therefore, by virtue of Theorem 3.5.9, the case where $n = 0$ is done. We may now focus on the case where $n > 0$.

Let us consider a commutative square of the following form.

$$\begin{array}{ccc} \Delta^1 \times \partial\Delta^n \cup \{1\} \times \Delta^n & \longrightarrow & k(B,X) \\ \big\downarrow & & \big\downarrow{\scriptstyle q} \\ \Delta^1 \times \Delta^n & \longrightarrow & k(A,X) \times_{k(A,Y)} k(B,Y) \end{array}$$

It gives rise to a commutative solid square of the form

$$\begin{array}{ccc} B\times\partial\Delta^n\cup A\times\Delta^n & \longrightarrow & h(\Delta^1,X) \\ \downarrow & \overset{l}{\nearrow} & \downarrow{\scriptstyle ev_1} \\ B\times\Delta^n & \longrightarrow & X\times_Y h(\Delta^1,Y) \end{array}$$

which admits a lift l by Theorem 3.5.9. It remains to check that the morphism $\tilde{l}\colon \Delta^1\times\Delta^n \to \underline{\mathrm{Hom}}(B,X)$, induced by l, factors through $k(B,X)$. But, by assumption, the restriction of $\tilde{l}$ to $\Delta^1\times\partial\Delta^n\cup\{1\}\times\Delta^n$ factors through $k(B,X)$. This means that, for each object b of B, in the commutative diagram of $\tau(X)$,

$$\tilde{l}(b) = \begin{array}{ccccccc} \tilde{l}_{0,0}(b) & \longrightarrow & \tilde{l}_{0,1}(b) & \longrightarrow & \cdots & \longrightarrow & \tilde{l}_{0,n}(b) \\ \downarrow & & \downarrow & & & & \downarrow \\ \tilde{l}_{1,0}(b) & \longrightarrow & \tilde{l}_{1,1}(b) & \longrightarrow & \cdots & \longrightarrow & \tilde{l}_{1,n}(b) \end{array}$$

all the vertical maps as well as all the maps of the second line are invertible. Therefore, since $n > 0$, all the maps of the first line are invertible, which readily implies that $\tilde{l}$ factors through $k(B,X)$, as required.

The map q is a Kan fibration: we have seen that it is a left fibration, and applying what precedes to $X^{\mathrm{op}} \to Y^{\mathrm{op}}$ shows that q also is a right fibration. The case $A = \varnothing$ and $Y = \Delta^0$ shows that $k(A,X)$ is a Kan complex. The Cartesian square

$$\begin{array}{ccc} k(A,X)\times_{k(A,Y)} k(B,Y) & \longrightarrow & k(B,Y) \\ \downarrow & & \downarrow \\ k(A,X) & \longrightarrow & k(A,Y) \end{array}$$

thus implies that $k(A,X)\times_{k(A,Y)} k(B,Y)$ is a Kan complex. □

Corollary 3.5.12 *For any ∞-category X and any simplicial set A, we have an equality:* $k(A,X) = k(\underline{\mathrm{Hom}}(A,X))$. *In other words, a natural transformation is invertible if and only if it is objectwise invertible.*

More generally, for any isofibration between ∞-categories $p\colon X \to Y$ *and any monomorphism of simplicial sets* $A \to B$, *we have*

$$k(A,X)\times_{k(A,Y)} k(B,Y) = k\big(\underline{\mathrm{Hom}}(A,X)\times_{\underline{\mathrm{Hom}}(A,Y)} \underline{\mathrm{Hom}}(B,Y)\big)\,.$$

Proof We obviously have $k(\underline{\mathrm{Hom}}(A,X)) \subset k(A,X)$. Since, by virtue of the preceding theorem, $k(A,X)$ is a Kan complex, we conclude with Corollary 3.5.3 that this inclusion is an equality.

The functor $k(-)$ sends ∞-categories to Kan complexes and isofibration between Kan complexes to Kan fibration. But it is also a right adjoint, hence it preserves representable limits (such as pull-backs of maps between ∞-categories along isofibrations). In particular it sends $\underline{\mathrm{Hom}}(A,X) \times_{\underline{\mathrm{Hom}}(A,Y)} \underline{\mathrm{Hom}}(B,Y)$ to $k(A,X) \times_{k(A,Y)} k(B,Y)$. □

Corollary 3.5.13 *Let $p \colon X \to Y$ be an isofibration between ∞-categories. For any anodyne extension $A \to B$, the induced map*

$$h(B,X) \to h(A,X) \times_{h(A,Y)} h(B,Y)$$

is a trivial fibration.

Proof We have the correspondence of lifting problems below.

$$\begin{array}{ccc} K & \longrightarrow & h(B,X) \\ \downarrow & \nearrow & \downarrow \\ L & \longrightarrow & h(A,X) \times_{h(A,Y)} h(B,Y) \end{array} \quad \leftrightsquigarrow \quad \begin{array}{ccc} A & \longrightarrow & k(L,X) \\ \downarrow & \nearrow & \downarrow \\ B & \longrightarrow & k(K,X) \times_{k(K,Y)} k(L,Y) \end{array}$$

Therefore, we conclude with Theorem 3.5.11 □

3.6 ∞-Categories as Fibrant Objects

Theorem 3.6.1 (Joyal) *A simplicial set is a fibrant object of the Joyal model category structure if and only if it is an ∞-category. A morphism of ∞-categories is a fibration of the Joyal model category structure if and only if it is an isofibration.*

Proof For any ∞-category X, any map $J \to X$ factors through $k(X)$, because $\tau(J)$ is the contractible groupoid with objects 0 and 1. Therefore, we have

$$\underline{\mathrm{Hom}}(J,X) = h(J,X)\,.$$

On the other hand, by Proposition 3.3.2, the inclusions $\{\varepsilon\} \to \Delta^1 \to J$ are anodyne extensions for $\varepsilon = 0,1$. Hence Corollary 3.5.13 implies that, for any isofibration between ∞-categories $p \colon X \to Y$, the evaluation at $\varepsilon = 0,1$

$$ev_\varepsilon \colon \underline{\mathrm{Hom}}(J,X) \to X \times_Y \underline{\mathrm{Hom}}(J,Y)$$

is a trivial fibration. Remarks 3.3.8 and 3.3.17 finish the proof. □

Proposition 3.6.2 *The class of weak categorical equivalences (i.e. of weak equivalences of the Joyal model category structure) is the smallest class of maps of simplicial sets W satisfying the following conditions.*

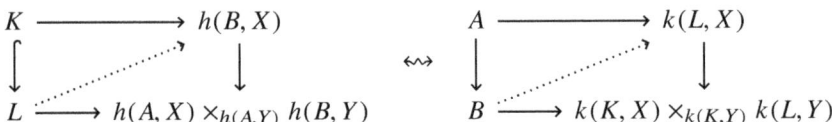

(i) *The class* W *has the two-out-of-three property.*
(ii) *Any inner anodyne extension is in* W.
(iii) *Any trivial fibration between* ∞*-categories is in* W.

Proof Let W be such a class of maps, and let us show that any weak categorical equivalence is in W. Let $f: X \to Y$ be a morphism of simplicial sets. We can form a commutative diagram of the form

$$\begin{array}{ccc} X & \xrightarrow{i} & X' \\ {\scriptstyle f}\downarrow & & \downarrow{\scriptstyle f'} \\ Y & \xrightarrow{j} & Y' \end{array}$$

in which i and j are inner anodyne morphisms, and X' and Y' both are ∞-categories. The proof of Ken Brown's lemma (Proposition 2.2.7), applied to $\mathcal{C} = (sSet)^{\mathrm{op}}$, shows that f' has a factorisation of the form $f' = kq$, where k is the section of a trivial fibration p, while q is an isofibration (i.e. a fibration of the Joyal model category structure). Therefore, the map f is a weak categorical equivalence if and only if q is a trivial fibration. Hence, $f \in W$ whenever f is a weak categorical equivalence. □

Corollary 3.6.3 *The class of weak categorical equivalences is closed under finite products.*

Proof Since the Cartesian product is symmetric and since weak categorical equivalences are stable under composition, it is sufficient to prove that the Cartesian product with a given simplicial set A preserves weak categorical equivalence. Let us consider the class of maps W which consists of the morphisms $X \to Y$ such that $A \times X \to A \times Y$ is a weak categorical equivalence. Then W has the two-out-of-three property, and it contains inner anodyne extensions (as a particular case of Corollary 3.2.4) as well as trivial fibrations (because the pull-back of a trivial fibration is a trivial fibration). Hence it contains the class of weak categorical equivalences. □

Corollary 3.6.4 *For any monomorphisms of simplicial sets* $i: K \to L$ *and* $j: U \to V$, *if either* i *or* j *is a weak categorical equivalence, so is the induced map* $K \times V \cup L \times U \to L \times V$.

For any trivial cofibration (cofibration) $K \to L$, *and for any fibration* $X \to Y$ *of the Joyal model category structure, the induced map*

$$\underline{\mathrm{Hom}}(L, X) \to \underline{\mathrm{Hom}}(K, X) \times_{\underline{\mathrm{Hom}}(K,Y)} \underline{\mathrm{Hom}}(L, Y)$$

is a trivial fibration (a fibration, respectively)

Proof The second assertion is a direct consequence of the first, which itself follows right away from the preceding corollary (and from the fact that trivial cofibrations are closed under push-outs). □

Remark 3.6.5 In Definition 1.6.10, one can find the notions of natural transformation and of equivalence of categories. It is clear that natural transformations are the morphisms in the ∞-categories of the form $\underline{\mathrm{Hom}}(A, X)$ (with X an ∞-category) and it follows from Corollary 3.5.12 that the invertible natural transformations precisely are the invertible maps in $\underline{\mathrm{Hom}}(A, X)$. Note finally that a natural transformation $\Delta^1 \to \underline{\mathrm{Hom}}(A, X)$ is invertible if and only if it factors through a map $J \to \underline{\mathrm{Hom}}(A, X)$, which corresponds to a homotopy $J \times A \to X$. In other words, two functors $u, v \colon A \to X$ are related by an invertible natural transformation if and only if they are J-homotopic. Hence the equivalences of ∞-categories precisely are the J-homotopy equivalences.

Corollary 3.6.6 *A functor between ∞-categories is an equivalence of ∞-categories if and only if it is a weak equivalence of the Joyal model category structure.*

Proof By virtue of Corollary 2.2.18 since any simplicial set is cofibrant, and since ∞-categories are fibrant, the weak equivalences between ∞-categories for the Joyal model category structures precisely are the homotopy equivalences. And by virtue of Lemma 2.2.12, we can pick any of our favourite cylinders to define the notion of homotopy; e.g. the Cartesian product with J. In that case, the preceding remark explains why the notion of equivalence of ∞-categories is nothing other than the notion of homotopy equivalence between cofibrant–fibrant objects in the Joyal model category structure. □

Corollary 3.6.7 *Let $p \colon X \to Y$ be an isofibration between ∞-categories. For any monomorphism $A \to B$, the induced map*

$$h(B, X) \to h(A, X) \times_{h(A,Y)} h(B, Y)$$

is an isofibration.

Proof For any trivial cofibration $K \to L$ of the Joyal model category structure, we have a Cartesian square of the form

$$\begin{array}{ccc}
k(L, X) & \hookrightarrow & \underline{\mathrm{Hom}}(L, X) \\
\downarrow & & \downarrow \\
k(K, X) \times_{k(K,Y)} k(L, Y) & \longrightarrow & \underline{\mathrm{Hom}}(K, X) \times_{\underline{\mathrm{Hom}}(K,Y)} \underline{\mathrm{Hom}}(L, Y)
\end{array}$$

in which the right-hand vertical map is a trivial fibration. We conclude as in the proof of Corollary 3.5.13. □

Theorem 3.6.8 *Let $f: X \to Y$ be a morphism of simplicial sets. The following conditions are equivalent.*

(i) The map f is a weak categorical equivalence.
(ii) For any ∞-category W, the map $f^*: \underline{\mathrm{Hom}}(Y, W) \to \underline{\mathrm{Hom}}(X, W)$ is an equivalence of ∞-categories.
(iii) For any ∞-category W, the functor $f^*: \tau\,\underline{\mathrm{Hom}}(Y, W) \to \tau\,\underline{\mathrm{Hom}}(X, W)$ is an equivalence of categories.
(iv) For any ∞-category W, the map $f^*: k(Y, W) \to k(X, W)$ is an equivalence of ∞-groupoids (or, equivalently, a weak homotopy equivalence).

Proof By definition of the Joyal model category structure, and, by virtue of Theorem 3.6.1, the map f is a weak categorical equivalence if and only if, for any ∞-category W, the induced map $f^*: [Y, W] \to [X, W]$ is bijective, where $[X, W]$ is the set of maps from X to W up to J-homotopy equivalence. Since $\underline{\mathrm{Hom}}(X, W)$ is an ∞-category, the explicit description of $\tau\,\underline{\mathrm{Hom}}(X, W)$ given by Theorem 1.6.6 shows that $[X, W]$ can alternatively be described as the set of isomorphism classes of objects in the category $\tau\,\underline{\mathrm{Hom}}(X, W)$. Remark 3.6.5 explains why a third description of the set $[X, W]$ is the one given by the set of connected components of the Kan complex $k(X, W)$. Therefore, it is clear that either of conditions (ii), (iii) or (iv) implies condition (i). By virtue of Proposition 3.3.14, the functor τ sends weak categorical equivalences (in particular, equivalences of ∞-categories) to equivalences of categories. Hence it is clear that condition (ii) implies condition (iii). Let us prove that condition (i) implies condition (ii): by virtue of Ken Brown's lemma (Proposition 2.2.7) and of Corollaries 3.6.6 and 3.6.4, the functor $\underline{\mathrm{Hom}}(-, W)$ sends weak categorical equivalences to equivalences of ∞-categories. It remains to prove that condition (ii) implies condition (iv). By virtue of Corollary 3.5.12, it is sufficient to prove that the functor k sends equivalences of ∞-categories to equivalences of ∞-groupoids. For this, using Ken Brown's lemma, it is sufficient to prove that the functor k sends trivial fibrations between ∞-categories to trivial fibrations between Kan complexes. Let $p: X \to Y$ be a trivial fibration, with Y an ∞-category. Then X is an ∞-category, and since p is obviously conservative (because it is an equivalence of ∞-categories), the commutative square

$$\begin{array}{ccc} k(X) & \longrightarrow & X \\ {\scriptstyle k(p)}\downarrow & & \downarrow{\scriptstyle p} \\ k(Y) & \longrightarrow & Y \end{array}$$

is Cartesian, which implies that the induced map $k(X) \to k(Y)$ is a trivial fibration. □

With similar arguments, one proves the next result.

Theorem 3.6.9 *Let $f\colon X \to Y$ be a functor between ∞-categories. The following conditions are equivalent.*

(i) The functor f is an equivalence of ∞-categories.
(ii) For any simplicial set A, the functor $\underline{\mathrm{Hom}}(A, X) \to \underline{\mathrm{Hom}}(A, Y)$ is an equivalence of ∞-categories.
(iii) For any simplicial set A, the functor $\tau\,\underline{\mathrm{Hom}}(A, X) \to \tau\,\underline{\mathrm{Hom}}(A, Y)$ is an equivalence of categories.
(iv) For any simplicial set A, the functor $k(A, X) \to k(A, Y)$ is an equivalence of ∞-groupoids.

3.7 The Boardman–Vogt Construction, Revisited

3.7.1 Let X be an ∞-category. Given two objects x and y in X, we form the ∞-category $X(x, y)$ of maps from x to y by considering the following pull-back square, in which s and t are the evaluation maps at 0 and 1, respectively.

$$\begin{array}{ccc} X(x,y) & \longrightarrow & \underline{\mathrm{Hom}}(\Delta^1, X) \\ \downarrow & & \downarrow{\scriptstyle (s,t)} \\ \Delta^0 & \xrightarrow{(x,y)} & X \times X \end{array} \tag{3.7.1.1}$$

By definition of $k(\Delta^1, X)$, and by virtue of Corollary 3.5.12, we have in fact two pull-back squares

$$\begin{array}{ccccc} X(x,y) & \longrightarrow & k(\underline{\mathrm{Hom}}(\Delta^1, X)) & \hookrightarrow & \underline{\mathrm{Hom}}(\Delta^1, X) \\ \downarrow & & \downarrow{\scriptstyle (k(s),k(t))} & & \downarrow{\scriptstyle (s,t)} \\ \Delta^0 & \xrightarrow{(x,y)} & k(X) \times k(X) & \hookrightarrow & X \times X \end{array} \tag{3.7.1.2}$$

in which the middle vertical map is a Kan fibration (hence the left vertical map as well). In other words, we have defined an ∞-groupoid of maps from x to y.

Proposition 3.7.2 *There is a canonical bijection*

$$\pi_0(X(x, y)) \simeq \mathrm{Hom}_{ho(X)}(x, y)$$

(where the right-hand side is the set of maps in the Boardman–Vogt homotopy category of X; see Theorem 1.6.6).

Proof The canonical morphism $X \to N(ho(X))$ induces a canonical morphism $X(x, y) \to N(ho(X))(x, y) = \mathrm{Hom}_{ho(X)}(x, y)$ (where we identify any set with the corresponding constant simplicial set), whence a canonical map

$$X(x, y)_0 \to \mathrm{Hom}_{ho(X)}(x, y)\,.$$

The domain of this map is the set of maps from x to y in X, and this map simply is the quotient map defining $\mathrm{Hom}_{ho(X)}(x, y)$. In particular, this map is surjective and obviously factors through $\pi_0(X(x, y))$ (since it comes from a morphism of simplicial sets). In other words, it is sufficient to check that, for any maps $f, g\colon x \to y$ in X, if f and g agree in $ho(X)$, then they belong to the same path component of the Kan complex $X(a, b)$. Let us choose a morphism $t\colon \Delta^2 \to X$ whose restriction to $\partial\Delta^2$ corresponds to the following diagram.

$$\begin{array}{ccccc} & & y & & \\ & {\scriptstyle f}\nearrow & & \searrow{\scriptstyle 1_y} & \\ x & & \xrightarrow[g]{} & & y \end{array}$$

Let $p\colon \Delta^1 \times \Delta^1 \to \Delta^2$ be the nerve of the surjective morphism which sends both $(0, 0)$ and $(0, 1)$ to 0. Then the composition $tp\colon \Delta^1 \times \Delta^1 \to X$ interprets the previous commutative triangle as a diagram of the form

$$\begin{array}{ccc} x & \xrightarrow{f} & y \\ {\scriptstyle 1_x}\big\downarrow & & \big\downarrow{\scriptstyle 1_y} \\ x & \xrightarrow{g} & y \end{array}$$

and induces by transposition a map $h\colon \Delta^1 \to X(x, y) \subset \underline{\mathrm{Hom}}(\Delta^1, X)$ which connects the points f and g. □

3.7.3 To understand the composition law better, we need the Grothendieck–Segal condition to hold for ∞-categories. Recall that, for $n \geq 2$, the spine Sp^n is the union of the images of the maps $u_i\colon \Delta^1 \to \Delta^n$ which send 0 to i and 1 to $i + 1$, for $0 \leq i < n$.

Proposition 3.7.4 *The inclusion $Sp^n \to \Delta^n$ is an inner anodyne extension (whence a weak categorical equivalence).*

For the proof, we will need the following lemma.

Lemma 3.7.5 *Let S be a subset of $\{0, \ldots, n\}$, $n \geq 2$. Assume that the complement of S is not an interval (i.e. that there exists a and c not in S and $b \in S$ such that $a < b < c$). Then the inclusion $\Lambda^n_S \to \Delta^n$ is an inner anodyne extension, where Λ^n_S is the union of the images of the faces δ^n_i for $i \notin S$.*

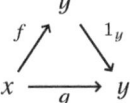

Proof We proceed by induction on n. For $n = 2$, we must have $S = \{1\}$, in which case the inclusion of $\Lambda^2_S = \Lambda^2_1$ into Δ^2 obviously is an inner anodyne extension. Assume that $n > 2$. We now proceed by induction on the cardinal of S. If S only has one element k, then we have $\Lambda^n_S = \Lambda^n_k$, and we are done (because, for the complement of $\{k\}$ not to be an interval, we need $0 < k < n$). Otherwise, we may assume that there exists a and c not in S and $b \in S$ such that $a < b < c$. Then we may choose $i \in S$ with $i \neq b$ (our inductive assumption implies that the cardinal of S is > 1). We shall write T for the set of elements j of $[n-1]$ such that $\delta^n_i(j) \in S$. Then, we have a biCartesian square of the following form.

$$\begin{array}{ccc} \Lambda^{n-1}_T & \hookrightarrow & \Delta^{n-1} \\ \downarrow & & \downarrow \delta^n_i \\ \Lambda^n_S & \hookrightarrow & \Lambda^n_{S-\{i\}} \end{array}$$

Moreover, the elements a, b and c are witness that the complement of $S - \{i\}$ is not an interval: a and c are not in S and distinct from i (because $i \in S$), while $b \in S - \{i\}$. Therefore, by induction on S, the map $\Lambda^n_{S-\{i\}} \to \Delta^n$ is an inner anodyne extension. It is thus sufficient to check that the map $\Lambda^{n-1}_T \to \Delta^{n-1}$ is an inner anodyne extension. By induction on n, it is sufficient to prove that the complement of T in $[n-1]$ is not an interval. Then, all elements distinct from i are in the image of δ^n_i. Let α, β and γ be the elements which are sent to a, b and c, respectively. We have $\alpha < \beta < \gamma$ with $\beta \in T$, but α and γ are not in T. □

Proof of Proposition 3.7.4 We first prove by induction that the inclusion

$$Im(\delta^n_0) \cup Sp^n \to \Delta^n$$

is an inner anodyne extension for $n \geq 2$. For $n = 2$, we have: $Im(\delta^n_0) \cup Sp^n = \Lambda^2_1$. For $n > 2$, the square

$$\begin{array}{ccc} Im(\delta^{n-1}_0) \cup Sp^{n-1} & \longrightarrow & Im(\delta^n_0) \cup Sp^n \\ \downarrow & & \downarrow \\ \Delta^{n-1} & \xrightarrow{\delta^n_n} & Im(\delta^n_0) \cup Im(\delta^n_n) \end{array}$$

is Cartesian, hence also coCartesian. Therefore, it is sufficient to prove that the inclusion of $Im(\delta^n_0) \cup Im(\delta^n_n)$ into Δ^n is an inner anodyne extension. Since, for $n > 1$, the subset $\{0, n\}$ obviously is not an interval in $[n]$, this readily follows from Lemma 3.7.5.

By duality (i.e. using the fact that the functor $X \mapsto X^{\mathrm{op}}$ preserves inner anodyne extensions), we have also proved that the inclusion $Im(\delta^n_n) \cup Sp^n \to \Delta^n$ is an inner anodyne extension for $n \geq 2$.

We now can prove the proposition by induction on n. The case $n = 2$ is obvious, and we may thus assume that $n > 2$. The Cartesian square

$$\begin{array}{ccc} Sp^{n-1} & \longrightarrow & Sp^n \\ \big\downarrow & & \big\downarrow \\ \Delta^{n-1} & \xrightarrow{\delta^n_n} & \Delta^n \end{array}$$

implies that the inclusion $Sp^n \to Im(\delta^n_n) \cup Sp^n$ is the push-out of the inclusion map $Sp^{n-1} \to \Delta^{n-1}$, whence is an inner anodyne extension. Therefore, the composition of the maps

$$Sp^n \to Im(\delta^n_n) \cup Sp^n \to \Delta^n$$

is an inner anodyne extension. □

Corollary 3.7.6 *For any isofibration between ∞-categories $X \to Y$, and for any integer $n \geq 2$, the canonical map*

$$\underline{\mathrm{Hom}}(\Delta^n, X) \to \underline{\mathrm{Hom}}(Sp^n, X) \times_{\underline{\mathrm{Hom}}(Sp^n, Y)} \underline{\mathrm{Hom}}(\Delta^n, Y)$$

is a trivial fibration. In particular, for any ∞-category X, the restriction map

$$\underline{\mathrm{Hom}}(\Delta^n, X) \to \underline{\mathrm{Hom}}(Sp^n, X)$$

is a trivial fibration.

Proof The first assertion is the particular case of Corollary 3.6.4 obtained from the trivial cofibration of Proposition 3.7.4. The second assertion is the particular case of the first when $Y = \Delta^0$. □

3.7.7 Let us consider an ∞-category X, together with an $(n+1)$-tuple of objects $(x_0, \ldots, x_n)$, for $n \geq 2$. We can then form the following pull-back squares.

$$\begin{array}{ccccc} X(x_0, \ldots, x_n) & \longrightarrow & k(\underline{\mathrm{Hom}}(\Delta^n, X)) & \hookrightarrow & \underline{\mathrm{Hom}}(\Delta^n, X) \\ \big\downarrow & & \big\downarrow & & \big\downarrow{\scriptstyle \text{evaluation}} \\ \Delta^0 & \xrightarrow{(x_0, \ldots, x_n)} & k(X)^{n+1} & \hookrightarrow & X^{n+1} \end{array} \tag{3.7.7.1}$$

The ∞-groupoid $X(x_0, \ldots, x_n)$ is the space of sequences of maps of the form

$x_0 \to \cdots \to x_n$ in X. We remark that we also have a pull-back square of the following form.

(3.7.7.2)

$$\begin{array}{ccccc} \prod_{0\leq i<n} X(x_i, x_{i+1}) & \longrightarrow & k(\underline{\mathrm{Hom}}(Sp^n, X)) & \hookrightarrow & \underline{\mathrm{Hom}}(Sp^n, X) \\ \downarrow & & \downarrow & & \downarrow{\scriptstyle \text{evaluation}} \\ \Delta^0 & \xrightarrow{(x_0,\ldots,x_n)} & k(X)^{n+1} & \hookrightarrow & X^{n+1} \end{array}$$

This implies that the diagram

(3.7.7.3)

$$\begin{array}{ccccc} X(x_0, \ldots, x_n) & \hookrightarrow & k(\underline{\mathrm{Hom}}(\Delta^n, X)) & \hookrightarrow & \underline{\mathrm{Hom}}(\Delta^n, X) \\ \downarrow & & \downarrow & & \downarrow \\ \prod_{0\leq i<n} X(x_i, x_{i+1}) & \hookrightarrow & k(\underline{\mathrm{Hom}}(Sp^n, X)) & \hookrightarrow & \underline{\mathrm{Hom}}(Sp^n, X) \end{array}$$

is made of Cartesian squares in which, by virtue of Corollary 3.7.6, all the vertical maps are trivial fibrations. The unique map $\gamma\colon \Delta^1 \to \Delta^n$ which sends 0 to 0 and 1 to n induces a morphism

(3.7.7.4) $$X(x_0, \ldots, x_n) \to X(x_0, x_n)\,.$$

The choice of a section of $X(x_0, \ldots, x_n) \to \prod_{0\leq i<n} X(x_i, x_{i+1})$ composed with the map (3.7.7.4) thus provides a composition law

(3.7.7.5) $$\prod_{0\leq i<n} X(x_i, x_{i+1}) \to X(x_0, x_n)\,.$$

Applying the functor π_0 to the map (3.7.7.5) gives a composition law

(3.7.7.6) $$\prod_{0\leq i<n} \mathrm{Hom}_{ho(X)}(x_i, x_{i+1}) \to \mathrm{Hom}_{ho(X)}(x_0, x_n),$$

which is nothing other than the composition law of Boardman and Vogt's homotopy category $ho(X)$.

3.8 Serre's Long Exact Sequence

3.8.1 Let X be an ∞-groupoid. For any object $x \in X_0$, we define the *loop space* of X at x as

(3.8.1.1) $$\Omega(X, x) = X(x, x)\,.$$

The *fundamental group* of X at the point x is $\pi_1(X, x) = \pi_0(\Omega(X, x))$ (the group structure is obtained, for instance, from the identification of $\pi_0(\Omega(X, x))$ with the group $\mathrm{Hom}_{ho(X)}(x, x)$ given by Proposition 3.7.2). There is a canonical element of $\Omega(X, x)_0$ which we write, by abuse of notation, and for historical reasons, x (but that really is 1_x). We thus can iterate this construction: we write $\Omega^1(X,x) = \Omega(X,x)$, and, for $n \geq 1$, we define

$$\Omega^{n+1}(X,x) = \Omega(\Omega^n(X,x),x)\,. \tag{3.8.1.2}$$

In particular, for each $n \geq 1$, we have defined a group

$$\pi_n(X,x) = \pi_0(\Omega^n(X,x)) \tag{3.8.1.3}$$

called the *nth homotopy group of X at the point x*.

Proposition 3.8.2 *For any integer $n \geq 2$, the group $\pi_n(X,x)$ is abelian.*

Proof It is sufficient to prove the case where $n = 2$. As a particular case of the map (3.7.7.5), we have a composition law

$$c_X\colon \Omega(X,x) \times \Omega(X,x) \to \Omega(X,x)$$

which, by functoriality of the loop space construction, gives a composition law

$$\Omega(c_X)\colon \Omega^2(X,x) \times \Omega^2(X,x) \to \Omega^2(X,x)\,.$$

On the other hand, we have the composition law

$$c_{\Omega(X,x)}\colon \Omega^2(X,x) \times \Omega^2(X,x) \to \Omega^2(X,x)\,.$$

Let us define

$$a \bullet b = \pi_0(\Omega(c_X))(a,b) \quad \text{and} \quad a \circ b = \pi_0(c_{\Omega(X,x)})(a,b)\,.$$

The pairing $a \circ b$ is the group law on $\pi_2(X,x)$, while $a \bullet b$ is another pairing which is a morphism of groups from $\pi_2(X,x) \times \pi_2(X,x)$ to $\pi_2(X,x)$. In particular, we have

$$(a \bullet b) \circ (c \bullet d) = (a \circ c) \bullet (b \circ d)\,.$$

We also have equalities of the form

$$a \bullet 1_x = a = 1_x \bullet a\,.$$

Let us prove that $a \bullet 1_x = a$ holds, for instance. There is the unique map $\gamma\colon \Delta^1 \to \Delta^2$ which sends 0 to 0 and 1 to 2, and its retraction $r\colon \Delta^2 \to \Delta^1$ which sends both 0 and 1 to 0. If s denotes the restriction of r on Λ^2_1, we then have a commutative square

$$\begin{array}{ccc}
\underline{\mathrm{Hom}}(\Delta^1, X) & \xrightarrow{r^*} & \underline{\mathrm{Hom}}(\Delta^2, X) \\
{\scriptstyle s^*}\downarrow & \overset{f}{\nearrow} & \downarrow \\
\underline{\mathrm{Hom}}(\Lambda^2_1, X) & = & \underline{\mathrm{Hom}}(\Lambda^2_1, X)
\end{array}$$

in which the vertical map on the left is a monomorphism, while the one on the right is a trivial fibration. Therefore, this square admits a lift f. The composition of f with $\gamma^*\colon \underline{\mathrm{Hom}}(\Delta^2, X) \to \underline{\mathrm{Hom}}(\Delta^1, X)$ is the composition law of two maps in X, i.e. c_X is homotopic to $\gamma^* f$ (because c_X is also induced by a composition of γ^* with a choice of section of the trivial fibration of the square above). Therefore, since Ω is a right Quillen functor, the map $\Omega(c_X)$ is also homotopic to $\Omega(\gamma^*)\Omega(f)$. But, by construction and by functoriality, the map

$$\Omega^2(X,x) \xrightarrow{(1_x, 1_{\Omega(X,x)})} \Omega^2(X,x) \times \Omega(X,x) \xrightarrow{\Omega(\gamma^*)\Omega(f)} \Omega^2(X,x)$$

is the identity. This shows that $a \bullet 1_x = a$. The equality $1_x \bullet a = a$ is obtained similarly, replacing r by the other retraction of γ.

With these properties, we now have

$$a \bullet b = (a \circ 1_x) \bullet (1_x \circ b) = (a \bullet 1_x) \circ (1_x \bullet b) = a \circ b,$$

and also

$$a \circ b = (1_x \bullet a) \circ (b \bullet 1_x) = (1_x \circ b) \bullet (a \circ 1_x) = b \bullet a,$$

from which we deduce that $a \bullet b = b \bullet a$. □

Proposition 3.8.3 *Let us consider a commutative triangle of simplicial sets.*

$$\begin{array}{ccc} X & \xrightarrow{f} & Y \\ & {\scriptstyle p}\searrow \quad \swarrow{\scriptstyle q} & \\ & S & \end{array}$$

(a) *If, for any map $\Delta^n \to S$, the induced morphism $\Delta^n \times_S X \to \Delta^n \times_S Y$ is a weak homotopy equivalence, then, for any map $S' \to S$, the induced morphism $X' = S' \times_S X \to Y' = S' \times_S Y$ is a weak homotopy equivalence (in particular, f itself is a weak homotopy equivalence).*

(b) *If p and q are Kan fibrations, the converse is true: if f is a weak homotopy equivalence, so are the maps $\Delta^n \times_S X \to \Delta^n \times_S Y$ for any map $\Delta^n \to S$.*

Proof Since the functors of the form $(-) \times_S X$ commute with small colimits and preserve cofibrations, by virtue of Corollaries 2.3.16, 2.3.18 and 2.3.29, the class of maps $S' \to S$ such that the induced morphism $S' \times_S X \to S' \times_S Y$ is a weak homotopy equivalence is saturated by monomorphisms in the sense of Definition 1.3.9 (when considered as a class of presheaves over the small category $\mathbf{\Delta}/S$). Corollary 1.3.10 for $A = \mathbf{\Delta}/S$ tells us that if ever it contains all maps of the form $\Delta^n \to S$, this class must be the one of all maps $S' \to S$. This proves the first assertion.[1]

Let us assume that both p and q are Kan fibrations. Then, for any map of the form $\Delta^n \to S$, we may consider the Cartesian squares

$$\begin{array}{ccc} \Delta^n \times_S X & \longrightarrow & X \\ \downarrow & & \downarrow p \\ \Delta^n & \longrightarrow & S \end{array} \quad \text{and} \quad \begin{array}{ccc} \Delta^n \times_S Y & \longrightarrow & Y \\ \downarrow & & \downarrow q \\ \Delta^n & \longrightarrow & S \end{array}$$

and see that both of them are homotopy Cartesian (by virtue of Corollary 2.3.28 and of Proposition 3.1.23 they have this property after we apply the finite limit preserving functor Ex^∞, and therefore the functorial trivial cofibration (3.1.22.5) gives our claim). This implies that the commutative square

$$\begin{array}{ccc} \Delta^n \times_S X & \longrightarrow & X \\ \downarrow & & \downarrow f \\ \Delta^n \times_S Y & \longrightarrow & Y \end{array}$$

is homotopy Cartesian. Therefore, the left-hand vertical map of the latter is a weak homotopy equivalence whenever f has this property. □

Corollary 3.8.4 *Let us consider a commutative square of simplicial sets*

$$\begin{array}{ccc} X' & \xrightarrow{u} & X \\ {\scriptstyle p'}\downarrow & & \downarrow {\scriptstyle p} \\ Y' & \xrightarrow{v} & Y \end{array}$$

in which the maps p and q are supposed to be Kan fibrations. Such a square is homotopy Cartesian if and only if, for any point $y' \in Y'_0$, if we put $y = v(y')$, the induced morphism between the fibres $X'_{y'} \to X_y$ is a weak homotopy equivalence.

Proof Since the map p is a Kan fibration, the homotopy pull-back of Y' and X over Y simply is the ordinary pull-back $Y' \times_Y X$. Therefore, this square is homotopy Cartesian if and only if the induced map $f : X' \to Y' \times_Y X$ is a weak homotopy equivalence. By virtue of the preceding proposition, this property is thus equivalent to the assertion that, for any map of the form $s' : \Delta^n \to Y'$, the induced map

$$\Delta^n \times_{Y'} X' \to \Delta^n \times_{Y'} (Y' \times_Y X) \simeq \Delta^n \times_Y X$$

[1] We remark that we have hardly used any specific property of the Kan–Quillen model category structure: the same proof applies to any model category structure on the category of presheaves of sets over an Eilenberg–Zilber category whose class of cofibrations is the class of monomorphisms.

is a weak homotopy equivalence. If we let y' and y be the image of 0 in Y' and Y, respectively, we have a commutative square of the form

$$\begin{array}{ccc} X'_{y'} & \longrightarrow & X_y \\ \downarrow & & \downarrow \\ \Delta^n \times_{Y'} X' & \longrightarrow & \Delta^n \times_Y X \end{array}$$

in which the vertical maps are strong deformation retracts (see Proposition 2.5.7). Therefore, it is sufficient to consider pull-backs along maps of the form $\Delta^0 \to Y'$ only. □

Corollary 3.8.5 *For any equivalence of ∞-groupoids $f\colon X \to Y$ and any point $x \in X_0$, if we put $y = f(x)$, the induced map $\pi_n(X, x) \to \pi_n(Y, y)$ is an isomorphism of groups for $n \geq 1$.*

Proof We have a commutative square of the form

$$\begin{array}{ccc} \underline{\mathrm{Hom}}(\Delta^1, X) & \xrightarrow{f_*} & \underline{\mathrm{Hom}}(\Delta^1, Y) \\ \downarrow{\scriptstyle (s,t)} & & \downarrow{\scriptstyle (s,t)} \\ X \times X & \xrightarrow[f \times f]{} & Y \times Y \end{array}$$

in which the horizontal maps are equivalences of ∞-groupoids (by virtue of Ken Brown's lemma, it is sufficient to check this latter property when f is a trivial fibration, in which case these maps are trivial fibrations as well). This commutative square is thus homotopy Cartesian. Therefore, the preceding corollary implies that it induces weak homotopy equivalences on fibres of the vertical maps. The latter weak homotopy equivalences precisely are the maps $\Omega(X, x) \to \Omega(Y, y)$ for all $x \in X_0$. Iterating the process gives weak equivalences $\Omega^n(X, x) \to \Omega^n(Y, y)$ for all $n \geq 1$. Since the functor π_0 sends weak homotopy equivalences to bijections, this achieves the proof. □

3.8.6 For an arbitrary simplicial set X, we have a canonical trivial cofibration $X \to Ex^\infty(X)$ which induces a bijection $X_0 \simeq Ex^\infty(X)_0$ (because $Sd(\Delta^0) \simeq \Delta^0$). By virtue of the preceding corollary, if we define

$$\pi_n(X, x) = \pi_n(Ex^\infty(X), x), \tag{3.8.6.1}$$

we extend the definition of homotopy groups to arbitrary simplicial sets in a way which is compatible with the definition given at the beginning of this chapter. Since the functor Ex^∞ sends weak homotopy equivalences to equivalences of ∞-groupoids, we can reformulate the previous corollary as follows.

Corollary 3.8.7 *For any weak homotopy equivalence of simplicial sets $f : X \to Y$ and any point $x \in X_0$, if we put $y = f(x)$, the induced map $\pi_n(X, x) \to \pi_n(Y, y)$ is an isomorphism of groups for $n \geq 1$.*

3.8.8 For any simplicial set X, one can freely add a base point by forming the pointed simplicial set $X_+ = X \amalg \Delta^0$ (where the base point corresponds to the new copy of Δ^0). This defines a left adjoint to the forgetful functor from pointed simplicial sets to simplicial sets.

Given two pointed simplicial sets X and Y, with base points x and y, respectively, the simplicial set of pointed maps from X to Y, denoted by $\underline{\mathrm{Hom}}_*(X, Y)$, is defined as the pull-back

(3.8.8.1)
$$\begin{array}{ccc} \underline{\mathrm{Hom}}_*(X,Y) & \longrightarrow & \underline{\mathrm{Hom}}(X,Y) \\ \downarrow & & \downarrow{\scriptstyle x^*} \\ \Delta^0 & \xrightarrow{\quad y \quad} & Y \end{array}$$

(where x^* is the evaluation at x).

We also define the smash product $X \wedge Y$ as the push-out

(3.8.8.2)
$$\begin{array}{ccc} X \vee Y & \hookrightarrow & X \times Y \\ \downarrow & & \downarrow \\ \Delta^0 & \longrightarrow & X \wedge Y \end{array}$$

where $X \vee Y$ is the union of $X \times \{y\}$ and $\{x\} \times Y$ in the Cartesian product $X \times Y$. For any three pointed simplicial sets X, Y and Z, we have a natural bijection

$$\mathrm{Hom}_*(X \wedge Y, Z) \simeq \mathrm{Hom}_*(X, \underline{\mathrm{Hom}}_*(Y, Z))$$

induced by the natural bijection $\mathrm{Hom}(X \times Y, Y) \simeq \mathrm{Hom}(X, \underline{\mathrm{Hom}}(Y, Z))$ (where Hom_* denotes the set of morphisms of pointed simplicial sets). Moreover the operation $\wedge$ defines a symmetric monoidal structure on the category of pointed simplicial sets with unit object $S^0 = (\Delta^0)_+$. In particular we have the following canonical isomorphisms (for all pointed simplicial sets X, Y and Z):

(3.8.8.3) $X \wedge (Y \wedge Z) \simeq (X \wedge Y) \wedge Z, \qquad X \wedge Y \simeq Y \wedge X, \qquad S^0 \wedge X \simeq X.$

One defines the *simplicial circle* S^1 with the following coCartesian square.

(3.8.8.4)
$$\begin{array}{ccc} \partial\Delta^1 & \hookrightarrow & \Delta^1 \\ \downarrow & & \downarrow \\ \Delta^0 & \longrightarrow & S^1 \end{array}$$

Then, for any integer $n \geq 2$ we define the *simplicial n-sphere* as

(3.8.8.5)
$$S^n = S^1 \wedge S^{n-1}.$$

One checks directly that, for each pointed Kan complex X, we have

$$\Omega^n(X, x) = \underline{\mathrm{Hom}}_*(S^n, X) \tag{3.8.8.6}$$

(and we really mean an equality here).

On the other hand, for any pointed simplicial set A and any pointed Kan complex X, the set $\pi_0(\underline{\mathrm{Hom}}_*(A, X))$ is canonically identified with the set of maps from A to X in the homotopy category $ho(\Delta^0\backslash sSet)$ (with respect to the Kan–Quillen model category structure): indeed, this is a particular case of Theorem 2.2.17, because, using Corollary 2.3.17, we see right away that $A \wedge (\Delta^1)_+$ is a cylinder of A in the Kan–Quillen model category structure. In particular, we get another proof of Corollary 3.8.5, but we also have the following property.

Proposition 3.8.9 *For any pointed weak homotopy equivalence $A \to B$ and any pointed Kan complex X, the induced map*

$$\pi_0(\underline{\mathrm{Hom}}_*(B, X)) \to \pi_0(\underline{\mathrm{Hom}}_*(A, X))$$

is bijective.

Proposition 3.8.10 *Let X be an ∞-groupoid, and $x \in X_0$ a point. For any integer $n \geq 1$, and for any choice of base point for $\partial\Delta^n$, there is a canonical bijection $\pi_0(\underline{\mathrm{Hom}}_*(\partial\Delta^{n+1}, X)) \simeq \pi_n(X, x)$ such that the constant map with value x corresponds to the unit 1_x of the homotopy group.*

Proof It is sufficient to prove that $\partial\Delta^{n+1}$ and S^n are isomorphic in the *unpointed* homotopy category of the Kan–Quillen model category structure. Indeed, if this is the case, there will exist a weak homotopy equivalence from $\partial\Delta^{n+1}$ to $Ex^\infty(S^n)$. Since S^n has exactly one 0-simplex (object) and since the trivial cofibration $S^n \to Ex^\infty(S^n)$ is bijective on objects, these maps will always be compatible with any choice of base point we make for $\partial\Delta^{n+1}$. We will conclude with the preceding proposition.

We prove this assertion by induction on n. If $n = 0$, this is clear: $S^0 = \partial\Delta^1$. Hence we may assume that $n > 0$. We have the (homotopy) coCartesian square below.

$$\begin{array}{ccc} \partial\Delta^n & \longrightarrow & \Lambda^{n+1}_{n+1} \\ \downarrow & & \downarrow \\ \Delta^n & \xrightarrow{\delta^{n+1}_{n+1}} & \partial\Delta^{n+1} \end{array}$$

Note that $X \wedge \Delta^0 = \Delta^0$ whatever X is. Therefore, we also have the following (homotopy) coCartesian square.

$$\begin{array}{ccc} S^{n-1}\wedge S^0 & \hookrightarrow & S^{n-1}\wedge\Delta^1 \\ \downarrow & & \downarrow \\ S^{n-1}\wedge\Delta^0 & \hookrightarrow & S^{n-1}\wedge S^1 \end{array} \quad = \quad \begin{array}{ccc} S^{n-1} & \hookrightarrow & S^{n-1}\wedge\Delta^1 \\ \downarrow & & \downarrow \\ \Delta^0 & \hookrightarrow & S^n \end{array}$$

By virtue of Corollary 2.3.17, the map $X \wedge \Delta^1 \to X \wedge \Delta^0 = \Delta^0$ is a weak homotopy equivalence for any pointed simplicial set X. In conclusion, the simplicial sets $\partial\Delta^{n+1}$ and S^n are homotopy push-outs of diagrams of the form

$$\Delta^0 \leftarrow \partial\Delta^n \to \Delta^0 \quad \text{and} \quad \Delta^0 \leftarrow S^{n-1} \to \Delta^0,$$

respectively. By induction, we see that these two diagrams are weakly equivalent, and therefore, by functoriality, so are their homotopy colimits. □

Proposition 3.8.11 *Let X be a simplicial set. The map $X \to \Delta^0$ is a weak homotopy equivalence if and only if the set $\pi_0(X)$ exactly has one element and if there exists a point $x_0 \in X_0$ such that, for any integer $n \geq 1$, the homotopy group $\pi_n(X, x_0)$ is trivial.*

Proof This clearly is a necessary condition, by Corollary 3.8.7. For the converse, we may assume that X is a Kan complex. Let us assume that X has exactly one path component as well as trivial homotopy groups $\pi_n(X, x_0)$ in any degree $n > 0$, for a specified point x_0. We first prove that, for any point x of X, the homotopy groups $\pi_n(X, x)$ are trivial. Indeed, since X is path-connected (and a Kan complex), there exists a map $h\colon \Delta^1 \to X$ such that $h(0) = x_0$ and $h(1) = x$. We can form the following (homotopy) Cartesian square.

$$\begin{array}{ccc} E & \longrightarrow & \underline{\mathrm{Hom}}(\Delta^1, X) \\ {\scriptstyle q}\downarrow & & \downarrow{\scriptstyle (s,t)} \\ \Delta^1 & \xrightarrow{(h,h)} & X \times X \end{array}$$

Since pulling back a weak homotopy equivalence along a Kan fibration is a weak homotopy equivalence, the two inclusions $\Omega(X, x_0) \to E \leftarrow \Omega(X, x)$ are weak homotopy equivalences. Iterating the process gives a zig-zag of weak homotopy equivalences relating the iterated loop spaces $\Omega^n(X, x_0)$ and $\Omega^n(X, x)$, hence isomorphisms $\pi_n(X, x_0) \simeq \pi_n(X, x)$. By virtue of Proposition 3.8.10, this means that, for $n > 0$ and for any map $a\colon \partial\Delta^n \to X$, there exists a morphism $h\colon \Delta^1 \times \partial\Delta^n \to X$ which is a homotopy from a to some constant map x. We thus get a morphism $(h, x)\colon \Delta^1 \times \partial\Delta^n \cup \{1\} \times \Delta^n \to X$, and, since X is a Kan complex, this morphism is the restriction of some morphism $k\colon \Delta^1 \times \Delta^n \to X$. The restriction of k to $\Delta^n \simeq \{0\} \times \Delta^n$ defines a morphism $b\colon \Delta^n \to X$ such that

$b_{|\partial\Delta^n} = a$. Since $X \neq \varnothing$, this means that $X \to \Delta^0$ is a trivial fibration, hence a weak homotopy equivalence. □

Theorem 3.8.12 *Let $p\colon X \to Y$ be a Kan fibration. We consider that we are given a point $x \in X_0$ and $y = p(x)$, and we let F be the fibre of p at y. Then there is a canonical long exact sequence of pointed sets (of groups, if we restrict to the part of degree ≥ 1) of the following form.*

$$\cdots \longrightarrow \pi_n(F,x) \longrightarrow \pi_n(X,x) \longrightarrow \pi_n(Y,y) \longrightarrow \pi_{n-1}(F,x) \longrightarrow \cdots \longrightarrow \pi_1(Y,y) \longrightarrow \pi_0(F) \longrightarrow \pi_0(X) \longrightarrow \pi_0(Y)$$

Proof We may always assume that Y is a Kan complex (if not, use Ex^∞). We first prove that, for any homotopy Cartesian square of simplicial sets of the form

$$\begin{array}{ccc} F & \longrightarrow & E \\ \downarrow & & \downarrow{\scriptstyle \pi} \\ P & \longrightarrow & B \end{array}$$

in which P is weakly contractible (i.e. the map $P \to \Delta^0$ is a weak homotopy equivalence), for any point x of E which is sent to the connected component of B receiving P, we have a short exact sequence of pointed sets:

$$\pi_0(F) \to \pi_0(E) \to \pi_0(B)$$

(where the base points correspond to the connected component containing (the image of) x). Indeed, we may assume that π is a Kan fibration between Kan complexes, and, by Corollary 3.8.4, we may also assume that $P = \Delta^0$ (because we can replace F by the fibre of $F \to P$). We then have to prove that if an object a of the ∞-groupoid X is sent to the connected component of y in Y, then there is a morphism $a \to b$ in X such that $p(b) = y$. This immediately follows from the fact that the map p has the right lifting property with respect to maps of the form $\{0\} \to \Delta^1$.

Therefore, we already have a short exact sequence of pointed sets

$$\pi_0(F) \to \pi_0(X) \to \pi_0(Y)\,.$$

We remark that, for any Kan complex X, the loop space $\Omega(X,x)$ fits in a diagram made of (homotopy) Cartesian squares of the form below.

$$\begin{array}{ccccc}
\Omega(X,x) & \longrightarrow & P(X) & \longrightarrow & \underline{\mathrm{Hom}}(\Delta^1, X) \\
\downarrow & & \downarrow & & \downarrow{\scriptstyle (s,t)} \\
\Delta^0 & \xrightarrow{x} & X & \xrightarrow{(1_X,x)} & X \times X \\
 & & \downarrow & & \downarrow{\scriptstyle pr_2} \\
 & & \Delta^0 & \xrightarrow{x} & X
\end{array}$$

Since the map $pr_2(s,t) = t$ is a trivial fibration, so is the map $P(X) \to \Delta^0$. This means that $\Omega(X,x)$ is the homotopy limit of the diagram $\Delta^0 \to X \leftarrow \Delta^0$. Using the fact that homotopy Cartesian squares are closed under composition, this implies that there is a canonical homotopy Cartesian square of the form

$$\begin{array}{ccc}
\Omega(Y,y) & \longrightarrow & P \\
\downarrow & & \downarrow \\
F & \hookrightarrow & X
\end{array}$$

where P is weakly contractible. Therefore, we also have a short exact sequence of pointed sets

$$\pi_1(Y,y) \to \pi_0(F) \to \pi_0(X)\,.$$

Applying what precedes to the Cartesian squares

$$\begin{array}{ccc}
\Omega^n(F,x) & \longrightarrow & \Omega^n(X,x) \\
\downarrow & & \downarrow \\
\Delta^0 & \hookrightarrow & \Omega^n(Y,y)
\end{array}$$

for $n \geq 1$ gives the rest of the long exact sequence. □

Remark 3.8.13 The natural identification comparing the homotopy pull-backs of the diagrams

$$\Delta^0 \to X \leftarrow P(X) \quad \text{and} \quad P(X) \to X \leftarrow \Delta^0$$

induces a bijection $\pi_1(X,x) \simeq \pi_1(X,x)$ which is not the identity: this is the inverse map $g \mapsto g^{-1}$. This is because this inversion can be seen in the construction of $\Omega(X,x) = \underline{\mathrm{Hom}}_*(\Delta^1/\partial\Delta^1, X)$: it corresponds to the unique isomorphism $\Delta^{1\mathrm{op}} \simeq \Delta^1$ which induces an isomorphism $S^{1\mathrm{op}} \simeq S^1$ and thus an isomorphism of groups

$$\pi_1(X,x) \simeq \pi_1(X^{\mathrm{op}},x) = \pi_1(X,x)^{\mathrm{op}}$$

which is the canonical one (where, for a group G, the opposite G^{op} is the group defined by the pairing $(g, h) \mapsto h^{-1}g^{-1}$).

Corollary 3.8.14 *A morphism of simplicial sets $f : X \to Y$ is a weak homotopy equivalence if and only if, it induces a bijection $\pi_0(X) \simeq \pi_0(Y)$ and if, for any integer $n \geq 1$ and any $x \in X_0$, if we put $y = f(x)$, the induced map $\pi_n(X, x) \to \pi_n(Y, y)$ is an isomorphism of groups.*

Proof This is a necessary condition, by Corollary 3.8.7. Conversely, let us assume that the map f induces a bijection at the levels of π_0 and of homotopy groups, for all base points. As already seen in the proof of Proposition 3.8.11, given a Kan complex K and two points k_0 and k_1 connected by a path γ, there are canonical bijections $\pi_n(K, k_0) \simeq \pi_n(K, k_1)$ for all $n \geq 1$. Therefore, since any map can be factored into a weak homotopy equivalence followed by a Kan fibration, Corollary 3.8.7 shows that it is sufficient to prove the case where f is a Kan fibration. The preceding theorem then ensures that the fibres of f are then Kan complexes F such that $\pi_0(F)$ has exactly one element and whose homotopy groups all are trivial. By virtue of Proposition 3.8.11, all the fibres of f are (weakly) contractible. Since strong deformation retracts are stable under pull-backs along Kan fibrations (see Proposition 2.5.7), we deduce that, for any map $\Delta^n \to Y$, the induced map $\Delta^n \times_Y X \to \Delta^n$ is a weak equivalence. Therefore, Proposition 3.8.3 tells us that f is a weak homotopy equivalence. □

3.9 Fully Faithful and Essentially Surjective Functors

Lemma 3.9.1 *Let X be an ∞-category. The formation of the ∞-groupoids $k(A, X)$ satisfies the following operations.*

(a) For any coCartesian square of simplicial sets

$$\begin{array}{ccc} A & \xrightarrow{a} & A' \\ \downarrow{\scriptstyle i} & & \downarrow{\scriptstyle i'} \\ B & \xrightarrow{b} & B' \end{array}$$

in which i is a monomorphism, the induced square

$$\begin{array}{ccc} k(B', X) & \longrightarrow & k(B, X) \\ \downarrow{\scriptstyle i'^*} & & \downarrow{\scriptstyle i^*} \\ k(A', X) & \longrightarrow & k(A, X) \end{array}$$

is Cartesian and its vertical maps are Kan fibrations.

(b) For any sequence of monomorphisms of simplicial sets

$$A_0 \to A_1 \to \cdots \to A_n \to A_{n+1} \to \cdots$$

the induced transition maps $k(A_{n+1}, X) \to k(A_n, X)$ *are Kan fibrations and the comparison map*

$$k(\varinjlim_n A_n, X) \to \varprojlim_n k(A_n, X)$$

is an isomorphism.

(c) For any small family of simplicial sets $(A_i)_{i \in I}$*, the induced map*

$$k(\coprod_i A_i, X) \to \prod_i k(A_i, X)$$

is an isomorphism.

Proof We know that $k(A, X) = k(\underline{\mathrm{Hom}}(A, X))$ is the maximal Kan complex contained in the ∞-category $\underline{\mathrm{Hom}}(A, X)$. On the other hand, we know from Theorem 3.6.1 and Corollary 3.6.4 that the functor $\underline{\mathrm{Hom}}(-, X)$ sends monomorphisms to isofibrations between ∞-categories. We also know from Theorem 3.5.11 that the functor k sends isofibrations between ∞-categories to Kan fibrations. On the other hand, since the functor k is a right adjoint of the inclusion functors of the category of ∞-groupoids into the category of ∞-categories, it preserves the limits which are representable in the category of ∞-categories. In conclusion, the properties listed in this lemma are direct consequences of the fact that the functor $\underline{\mathrm{Hom}}(-, X)$ preserves small limits. □

Theorem 3.9.2 *Let* $f : X \to Y$ *be a functor of* ∞*-categories. The following conditions are equivalent.*

(i) The induced maps $k(\Delta^n, X) \to k(\Delta^n, Y)$ *are equivalences of* ∞*-groupoids for* $n = 0$ *and* $n = 1$.
(ii) The induced maps $k(\Delta^n, X) \to k(\Delta^n, Y)$ *are equivalences of* ∞*-groupoids for all non-negative integers* n.
(iii) The map f *is an equivalence of* ∞*-categories.*

Proof The dual versions of Corollaries 2.3.16, 2.3.18 and 2.3.29 (applied to the opposite of the category of simplicial sets, endowed with the Kan–Quillen model category structure) together with the preceding lemma show that, given such a functor $f : X \to Y$, the class of simplicial sets A such that the map $k(A, X) \to k(A, Y)$ is a homotopy equivalence between Kan complexes is saturated by monomorphisms. Therefore, applying Corollary 1.3.10 (for $A = \mathbf{\Delta}$) and Theorem 3.6.9, we see that condition (ii) is equivalent to condition (iii).

Therefore, it is sufficient to prove that condition (i) implies condition (ii). For $n > 1$, we have coCartesian squares of the form

$$\begin{array}{ccc} \Delta^0 & \longrightarrow & Sp^{n-1} \\ \downarrow & & \downarrow \\ \Delta^1 & \longrightarrow & Sp^n \end{array}$$

(where we put $Sp^1 = \Delta^1$). Hence condition (i) and the first assertion of the preceding lemma imply that, for any integer $n > 1$, the induced map $k(Sp^n, X) \to k(Sp^n, Y)$ is a weak homotopy equivalence. On the other hand, it follows from Proposition 3.7.4 that the inclusions $Sp^n \to \Delta^n$ induce a commutative diagram

$$\begin{array}{ccc} k(\Delta^n, X) & \longrightarrow & k(\Delta^n, Y) \\ \downarrow & & \downarrow \\ k(Sp^n, X) & \longrightarrow & k(Sp^n, Y) \end{array}$$

in which the vertical maps are trivial fibrations (they are Kan fibrations by Theorem 3.5.11, as well as weak homotopy equivalences by Theorem 3.6.8). Hence, condition (ii) is fulfilled. □

Definition 3.9.3 Let $f\colon X \to Y$ be a functor between ∞-categories.

We say that f is *fully faithful* if, for any objects x and y in X, the induced map

$$X(x, y) \to Y(f(x), f(y))$$

is an equivalence of ∞-groupoids.

We say that f is *essentially surjective* if, for any object y in Y, there exists an object x in X as well as an invertible morphism from $f(x)$ to y in Y.

Remark 3.9.4 The explicit description of $\tau(X)$ given by Boardman and Vogt (Theorem 1.6.6) implies that the functor f is essentially surjective if and only if the induced functor $\tau(f)\colon \tau(X) \to \tau(Y)$ is essentially surjective. Proposition 3.7.2 also implies that the functor $\tau(f)$ is fully faithful whenever f has this property.

Example 3.9.5 Given an ∞-category X and a subset $A \subset X_0$, we define the *full subcategory generated by A in X* as the subcomplex $X_A \subset X$ whose simplices are those maps $f\colon \Delta^n \to X$ such that $f(i)$ belongs to A for all i, $0 \leq i \leq n$. If $ho(X)_A$ denotes the full subcategory of $ho(X)$ whose objects are in A, there is a canonical Cartesian square

$$\begin{array}{ccc} X_A & \longrightarrow & X \\ \downarrow & & \downarrow \\ N(ho(X)_A) & \longrightarrow & N(ho(X)) \end{array}$$

in which both vertical maps are conservative isofibrations. It is thus clear that X_A is an ∞-category. Moreover, by construction, for any objects a and b in A, the induced map

$$X_A(a,b) \to X(a,b)$$

is the identity. Therefore, the inclusion map $X_A \to X$ is fully faithful.

Proposition 3.9.6 *A functor between ∞-categories $f\colon X \to Y$ is fully faithful if and only if the induced commutative square of ∞-groupoids*

$$\begin{array}{ccc} k(\Delta^1, X) & \longrightarrow & k(\Delta^1, Y) \\ {\scriptstyle (s,t)}\downarrow & & \downarrow{\scriptstyle (s,t)} \\ k(X) \times k(X) & \longrightarrow & k(Y) \times k(Y) \end{array}$$

is homotopy Cartesian.

Proof Since both vertical maps are Kan fibrations between Kan complexes (Theorem 3.5.11), and since both squares of diagram (3.7.1.2) are Cartesian, this is a particular case of Corollary 3.8.4. □

Theorem 3.9.7 *A functor between ∞-categories is an equivalence of ∞-categories if and only if it is fully faithful and essentially surjective.*

Proof Let $f\colon X \to Y$ be such a functor.

If f is an equivalence of ∞-categories, then the functor $\tau(f)\colon \tau(X) \to \tau(Y)$ is an equivalence of categories (this is implied by Proposition 3.3.14), hence f is essentially surjective. Moreover, the commutative square of Proposition 3.9.6 obviously is homotopy Cartesian because its horizontal maps are weak homotopy equivalences (Theorem 3.6.9), hence f is fully faithful.

Conversely, let us assume that f is fully faithful and essentially surjective. To prove that f is an equivalence of ∞-categories, by virtue of Theorem 3.9.2 and of Proposition 3.9.6, it is sufficient to prove that the induced morphism of Kan complexes $k(X) \to k(Y)$ is a weak homotopy equivalence. We have a bijection $\pi_0(k(X)) \to \pi_0(k(Y))$ because we have an equivalence of categories $\tau(X) \simeq \tau(Y)$ whence a bijection at the level of sets of isomorphism classes of objects. Therefore, by virtue of Corollary 3.8.14, it is sufficient to prove that, for any object x in X, if we put $y = f(x)$, the induced map $\Omega(k(X), x) \to$

$\Omega(k(Y), y)$ is a weak homotopy equivalence. Since f is conservative (because $\tau(f)$ is, being an equivalence of categories), the commutative square

$$\begin{array}{ccc} k(X) & \longrightarrow & k(Y) \\ \downarrow & & \downarrow \\ X & \xrightarrow{f} & Y \end{array}$$

is Cartesian. In particular, we also have a Cartesian square in the category of ∞-groupoids

$$\begin{array}{ccc} \Omega(k(X), x) & \longrightarrow & \Omega(k(Y), y) \\ \downarrow & & \downarrow \\ X(x, x) & \xrightarrow{f} & Y(y, y) \end{array}$$

in which the vertical maps are both monomorphisms and Kan fibrations: the Kan complex $\Omega(k(X), x)$ is the union of the connected components of the Kan complex $X(x, x)$ corresponding to the elements of $\pi_0(X(x, x)) = \mathrm{Hom}_{\tau(X)}(x, x)$ which are the invertible automorphisms of x in the category $\tau(X)$. Since the map $X(x, x) \to Y(y, y)$ is a weak homotopy equivalence, so is its (homotopy) pull-back $\Omega(k(X), x) \to \Omega(k(Y), y)$. □

Corollary 3.9.8 *The class of weak categorical equivalences is stable under small filtered colimits.*

Proof Let us check first that the class of homotopy equivalences between Kan complexes is closed under filtered colimits. This makes sense because the class of Kan complexes is stable under filtered colimits (Lemma 3.1.24). For any pointed finite simplicial set K (such as Δ^1 or $\partial\Delta^1$), the functor $\underline{\mathrm{Hom}}_*(K, -)$ preserves filtered colimits: since filtered colimits commute with finite limits, the pull-back square (3.8.8.1) shows that it is sufficient to prove that the functor $\underline{\mathrm{Hom}}(K, -)$ preserves filtered colimits, which follows from the fact that each product $\Delta^n \times K$ has finitely many non-degenerate simplices, applying Corollary 1.3.12. Therefore, the formation of iterated loop spaces is compatible with filtered colimits. Since the functor π_0 obviously preserves filtered colimits as well, this proves that the formation of the groups $\pi_n(X, x)$ is compatible with filtered colimits. The characterisation of weak equivalences between Kan complexes given by Corollary 3.8.14 thus implies that weak equivalences between Kan complexes are closed under filtered colimits.

The proof of Lemma 3.1.24 also shows that the class of ∞-categories is closed under filtered colimits. Since the functor $\underline{\mathrm{Hom}}(\Delta^1, -)$ preserves filtered colimits and since filtered colimits commute with finite limits in sets, the formation of

∞-groupoids of morphisms $X(x, y)$ is compatible with filtered colimits of ∞-categories. The stability of equivalences of ∞-groupoids by filtered colimits implies the stability of fully faithful functors between ∞-categories by filtered colimits. It is an easy exercise to check that essentially surjective functors are stable under filtered colimits. Therefore, by virtue of Theorem 3.9.7, the class of equivalences of ∞-categories is closed under filtered colimits.

Finally, since the functor $\mathrm{Hom}(K, -)$ commutes with filtered colimits for any finite simplicial set K, the explicit construction of the small object argument shows that there is a functorial inner anodyne extension $X \to L(X)$ such that the functor L takes its values in the category of ∞-categories and commutes with filtered colimits. Therefore, the class of weak categorical equivalences is the class of maps whose image by L is an equivalence of ∞-categories. This implies the theorem. □

4

Presheaves: Externally

The first paragraph of the first section of this chapter is heuristic: it explains why it is natural to see right fibrations $X \to C$ as presheaves over C. The rest of this first section is devoted to the construction of the homotopy theory of right fibrations with fixed codomain C, through the construction of the contravariant model category structure over C. This is achieved as an interpretation of the last section of Chapter 2. This is also the opportunity to introduce one of the most fundamental classes of maps in the theory of ∞-categories: the one of final maps. One then proves that a map between right fibrations over C is a weak equivalence if and only if it induces a fibrewise equivalence of ∞-groupoids. In the second section, using an alternative construction of the join operation, we prove that the homotopy fibre at x of the slice fibration $X/y \to X$ is the mapping space of maps from x to y in the ∞-category X. This is used in the third section to study final objects.

Section 4.4 revisits Quillen's famous theorem A (see Corollary 4.4.32 below), after Grothendieck, Joyal and Lurie, introducing the notions of proper functors and of smooth functors. In particular, it provides useful computational tools, that will ramify in various forms all through the rest of the book. After a technical section on fully faithful and essentially surjective functors through the lenses of the covariant and contravariant model category structures, we finally devote Section 4.6 to Quillen's Theorem B, or, in other words, to locally constant presheaves.

4.1 Catégories Fibrées en ∞-Groupoïdes

4.1.1 Let C be a small category. The category of presheaves of sets over C can be embedded in the category Cat/C of categories over C as follows. For a presheaf $F\colon C^{\mathrm{op}} \to Set$ one considers the category of elements C/F

(see Definition 1.1.7), which comes equipped with a canonical projection map $C/F \to C$. This defines a functor

$$\widehat{C} \to Cat/C\,, \qquad F \mapsto C/F\,. \tag{4.1.1.1}$$

This functor is fully faithful and admits a left adjoint as well as a right adjoint. The left adjoint is easy: it associates to a functor $\varphi\colon I \to C$ the colimit of the functor

$$I \xrightarrow{\varphi} C \xrightarrow{h} \widehat{C}$$

(where h is the Yoneda embedding). The right adjoint is less enlightening (at least for the author of this book) but has the merit of being explicit: it associates to a functor $\varphi\colon I \to C$ the presheaf whose evaluation at c is the set of functors $u\colon C/c \to I$ such that $\varphi u\colon C/c \to C$ is the canonical projection.

Therefore, there is a way to think of presheaves as a full subcategory of the category Cat/C. The aim of this section is to introduce a candidate for an analogous subcategory of the category of ∞-categories over a given ∞-category C. Well, in fact, over *any* object of the Joyal model category structure, i.e. over any simplicial set.

For this, we need to characterise functors isomorphic to functors of the form $C/F \to C$: these are the functors $p\colon X \to C$ such that, for any object x in X, if we put $c = p(x)$, the induced functor $X/x \to C/c$ is an isomorphism. Since the nerve of any functor between small categories is an inner fibration, the following proposition is enlightening.

Proposition 4.1.2 *A morphism of simplicial sets $p\colon X \to C$ is a right fibration if and only if it is an inner fibration such that, for any object x in X, if we put $c = p(x)$, the induced functor $X/x \to C/c$ is a trivial fibration.*

Proof The identification $\Lambda^n_n = \partial\Delta^{n-1} * \Delta^0$, for $n > 0$, and Lemma 3.4.20 show that lifting problems of the form

$$\begin{array}{ccc} \partial\Delta^{n-1} & \longrightarrow & X/x \\ \downarrow & \nearrow & \downarrow \\ \Delta^{n-1} & \longrightarrow & C/c \end{array} \qquad \text{(with } c = p(x)\text{)}$$

correspond to lifting problems of the form

$$\begin{array}{ccc} \Lambda^n_n & \xrightarrow{u} & X \\ \downarrow & \nearrow & \downarrow \\ \Delta^n & \xrightarrow{v} & C \end{array} \qquad \text{(with } u(n) = x\text{)}.$$

Since right fibrations can be characterised as the inner fibrations with the right lifting property with respect to inclusions of the form $\Lambda^n_n \subset \Delta^n$ for $n > 0$, this proves the proposition. □

Corollary 4.1.3 *The nerve of any Grothendieck fibration with discrete fibres between small categories is a right fibration.*

Proof By virtue of the preceding proposition, this follows from the canonical identification $N(C)/x \simeq N(C/x)$ for any small category C equipped with an object x, and from the characterisation of Grothendieck fibrations with discrete fibres recalled at the end of paragraph 4.1.1. □

4.1.4 Let C be a simplicial set. We denote by $P(C)$ the full subcategory of $sSet/C$ whose objects are the right fibrations of the form $p_F : F \to C$. The objects of $P(C)$ will be called the *right fibrant objects over* C. Most of the time, the structural map p_F will be implicitly given, and we shall speak of the right fibrant object F over C. A morphism $F \to G$ of right fibrant objects over C is a *fibrewise equivalence* if, for any object c of C, the induced map $F_c \to G_c$ is an equivalence of ∞-groupoids. Here, F_c is the pull-back of the map $c : \Delta^0 \to C$ along the structural map p_F; such a fibre F_c always is an ∞-groupoid, by Corollary 3.5.6.

Theorem 4.1.5 (Joyal) *There is a unique model category structure on the category $sSet/C$ whose cofibrations are the monomorphisms, and whose fibrant objects are the right fibrant objects over C. Moreover, a morphism between right fibrant objects over C is a fibration if and only if it is a right fibration.*

The proof of the theorem will be given below, after Proposition 4.1.7. This model category structure will be called the *contravariant model category structure over C*.

4.1.6 We observe the following fact: Proposition 3.5.5 implies that, for $C = \Delta^0$, the contravariant model category structure must coincide with the Kan–Quillen model category structure. More generally, when C is an ∞-groupoid, the contravariant model category structure over C must coincide with the model category structure induced by the Kan–Quillen model category structure. This is why, in order to construct the contravariant model category structure over C, we shall forget C temporarily, for studying another presentation of the Kan–Quillen model category structure on the category of simplicial sets.

Let us consider the interval J' defined as the nerve of the contractible groupoid with set of objects $\{0, 1\}$. This is a contractible Kan complex: this is a Kan complex because this is an ∞-groupoid in which all morphisms are invertible, and it is contractible because the map $J' \to \Delta^0$ is a simplicial homotopy equivalence. Therefore, the projection map $J' \times X \to X$ is a trivial fibration for any simplicial set X. The interval J' thus defines an exact cylinder $X \mapsto J' \times X$. Let us consider the smallest class $An^{right}_{J'}$ of J'-anodyne extensions which contains the inclusions of the form $\Lambda^n_k \to \Delta^n$ with $n \geq 1$ and $0 < k \leq n$; see Example 2.4.13. Using Lemma 3.1.3, we see that it coincides with the smallest saturated class of morphisms of simplicial sets containing the following two kinds of maps.

1. Inclusions of the form $J' \times \partial\Delta^n \cup \{\varepsilon\} \times \Delta^n \to J' \times \Delta^n$ for $n \geq 0$ and $\varepsilon = 0, 1$.
2. Inclusions of the form $\Lambda^n_k \to \Delta^n$ with $n \geq 1$ and $0 < k \leq n$.

However, according to the next proposition, this description can be dramatically simplified.

Proposition 4.1.7 *The class $An^{right}_{J'}$ is the class of right anodyne extensions.*

Proof It is clear that any right anodyne extension belongs to $An^{right}_{J'}$. To prove the converse, it is sufficient to prove that any right fibration has the right lifting property with respect to the generators of the class $An^{right}_{J'}$. These generators are all of the form $i: A \to B$, where B is the nerve of a small category (in particular, an ∞-category). Let $p: X \to Y$ be a right fibration. To solve lifting problems of the form

$$\begin{array}{ccc} A & \xrightarrow{a} & X \\ {\scriptstyle i}\downarrow & & \downarrow{\scriptstyle p} \\ B & \xrightarrow{b} & Y \end{array}$$

we may factor the given commutative square into

$$\begin{array}{ccccc} A & \xrightarrow{\alpha} & B \times_Y X & \xrightarrow{\alpha} & X \\ {\scriptstyle i}\downarrow & & \downarrow{\scriptstyle q} & & \downarrow{\scriptstyle p} \\ B & = & B & \xrightarrow{b} & Y \end{array}$$

(where the right-hand square is Cartesian). The map q is now a right fibration whose codomain is an ∞-category. In other words, it is sufficient to prove that any right fibration whose codomain is an ∞-category has the right lifting property with respect to the generators of the class $An^{right}_{J'}$. And here is a final translation of the latter problem: we must prove that, for any right fibration $p: X \to Y$ with Y an ∞-category, the evaluation at ε map

$$\underline{\mathrm{Hom}}(J', X) \to X \times_Y \underline{\mathrm{Hom}}(J', Y)$$

is a trivial fibration for $\varepsilon = 0, 1$. But such morphism p is always an isofibration between ∞-categories (Proposition 3.4.8) hence a fibration of the Joyal model category structure (Theorem 3.6.1). In particular, it must have the required lifting property: since J' is an ∞-groupoid, we have $h(J', C) = \underline{\mathrm{Hom}}(J', C)$ for any ∞-category C, so that we can apply Corollary 3.5.13. □

Proof of Theorem 4.1.5 Thanks to Proposition 4.1.7, this is a particular case of the construction of paragraph 2.5.1, which allows one to apply Theorem 2.4.19 for $A = \Delta/C$. □

In fact, Proposition 4.1.7 has several other consequences.

Definition 4.1.8 A morphism of simplicial sets $u\colon A \to B$ is *final* if, for any simplicial set C and any morphism $p\colon B \to C$, the map $u\colon (A, pu) \to (B, p)$ is a weak equivalence of the contravariant model category structure over C.

Corollary 4.1.9 *A monomorphism of simplicial sets is a right anodyne extension if and only if it is a final map. A morphism of simplicial sets is final if and only if it admits a factorisation into a right anodyne extension followed by a trivial fibration.*

The class of final maps is the smallest class C *of morphisms of simplicial sets satisfying the following properties:*

(a) the class C *is closed under composition;*
(b) for any pair of composable morphisms $f\colon X \to Y$ *and* $g\colon Y \to Z$*, if both* f *and* gf *are in* C*, so is* g*;*
(c) any right anodyne extension belongs to C*.*

Proof This is a particular case of Propositions 2.5.3 and 2.5.4. □

Corollary 4.1.10 *Let us consider a pair of composable monomorphisms* $f\colon X \to Y$ *and* $g\colon Y \to Z$*, and assume that* f *is a right anodyne extension. Then the map* g *is a right anodyne extension if and only if the map* gf *has the same property.*

A useful sufficient condition for being final is given by the following statement.

Proposition 4.1.11 *Let* $f\colon X \to Y$ *be a morphism of simplicial sets, and* $p\colon Y \to C$ *a right fibration. Then* f *is final if and only if the map* p *turns* f *into a weak equivalence of the contravariant model category structure over* C*.*

Proof This is a special case of Proposition 2.5.6. □

4.1.12 Let $p_F : F \to C$ and $p_G : G \to C$ be two morphisms of simplicial sets. We define $\mathrm{Map}_C(F, G)$ by forming the following Cartesian square.

$$\begin{array}{ccc} \mathrm{Map}_C(F,G) & \hookrightarrow & \underline{\mathrm{Hom}}(F,G) \\ \downarrow & & \downarrow{\scriptstyle (p_G)_*} \\ \Delta^0 & \xrightarrow{p_F} & \underline{\mathrm{Hom}}(F,C) \end{array} \tag{4.1.12.1}$$

If ever G is right fibrant over C, then $\mathrm{Map}_C(F, G)$ is the fibre of a right fibration (by Proposition 3.4.5), and therefore is a Kan complex, by Corollary 3.5.6. In the case of an object c of C, if we write $h(c)$ for the image of the corresponding morphism $c\colon \Delta^0 \to C$, the canonical isomorphism $\underline{\mathrm{Hom}}(\Delta^0, X) \simeq X$ induces a canonical isomorphism

$$\mathrm{Map}_C(h(c), G) \simeq G_c. \tag{4.1.12.2}$$

Proposition 4.1.13 *For any monomorphism $F \to F'$ of simplicial sets over C, and for any right fibrant object G over C, the induced map*

$$\mathrm{Map}_C(F, G) \to \mathrm{Map}_C(F', G)$$

is a Kan fibration.

Proof Since the source and target of this map are Kan complexes, by virtue of Proposition 3.5.5 it is sufficient to prove that it is a right Kan fibration. Since right fibrations are stable under base change, by virtue of Proposition 3.4.5, the Cartesian square

$$\begin{array}{ccc} \mathrm{Map}_C(F,G) & \hookrightarrow & \underline{\mathrm{Hom}}(F,G) \\ \downarrow & & \downarrow \\ \mathrm{Map}_C(F',G) & \hookrightarrow & \underline{\mathrm{Hom}}(F',G) \times_{\underline{\mathrm{Hom}}(F',C)} \underline{\mathrm{Hom}}(F,C) \end{array} \tag{4.1.13.1}$$

thus implies this proposition. □

Proposition 4.1.14 *For any weak equivalence of the contravariant model category structure $F \to F'$ over C, and for any right fibrant object G over C, the induced map*

$$\mathrm{Map}_C(F, G) \to \mathrm{Map}_C(F', G)$$

is an equivalence of ∞-groupoids.

Proof By a variation on the proof of the preceding proposition, one checks that, for any right fibrant object G over C, the functor $\mathrm{Map}_C(-, G)$ sends right anodyne extensions to trivial fibrations. By virtue of Propositions 2.4.40 and 4.1.7, this implies that the functor $\mathrm{Map}_C(-, G)$ is a left Quillen functor from the contravariant model category structure to the opposite of the Kan–Quillen model category structure. In particular, this functor preserves weak equivalences between cofibrant objects (which are all simplicial sets over C), whence the proposition. □

Lemma 4.1.15 *For any integer $n \geq 0$, the inclusion $\{n\} \to \Delta^n$ is a right anodyne extension.*

Proof Let $h \colon \Delta^1 \times \Delta^n \to \Delta^n$ be the morphism induced by the map

$$(\varepsilon, x) \to \begin{cases} n & \text{if } \varepsilon = 1, \\ x & \text{otherwise.} \end{cases}$$

Identifying Δ^n with $\{0\} \times \Delta^n$, we have the following commutative diagram

$$\begin{array}{ccccc} \{n\} & \longrightarrow & \Delta^1 \times \{n\} \cup \{1\} \times \Delta^n & \longrightarrow & \{n\} \\ \downarrow & & \downarrow & & \downarrow \\ \Delta^n & \longrightarrow & \Delta^1 \times \Delta^n & \xrightarrow{h} & \Delta^n \end{array}$$

which turns the inclusion $\{n\} \to \Delta^n$ into a retract of the right anodyne extension $\Delta^1 \times \{n\} \cup \{1\} \times \Delta^n \to \Delta^1 \times \Delta^n$ (see Proposition 3.4.3). This proves the lemma. □

Theorem 4.1.16 *Let $\varphi \colon F \to G$ be a morphism between right fibrant objects over C. The following conditions are equivalent.*

(i) The morphism φ is a weak equivalence of the contravariant model category structure.
(ii) The morphism φ is a fibrewise equivalence.
(iii) For any simplicial set X over C, the induced map

$$\mathrm{Map}_C(X, F) \to \mathrm{Map}_C(X, G)$$

is an equivalence of ∞-groupoids.

Proof By virtue of the identification (4.1.12.2), it is clear that condition (iii) implies condition (ii). The Cartesian square (4.1.13.1) shows that the functors $\mathrm{Map}_C(X, -)$ preserve trivial fibrations, whence weak equivalences between fibrant objects. In other words, condition (i) implies condition (iii). One deduces from Propositions 4.1.13 and 4.1.14 that, for any simplicial set X over C and any right fibrant object F over C, there is a canonical identification between the set $\pi_0(\mathrm{Map}_C(X, F))$ and the set of homotopy classes of maps from X to F: this is because the functor $\mathrm{Map}_C(-, F)$ will send any cylinder of X (in the sense of the contravariant model category structure) into a path object of the

Kan complex $\operatorname{Map}_C(X, F)$. This implies that condition (iii) implies condition (i). It now remains to prove that condition (ii) implies condition (iii). For any integer $n \geq 0$, by virtue of the preceding lemma and of Proposition 4.1.14, for any map $s\colon \Delta^n \to C$, there is a commutative square of the form

$$\begin{array}{ccc} \operatorname{Map}_C((\Delta^n, s), F) & \longrightarrow & F_{s(n)} \\ {\scriptstyle \varphi_*}\downarrow & & \downarrow{\scriptstyle \varphi_{s(n)}} \\ \operatorname{Map}_C((\Delta^n, s), G) & \longrightarrow & G_{s(n)} \end{array}$$

in which the horizontal maps are equivalences of ∞-groupoids. Therefore, condition (ii) is equivalent to the condition that, for any map $s\colon \Delta^n \to C$, the induced morphism

$$\varphi_*\colon \operatorname{Map}_C((\Delta^n, s), F) \to \operatorname{Map}_C((\Delta^n, s), G)$$

is an equivalence of ∞-groupoids. On the other hand, since the functors of the form $\operatorname{Map}_C(-, F)$ send small colimits of simplicial sets over C to limits of simplicial sets, Proposition 4.1.13 and Corollaries 2.3.16, 2.3.18 and 2.3.29 (all three applied to the opposite of the Kan–Quillen model category) imply that the class of simplicial sets X over C such that the induced map

$$\operatorname{Map}_C(X, F) \to \operatorname{Map}_C(X, G)$$

is an equivalence of ∞-groupoids is saturated by monomorphisms. Applying Corollary 1.3.10 in the case of $A = \mathbf{\Delta}/C$ thus proves that condition (ii) implies condition (iii). □

Corollary 4.1.17 *The class of weak equivalences of the contravariant model category structure over C is closed under small filtered colimits.*

Proof The proof of Lemma 3.1.24 shows that the class of right fibrations is closed under small filtered colimits, and that the fibrant resolution functor of the contravariant model category structure over C preserves small filtered colimits. Therefore, it is sufficient to prove that the class of weak equivalences between fibrant objects of the contravariant model category structure over C is closed under small filtered colimits. By virtue of the preceding theorem, this amounts to prove that fibrewise equivalences are closed under filtered colimits. Since filtered colimits are exact, they are compatible with the formation of fibres of right fibrations. Finally, we simply have to check that the class of equivalences of ∞-groupoids is closed under small filtered colimits, which follows right away from Corollary 3.9.8 (and was in fact used in its proof). □

Remark 4.1.18 In the case where C is an ∞-category, if $p\colon X \to C$ and

$q\colon Y \to C$ are right fibrations, a map $f\colon X \to Y$ over C is a fibrewise equivalence if and only if it is an equivalence of ∞-categories (or, equivalently, a weak categorical equivalence). Indeed, by virtue of general computations in model categories such as Lemma 2.2.12 and Theorem 2.2.17, these properties of f are also equivalent to the assertion that f is a J'-homotopy equivalence (over C).

4.2 Mapping Spaces as Fibres of Slices

4.2.1 For two simplicial sets X and Y, one defines $X \diamond Y$ by the following push-out square:

$$\begin{array}{ccc} X \times \partial\Delta^1 \times Y & \longrightarrow & X \amalg Y \\ \downarrow & & \downarrow \\ X \times \Delta^1 \times Y & \longrightarrow & X \diamond Y \end{array} \qquad (4.2.1.1)$$

where the upper horizontal map is the disjoint union of the two canonical projections $X \times Y \to X$ and $X \times Y \to Y$ interpreted through the identification

$$X \times \partial\Delta^1 \times Y \simeq (X \times Y) \amalg (X \times Y)\,.$$

One checks that the functor $sSet \to Y\backslash sSet$, $X \mapsto (Y \to X \diamond Y)$ has a right adjoint $Y\backslash sSet \to sSet$, $(t\colon Y \to W) \mapsto W//t$. Indeed, the functor $(-) \diamond Y$ preserves connected colimits (because one directly checks that it preserves push-outs as well as filtered colimits). The object $W//t$ has to be distinguished from the slice W/t that was introduced earlier (see paragraph 3.4.14). However, this chapter is all about comparing them. If there is no ambiguity on t, we will also write $W//T = W//t$.

We remark that we have $\Delta^1 = \Delta^0 \diamond \Delta^0 = \Delta^0 * \Delta^0$, so that there are canonical maps $X \diamond Y \to \Delta^1$ and $X * Y \to \Delta^1$. For either of these maps, the fibre at 0 is X, and the fibre at 1 is Y.

Proposition 4.2.2 *There is a unique natural map $\gamma_{X,Y}\colon X \diamond Y \to X * Y$ such that the diagram below commutes.*

$$\begin{array}{ccc} X \amalg Y & \longrightarrow & X * Y \\ \downarrow & \overset{\gamma_{X,Y}}{\nearrow} & \downarrow \\ X \diamond Y & \longrightarrow & \Delta^1 \end{array}$$

Proof Recall that, for $n \geq 0$, we have

$$(X * Y)_n = \coprod_{i+1+j=n} X_i \times Y_j\,.$$

A map $\Delta^n \to X \times \Delta^1 \times Y$ consists of elements $x \in X_n$ and $y \in Y_n$ together with an integer i, which correspond to the map $u\colon \Delta^n \to \Delta^1$ (the integer i is the smallest integer such that $u(i+1) = 1$). There is a unique $\xi\colon \Delta^i \to X$ which is the restriction of x to Δ^i, as well as a unique $\eta\colon \Delta^j \to Y$, with $j = n - i - 1$, which corresponds to the restriction of y to $\Delta^{\{i+1,\dots,n\}} \simeq \Delta^j$. One associates to such a triple (x, i, y) the element (ξ, η). This defines a commutative square

$$\begin{array}{ccc} X \times \partial\Delta^1 \times Y & \longrightarrow & X \amalg Y \\ \downarrow & & \downarrow \\ X \times \Delta^1 \times Y & \xrightarrow{c_{X,Y}} & X * Y \end{array}$$

over Δ^1, and thus a map $\gamma_{X,Y}\colon X \diamond Y \to X * Y$. This proves the existence. The uniqueness is clear. □

Proposition 4.2.3 (Joyal) *The map $X \diamond Y \to X * Y$ is a weak categorical equivalence.*

Proof The map $\gamma_{X,Y}$ is the identity whenever X or Y is empty. Since the functors $(-)\diamond Y$ and $(-)*Y$ preserve connected colimits as well as monomorphisms, we deduce from Corollaries 2.3.16, 2.3.18 and 2.3.29 that the class of simplicial sets X such that the map $\gamma_{X,Y}$ is a weak categorical equivalence is saturated by monomorphisms; similarly for the class of sets Y such that $\gamma_{X,Y}$ is a weak categorical equivalence. By Corollary 1.3.10, it is thus sufficient to prove the case where $X = \Delta^m$ and $Y = \Delta^n$. In this case, there is a section s of the map $c = c_{\Delta^m,\Delta^n}\colon \Delta^m \times \Delta^1 \times \Delta^n \to \Delta^m * \Delta^n = \Delta^{m+1+n}$ above Δ^1, defined as follows:

$$s(x) = \begin{cases} (x, 0, 0) & \text{if } x \le m, \\ (m, 1, x - m - 1) & \text{if } x > m. \end{cases}$$

This defines a section σ of the map $\gamma = \gamma_{\Delta^m,\Delta^n}$. It is sufficient to prove that $\sigma\gamma$ is a weak categorical equivalence.

For this purpose, we also define a morphism

$$p\colon \Delta^m \times \Delta^1 \times \Delta^n \to \Delta^m \times \Delta^1 \times \Delta^n$$

by the formula $p(x, u, y) = (x, u, uy)$. There is a natural transformation from p to the identity and a natural transformation from p to sc. The map p induces a map $\pi\colon X \diamond Y \to X \diamond Y$, and in $\underline{\mathrm{Hom}}(\Delta^m \diamond \Delta^n, \Delta^m \diamond \Delta^n)$, we have a morphism h from π to the identity and a morphism k from π to $\sigma\gamma$. These natural transformations are the identity on the image of s. Therefore, after applying the functor τ, they are equal to a single invertible natural transformation from $\tau(\pi)$ to the identity (which is equal to $\tau(\sigma)\tau(\gamma)$ because $\tau(\gamma)$ is an isomorphism). Therefore, for any ∞-category C and any functor $f\colon \Delta^m \diamond \Delta^n \to C$, there are natural transformations $fh\colon f\pi \to f$ and $fk\colon \pi \to \sigma\gamma$; both fh and fk are in fact invertible natural transformations because it is sufficient to check this after composing with the canonical conservative functor $C \to N(\tau(C))$. Let $f\colon \Delta^m \diamond \Delta^n \to C$ be a weak categorical equivalence with C an ∞-category. Then h and k induce invertible morphisms from $f\pi$ to f, and from $f\pi$ to $f\sigma\gamma$, respectively, in the ∞-category $\underline{\mathrm{Hom}}(\Delta^m \diamond \Delta^n, C)$. Therefore, the maps $f\pi$, f and $f\sigma\gamma$ must be J-homotopic (see Remark 3.6.5). Since f is a weak categorical equivalence, this proves that $\sigma\gamma$ is a weak categorical equivalence. □

Proposition 4.2.4 *The functors $X \diamond (-)$ and $(-) \diamond Y$ preserve weak categorical equivalences.*

Proof This follows right away from Corollary 2.3.29, applied to the coCartesian squares of the form (4.2.1.1). □

Corollary 4.2.5 *The functors $X * (-)$ and $(-) * Y$ preserve weak categorical equivalences.*

Proof This follows right away from the preceding two propositions. □

Remark 4.2.6 One can prove the preceding corollary without using Proposition 4.2.2. Here is another proof. By Ken Brown's lemma, it is sufficient to prove that these functors preserve trivial cofibrations of the Joyal model structure. Let $K \to L$ be such a trivial cofibration. To check that $K * Y \to L * Y$ is a trivial cofibration, it is sufficient to prove that it has the right lifting property with respect to isofibrations between ∞-categories (see Proposition 2.4.30). But, for such an isofibration between ∞-categories $E \to B$, for any map $Y \to E$, the induced map $E/Y \to E \times_B B/Y$ is a right fibration between ∞-categories (Theorem 3.4.22) whence an isofibration (Proposition 3.4.8). Since such an isofibration has the right lifting property with respect to the trivial cofibrations of the Joyal model category structure (Theorem 3.6.1), we conclude from Lemma 3.4.20 that the map $K * Y \to L * Y$ has the left lifting property with respect to $E \to B$.

The case of the functor $X * (-)$ is deduced from the case of $(-) * X^{\mathrm{op}}$.

Proposition 4.2.7 *Let $p\colon X \to Y$ be a fibration of the Joyal model category (e.g. an isofibration between ∞-categories). We consider a monomorphism $S \to T$, together with a morphism $t\colon T \to X$. Then the canonical projection*

$$X//T \to X//S \times_{Y//S} Y//T$$

is a right fibration. Furthermore, if ever Y is an ∞-category, then so are $X//T$ as well as the fibre product $X//S \times_{Y//S} Y//T$.

Proof As in the proof of Theorem 3.4.22, it is sufficient to prove that, for any right anodyne extension $U \to V$, the induced inclusion

$$V \diamond S \cup U \diamond T = V \diamond S \amalg_{U \diamond S} U \diamond T \to V \diamond T$$

is a trivial cofibration of the Joyal model category structure. But we have the commutative square

$$\begin{array}{ccc} V \diamond S \cup U \diamond T & \hookrightarrow & V \diamond T \\ \downarrow & & \downarrow \\ V * S \cup U * T & \hookrightarrow & V * T \end{array}$$

of which the vertical maps are weak categorical equivalences, by a repeated use of Proposition 4.2.3 (and by Corollary 2.3.29). Therefore, it is sufficient to prove that the lower horizontal map of the commutative square above is a trivial cofibration of the Joyal model category structure. But, by virtue of Lemma 3.4.20 and Theorem 3.4.22, we know that it has the left lifting property with respect to any inner fibration (hence with respect to any fibration of the Joyal model category structure). □

4.2.8 For Y fixed, the comparison map γ_X of Proposition 4.2.2 is a natural transformation between left adjoints and thus induces a natural transformation of their right adjoints

$$X/T \to X//T\,.$$

Proposition 4.2.9 *For any ∞-category X and any map $t\colon T \to X$, the canonical comparison map $X/T \to X//T$ is an equivalence of ∞-categories.*

Proof Both functors $(-) \diamond T$ and $(-) * T$ are left Quillen functors (they preserve monomorphisms as well as weak equivalences by Proposition 4.2.4 and Corollary 4.2.5) and the natural transformation $\gamma_{(-),T}$ is a termwise weak equivalence, by Proposition 4.2.3. Therefore, their right adjoints are right Quillen functors and the corresponding transposed map from $(-)/T$ to $(-)//T$ is a weak equivalence on fibrant objects. This is because the total left derived functors of $(-) \diamond T$ and $(-) * T$ are isomorphic, so that their right adjoints, the total right derived functors of $(-)/T$ and $(-)//T$, must be isomorphic as well. □

Corollary 4.2.10 *Let X be an ∞-category, and $x, y\colon \Delta^0 \to X$ two objects. There is a canonical equivalence of ∞-groupoids from the fibre of the right fibration $X/y \to X$ over x to the ∞-groupoid of maps $X(x, y)$.*

Proof Since both maps $X/y \to X$ and $X//y \to X$ are right fibrations, by virtue of Theorem 4.1.16, the weak equivalence of Proposition 4.2.9 induces equivalences of ∞-groupoids on their fibres. Therefore, we can replace X/y by $X//y$. By definition, the simplicial set $X//y$ fits in the Cartesian square below.

$$\begin{array}{ccc} X//y & \hookrightarrow & \underline{\mathrm{Hom}}(\Delta^1, X) \\ \downarrow & & \downarrow{\scriptstyle ev_1} \\ \Delta^0 & \longrightarrow & X \end{array}$$

Since the square

$$\begin{array}{ccc} X & \xrightarrow{(1_X, y)} & X \times X \\ \downarrow & & \downarrow{\scriptstyle pr_2} \\ \Delta^0 & \xrightarrow{y} & X \end{array}$$

is Cartesian, we thus have a Cartesian square of the following form.

$$\begin{array}{ccc} X//y & \hookrightarrow & \underline{\mathrm{Hom}}(\Delta^1, X) \\ \downarrow & & \downarrow \\ X & \longrightarrow & X \times X \end{array}$$

This immediately implies, by construction of $X(x, y)$, that the fibre at x of the right fibration $X//Y \to X$ is $X(x, y)$. □

Remark 4.2.11 In his monograph [Lur09], Lurie denotes by $\mathrm{Hom}^L_A(x, y)$ and $\mathrm{Hom}_A(x, y)$ the fibres at x of the canonical projections $A/y \to A$ and $A//y \to A$, respectively. However, we will use the notation $\mathrm{Hom}_A(x, y)$ rather differently: for the fully functorial version of the ∞-groupoid of maps from x to y; see paragraph 5.8.1 below.

Proposition 4.2.12 *Let X be an ∞-category endowed with an object x. For any simplicial set A, the canonical functor $\underline{\mathrm{Hom}}(A, X/x) \to \underline{\mathrm{Hom}}(A, X)/x$ is an equivalence of ∞-categories (where we also denote by x the constant functor $A \to X$ with value x).*

Proof The canonical functor $\underline{\mathrm{Hom}}(A, X/x) \to \underline{\mathrm{Hom}}(A, X)$ sends 1_x to x and thus induces a canonical functor $\underline{\mathrm{Hom}}(A, X/x) \to \underline{\mathrm{Hom}}(A, X)/x$. We have a commutative square

$$\begin{array}{ccc} \underline{\mathrm{Hom}}(A, X/x) & \longrightarrow & \underline{\mathrm{Hom}}(A, X)/x \\ \downarrow & & \downarrow \\ \underline{\mathrm{Hom}}(A, X//x) & \longrightarrow & \underline{\mathrm{Hom}}(A, X)//x \end{array}$$

whose vertical maps are equivalences of ∞-categories (Theorem 3.6.9 and Proposition 4.2.9). But we also have the Cartesian square

$$\begin{array}{ccc} X//x & \hookrightarrow & \underline{\mathrm{Hom}}(\Delta^1, X) \\ \downarrow & & \downarrow{\scriptstyle ev_1} \\ \Delta^0 & \longrightarrow & X \end{array}$$

and since $\underline{\mathrm{Hom}}(A, \underline{\mathrm{Hom}}(B, X)) \simeq \underline{\mathrm{Hom}}(B, \underline{\mathrm{Hom}}(A, X))$ for all simplicial sets A and B, we conclude that the canonical map

$$\underline{\mathrm{Hom}}(A, X//x) \to \underline{\mathrm{Hom}}(A, X)//x$$

is an isomorphism. □

4.3 Final Objects

Definition 4.3.1 An object x of a simplicial set X is *final* if the corresponding morphism $x\colon \Delta^0 \to X$ is final (see Definition 4.1.8).

Remark 4.3.2 An object x of X is final if and only if the map $x\colon \Delta^0 \to X$ is a right anodyne extension (this is a particular case of the first assertion of Corollary 4.1.9).

Proposition 4.3.3 *Let $f\colon X \to Y$ be a morphism of simplicial sets, and assume that X has a final object x. Then the morphism f is final if and only if the object $f(x)$ is final in Y.*

Proof This is a special case of properties (a) and (b) of Corollary 4.1.9. □

4.3.4 When we see the category of simplicial sets as a monoidal category with the join operation $X * Y$, it acts naturally on the category of pointed simplicial sets as follows. Given a pointed simplicial set (X, x) and a simplicial set S, we define the pointed simplicial set $X *_x S$ by forming the push-out square below.

$$\begin{array}{ccc} \Delta^0 * S & \longrightarrow & \Delta^0 \\ {\scriptstyle x*1_S}\downarrow & & \downarrow{\scriptstyle x'} \\ X * S & \longrightarrow & X *_x S \end{array}$$

The fact that this actually defines an action (in particular, that this operation is associative up to a coherent isomorphism) follows from the property of the join operation $(-) * S$ of commuting with connected colimits.

We define

$$C(X) = X *_x \Delta^0 \,.$$

Lemma 4.3.5 *For any pointed simplicial set (X, x), the object x' is final in $C(X)$.*

Proof The map $x' \colon \Delta^0 \to C(X)$ is the push-out of the map of joins $x * 1_{\Delta^0} \colon \Delta^0 * \Delta^0 \to X * \Delta^0$. It is thus sufficient to prove that the latter is a right anodyne extension. By virtue of Proposition 4.1.2, an inner fibration $p \colon A \to B$ is a right fibration if and only if, for any object a in A, the induced map $A/a \to B/p(a)$ is a trivial fibration. In particular, for any monomorphism $i \colon X' \to X$, the induced inclusion from $X' * \Delta^0$ into $X * \Delta^0$ is a right anodyne extension. The particular case where $X' = \Delta^0$ thus proves the lemma. □

4.3.6 For a pointed simplicial set (Y, y), the map $1_y \colon \Delta^1 = \Delta^0 * \Delta^0 \to Y$ defines an object $1_y \colon \Delta^0 \to Y/y$. The association $(Y, y) \to (Y/y, 1_y)$ is functorial and actually defines a right adjoint to the functor C. In other words, pointed maps $X \to Y/y$ correspond to pointed maps $C(X) \to Y$.

Proposition 4.3.7 *Let X be a simplicial set, and x an object of X. If the canonical map $X/x \to X$ has a section s such that $s(x) = 1_x$, then x is a final object of X. In the case where X is an ∞-category, the converse is true.*

Proof If there is such a section, by transposition, the inclusion $X \subset C(X)$ has a retraction $r \colon C(X) \to X$ such that $r(x') = x$. This means that the map $x \colon \Delta^0 \to X$ is a retract of the final map $x' \colon \Delta^0 \to C(X)$, whence that x is a final object.

If X is an ∞-category and if x is a final object of X, then the commutative square

$$\begin{array}{ccc} \Delta^0 & \xrightarrow{1_x} & X/x \\ {\scriptstyle x}\downarrow & & \downarrow \\ X & = & X \end{array}$$

has a filler because $X/x \to X$ is a right fibration, while the left-hand vertical map is a right anodyne extension (see Remark 4.3.2). □

Corollary 4.3.8 *For any object x in a simplicial set X, the object 1_x is final in the slice X/x.*

Proof The object Δ^0 is a unital associative monoid with respect to the tensor product defined by the join operation: the unit is $\varnothing \to \Delta^1$, while the multiplication is the unique map $\Delta^1 \to \Delta^0$. Since the operation $X *_x S$ defines a unital and associative action of simplicial sets, this implies that the functor C has a natural structure of a monad. By transposition, its right adjoint has a structure of a comonad. The map $(X/x)/1_x \to X/x$ is the comultiplication of this comonad, and thus has a pointed section s (i.e. a section such that $s(1_x) = 1_{1_x}$). Therefore, the preceding proposition implies that 1_x is a final object in X/x. □

We just proved the following important theorem (one of the many incarnations of the Yoneda lemma).

Theorem 4.3.9 *Let x be an object in an ∞-category X. The map $\Delta^0 \to X$, seen as an object of $sSet/X$, has a canonical fibrant replacement in the contravariant model category structure over X, namely the canonical right fibration $X/x \to X$. In particular, if y is another object of X, there is a canonical equivalence of ∞-groupoids*

$$\mathrm{Map}_X(X/x, X/y) \to X(x, y).$$

Proof The first assertion directly follows from the previous corollary. This implies that the map $1_x \colon \Delta^0 \to X/x$ induces a trivial fibration from the Kan complex $\mathrm{Map}_X(X/x, X/y)$ to the fibre of $X/y \to X$ at x (see Propositions 4.1.13 and 4.1.14). We conclude with Corollary 4.2.10. □

Proposition 4.3.10 *Let X be an ∞-category, and ω an object of X. We assume that there is a natural transformation a from the identity of X to the constant functor with value ω, such that the induced morphism $a_\omega \colon \omega \to \omega$ is the identity in $\mathrm{ho}(X)$. Then ω is a final object of X.*

Proof The natural transformation a determines a homotopy

$$h \colon \Delta^1 \times X \to X.$$

We can modify h so that its restriction to $\Delta^1 \times \{\omega\}$ is constant with value ω. Indeed, by assumption on a, the restriction of h on $\Delta^1 \simeq \Delta^1 \times \{\omega\}$ is the identity of ω in $\mathrm{ho}(X)$. This means that there exists a map $\mu \colon \Delta^2 \to X$ whose boundary is a triangle of the following shape.

$$\begin{array}{ccc} & \omega & \\ {}^{1_\omega}\nearrow & & \searrow{}^{a_\omega} \\ \omega & \xrightarrow[1_\omega]{} & \omega \end{array}$$

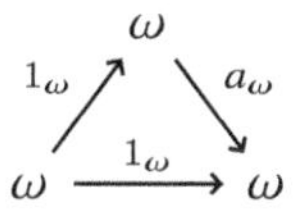

We consider the map

$$\tilde{h} \colon \Lambda^2_1 \times X \to X$$

whose restriction to $\Delta^{\{0,1\}} \times X$ is the constant homotopy with value 1_X and whose restriction to $\Delta^{\{1,2\}} \times X$ is h. This defines a map

$$(\mu, \tilde{h}) \colon \Delta^2 \times \{\omega\} \cup \Lambda^2_1 \times X \to X$$

which is the restriction of some map of the form

$$k\colon \Delta^2 \times X \to X$$

(because the inclusion of $\Delta^2 \times \{\omega\} \cup \Lambda^2_1 \times X$ into the product $\Delta^2 \times X$ is inner anodyne). The restriction of k to $\Delta^{\{0,2\}} \times X$ thus defines a homotopy

$$H\colon \Delta^1 \times X \to X$$

from the identity of X to the constant map with value ω, whose restriction on $\Delta^1 \times \{\omega\}$ is constant with value ω. In other words, the map H defines by transposition a map

$$\tilde{H}\colon X \to X//\omega \subset \underline{\mathrm{Hom}}(\Delta^1, X)$$

which is a section of the canonical projection $X//\omega \to X$ and which sends ω to 1_ω. We observe that the canonical equivalence (whence final functor) from X/ω to $X//\omega$ provided by Proposition 4.2.9 sends the final object of X/ω to the identity of ω. In particular, 1_ω is a final object of $X//\omega$. The map $\omega\colon \Delta^0 \to X$ being a retract of the right anodyne extension $1_\omega\colon \Delta^0 \to X//\omega$, it is a right anodyne extension as well. Therefore, ω is a final object of X. □

Theorem 4.3.11 (Joyal) *Let X be an ∞-category. We consider an object ω in X, and write $\pi\colon X/\omega \to X$ for the canonical projection. The following conditions are equivalent.*

(i) The object ω is final in X.
(ii) For any object x in X, the ∞-groupoid of maps $X(x, \omega)$ is contractible.
(iii) The map $\pi\colon X/\omega \to X$ is a trivial fibration.
(iv) The map $\pi\colon X/\omega \to X$ is an equivalence of ∞-categories.
(v) The map $\pi\colon X/\omega \to X$ has a section which sends ω to 1_ω.
(vi) Any morphism $\partial\Delta^n \to X$, such that $n > 0$ and $u(n) = \omega$, is the restriction of a morphism $v\colon \Delta^n \to X$.

Proof We already know that conditions (i) and (v) are equivalent, by Proposition 4.3.7. Since π is a fibration between fibrant objects of the Joyal model category structure, it is a weak equivalence of this model structure if and only if it is a trivial fibration. Therefore, conditions (iii) and (iv) are equivalent. On the other hand, the map π is also a fibration between fibrant objects of the contravariant model category structure over X. Hence it is a trivial fibration if and only if it is a weak equivalence of the contravariant model category structure. By virtue of Theorem 4.1.16 and of Corollary 4.2.10, conditions (ii) and (iii) are equivalent. Condition (vi) is equivalent to condition (iii): this is because $\partial\Delta^n = \Delta^{n-1} * \varnothing \cup \partial\Delta^{n-1} * \Delta^0$, using Lemma 3.4.20.

If π is a trivial fibration, then it is final, and since final maps are stable by composition, the map $\omega\colon \Delta^0 \to X$ is final, by Corollary 4.3.8. Conversely, if ω is a final object, then condition (vi) holds, which implies that the map π has a section which sends ω to a final object (namely 1_ω, by Corollary 4.3.8). By virtue of Proposition 4.3.3, such a section is a right anodyne extension. This means that π is a weak equivalence of the contravariant model category structure over X, hence a trivial fibration as well. Therefore, conditions (i) and (iii) are equivalent. □

Corollary 4.3.12 *If ω is a final object in an ∞-category X, then it is a final object in the category $\tau(X)$.*

Proof This follows right away from Theorem 4.3.11(ii) and from Proposition 3.7.2. □

Corollary 4.3.13 *The final objects of an ∞-category X form an ∞-groupoid which is either empty or equivalent to the point.*

Proof Let F be the full subcategory of X which consists of final objects: this is the subobject of the simplicial set X whose simplices are the maps $s\colon \Delta^n \to X$ such that $s(i)$ is a final object for $0 \le i \le n$. Theorem 4.3.11(ii) says that the map $F \to \Delta^0$ is fully faithful. Therefore, if F is not empty, such a functor is an equivalence of ∞-categories. □

Corollary 4.3.14 *Let x be a final object in an ∞-category X. For any simplicial set A, the constant functor $A \to X$ with value x is a final object of $\underline{\mathrm{Hom}}(A, X)$.*

Proof This follows from Proposition 4.2.12, by using the characterisation of final objects given by Theorem 4.3.11(iv). □

Lemma 4.3.15 *Let X be a simplicial set such that, for any finite partially ordered set E, any map $N(E) \to X$ is Δ^1-homotopic to a constant map. Then the map $X \to \Delta^0$ is a weak homotopy equivalence.*

Proof It follows from Lemmas 3.1.25 and 3.1.26 that, for any integers $n \ge 0$ and $i > 0$, the iterated subdivision $Sd^i(\partial\Delta^n)$ is the nerve of a partially ordered set. Therefore, any map $\partial\Delta^n \to Ex^\infty(X)$ is homotopic to a constant map: indeed, such a map factors through a map $\partial\Delta^n \to Ex^i(X)$ with $i > 0$, which corresponds by adjunction to a map of the form $Sd^i(\Delta^n) \to X$. In particular, $\pi_0(X)$ has exactly one element.[1] This also implies that the homotopy groups $\pi_n(X, x)$ are all trivial (they do not depend on the base point, since all objects are

[1] The map $u\colon \varnothing \to Ex^\infty X$ is homotopic to a constant map. This means that there exists an element x_0 of X_0 such that $u(x) = x_0$ for any element x of $\varnothing$. In particular, X is not empty.

isomorphic in the groupoid $\tau(Ex^\infty(X))$). Therefore, the Kan complex $Ex^\infty(X)$ is contractible, by Propositions 3.8.10 and 3.8.11. Since we have a canonical trivial cofibration $X \to Ex^\infty(X)$, this achieves the proof. □

Theorem 4.3.16 *Let ω be an object in an ∞-category X. The following conditions are equivalent.*

(i) The object ω is final.
(ii) For any simplicial set A, the constant functor from A to X with value ω is a final object in the category $\tau(\underline{\mathrm{Hom}}(A, X))$.
(iii) For any simplicial set A which is isomorphic to the nerve of a finite partially ordered set, the constant functor from A to X with value ω is a final object in the category $\tau(\underline{\mathrm{Hom}}(A, X))$.

Proof Corollaries 4.3.12 and 4.3.14 show that condition (i) implies condition (ii). Since condition (iii) is obviously a particular case of condition (ii), it remains to prove that condition (iii) implies that ω is a final object of X. Let us assume that ω (or the constant functor with value ω) is a final object of $\tau(\underline{\mathrm{Hom}}(A, X))$ for any A of the form $A = N(E)$, with E a finite partially ordered set. We want to prove that, for any object x in X, the ∞-groupoid $X(x, \omega)$ is equivalent to the point. For any simplicial set A, since it commutes with limits and preserves Kan fibrations as well as trivial fibrations, the functor $\underline{\mathrm{Hom}}(A, -)$ commutes with the formation of homotopy fibres of Kan fibrations between Kan complexes. Therefore, it follows from Corollary 4.2.10 that the Kan complex $\underline{\mathrm{Hom}}(A, X(x, \omega))$ is the homotopy fibre of the canonical map

$$\underline{\mathrm{Hom}}(A, X/\omega) \to \underline{\mathrm{Hom}}(A, X)\,.$$

Hence, by virtue of Proposition 4.2.12, the Kan complex $\underline{\mathrm{Hom}}(A, X(x, \omega))$ is homotopy equivalent to the Kan complex $\underline{\mathrm{Hom}}(A, X)(x, \omega)$ (where x and ω also denote the corresponding constant functors). Therefore, the set of connected components $\pi_0(\underline{\mathrm{Hom}}(A, X(x, \omega)))$ has exactly one element (being the set of maps from x to ω in $\tau(\underline{\mathrm{Hom}}(A, X))$). The preceding lemma implies that $X(x, \omega)$ is contractible. We conclude thanks to Theorem 4.3.11. □

Remark 4.3.17 There is a 2-category of ∞-categories: the objects are the ∞-categories, and the categories of morphisms $\mathrm{Hom}(A, B)$ are the categories $\tau(\underline{\mathrm{Hom}}(A, B))$. The preceding theorem means that the property of having a final object may be detected in this 2-category. This is a particular case of a more general feature of the theory of ∞-categories: the notion of adjunction is 2-categorical; see Theorem 6.1.23 below (we recall that having a final object is equivalent to the property that the map to the final category has a right adjoint).

We also observe that the notion of equivalence of ∞-categories is 2-categorical as well: this follows from Theorems 3.6.8 and 3.6.9.

4.4 Grothendieck Base Change Formulas and Quillen's Theorem A

Definition 4.4.1 (Grothendieck) A morphism of simplicial sets $p\colon A \to B$ is *proper* if, for any Cartesian squares of the form

$$\begin{array}{ccccc} A'' & \xrightarrow{u'} & A' & \xrightarrow{u} & A \\ \downarrow{\scriptstyle p''} & & \downarrow{\scriptstyle p'} & & \downarrow{\scriptstyle p} \\ B'' & \xrightarrow{v'} & B' & \xrightarrow{v} & B \end{array} \tag{4.4.1.1}$$

if v' is final, so is u'.

The proof of the following proposition is an easy exercise left to the reader.

Proposition 4.4.2 *The class of proper morphisms is stable under composition and under base change.*

Lemma 4.4.3 *For any $n \geq 1$ and $0 < k \leq n$, the inclusion of $\{n\}$ in Λ^n_k is a right anodyne extension.*

Proof In the proof of Lemma 3.1.3, we saw that Λ^n_k is a retract of $\Delta^1 \times \Lambda^n_k \cup \{1\} \times \Delta^n$. We observe that this is a retraction of pointed simplicial sets, where the base points are n and $(1, n)$, respectively. Since right anodyne extensions are stable under retracts, this shows that it is sufficient to check that $(1, n)$ is a final object of $\Delta^1 \times \Lambda^n_k \cup \{1\} \times \Delta^n$. By definition, we have a coCartesian square of the form

$$\begin{array}{ccc} \{1\} \times \Lambda^n_k & \hookrightarrow & \{1\} \times \Delta^n \\ \downarrow & & \downarrow \\ \Delta^1 \times \Lambda^n_k & \hookrightarrow & \Delta^1 \times \Lambda^n_k \cup \{1\} \times \Delta^n \end{array}$$

whose first vertical arrow is a right anodyne extension. Therefore, the second vertical arrow is a right anodyne extension as well, and it remains to check that $(1, n)$ is a final object of $\{1\} \times \Delta^n$, which is obvious. □

Proposition 4.4.4 *For a morphism of simplicial sets $p\colon A \to B$, the following conditions are equivalent.*

(i) The morphism p is proper.

(ii) *For any Cartesian squares of the form* (4.4.1.1)*, if B'' has a final object whose image by v' is a final object of B', the map u' is final.*
(iii) *For any Cartesian squares of the form* (4.4.1.1)*, if $B'' = \Delta^0$ and if v' is a final object of B', the map u' is final.*
(iv) *For any Cartesian squares of the form* (4.4.1.1)*, if v' is a right anodyne extension, so is u'.*
(v) *For any Cartesian squares of the form* (4.4.1.1) *in which the map v' is the canonical inclusion of Λ^n_k into Δ^n, for $n \geq 1$ and $0 < k \leq n$, the map u' is a right anodyne extension.*

Proof A monomorphism is final if and only if it is a right anodyne extension (see Corollary 4.1.9). Therefore, condition (i) implies condition (iv). We also know that a morphism of simplicial sets is final if and only if it admits a factorisation into a right anodyne extension followed by a trivial fibration. Since the class of trivial fibrations is stable under base change, one deduces that condition (iv) implies (i). It is clear that condition (iv) implies condition (v). Since any map of the form $1_b : \Delta^0 \to B/b$ is final (Corollary 4.3.8), we see that condition (iii) is equivalent to condition (ii) as follows. It is clear that condition (ii) implies condition (iii). Conversely, for any Cartesian squares of the form (4.4.1.1), if B'' has a final object ω whose image by v' is a final object of B', if we write A''_ω for the fibre of p'' at ω, there is a commutative triangle of the form

$$\begin{array}{ccc} & A'_\omega & \\ {}^{i'}\swarrow & & \searrow{}^{i} \\ A'' & \xrightarrow{u'} & A' \end{array}$$

in which, assuming (iii), the maps i and i' are right anodyne extensions. Corollary 4.1.9 thus implies that u' is a right anodyne extension. Condition (ii) implies (iv): by virtue of the preceding lemma, for $n \geq 1$ and $0 < k \leq n$, the object n is final in both Λ^n_k and Δ^n. Since condition (i) clearly implies condition (iii), it is now sufficient to prove that condition (v) implies condition (i).

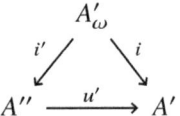

Let C be the class of maps $v' : B'' \to B'$ such that, for any map $v : B' \to B$, if we form the Cartesian squares as in diagram (4.4.1.1), the induced map $u' : A'' \to A'$ is final. Since pulling back along p preserves monomorphisms and commutes with colimits over A, we see that the class of monomorphisms which are in C is saturated. Assuming condition (v), the class C contains all inclusions of the form Λ^n_k into Δ^n, for $n \geq 1$ and $0 < k \leq n$. Therefore, it contains all right anodyne extensions. The class C is closed under composition, and, for any pair of morphisms $f : X \to Y$ and $g : Y \to Z$, if both f and gf are in C, so is g. These latter conditions are verified because the class of final maps has these properties. And since the class of final maps is the smallest class with these stability properties containing right anodyne extensions, the second part of Corollary 4.1.9 implies that any final map is in C. Hence the map p is proper. □

4.4.5 For any morphism of simplicial sets $p\colon A \to B$, the functor

$$p_!\colon sSet/A \to sSet/B\,, \quad (X \to A) \mapsto (X \to B) \tag{4.4.5.1}$$

has a right adjoint

$$p^*\colon sSet/B \to sSet/A\,, \quad (X \to B) \mapsto (X \times_B A \to A)\,. \tag{4.4.5.2}$$

Since it commutes with small colimits, the latter functor also has a right adjoint

$$p_*\colon sSet/A \to sSet/B\,. \tag{4.4.5.3}$$

Applying Proposition 2.4.40 twice, we obtain the following two statements.

Proposition 4.4.6 *For any morphism of simplicial sets $p\colon A \to B$, the pair $(p_!, p^*)$ is a Quillen adjunction for the contravariant model category structures.*

Proposition 4.4.7 *If a morphism of simplicial sets $p\colon A \to B$ is proper, the pair (p^*, p_*) is a Quillen adjunction for the contravariant model category structures.*

4.4.8 Let $RFib(A)$ be the homotopy category of the contravariant model category structure on $sSet/A$. As seen in the proof of Theorem 4.1.16, it can be described as the category whose objects are the right fibrant objects over A, and such that, for two such right fibrant objects F and G,

$$\mathrm{Hom}_{RFib(A)}(F, G) = \pi_0(\mathrm{Map}_A(F, G))\,. \tag{4.4.8.1}$$

For any morphism of simplicial sets $p\colon A \to B$, we thus have a derived adjunction

$$p_! = \mathbf{L}p_!\colon RFib(A) \rightleftarrows RFib(B) :\mathbf{R}p^*\,. \tag{4.4.8.2}$$

For any map $f\colon X \to A$, we have $p_!(X \to A) = (X \to B)$, and, for any right fibration $g\colon Y \to B$, we have $\mathbf{R}p^*(Y \to X) = (A \times_B Y \to A)$.

If the map p is proper we also have a derived adjunction

$$p^* = \mathbf{L}p^*\colon RFib(B) \rightleftarrows RFib(A) :\mathbf{R}p_*\,, \tag{4.4.8.3}$$

where $p^* = \mathbf{L}p^* = \mathbf{R}p^*$. Proposition 4.4.7 can be interpreted as a base change formula. Indeed, for any commutative square of simplicial sets

$$
\begin{array}{ccc}
A' & \xrightarrow{u} & A \\
{\scriptstyle q}\downarrow & & \downarrow{\scriptstyle p} \\
B' & \xrightarrow{v} & B
\end{array}
\tag{4.4.8.4}
$$

there is a canonical base change map

$$\mathbf{L}q_!\mathbf{R}u^*(X) \to \mathbf{R}v^*\mathbf{L}p_!(X) \tag{4.4.8.5}$$

defined as follows, for any object X of $RFib(A)$. By adjunction, it corresponds to a map

$$
\begin{aligned}
\mathbf{R}u^*(X) \to \mathbf{R}q^*\mathbf{R}v^*\mathbf{L}p_!(X) &\simeq \mathbf{R}(vq)^*\mathbf{L}p_!(X) \\
&= \mathbf{R}(pu)^*\mathbf{L}p_!(X) \\
&\simeq \mathbf{R}u^*\mathbf{R}p^*\mathbf{L}p_!(X)
\end{aligned}
$$

which is the pull-back along u of the unit map $X \to \mathbf{R}p^*\mathbf{L}p_!(X)$. Similarly, there is a base change map

$$\mathbf{L}u_!\mathbf{R}q^*(Y') \to \mathbf{R}p^*\mathbf{L}v_!(Y') \tag{4.4.8.6}$$

for any object Y' of $RFib(B')$.

Corollary 4.4.9 *For any Cartesian square of the form* (4.4.8.4) *in which the map p is proper, the base change map* (4.4.8.6) *is an isomorphism for all objects Y' of $RFib(B')$.*

Proof Since $\mathbf{L}p^* = \mathbf{R}p^*$ whenever p is proper, since the formation of total left derived functors commutes with composition of left Quillen functors, it is sufficient to prove the analogue of this assertion in the non-derived case, which consists essentially in seeing that Cartesian squares are stable by composition. □

Lemma 4.4.10 *Let $i\colon K \to L$ be a monomorphism of simplicial sets. Assume that there exists a retraction $r\colon L \to K$ of i as well as a simplicial homotopy $h\colon \Delta^1 \times L \to L$ which is constant on K, from 1_L to ir (i.e. such that $h(1_{\Delta^1} \times i)$ is the composition of the projection from $\Delta^1 \times K$ to K with the map i, and such that $h(\{0\} \times L) = 1_L$ and $h(\{1\} \times L) = ir$). Then the map i is a right anodyne extension.*

Proof We have a commutative diagram

$$
\begin{array}{ccccc}
K & \xrightarrow{u} & \Delta^1 \times K \cup \{1\} \times L & \xrightarrow{k} & K \\
{\scriptstyle i}\downarrow & & \downarrow & & \downarrow{\scriptstyle i} \\
L & \xrightarrow{v} & \Delta^1 \times L & \xrightarrow{h} & L
\end{array}
$$

in which v corresponds to the identification of L with $\{0\} \times L$, and u and k are the restrictions of v and h, respectively. The map i thus is a retract of the middle vertical inclusion, which is a right anodyne extension, by Proposition 3.4.3. □

A good supply of proper morphisms comes from the following two propositions.

Proposition 4.4.11 *Any left fibration is proper.*

Proof Let C be the class of monomorphisms $j \colon Y' \to Y$ such that, for any Cartesian square of the form

$$\begin{array}{ccc} X' & \xrightarrow{i} & X \\ {\scriptstyle q}\big\downarrow & & \big\downarrow{\scriptstyle p} \\ Y' & \xrightarrow{j} & Y \end{array}$$

in which p is a left fibration, the map i is a right anodyne extension. This class C is saturated, and for any monomorphisms $f \colon Y' \to Y$ and $f' \colon Y'' \to Y'$ such that f' is in C, we have f' in C if and only if ff' is in C. Using Lemma 3.1.3 and Corollary 4.1.10, we see that it is sufficient to prove that the class C contains the inclusions of $\{1\} \times Y$ into $\Delta^1 \times Y$ for $Y = \partial\Delta^n$ or $Y = \Delta^n$. Let us consider a Cartesian square of the form

$$\begin{array}{ccc} X' & \xrightarrow{i} & X \\ {\scriptstyle q}\big\downarrow & & \big\downarrow{\scriptstyle p} \\ Y & \xrightarrow{j} & \Delta^1 \times Y \end{array}$$

in which p is a left fibration, and j is the inclusion of $\{1\} \times Y$ into $\Delta^1 \times Y$. Let $r \colon \Delta^1 \times Y \to Y$ be the projection, and $h \colon \Delta^1 \times \Delta^1 \times Y \to \Delta^1 \times Y$ be the product of Y with the homotopy from the identity of Δ^1 to the constant map with value 1. The proof of Proposition 2.5.7 can be repeated *mutatis mutandis* and we see that the map i satisfies the hypothesis of Lemma 4.4.10, hence is right anodyne. □

Proposition 4.4.12 *For any simplicial sets A and B, the canonical projection map $A \times B \to B$ is proper.*

Proof Since the class of proper morphisms is stable under pull-back, we may assume, without loss of generality, that $B = \Delta^0$. Using property (v) of Proposition 4.4.4, it is now sufficient to observe that, by Proposition 3.4.3, for any $n \geq 1$ and $0 < k \leq n$, the induced map $A \times \Lambda^n_k \to A \times \Delta^n$ is a right anodyne extension. □

In order to go further, we need to consider the dual version of the notions of final maps and of proper maps.

Definition 4.4.13 A morphism of simplicial sets $X \to Y$ is *cofinal* if the induced morphism $X^{\mathrm{op}} \to Y^{\mathrm{op}}$ is final. An object x of a simplicial set X is *initial* if the corresponding map $x\colon \Delta^0 \to X$ is cofinal.

A *left fibrant object* over a simplicial set C is a left fibration of the form $F \to C$, seen as an object of $sSet/C$. A morphism of left fibrant objects $F \to G$ over C is a *fibrewise equivalence* if, for any object c of C, the induced maps on the fibres $F_c \to G_c$ is an equivalence of ∞-groupoids.

Theorem 4.4.14 (Joyal) *There is a unique model category structure on the category $sSet/C$ whose cofibrations are the monomorphisms, and whose fibrant objects are the left fibrant objects over C. Moreover, a morphism between left fibrant objects over C is a fibration if and only if it is a right fibration. Finally, a morphism between left fibrant objects is a weak equivalence if and only if it is a fibrewise equivalence.*

This model category structure will be called the *covariant model category structure over C.*

Proof This comes from translating Theorems 4.1.5 and 4.1.16 through the equivalence of categories $X \mapsto X^{\mathrm{op}}$. □

Definition 4.4.15 A morphism of simplicial sets $p\colon A \to B$ is *smooth* if the induced morphism $A^{\mathrm{op}} \to B^{\mathrm{op}}$ is proper.

Remark 4.4.16 The dual version of Proposition 4.4.4 holds; we leave to the reader the task of translating the precise statement and proof. Similarly, the dual versions of Propositions 4.4.11 and 4.4.12 give us examples of smooth maps: any right fibration is smooth, and so is any Cartesian projection $A \times B \to B$.

Proposition 4.4.17 *For any morphism of simplicial sets $p\colon A \to B$, the pair $(p_!, p^*)$ is a Quillen adjunction for the covariant model category structures.*

Proposition 4.4.18 *If a morphism of simplicial sets $p\colon A \to B$ is smooth, the pair (p^*, p_*) is a Quillen adjunction for the covariant model category structures.*

4.4.19 Let $LFib(A)$ be the homotopy category of the covariant model category structure on $sSet/A$. Since the functor $X \mapsto X^{\mathrm{op}}$ is a Quillen equivalence from the contravariant model structure over A^{op} to the covariant model category structure over A, we have a canonical equivalence of categories

$$RFib(A^{\mathrm{op}}) \simeq LFib(A)\,. \tag{4.4.19.1}$$

The constructions of paragraph 4.4.8 correspond to the following ones. For any morphism of simplicial sets $p\colon A \to B$, we thus have a derived adjunction

$$p_! = \mathbf{L}p_! \colon LFib(A) \rightleftarrows LFib(B) \colon \mathbf{R}p^* \,. \tag{4.4.19.2}$$

If the map p is smooth we also have a derived adjunction

$$p^* = \mathbf{L}p^* \colon RFib(B) \rightleftarrows RFib(A) \colon \mathbf{R}p_* \,, \tag{4.4.19.3}$$

where $p^* = \mathbf{L}p^* = \mathbf{R}p^*$.

For any commutative square of simplicial sets

$$\begin{array}{ccc} A' & \xrightarrow{u} & A \\ {\scriptstyle q}\downarrow & & \downarrow{\scriptstyle p} \\ B' & \xrightarrow{v} & B \end{array} \tag{4.4.19.4}$$

there is a canonical base change map

$$\mathbf{L}q_!\mathbf{R}u^*(X) \to \mathbf{R}v^*\mathbf{L}p_!(X) \tag{4.4.19.5}$$

for any object X of $LFib(A)$. There is also a base change map

$$\mathbf{L}u_!\mathbf{R}q^*(Y') \to \mathbf{R}p^*\mathbf{L}v_!(Y') \tag{4.4.19.6}$$

for any object Y' of $LFib(B')$.

Corollary 4.4.20 *For any Cartesian square of the form* (4.4.19.4) *in which the map v is smooth, the base change map* (4.4.19.6) *is an isomorphism for all objects X of $LFib(A)$.*

4.4.21 For a simplicial set X and an object F of $LFib(X)$, we define

$$\int_A F = \mathbf{L}p_!(F), \tag{4.4.21.1}$$

where $p\colon A \to \Delta^0$ denotes the structural map. We define the category *ho(sSet)* as the homotopy category of the Kan–Quillen model category structure. In other words, we have

$$ho(sSet) = LFib(\Delta^0) = RFib(\Delta^0)\,. \tag{4.4.21.2}$$

The object $\int_A F$ is thus an object of *ho(sSet)*.

Lemma 4.4.22 *Let A be a simplicial set endowed with a final object a. Then there is a canonical isomorphism $\mathbf{R}a^*(F) \simeq \int_A F$ in the homotopy category ho(sSet).*

Proof Up to isomorphism, we may assume that F is represented by a left fibration $p\colon X \to A$ whose fibre over a will be written X_a. Since, by virtue of Proposition 4.4.11, p is proper, and since $a\colon \Delta^0 \to A$ is a right anodyne extension, the induced map $X_a \to X$ is a right anodyne extension, hence an anodyne extension, and therefore defines an invertible morphism from X_a to X in *ho*(*sSet*). But, by construction, we have $\int_A F = X$ and $\mathbf{R}a^*(F) = X_a$. □

Lemma 4.4.23 *Let $p\colon X \to A$ be a proper morphism, and a an object of A. For any object F of LFib(X), the base change morphism*

$$\int_{X_a} F_{|X_a} \to \mathbf{R}a^*\mathbf{L}p_!(F)$$

is invertible, where $F_{|X_a} = \mathbf{R}i^*(F)$, with $i\colon X_a \to X$ the embedding of the fibre of p over a.

Proof Let us choose a fibrant replacement $\tilde{a}\colon \tilde{A}_{/a} \to A$ of $a\colon \Delta^0 \to A$ in the contravariant model category structure over A. In particular, the map a factors through a final map $\ell_a\colon \Delta^0 \to \tilde{A}_{/a}$. We form the following Cartesian squares.

$$\begin{array}{ccccc} X_a & \xrightarrow{i} & \tilde{A}_{/a} \times_A X & \xrightarrow{\xi} & X \\ \downarrow & & \downarrow{\scriptstyle q} & & \downarrow{\scriptstyle p} \\ \Delta^0 & \xrightarrow{\ell_a} & \tilde{A}_{/a} & \xrightarrow{\tilde{a}} & A \end{array}$$

Since $\tilde{a}$ is a right fibration, it is smooth. Therefore, the base change map

$$\mathbf{L}q_!\mathbf{R}\xi^*(F) \to \mathbf{R}\tilde{a}^*\mathbf{L}p_!(F)$$

is invertible in *LFib*$(\tilde{A}_{/a})$. Hence we have an invertible map in *ho*(*sSet*) of the form

$$\int_{\tilde{A}_{/a}\times_A X} \mathbf{R}\xi^*(F) \simeq \int_{\tilde{A}_{/a}} \mathbf{L}q_!\mathbf{R}\xi^*(F) \to \int_{\tilde{A}_{/a}} \mathbf{R}\tilde{a}^*\mathbf{L}p_!(F)\,.$$

The preceding lemma provides a canonical isomorphism

$$\mathbf{R}a^*\mathbf{L}p_!(F) \simeq \mathbf{R}\ell_a^*\mathbf{R}\tilde{a}^*\mathbf{L}p_!(F) \to \int_{\tilde{A}_{/a}} \mathbf{R}\tilde{a}^*\mathbf{L}p_!(F)\,.$$

On the other hand, we may assume that F is represented by a left fibration of the form $\pi\colon Y \to X$. Then $p\pi\colon Y \to A$ is proper (as the composition of two proper maps), which implies that the inclusion $X_a \times_X Y \to \tilde{A}_{/a} \times_A Y$ is a (right) anodyne extension, hence an invertible map in *ho*(*sSet*). This map thus provides an invertible map of the form

$$\int_{X_a} F_{|X_a} \to \int_{\tilde{A}_{/a}\times_A X} \mathbf{R}\xi^*(F)$$

in $ho(sSet)$. Since the square

$$\begin{array}{ccc} \int_{X_a} F_{|X_a} & \xrightarrow{\sim} & \int_{\tilde{A}/a \times_A X} \mathbf{R}\xi^*(F) \\ \downarrow & & \downarrow \wr \\ \mathbf{R}a^*\mathbf{L}p_!(F) & \xrightarrow{\sim} & \int_{\tilde{A}/a} \mathbf{R}\tilde{a}^*\mathbf{L}p_!(F) \end{array}$$

commutes in $ho(sSet)$ (because it is the image of a commutative square in $sSet$), this proves the lemma. □

Theorem 4.4.24 *For any Cartesian square of the form* (4.4.19.4) *in which the map p is proper, the base change map* (4.4.19.5) *is an isomorphism for all objects X of $LFib(A)$.*

Proof A map in $LFib(B')$ is invertible if and only if its image by the functor $\mathbf{R}x^*$ is invertible in $LFib(\Delta^0) = ho(sSet)$ for any object x of B'. Indeed, we may assume that such a map comes from a morphism between left fibrant objects over B', and we see that we are reformulating the last assertion of Theorem 4.4.14. Therefore, to prove that the base change map (4.4.8.6) is an isomorphism, it is sufficient to prove that, for any object x in B', the induced map

$$\mathbf{R}x^*\mathbf{L}q_!\mathbf{R}u^*(X) \to \mathbf{R}x^*\mathbf{R}v^*\mathbf{L}p_!(X) \simeq \mathbf{R}v(x)^*\mathbf{L}p_!(X)$$

is invertible. But, applying the preceding lemma twice, we see that this map is isomorphic to the identity of $\int_{A_{v(x)}} X_{|A_{v(x)}}$. □

Corollary 4.4.25 *Let $p\colon A \to B$ be a proper morphism, and consider a commutative square of simplicial sets of the form*

$$\begin{array}{ccc} X & \xrightarrow{\phi} & A \\ {\scriptstyle f}\downarrow & & \downarrow{\scriptstyle p} \\ Y & \xrightarrow{\psi} & B \end{array}$$

in which both ϕ and ψ are left fibrations and f is a cofinal map. Then, for any map $v\colon B' \to B$, the induced map

$$g\colon X' = B' \times_B X \to Y' = B' \times_B Y$$

is cofinal.

Proof We form the pull-back square

$$\begin{array}{ccc} A' & \xrightarrow{u} & A \\ {\scriptstyle q}\downarrow & & \downarrow{\scriptstyle p} \\ B' & \xrightarrow{v} & B \end{array}$$

and then apply the preceding theorem: the base change map

$$\mathbf{L}q_!\mathbf{R}u^*(X) \to \mathbf{R}v^*\mathbf{L}p_!(X)$$

is an isomorphism in $LFib(B')$. Unpacking the construction of the involved derived functors, we see that this base change map is the image in $LFib(B')$ of the map g. Therefore, the morphism g is a weak equivalence with fibrant codomain of the covariant model category structure over B'. The dual version of Proposition 4.1.11 tells us that g must be cofinal. □

Dually, we have the following two statements.

Theorem 4.4.26 *For any Cartesian square of the form* (4.4.8.4) *in which the map v is smooth, the base change map* (4.4.8.5) *is an isomorphism for all objects X of $RFib(A)$.*

Corollary 4.4.27 *Let $p\colon A \to B$ be a smooth morphism, and consider a commutative square of simplicial sets of the form*

$$\begin{array}{ccc} X & \xrightarrow{\phi} & A \\ {\scriptstyle f}\downarrow & & \downarrow{\scriptstyle p} \\ Y & \xrightarrow{\psi} & B \end{array}$$

in which both ϕ and ψ are right fibrations and f is a final map. Then, for any map $v\colon B' \to B$, the induced map

$$g\colon X' = B' \times_B X \to Y' = B' \times_B Y$$

is final.

Corollary 4.4.28 *Let us consider the commutative triangle of simplicial sets below.*

$$\begin{array}{ccc} X & \xrightarrow{f} & Y \\ & {\scriptstyle p}\searrow \quad \swarrow{\scriptstyle q} & \\ & A & \end{array}$$

We assume that p and q are proper (smooth, respectively). Then the map f is a weak equivalence of the covariant model category structure (of the contravariant model category structure, respectively) if and only if it is a fibrewise weak homotopy equivalence.

Proof We consider the case where p and q are proper (the smooth case will follow by duality). We factor the map q into a cofinal map $j\colon Y \to Y'$ followed by a left fibration $q'\colon Y' \to A$. Similarly, we factor the map jf into a cofinal map $i\colon X \to X'$ followed by a left fibration $p'\colon X' \to Y'$. For each object a of A, we obtain a commutative square of the form

$$\begin{array}{ccc} X_a & \xrightarrow{f_a} & Y_a \\ {\scriptstyle i_a}\downarrow & & \downarrow{\scriptstyle j_a} \\ X'_a & \xrightarrow{p'_a} & Y'_a \end{array}$$

in which the vertical maps are cofinal, by Corollary 4.4.25. Therefore, the map f is a fibrewise weak homotopy equivalence over A if and only if the map p' has the same property. We also have the property that f is a weak equivalence of the covariant model structure over A if and only if p' is a weak equivalence. Therefore, it is sufficient to prove the corollary when p and q are left fibrations, in which case this is already known: this is the last assertion of Theorem 4.4.14. □

4.4.29 Let A be a simplicial set. Given two objects E and F of $sSet/A$, corresponding to maps $p\colon X \to A$ and $q\colon Y \to A$, respectively, we define

$$\langle E, F\rangle = X \times_A Y\,. \tag{4.4.29.1}$$

This defines a functor

$$\langle -, -\rangle\colon sSet/A \times sSet/A \to sSet\,. \tag{4.4.29.2}$$

If E is a right fibration, then the functor $\langle E, -\rangle$ is a left Quillen functor from the covariant model category structure over A to the Kan–Quillen model category structure; similarly, if F is a left fibration, the functor $\langle -, F\rangle$ is a left Quillen functor from the contravariant model category structure over A to the Kan–Quillen model category structure. In particular, the functor (4.4.29.2) preserves weak equivalences between fibrant objects on the product $sSet/A \times sSet/A$, where we consider the contravariant model category structure on the first factor, and the covariant model structure on the second factor. Therefore, we have a total right derived functor

$$\mathbf{R}\langle -, -\rangle\colon RFib(A) \times LFib(A) \to ho(sSet)\,. \tag{4.4.29.3}$$

When either E is a right fibration or F is a left fibration, we have a canonical isomorphism

$$\langle E, F\rangle \simeq \mathbf{R}\langle E, F\rangle \tag{4.4.29.4}$$

in $ho(sSet)$. Given an object a of A and an object F of $LFib(A)$, we define

$$F_{/a} = \mathbf{R}\langle(\Delta^0, a), F\rangle\,. \tag{4.4.29.5}$$

Dually, for an object E of $RFib(A)$, we define

$$E_{a/} = \mathbf{R}\langle E, (\Delta^0, a)\rangle\,. \tag{4.4.29.6}$$

In other words, we may construct $F_{/a} = \tilde{A}_{/a} \times_A Y$ and $E_{a/} = X \times_A \tilde{A}_{a/}$, where $\tilde{A}_{/a} \to A$ and $\tilde{A}_{a/} \to A$ are fibrant replacements of $a\colon \Delta^0 \to A$ in the contravariant and covariant model category structure over A, respectively. We observe that, for an object F of $LFib(A)$ (for an object E of $RFib(A)$), there is a canonical isomorphism

$$F_{/a} \simeq \mathbf{R}a^*(F) \quad (E_{a/} \simeq \mathbf{R}a^*(E), \text{respectively}) \tag{4.4.29.7}$$

in $ho(sSet)$. Indeed, if F corresponds to a map $p\colon X \to A$, we may choose a factorisation of p into a cofinal map $i\colon X \to Y$ followed by a left fibration $q\colon Y \to A$. Then the map i induces a cofinal map $X \times_A \tilde{A}_{/a} \to Y \times_A \tilde{A}_{/a}$ because the map $\tilde{A}_{/a} \to A$ is a right fibration whence is smooth. Similarly the inclusion map $Y_a \to Y \times_A \tilde{A}_{/a}$ is right anodyne since q is a left fibration whence proper. Finally, Y_a and $X \times_A \tilde{A}_{/a}$ are isomorphic in $ho(sSet)$. The case of an object of $RFib(A)$ is obtained by duality.

Proposition 4.4.30 (Joyal) *Let us consider a morphism $\varphi\colon F \to G$ in $LFib(A)$. The following conditions are equivalent.*

(i) The morphism φ is an isomorphism in $LFib(A)$.
(ii) For any object E of $RFib(A)$, the induced morphism $\mathbf{R}\langle E, F\rangle \to \mathbf{R}\langle E, G\rangle$ is an isomorphism in $ho(sSet)$.
(iii) For any object a of A, the induced map $\varphi_{/a}\colon F_{/a} \to G_{/a}$ is an isomorphism in $ho(sSet)$.

Proof By virtue of identification (4.4.29.7), the equivalence between conditions (i) and (iii) is a reformulation of Theorem 4.1.16. Since the implications (i) $\Rightarrow$ (ii) $\Rightarrow$ (iii) are obvious, this proves the proposition. □

Corollary 4.4.31 *A functor between ∞-categories $u\colon A \to B$ is cofinal if and only if, for any object b of B, the ∞-category $A/b = B/b \times_B A$ has the weak homotopy type of the point.*

Proof Let F be the object of $LFib(B)$ defined as $F = (A, u)$. The map u is final if and only if F is a final object of $LFib(B)$. By virtue of the preceding proposition, this is equivalent to saying that $F_{/b}$ is a final object of $ho(sSet)$ for any object b of B. But, by virtue of Theorem 4.3.9, a canonical fibrant replacement of (Δ^0, b) in the contravariant model category structure over B is the canonical map $B/b \to B$. Therefore, we have a canonical isomorphism $A/b = F_{/b}$ in the homotopy category $ho(sSet)$. □

Corollary 4.4.32 (Quillen's theorem A) *Let $u\colon A \to B$ be a functor between small categories. Assume that, for any object b of B, the nerve of $A/b = B/b \times_B A$ has the weak homotopy type of the point. Then the nerve of u is a weak homotopy equivalence.*

Proof We have $N(B)/b = N(B/b)$. Therefore, the hypothesis implies that $N(u)\colon N(A) \to N(B)$ is a cofinal map, hence a weak equivalence of the Kan–Quillen model structure. □

Proposition 4.4.33 *Let $u\colon A \to B$ be a morphism of simplicial sets. For any objects E in $RFib(A)$ and F in $LFib(B)$, there is a canonical isomorphism*

$$\mathbf{R}\langle E, \mathbf{R}u^*(F)\rangle \simeq \mathbf{R}\langle \mathbf{L}u_!(E), F\rangle$$

in the homotopy category $ho(sSet)$. Similarly, for any object F' in $RFib(B)$ and any object E' in $LFib(A)$, there is a canonical isomorphism

$$\mathbf{R}\langle F', \mathbf{L}u_!(E')\rangle \simeq \mathbf{R}\langle \mathbf{R}u^*(F'), E'\rangle\,.$$

Proof Cartesian diagrams of the form

$$\begin{array}{ccccc} X \times_B Y & \longrightarrow & A \times_B Y & \longrightarrow & Y \\ \downarrow & & \downarrow & & \downarrow{\scriptstyle q} \\ X & \xrightarrow{\;\;p\;\;} & A & \xrightarrow{\;\;u\;\;} & B \end{array}$$

show that, for $E = (X, p)$ and $F = (Y, q)$, we have

$$\langle E, u^*(F)\rangle = \langle u_!(E), F\rangle\,.$$

When q is a left fibration, this proves the first isomorphism. Using the identification

$$\langle E, F\rangle^{\mathrm{op}} \simeq \langle F^{\mathrm{op}}, E^{\mathrm{op}}\rangle,$$

where $E^{\mathrm{op}} = (X^{\mathrm{op}}, p^{\mathrm{op}})$, this also implies the second one. □

Proposition 4.4.34 *Let $p\colon A \to B$ be a morphism of simplicial sets. The following conditions are equivalent.*

(i) The morphism p is proper.

(ii) For any pull-back squares of the form

$$\begin{array}{ccccc} A'' & \xrightarrow{\;u'\;} & A' & \xrightarrow{\;u\;} & A \\ \downarrow{\scriptstyle p''} & & \downarrow{\scriptstyle p'} & & \downarrow{\scriptstyle p} \\ B'' & \xrightarrow{\;v'\;} & B' & \xrightarrow{\;v\;} & B \end{array} \qquad (4.4.34.1)$$

and for any object F of LFib(A′), the base change map

$$\mathbf{L}p''_!\mathbf{R}u''^*(F) \to \mathbf{R}v''^*\mathbf{L}p'_!(F)$$

is an isomorphism in LFib(B″).

(iii) For any pull-back squares of the form (4.4.34.1), *and for any object E in RFib(B″), the base change map*

$$\mathbf{L}u'_!\mathbf{R}p''^*(E) \to \mathbf{R}p'^*\mathbf{L}v'_!(E)$$

is an isomorphism in RFib(A′).

Proof For E and F corresponding to a right fibration $X \to A$ and a left fibration $Y \to A$, respectively, we have the following commutative diagram.

$$\begin{array}{ccc} \mathbf{R}\langle E, \mathbf{L}p''_!\mathbf{R}u''^*(F)\rangle & \longrightarrow & \mathbf{R}\langle E, \mathbf{R}v''^*\mathbf{L}p'_!(F)\rangle \\ \wr\downarrow & & \downarrow\wr \\ \mathbf{R}\langle \mathbf{L}u'_!\mathbf{R}p''^*(E), F\rangle & \longrightarrow & \mathbf{R}\langle \mathbf{R}p'^*\mathbf{L}v'_!(E), F\rangle \end{array}$$

Using Propositions 4.4.30 and 4.4.33, we see easily that conditions (ii) and (iii) are equivalent. We already know that condition (i) implies conditions (ii) (by Corollary 4.4.9) and (iii) (by Theorem 4.4.24).[2] It is thus sufficient to prove that condition (ii) implies condition (i).

Let us consider Cartesian squares of the form (4.4.34.1) in which we have $B'' = \Delta^0$ with v' final. It is sufficient to prove that the map u is final, by Proposition 4.4.4. In other words, we must show that the map u is a weak equivalence of the contravariant model category structure over A'. The dual version of Proposition 4.4.30 shows that it is sufficient to prove that, for any object a' of A', if $\tilde{A}'_{a'/} \to A'$ denotes a fibrant replacement of $a' : \Delta^0 \to A'$ with respect to the covariant model category structure over A', then the projection from $A''_{a'/} = A'' \times_{A'} \tilde{A}'_{a'/}$ to $\tilde{A}'_{a'/}$ is a weak equivalence of the Kan–Quillen model category structure. Let F be the object of $LFib(A')$ corresponding to the map $\tilde{A}'_{a'/} \to A'$. Then there is a canonical isomorphism $\int F \simeq \int \mathbf{L}p'_!(F)$, and, since v' is final, Lemma 4.4.22 ensures that we have a canonical isomorphism $\mathbf{R}v''^*\mathbf{L}p'_!(F) \simeq \int F$. On the other hand, assuming condition (ii), we have a base change isomorphism in $ho(sSet)$:

$$A''_{a'/} \simeq \int \mathbf{R}u''^*(F) = \mathbf{L}p''_!\mathbf{R}u''^*(F) \to \mathbf{R}v''^*\mathbf{L}p'_!(F) \simeq \int F \simeq \tilde{A}'_{a'/}.$$

This ends the proof. □

[2] We remark that (i) ⇒ (iii) ⇒ (ii) gives another proof of Theorem 4.4.24.

4.4.35 Let $p\colon A \to B$ be a morphism of simplicial sets. We consider a morphism $v\colon b_0 \to b_1$ in B, as well as an object a_0 of A such that $p(a_0) = b_0$. We may see the map v as a morphism from v to 1_{b_1} in the slice B/b_1, hence as a morphism of simplicial sets $v\colon \Delta^1 \to B/b_1$. We then form the following pull-back squares.

(4.4.35.1)
$$\begin{array}{ccccccc} A_{b_1} & \hookrightarrow & A' & \longrightarrow & A/b_1 & \longrightarrow & A \\ \downarrow & & \downarrow & & \downarrow & & \downarrow{\scriptstyle p} \\ \{1\} & \hookrightarrow & \Delta^1 & \xrightarrow{v} & B/b_1 & \longrightarrow & B \end{array}$$

We denote by (v, b_0) the object of B/b_1 corresponding to $v\colon \Delta^0 * \Delta^0 = \Delta^1 \to B$, and we write $(v, a_0) = ((v, b_0), a_0)$ for the corresponding object in the fibre product $B/b_1 \times_B A = A/b_1$. Similarly, we have the object $a' = (0, a_0)$ of $A' = \Delta^1 \times_B A$. Since the map u sends a' to (v, a_0), we have a canonical commutative square of the following form.

(4.4.35.2)
$$\begin{array}{ccc} a'\backslash A' & \longrightarrow & (v, a_0)\backslash(A/b_1) \\ \downarrow & & \downarrow \\ 0\backslash\Delta^1 & \longrightarrow & (v, b_0)\backslash(B/b_1) \end{array}$$

This square is Cartesian, since the functor $(X, x) \mapsto x\backslash X$ is right adjoint to the functor $C \mapsto \Delta^0 * C$, and, therefore, commutes with small limits. Since we also have Cartesian squares of the form

(4.4.35.3)
$$\begin{array}{ccccccc} A_{b_1} & \hookrightarrow & 0\backslash A' & \longrightarrow & (v, b_0)\backslash(A/b_1) & \longrightarrow & A/b_1 \\ \downarrow & & \downarrow & & \downarrow & & \downarrow \\ \{1\} & \xrightarrow{0<1} & 0\backslash\Delta^1 & \longrightarrow & (v, b_0)\backslash(B/b_1) & \longrightarrow & B/b_1 \end{array}$$

we deduce that there is the Cartesian square below.

(4.4.35.4)
$$\begin{array}{ccc} a'\backslash A' & \longrightarrow & (v, a_0)\backslash(A/b_1) \\ \downarrow & & \downarrow \\ 0\backslash A' & \longrightarrow & (v, b_0)\backslash(A/b_1) \end{array}$$

In particular, the fibres of the two vertical maps of the latter are isomorphic, which implies that there is a canonical isomorphism

(4.4.35.5)
$$a'\backslash A_{b_1} = A_{b_1} \times_{A'} (a'\backslash A') \simeq A_{b_1} \times_{A/b_1} ((v, a_0)\backslash(A/b_1)) = (v, a_0)\backslash A_{b_1}\,.$$

Theorem 4.4.36 (Grothendieck) *Let $p\colon A \to B$ be an inner fibration between ∞-categories. The following conditions are equivalent.*

(i) *The functor p is proper.*
(ii) *For any object b of B, the canonical map $A_b \to A/b$ is final.*
(iii) *For any Cartesian squares of the form*

$$\begin{array}{ccccc} A'' & \xrightarrow{u'} & A' & \xrightarrow{u} & A \\ \downarrow & & \downarrow & & \downarrow{\scriptstyle p} \\ \{1\} & \hookrightarrow & \Delta^1 & \xrightarrow{v} & B \end{array}$$

the functor u′ is final.
(iv) *For any Cartesian squares of the form*

$$\begin{array}{ccccc} A'' & \xrightarrow{u'} & A' & \xrightarrow{u} & A \\ \downarrow{\scriptstyle p''} & & \downarrow{\scriptstyle p'} & & \downarrow{\scriptstyle p} \\ \Delta^0 & \xrightarrow{v'} & B' & \xrightarrow{v} & B \end{array}$$

in which B′ is an ∞-category, if v′ is final, so is u′.

Proof Proposition 4.4.4 shows that condition (i) implies both conditions (ii) and (iii): the map u' is the pull-back along p of the final map $\{1\} \to \Delta^1$, while $A_b \to A/b$ is the inverse image by p of the final map $b\colon \Delta^0 \to B/b$. Let us check that conditions (ii) and (iii) are equivalent.

Let $v\colon \Delta^1 \to B$ be a morphism from b_0 to b_1 in B. An object a' of $A' = \Delta^1 \times_B A$ is a pair (ε, a) where a is an object of A, and $\varepsilon = 0, 1$, such that $p(a) = b_\varepsilon$. The object a' belongs to $A'' = A_{b_1}$ if and only if $\varepsilon = 1$. For any object a such that $p(a) = b_1$, since $a' = (1, a)$ belongs to A'', the fibre product $A'' \times_{A'} a'\backslash A'$ is isomorphic to the slice $a'\backslash A''$, and thus has an initial object. This proves in particular that the ∞-category $A'' \times_{A'} (a'\backslash A')$ has an initial object and thus is weakly equivalent to the point. In other words, by virtue of Corollary 4.4.31, the map $A'' \to A'$ is final if and only if, for any object a_0 of A such that $p(a_0) = b_0$, if we put $a' = (0, a_0)$, the ∞-category $A'' \times_{A'} a'\backslash A'$ is weakly contractible. The pair (v, b_0) can be seen as an object of $A/b_1 = B/b_1 \times_B A$. Therefore, condition (ii) is equivalent to the property that the ∞-category $A_{b_1} \times_{A/b_1} ((v, b_0)\backslash(A/b_1))$ is weakly contractible for any v and a_0 as above. The isomorphism (4.4.35.5) thus proves that conditions (ii) and (iii) are equivalent.

We now prove that condition (iii) implies condition (iv). Let us consider Cartesian squares of the form

$$\begin{array}{ccccc} A'' & \xrightarrow{u'} & A' & \xrightarrow{u} & A \\ \downarrow{\scriptstyle p''} & & \downarrow{\scriptstyle p'} & & \downarrow{\scriptstyle p} \\ \Delta^0 & \xrightarrow{v'} & B' & \xrightarrow{v} & B \end{array}$$

in which B' is assumed to be an ∞-category, with v' final. If p satisfies condition (iii), so does p'. Therefore, the functor p' also satisfies condition (ii). In other words, the map $A'' = A'_v \to A'/v$ is final. Since the map $B'/v \to B'$ is a trivial fibration, so is the induced map $A'/v = B'/v \times_{B'} A' \to A'$. As any trivial fibration is final, the composed map u' is final.

To finish the proof, it remains to prove that condition (iv) implies that p is proper. Let C be the class of monomorphisms $v' \colon B'' \to B'$ such that, for any map $v \colon B' \to B$, the induced map $u' \colon B'' \times_B A \to B' \times_B A$ is a right anodyne extension. The class C is saturated. Moreover, for any pair of monomorphisms $f \colon X \to Y$ and $g \colon Y \to Z$, if both f and gf are in C, so is g: this follows right away from the fact that the class of right anodyne extensions has this property; see Corollary 4.1.10. In particular, condition (iv) implies that the class C contains any monomorphism $v' \colon B'' \to B'$ such that both B' and B'' are ∞-categories, and such that there exists a final object in B'' whose image by v is a final object of B'. Therefore, it contains all the inclusions of the form $\{1\} \times \Delta^n \subset \Delta^1 \times \Delta^n$, for $n \geq 0$. This implies that it contains all inclusions of the form $\{1\} \times X \to \Delta^1 \times X$, for any simplicial set X. To prove this, by Corollary 1.3.10, it is sufficient to prove that the class of such simplicial sets is saturated by monomorphisms. The stability by small sums is obvious. Let us consider a coCartesian square of the form

$$\begin{array}{ccc} X_0 & \longrightarrow & X_1 \\ \downarrow & & \downarrow \\ X_2 & \longrightarrow & X_3 \end{array}$$

and assume that the map $\{1\} \times X_i \to \Delta^1 \times X_i$ is in C for $i = 0, 1, 2$. For any map $\Delta^1 \times X_3 \to B$, by virtue of Corollary 2.3.17, the induced map

$$(\{1\} \times X_3) \times_B A \to (\Delta^1 \times X_3) \times_B A$$

is a trivial cofibration of the contravariant model category structure over the fibre product $(\Delta^1 \times X_3) \times_B A$, and is thus a final monomorphism, hence a right anodyne extension. The case of a countable union is proved similarly.

For any integer $n \geq 0$, the inclusion of $\{1\} \times \Delta^n$ into $\Delta^1 \times \partial\Delta^n \cup \{1\} \times \Delta^n$ is in C, because it is the push-out of the inclusion of $\{1\} \times \partial\Delta^n$ into $\Delta^1 \times \partial\Delta^n$. Since $\{1\} \times \Delta^n \subset \Delta^1 \times \Delta^n$ is in C, this shows that the inclusion of $\Delta^1 \times \Delta^n \cup \{1\} \times \Delta^n$ into $\Delta^1 \times \Delta^n$ is in C. Lemma 3.1.3 thus implies that inclusions $\Lambda^n_k \to \Delta^n$ are all in C for $0 < k \leq n$. This shows that p satisfies condition (v) of Proposition 4.4.4, hence is proper. □

Remark 4.4.37 The proof that conditions (i) and (iv) of Theorem 4.4.36 are equivalent does not require that A and B are ∞-categories. It would deserve to be inserted in Proposition 4.4.4.

4.5 Fully Faithful and Essentially Surjective Functors, Revisited

Proposition 4.5.1 *Let $u\colon A \to B$ be a morphism of simplicial sets. The following conditions are equivalent.*

(i) For any objects E in $RFib(A)$, the unit map $E \to \mathbf{R}u^*\mathbf{L}u_!(E)$ is an isomorphism.
(ii) For any object a of A, the unit map $h_a \to \mathbf{R}u^*\mathbf{L}u_!(h_a)$ is an isomorphism in $RFib(A)$, where $h_a = (\Delta^0, a)$.
(iii) For any objects F in $LFib(A)$, the unit map $F \to \mathbf{R}u^*\mathbf{L}u_!(F)$ is an isomorphism.

Proof We shall use Propositions 4.4.30 and 4.4.33 repeatedly. Condition (i) is equivalent to the condition that, for any objects E in $RFib(A)$ and F in $LFib(A)$, the induced map

$$\mathbf{R}\langle E, F\rangle \to \mathbf{R}\langle \mathbf{R}u^*\mathbf{L}u_!(E), F\rangle \simeq \mathbf{R}\langle \mathbf{L}u_!(E), \mathbf{L}u_!(F)\rangle \simeq \mathbf{R}\langle E, \mathbf{R}u^*\mathbf{L}u_!(F)\rangle$$

is invertible in $ho(sSet)$. This is equivalent to the assertion that, for any object a of A and any object F in $LFib(A)$, the induced map

$$\mathbf{R}\langle h_a, F\rangle \to \mathbf{R}\langle h_a, \mathbf{R}u^*\mathbf{L}u_!(F)\rangle \simeq \mathbf{R}\langle \mathbf{R}u^*\mathbf{L}u_!(h_a), F\rangle$$

is invertible. □

Proposition 4.5.2 *Let $u\colon A \to B$ be a functor of ∞-categories. The following conditions are equivalent.*

(i) The functor u is fully faithful.
(ii) The functor $\mathbf{L}u_!\colon RFib(A) \to RFib(B)$ is fully faithful.
(iii) The functor $\mathbf{L}u_!\colon LFib(A) \to LFib(B)$ is fully faithful.

Proof We already know from Proposition 4.5.1 that conditions (ii) and (iii) are equivalent. To prove that condition (i) is equivalent to condition (ii), we shall verify that u is fully faithful if and only if condition (ii) of Proposition 4.5.1 is satisfied. Let a be an object of A. Then $\mathbf{L}u_!(h_a) = h_{u(a)}$. Since A and B are ∞-categories, fibrant replacements of h_a and $h_{u(a)}$ are given by A/a and $B/u(a)$ in the contravariant model category structures over A and B, respectively. The map $h_a \to \mathbf{R}u^*\mathbf{L}u_!(h_a)$ is thus the natural morphism $A/a \to A \times_B B/u(a)$. By Corollary 4.2.10, for any object x of A, the induced map on the homotopy fibres is the natural morphism $A(x, a) \to B(u(x), u(a))$. Therefore, u is fully faithful if and only if the functor $\mathbf{L}u_!$ is fully faithful on $RFib(A)$. □

Remark 4.5.3 Let A be a simplicial set, and a an object of A. For any object F of $RFib(A)$, there is a canonical isomorphism

$$\mathbf{R}\langle F, h_a\rangle \simeq \mathbf{R}a^*(F) \tag{4.5.3.1}$$

in $ho(sSet)$. Indeed, it is sufficient to prove this for F fibrant, i.e. for F corresponding to a right fibration $p\colon X \to A$. In this case, $\mathbf{R}a^*(F) = X_a = \langle F, h_a\rangle$ is simply the fibre of X over a.

Any map $u\colon a \to b$ in A induces a natural morphism $u^*\colon \mathbf{R}b^*(F) \to \mathbf{R}a^*(F)$. Indeed, for F fibrant as above, one can also describe $\mathbf{R}a^*(F)$ as the mapping space $\mathrm{Map}_A(h_a, X) \simeq X_a$, where $h_a = (\Delta^0, a)$. Since the inclusion $\{1\} \to \Delta^1$ is a right anodyne extension, the canonical map

$$\mathrm{Map}_A((\Delta^1, u), X) \to X_b$$

is a trivial fibration. The choice of a section of the latter, composed with the map $\mathrm{Map}_A((\Delta^1, u), X) \to X_a$, defines a morphism

$$u^*\colon \mathbf{R}b^*(F) \simeq \mathrm{Map}_A(h_b, X) \to \mathrm{Map}_A(h_a, X) \simeq \mathbf{R}a^*(F) \tag{4.5.3.2}$$

in $ho(sSet)$.

Lemma 4.5.4 *Let us assume that A is an ∞-category. If the map $u\colon a \to b$ is invertible in A, for any object F of $RFib(A)$, the induced map $u^*\colon \mathbf{R}b^*(F) \to \mathbf{R}a^*(F)$ is an isomorphism in $ho(sSet)$.*

Proof Since $\Delta^2 = \Delta^1 * \Delta^0 = \Delta^0 * \Delta^1$, the map

$$v\colon \Delta^2 \xrightarrow{\sigma_0^1} \Delta^1 \xrightarrow{u} A$$

can be interpreted in two ways: as a map $\tilde{u}\colon \Delta^1 \to A/b$, and as a map $\tilde{a}\colon \Delta^0 \to A/u$. We then have the following commutative diagram.

$$\begin{array}{ccccc} & & \{0\} & \xrightarrow{\tilde{a}} & A/u & \longrightarrow & A/a \\ & & \downarrow & & \downarrow & & \\ \{1\} & \hookrightarrow & \Delta^1 & \xrightarrow{\tilde{u}} & A/b & & \end{array}$$

The map $A/u \to A/a$ is a trivial fibration because the inclusion $\{0\} \to \Delta^1$ is a left anodyne extension; see Lemma 3.4.21. Therefore, $\tilde{a}$ is a final object of A/u, because its image in A/a is final. Similarly, the map $\tilde{u}$ is final because it sends the final object 1 of Δ^1 to the final object 1_b of A/b. It is thus sufficient to prove that the canonical map $A/u \to A/b$ has contractible fibres over A (this map is a morphism of right fibrations over A, so that this will ensure that it is a weak equivalence of the contravariant model category structure over A, by Theorem 4.1.16, and this will allow us to apply Proposition 4.1.14). Proposition 4.2.9 shows that we may as well consider the fibres of the canonical map $A//u \to A//b$ over A. For each object x of A, we have a trivial fibration $A(x,a) \leftarrow (A//u)_x$, and a canonical map $(A//u)_x \to A(x,b)$. Therefore, we obtain a map

$$\mathrm{Hom}_{ho(A)}(x,a) = \pi_0(A(x,a)) \to \pi_0(A(x,b)) = \mathrm{Hom}_{ho(A)}(x,b)$$

which is nothing other than the composition by f in $ho(A) \simeq \tau(A)$. Since u is invertible, the latter map is bijective. But, since the image of an invertible map is invertible, we also know that u, seen as a morphism of constant diagrams, is invertible in the ∞-category $\underline{\mathrm{Hom}}(K,A)$ for any simplicial set K. Therefore, since $\underline{\mathrm{Hom}}(K,A)(x,a) = \underline{\mathrm{Hom}}(K,A(x,a))$ for x and a as above, seen as constant diagrams, what precedes also proves that the induced map

$$\pi_0(\underline{\mathrm{Hom}}(K,A(x,a))) \simeq \pi_0\,\underline{\mathrm{Hom}}(K,(A//u)_x) \to \pi_0(\underline{\mathrm{Hom}}(K,A(x,b)))$$

is bijective for any simplicial set K. For any simplicial set K and any Kan complex T, there is a canonical bijection $\mathrm{Hom}_{ho(sSet)}(K,T) \simeq \pi_0(\underline{\mathrm{Hom}}(K,T))$. Therefore, the Yoneda lemma applied to the homotopy category $ho(sSet)$ implies that the map $(A//u)_x \to A(x,b)$ is a simplicial homotopy equivalence. □

Proposition 4.5.5 *If a functor of ∞-categories $u\colon A \to B$ is essentially surjective, then the derived pull-back functor $\mathbf{R}u^*\colon RFib(B) \to RFib(A)$ is conservative.*

Proof If a map $E \to F$ in $RFib(A)$ induces an isomorphism $\mathbf{R}u^*(E) \simeq \mathbf{R}u^*(F)$, then it induces an isomorphism

$$\mathbf{R}u(a)^*(E) \simeq \mathbf{R}a^*\mathbf{R}u^*(E) \simeq \mathbf{R}a^*\mathbf{R}u^*(F) \simeq \mathbf{R}u(a)^*(F)$$

for any object a of A. But, since u is essentially surjective, any object b of B is equivalent to an object of the form $u(a)$. Hence the functor $\mathbf{R}b^*$ is isomorphic to $\mathbf{R}u(a)^*$, by the preceding lemma. Therefore, Theorem 4.1.16 implies that the map $E \to F$ is an isomorphism. □

Corollary 4.5.6 *Any equivalence of ∞-categories $u\colon A \to B$ induces equivalences of categories*

$$\mathbf{L}u_!\colon RFib(A) \to RFib(B) \quad \textit{and} \quad \mathbf{L}u_!\colon LFib(A) \to LFib(B)\,.$$

Proof If u is an equivalence of ∞-categories, it is fully faithful and essentially surjective. Therefore, by virtue of Propositions 4.5.2 and 4.5.5, the functor $\mathbf{L}u_!$

is fully faithful on $RFib(A)$, and its right adjoint $\mathbf{R}u^*$ is conservative, hence both functors are equivalences of categories and are quasi-inverse to each other. The other equivalence comes from the natural identification $RFib(A) \simeq LFib(A^{op})$. □

Remark 4.5.7 There is a much more direct way to prove Corollary 4.5.6: one proves rather easily that any trivial fibration of the Joyal model category structure $u: A \to B$ induces an equivalence of categories $\mathbf{L}u_!: RFib(A) \simeq RFib(B)$ and we use Ken Brown's lemma (Proposition 2.2.7) for the Joyal model category structure. Although this is non-trivial, it will be seen later that there is no need to restrict ourselves to equivalences between ∞-categories; see Theorem 5.2.14.

4.6 Locally Constant Functors and Quillen's Theorem B

For a simplicial set X, we have three model category structures on the category $sSet/X$: the contravariant model category structure, the covariant model category structure, and the model category structure induced by the Kan–Quillen model category structure. Since we want to interpret the contravariant model category structure as a theory of presheaves over X, and since any Kan fibration of codomain X is a fibrant object in the contravariant model category structure, it is natural to ask what is the meaning of Kan fibrations in this semantic interpretation of the theory of presheaves. This is what we shall investigate in this section.

Proposition 4.6.1 *Let W be a class of morphisms of simplicial sets. We assume that W has the following properties.*

(i) For any morphisms of the form $f: X \to Y$ and $g: Y \to Z$, with f in W, then g is in W if and only if gf is in W.
(ii) The class of monomorphisms which are in W is saturated.
(iii) Any trivial fibration is in W.
(iv) For any integer $n \geq 0$, any map $\Delta^0 \to \Delta^n$ is in W.

Then any weak homotopy equivalence is in W.

Proof We first prove that any inclusion of the form $\Lambda^n_k \to \Delta^n$, $n \geq 1$, $0 \leq k \leq n$, is in W. In fact, we have already seen a proof: we just have to use the same combinatorial arguments as in the proof of Proposition 3.1.18 (this only requires conditions (i), (ii) and (iv)). Condition (ii) thus implies that any anodyne extension is in W. Since any weak homotopy equivalence is the composition of an anodyne extension with a trivial fibration, condition (iv) implies that any weak homotopy equivalence is in W. □

Proposition 4.6.2 *Let $p\colon X \to Y$ be a morphism of simplicial sets. The following conditions are equivalent.*

(i) For any Cartesian squares of the form

$$
\begin{array}{ccccc}
X'' & \xrightarrow{u'} & X' & \xrightarrow{u} & X \\
\downarrow{\scriptstyle p''} & & \downarrow{\scriptstyle p'} & & \downarrow{\scriptstyle p} \\
Y'' & \xrightarrow{v'} & Y' & \xrightarrow{v} & Y
\end{array}
\tag{4.6.2.1}
$$

if v' is a weak homotopy equivalence, so is u'.

(ii) As in (i), but only when $Y'' = \Delta^0$.

(iii) As in (ii), but only when $Y' = \Delta^n$.

(iv) As in (i), but only when v' is an inclusion of the form $\Lambda^n_k \to \Delta^n$ for $n \geq 1$ and $0 \leq k \leq n$.

(v) Any Cartesian square of the form

$$
\begin{array}{ccc}
X' & \xrightarrow{u} & X \\
\downarrow{\scriptstyle p'} & & \downarrow{\scriptstyle p} \\
Y' & \xrightarrow{v} & Y
\end{array}
\tag{4.6.2.2}
$$

is homotopy Cartesian in the Kan–Quillen model category structure.

Proof We prove first that conditions (i) and (v) are equivalent. If condition (v) is verified, then any Cartesian squares of the form (4.6.2.1) are homotopy Cartesian, and, since weak equivalences are stable under homotopy pull-backs, this proves that (v) $\Rightarrow$ (i). The converse follows right away from Corollary 3.8.4. Let us assume condition (i). Given any Cartesian square of the form (4.6.2.2), we choose a factorisation of the map v into an anodyne extension $i\colon Y' \to T$ followed by a Kan fibration $s\colon T \to Y$, and we form the Cartesian squares below.

$$
\begin{array}{ccccc}
X' & \xrightarrow{i} & U & \xrightarrow{r} & X \\
\downarrow & & \downarrow & & \downarrow{\scriptstyle p} \\
Y'' & \xrightarrow{j} & T & \xrightarrow{s} & Y
\end{array}
$$

Then both maps i and j are weak homotopy equivalences, and the right-hand square is homotopy Cartesian, which implies that the commutative square (4.6.2.2) has the same property.

Let W be the class of morphisms of simplicial sets $A \to B$ such that, for any map $B \to Y$, the induced map $A \times_Y X \to B \times_Y X$ is an anodyne extension. Since pulling back along p preserves monomorphisms and colimits, we check that the class of monomorphisms which are in W is saturated. Finally, the class W satisfies the two-out-of-three property and contains the class of trivial fibrations. Since any weak homotopy equivalence is the composition of an anodyne extension and a trivial fibration, condition (i) is equivalent to the assertion that the pull-back of any anodyne extension along p is a trivial cofibration of the Kan–Quillen model category structure. Since, by definition, the smallest saturated class of morphisms of simplicial sets which contains horns $\Lambda^n_k \to \Delta^n$ for $n \geq 1$ and $0 \leq k \leq n$ is the class of anodyne extensions, we see that conditions (i) and (iv) are equivalent.

It is clear that we have (i) $\Rightarrow$ (ii) $\Rightarrow$ (iii). Proposition 4.6.1 shows that condition (iii) implies condition (iv). □

Definition 4.6.3 A morphism of simplicial sets is *locally constant* if it satisfies the equivalent conditions of Proposition 4.6.2.

Proposition 4.6.4 *A proper morphism of simplicial sets $p \colon X \to Y$ is locally constant if and only if, for any map $v \colon \Delta^n \to Y$, the inclusion $X_{v(0)} \to \Delta^n \times_Y X$ is a weak homotopy equivalence.*

Proof This is clearly a necessary condition. For the converse, we shall prove that a proper morphism p satisfying this extra condition satisfies condition (iii) of Proposition 4.6.2. Let us consider Cartesian squares of the form (4.6.2.1) such that $Y'' = \Delta^0$. We want to prove that the map u' is a weak equivalence. We proceed by induction on n. Since the case $n = 0$ is trivial, we may assume that $n > 0$. If v' is initial, the property we seek is true by assumption. If not, we may assume that v' factors through the final map $w \colon \Delta^{n-1} \to \Delta^n$ defined by $w(i) = i + 1$. Since p is proper, the induced map $\Delta^{n-1} \times_Y X \to \Delta^n \times_Y X$ is final, hence a weak homotopy equivalence, and the induced map $\Delta^0 \times_Y X \to \Delta^{n-1} \times_Y X$ is a weak homotopy equivalence by induction on n. Therefore, u' is the composition of two weak homotopy equivalences. □

Corollary 4.6.5 *Any morphism of simplicial sets which is both smooth and proper is locally constant.*

Proof If $p \colon X \to Y$ is smooth then, for any map $v \colon \Delta^n \to Y$, the map $X_{v(0)} \to \Delta^n \times_Y X$ is cofinal, hence a weak homotopy equivalence. Therefore, if p is also proper, the preceding proposition ensures that it is locally constant. □

Corollary 4.6.6 *Any Kan fibration is locally constant.*

Proof Any right fibration is smooth, and any left fibration is proper. □

Remark 4.6.7 The proof above is a little pedantic, since, for Kan fibrations, one can check several of the conditions of Proposition 4.6.2 directly. For instance, as explained in the proof of that proposition, condition (v) follows right away from Corollary 3.8.4 (which is itself a rather direct consequence of the good properties of the functor Ex^∞).

Definition 4.6.8 Let A be a simplicial set. An object F of $RFib(A)$ is *locally constant* if, for any map $u\colon a \to b$ in A, the induced map $u^*\colon \mathbf{R}b^*(F) \to \mathbf{R}a^*(F)$ (as defined in Remark 4.5.3) is an isomorphism in $ho(sSet)$.

Lemma 4.6.9 *Let $p\colon X \to A$ be a right fibration, and $u\colon a \to b$ be a morphism in A. We form the following Cartesian squares.*

$$\begin{array}{ccccc} X_b & \xrightarrow{i} & X_u & \longrightarrow & X \\ \downarrow & & \downarrow & & \downarrow{\scriptstyle p} \\ \Delta^0 & \xrightarrow{1} & \Delta^1 & \xrightarrow{u} & A \end{array}$$

If $F = (X, p)$ denotes the object of $RFib(A)$ associated to p, then there are canonical isomorphisms

$$X_b \simeq \mathbf{R}b^*(F) \quad \textit{and} \quad X_u \simeq \mathbf{R}a^*(F)$$

in $ho(sSet)$. Furthermore, under these identifications, the map i corresponds in $ho(sSet)$ to the canonical map $\mathbf{R}b^*(F) \to \mathbf{R}a^*(F)$ of Remark 4.5.3.

Proof Since 0 is an initial object of Δ^1 and since p is smooth, the inclusion $j\colon X_a \to X_u$ is cofinal, hence a weak homotopy equivalence. It remains to check the compatibility of the map i with the construction of Remark 4.5.3. We want to check that the diagram

$$\begin{array}{ccc} X_b & \xrightarrow{u^*} & X_a \\ & {\scriptstyle i}\searrow \quad \swarrow{\scriptstyle j} & \\ & X_u & \end{array}$$

commutes in $ho(sSet)$.

In other words, we want to check that, for any simplicial set K, the diagram

$$\begin{array}{ccc} \pi_0(\underline{\mathrm{Hom}}(K, X_b)) & \xrightarrow{u^*} & \pi_0(\underline{\mathrm{Hom}}(K, X_a)) \\ & {\scriptstyle i}\searrow \qquad \swarrow{\scriptstyle j} & \\ & \pi_0(\underline{\mathrm{Hom}}(K, X_u)) & \end{array}$$

commutes in the category of sets.

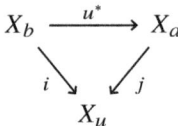

For a given simplicial set K, let $q\colon K \times A \to A$ be the second projection. Since q is proper, the pair (q^*, q_*) is a Quillen adjunction for the contravariant model category structures, and we write $p^K\colon X^K \to A$ for the map whose corresponding object of $sSet/A$ is $q_*q^*(X, p)$. There is a Cartesian square of simplicial sets of the form

$$\begin{array}{ccc} X^K & \longrightarrow & \underline{\mathrm{Hom}}(K, X) \\ {\scriptstyle p^K}\downarrow & & \downarrow{\scriptstyle p_*} \\ A & \longrightarrow & \underline{\mathrm{Hom}}(K, A) \end{array}$$

where the lower horizontal map is the one induced by $K \to \Delta^0$. We remark that, for any map $w\colon A' \to A$, if $p'\colon X' = A' \times_A X \to A'$ denotes the pull-back of p along w, then we have a canonical Cartesian square of the following form.

$$\begin{array}{ccc} X'^K & \longrightarrow & X^K \\ {\scriptstyle p'^K}\downarrow & & \downarrow{\scriptstyle p^K} \\ A' & \longrightarrow & A \end{array}$$

In particular, the fibre of p^K at $a \in A_0$ is $\underline{\mathrm{Hom}}(K, X_a)$. We also have, for any simplicial set E over A, a canonical isomorphism of simplicial sets

$$\underline{\mathrm{Hom}}(K, \mathrm{Map}_A(E, X)) \simeq \mathrm{Map}_A(E, X^K)\,.$$

This shows that the construction of the map u^* is compatible with the operation $X \mapsto X^K$. We remark finally that the map $X_u^K \to \underline{\mathrm{Hom}}(K, X_u)$ is a weak homotopy equivalence: this is the inverse image of the inclusion $\Delta^1 \to \underline{\mathrm{Hom}}(K, \Delta^1)$ along the right fibration $\underline{\mathrm{Hom}}(K, X_u) \to \underline{\mathrm{Hom}}(K, \Delta^1)$. Since any right fibration is smooth, it is sufficient to prove that $\Delta^1 \to \underline{\mathrm{Hom}}(K, \Delta^1)$ is cofinal. This follows from the fact that the object 0 is initial in both Δ^1 and $\underline{\mathrm{Hom}}(K, \Delta^1)$. This shows that, up a canonical weak homotopy equivalence, the formation of the maps i and j is compatible with the operation $X \mapsto X^K$.

In other words, we may replace X by X^K, and it is sufficient to prove that $\pi_0(j)\pi_0(u^*) = \pi_0(i)$. In fact, both $\pi_0(u^*)$ and $\pi_0(j)^{-1}\pi_0(i)$ have the following explicit description. Given a connected component C of X_b, we choose an element $x \in C$. We then have the solid commutative square below

$$\begin{array}{ccc} \{1\} & \xrightarrow{\;x\;} & X \\ \downarrow & \overset{h}{\nearrow} & \downarrow{\scriptstyle p} \\ \Delta^1 & \xrightarrow{\;u\;} & A \end{array}$$

a dotted filler of which always exists (because 1 is final in Δ^1 and p is a right fibration), and the image of C in $\pi_0(X_a)$ is the connected component of $h(0)$. Each of the maps $\pi_0(u^*)$ and $\pi_0(j)^{-1}\pi_0(i)$ is determined by a choice, for each connected component C, of a point x in C, and of such a lift h.

Nevertheless, the filler h is unique in the following sense. If we endow the category of pointed simplicial sets over A with the model structure induced by the contravariant model category structure over A, the map h is a morphism from the cofibrant object $(\Delta^1, 1, u)$ to the fibrant object (X, x, p). Since $(\Delta^1, 1, u)$ is weakly equivalent to the initial object, such a map h is unique up to homotopy in the category of pointed simplicial sets over A. In particular, the connected component of $h(0)$ in the fibre X_a does not depend on the choice of the lift h. It remains to check that it does not depend on the choice of the point x in C. Since C is the connected component of a Kan complex, if there are two points x and x' in C, there is a map $\xi\colon x' \to x$ in C. The pair (ξ, h) can be seen as a diagram of the form

$$\begin{array}{ccc} & & x' \\ & & \downarrow{\scriptstyle \xi} \\ h(0) & \xrightarrow{\ h\ } & x \end{array}$$

whence as a map $(\xi, h)\colon \Lambda^2_2 \to X$. The obvious commutative triangle

$$\begin{array}{ccc} & b & \\ {\scriptstyle u}\nearrow & & \searrow{\scriptstyle 1_b} \\ a & \xrightarrow{\ u\ } & b \end{array}$$

given by the map $u\sigma^1_2\colon \Delta^2 \to A$ thus provides a solid commutative square of simplicial sets

$$\begin{array}{ccc} \Lambda^2_2 & \xrightarrow{(\xi,h)} & X \\ \big\downarrow & \overset{t}{\nearrow} & \big\downarrow{\scriptstyle p} \\ \Delta^2 & \xrightarrow{\sigma^1_2 u} & A \end{array}$$

which admits a filler t. The restriction of t to $\Delta^{\{0,1\}}$ defines a map $h'\colon h(0) \to x'$ in X such that $p(h') = u$, which shows that the choice of a point of C does not affect the final result in $\pi_0(X_a)$. □

Theorem 4.6.10 *Let $p\colon X \to A$ be a right fibration, and $F = (X, p)$ the corresponding object in $RFib(A)$. The following conditions are equivalent.*

(i) The map p is locally constant.
(ii) The object F of $RFib(A)$ is locally constant.
(iii) The map p is a Kan fibration.

Proof We already know that (iii) ⇒ (i). It follows right away from Lemma 4.6.9 that (i) ⇒ (ii). It remains to prove that (ii) ⇒ (iii).

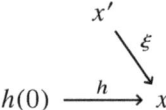

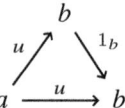

Let us assume that F is locally constant. For any map $u\colon \Delta^n \to A$, and any element i of $[n]$, the induced map

$$\mathrm{Map}_A((\Delta^n, u), F) \to \mathrm{Map}_A((\Delta^0, u(i)), F) = X_{u(i)}$$

is a trivial fibration. Indeed, by Proposition 4.1.13, this map is always a Kan fibration, and, by Proposition 4.1.14, the map

$$\mathrm{Map}_A((\Delta^n, u), F) \to \mathrm{Map}_A((\Delta^0, u(n)), F) = X_{u(n)}$$

is a trivial fibration. Under the latter identification, the corresponding map

$$X_{u(n)} = \mathbf{R}u(n)^*(F) \to \mathbf{R}u(i)^*(F) = X_{u(i)}$$

in $ho(sSet)$ is the canonical map induced by the map $u(i) \to u(n)$ in A. Let W be the class of maps $u\colon K \to L$ such that, for any map $v : L \to A$, the induced morphism

$$u^*\colon \mathrm{Map}_A((L, v), F) \to \mathrm{Map}_A((K, vu), F)$$

is an equivalence of ∞-groupoids. This class satisfies the hypothesis of Proposition 4.6.1 and thus contains all weak homotopy equivalences. In particular, for any anodyne extension $u\colon K \to L$ and any map $v : L \to A$, the map

$$u^*\colon \mathrm{Map}_A((L, v), F) \to \mathrm{Map}_A((K, vu), F)$$

is a trivial fibration. A section of the latter, at the level of objects, provides a lift in any solid commutative square of the form below.

$$\begin{array}{ccc} K & \longrightarrow & X \\ {\scriptstyle u}\downarrow & \nearrow & \downarrow{\scriptstyle p} \\ L & \xrightarrow{\ v\ } & A \end{array}$$

This implies that p is a Kan fibration. □

Proposition 4.6.11 *Let $u\colon A \to B$ be a morphism of simplicial sets. The following conditions are equivalent.*

(i) The pair (A, u) is locally constant as an object of $RFib(B)$.
(ii) For any Cartesian squares of the form

$$\begin{array}{ccccc} A'' & \xrightarrow{f'} & A' & \xrightarrow{f} & A \\ \downarrow{\scriptstyle u''} & & \downarrow{\scriptstyle u'} & & \downarrow{\scriptstyle u} \\ B'' & \xrightarrow{g'} & B' & \xrightarrow{g} & B \end{array} \qquad (4.6.11.1)$$

if g and gg' are left fibrations and if both B' and B'' have initial objects, then the map f' is a weak homotopy equivalence.

(iii) Any Cartesian square of the form

$$
\begin{array}{ccc}
A' & \xrightarrow{f} & A \\
\downarrow{\scriptstyle u'} & & \downarrow{\scriptstyle u} \\
B' & \xrightarrow{g} & B
\end{array}
\tag{4.6.11.2}
$$

in which the map g is a left fibration is homotopy Cartesian in the Kan–Quillen model category structure.

Proof In a Cartesian square of the form (4.6.11.2), if g is a left fibration and if B' has an initial object whose image in B is denoted by b, then $\mathbf{R}b^*(A, u)$ is canonically isomorphic to A' in $ho(sSet)$. Using Lemma 4.6.9, one sees that conditions (i) and (ii) are equivalent. Since the homotopy pull-back of a weak equivalence is a weak equivalence, condition (iii) implies condition (ii). Let $i \colon A \to C$ be a right anodyne extension followed by a right fibration $p \colon C \to B$ such that $u = pi$. Since (A, u) and (C, p) are isomorphic in $RFib(B)$, one is locally constant if and only if the other has the same property. By virtue of Theorem 4.6.10, condition (i) is equivalent to the condition that p is a Kan fibration. Let us consider a Cartesian square of the form (4.6.11.2). If we put $C' = B' \times_B C$, the map $i' \colon A' \to C'$ induced by i is a right anodyne extension because it is the pull-back of i by the left fibration g, hence by a proper map. Since C' is the homotopy pull-back of C and B' over B (because p is a Kan fibration), this proves that the square (4.6.11.2) is homotopy Cartesian. □

Corollary 4.6.12 (Quillen's theorem B) *Let $u \colon A \to B$ be a functor between small categories. Assume that, for any map $b_0 \to b_1$ in B, the nerve of the induced functor $b_1\backslash A \to b_0\backslash A$ is a weak homotopy equivalence. Then, for any object b of B, the Cartesian square*

$$
\begin{array}{ccc}
N(b\backslash A) & \longrightarrow & N(A) \\
\downarrow & & \downarrow{\scriptstyle N(u)} \\
N(b\backslash B) & \longrightarrow & N(B)
\end{array}
$$

is homotopy Cartesian in the Kan–Quillen model category structure.

Proof For any left fibration $B' \to N(B)$, if B' has an initial object b' whose image in B is denoted by b, the canonical maps $b'\backslash B' \to N(b\backslash B) = b\backslash N(B)$ and $b'\backslash B' \to B'$ are trivial fibrations. Therefore, the pull-backs of any of these maps along any morphism $X \to N(B)$ will remain a weak homotopy equivalence. One deduces from there that the nerve of u satisfies condition (ii) of the previous Proposition 4.6.11, hence also condition (iii). □

5

Presheaves: Internally

Our aim here is to construct the ∞-category of ∞-groupoids $\mathcal{S}$ (with a smallness condition determined by a given universe, to keep $\mathcal{S}$ small itself). The way we will define $\mathcal{S}$ will be so that any ∞-groupoid can be interpreted tautologically as an object of $\mathcal{S}$. This gives rise to the question of turning the assignment

$$(x, y) \mapsto A(x, y)$$

into a functor $A^{op} \times A \to \mathcal{S}$ for any ∞-category A. Equivalently, the question is one of defining the Yoneda embedding $h_A : A \to \widehat{A}$ for any ∞-category A (with appropriate smallness assumptions). A related question consists in interpreting each left fibration $p : X \to A$ as a functor $F : A \to \mathcal{S}$ (the value of F at a being the fibre of p at a). The way we shall define $\mathcal{S}$ will make the latter correspondence true by definition. Significant efforts will be necessary to prove that this defines an ∞-category (as opposed to a mere simplicial set), and then to prove that this correspondence is not only syntactic, but also homotopy-theoretic: given two left fibrations $p : X \to A$ and $q : Y \to A$ corresponding to functors $F, G : A \to \mathcal{S}$, we shall have to compare the mapping space $\mathrm{Map}_A(X, Y)$ with the ∞-groupoid $\underline{\mathrm{Hom}}(A, \mathcal{S})(F, G)$.

The first section is a complement to Section 2.4 of Chapter 2: this is a general theory of minimal fibrations in a model category on a category of presheaves, under appropriate combinatorial assumptions. One proves that any fibration may be approximated by a minimal fibration, and that weak equivalences between minimal fibrations always are isomorphisms. These properties mean that some coherence problems can be solved whenever they have solutions up to a weak equivalence. This will be used in Section 5.2, where we define the universal left fibration with small fibres $p_{univ} : \mathcal{S}_\bullet \to \mathcal{S}$. We also prove that the homotopy theory of left fibrations over a simplicial set X is invariant under weak categorical equivalences $X \to Y$ (Theorem 5.2.14).

We then prove in Section 5.3 that the correspondence between left fibrations over A and functors $A \to \mathcal{S}$ is also homotopy-theoretic in the sense sketched above. This involves a correspondence between the ∞-groupoid of invertible maps between two functors with values in $\mathcal{S}$ and the space of fibre-wise equivalences between the associated left fibrations. Section 5.4 extends this correspondence to possibly non-invertible morphisms.

In order to define the Yoneda embedding, we develop in the fifth section a homotopy theory of left bifibrations in the category of bisimplicial sets. This section is quite technical and may be avoided at first: it is used only twice in Section 5.6, in the proof (but not in the formulation) of Propositions 5.6.2 and 5.6.5, which explain how to see mapping spaces in families, using the language of left fibrations. However, this theory of left bifibrations might be useful for other purposes as well: it relies on a generalisation to Joyal's covariant model structures, of the well-known fact that the diagonal of a levelwise weak homotopy equivalence of bisimplicial sets is a weak homotopy equivalence. It also provides a source of left fibrations under the form of diagonals of left bifibrations (and it is this latter property that we really use in the text).

Section 5.7 compares various versions of the notion of locally small ∞-category. Discussing such a set-theoretic issue is essential, simply to properly formulate and use the Yoneda lemma. Practical criteria for local smallness will also be provided later in Section 7.10.

Finally, Section 5.8 is devoted to the construction of the Yoneda embedding and to the proof of the Yoneda lemma itself: Theorem 5.8.13. We emphasise that, as can be seen in its proof, the Yoneda lemma really is a cofinality statement; this is explicitly formulated in Lemma 5.8.11 below.

5.1 Minimal Fibrations

5.1.1 In this section, we fix once and for all an *Eilenberg–Zilber category* A; see Definition 1.3.1 (in practice, A will be of the form $\mathbf{\Delta}/Y$ for a fixed simplicial set Y). Given a representable presheaf a on A, we shall write ∂a for the maximal proper subobject of a.

We consider that we are given a model category structure on $\widehat{A}$, whose cofibrations are precisely the monomorphisms.

We also choose an exact cylinder I such that the projection $I \otimes X \to X$ is a weak equivalence for any presheaf X on A (e.g. we can take for I the Cartesian product with the subobject classifier of the topos $\widehat{A}$). As usual, we write $\partial I = \{0\} \amalg \{1\} \subset I$ for the inclusion of the two end-points of I.

5.1.2 Let $h \colon I \otimes X \to Y$ be a homotopy. For $e = 0, 1$, we write h_e for the composite

$$X = \{e\} \otimes X \to I \otimes X \xrightarrow{h} Y .$$

Given a subobject $S \subset X$, we say that h is *constant on* S if the restriction $h|_{I \otimes S}$ is the canonical projection $I \otimes S \to S$.

Given a presheaf X, a *section of* X is a map $x\colon a \to X$ with a a representable presheaf. The *boundary* of such a section x is the map

$$\partial x\colon \partial a \to a \xrightarrow{x} X .$$

Definition 5.1.3 Let X be an object of $\widehat{A}$.

Two sections $x, y\colon a \to X$ are *∂-equivalent* if the following conditions are satisfied.

(i) These have the same boundaries: $\partial x = \partial y$.
(ii) There exists a homotopy $h\colon I \otimes a \to X$ which is constant on ∂a, and such that $h_0 = x$ and $h_1 = y$.

We write $x \sim y$ whenever x and y are ∂-equivalent.

A *minimal complex* is a fibrant object S such that, for any two sections $x, y\colon a \to S$, if x and y are ∂-equivalent, then $x = y$.

A *minimal model* of X is a trivial cofibration $S \to X$ with S a minimal complex.

Proposition 5.1.4 *Let X be a fibrant object. The ∂-equivalence relation is an equivalence relation on the set of sections of X, and this relation does not depend on the choice of the exact cylinder I.*

Proof Let a be a representable presheaf on A, and let x and y be two sections of X over a such that $\partial x = \partial y$. One can see x and y as maps from the cofibrant object a to the fibrant object X of $\partial a\backslash\widehat{A}$, with the obvious induced model category structure. Since the relation of homotopy between maps from a cofibrant object to a fibrant object is always an equivalence relation which is independent of the choice of a cylinder object, this shows the proposition. □

Proposition 5.1.5 *Let $\varepsilon \in \{0, 1\}$, and consider a fibrant object X, together with two maps $h, k\colon I \otimes a \to X$, with a representable, such that the restrictions of k and h coincide on $I \otimes \partial a \cup \{1-\varepsilon\} \otimes a$. If, furthermore, we have $\partial h_\varepsilon = \partial k_\varepsilon$, then the sections h_ε and k_ε are ∂-equivalent.*

Proof Put $z = h_{1-\varepsilon} = k_{1-\varepsilon}$; we also write ℓ for the restriction of h (and of k) on $I \otimes \partial a$. We define $\varphi\colon I \otimes (I \otimes \partial a \cup \{1-\varepsilon\} \otimes a) \to X$ as the constant homotopy at the map $(\ell, z)\colon I \otimes \partial a \cup \{1-\varepsilon\} \otimes a \to X$, and $\psi\colon \partial I \otimes (I \otimes a) \to X$ as the map whose restrictions to $\{0\} \otimes I \otimes a$ and $\{1\} \otimes I \otimes a$ are h and k, respectively. Since $\partial h_\varepsilon = \partial k_\varepsilon$, this defines a map

$$f = (\varphi, \psi) \colon (I \otimes (I \otimes \partial a \cup \{1 - \varepsilon\} \otimes a)) \cup (\partial I \otimes (I \otimes a)) \to X \,.$$

Since the embedding of the source of f into $I \otimes I \otimes a$ is a trivial cofibration, the map f is the restriction of some map $F \colon I \otimes I \otimes a \to X$. We define a morphism $H \colon I \otimes a \to X$ as the restriction of F on $I \otimes \{\varepsilon\} \otimes a$. This is a homotopy from h_ε to k_ε which is constant on ∂a. □

Lemma 5.1.6 *Let X be any presheaf on A, and $x_0, x_1 \colon a \to X$ two degenerate sections. If x_0 and x_1 are ∂-equivalent, then they are equal.*

Proof For $\varepsilon = 0, 1$, there is a unique couple $(p_\varepsilon, y_\varepsilon)$, where $p_\varepsilon \colon a \to b_\varepsilon$ is a split epimorphism in A and $y_\varepsilon \colon b_\varepsilon \to X$ is a non-degenerate section of X such that $x_\varepsilon = y_\varepsilon p_\varepsilon$. Let us choose a section s_ε of p_ε. As x_0 and x_1 are degenerate and since $\partial x_0 = \partial x_1$, we have $x_0 s_0 = x_1 s_0$ and $x_0 s_1 = x_1 s_1$. On the other hand, we have $y_\varepsilon = x_\varepsilon s_\varepsilon$. We thus have the equalities $y_0 = y_1 p_1 s_0$ and $y_1 = y_0 p_0 s_1$. These imply that the maps $p_\varepsilon s_{1-\varepsilon} \colon b_{1-\varepsilon} \to b_\varepsilon$ are in A_+ and that b_0 and b_1 have the same dimension. This means that $p_\varepsilon s_{1-\varepsilon}$ is the identity for $\varepsilon = 0, 1$. In other words, we have $b_0 = b_1$ and $y_0 = y_1$, and we also have proven that p_0 and p_1 have the same sections, whence are equal, by Condition EZ3 of Definition 1.3.1. □

Theorem 5.1.7 *Any fibrant object has a minimal model.*

Proof Let X be a fibrant object. A subobject S of X will be called *thin* if it satisfies the following two conditions.

(a) If x and y are two sections of S which are ∂-equivalent as sections of X, then $x = y$.
(b) If x is a section of S whose image in X is ∂-equivalent to a degenerate section of X, then x is degenerate.

Let E be the set of thin subobjects of X. We observe that E is not empty: it follows from Lemma 5.1.6 that the 0-skeleton of X is an element of E (but the empty subobject is good as well). By Zorn's lemma, we can choose a maximal element S of E (with respect to inclusion). We shall first observe that any section $x \colon a \to X$ whose boundary ∂x factors through S must be ∂-equivalent to a section of S. Indeed, if x is degenerate, then it factors through ∂a hence through S. Otherwise, let us consider $S' = S \cup \mathrm{Im}(x)$. A non-degenerate section of S' must either factor through S or be precisely equal to x. If S' is thin, then the maximality of S implies that $S = S'$. Otherwise S' is not thin. In this case, if y is a section of S' ∂-equivalent to a degenerate section z of X, with $y \neq z$, then y does not belong to S. This means that $y = x\sigma$ where $\sigma\colon b \to a$ is a map in A_+. By Lemma 5.1.6, the section y must also be non-degenerate. In other words, we must have $x = y$. But since $\partial x = \partial z$ factors through S with z degenerate, we must have z in S. This means that, for S' not to be thin, either x is ∂-equivalent to a degenerate section of S, or we have the existence of two ∂-equivalent sections y_0 and y_1 of S' which are not equal. In the second situation, condition (a) for S implies that one of the sections, say y_0, must be out of S. This implies as above that y_0 must be of the form $x\sigma$, with $\sigma\colon a_0 \to a$ in A_+. If σ is not an identity, then condition (b) implies that y_1 is degenerate, and thus, by virtue of Lemma 5.1.6, we must have $y_0 = y_1$, which is not possible, or that y_1 is not in S and non-degenerate, i.e. that $y_1 = x$. But then, since y_0 and y_1 must have the same domain, we must also have $y_0 = x$, which contradicts the fact that $y_0 \neq y_1$. Therefore, we must have $y_0 = x$ and y_1 in S, which proves that x is ∂-equivalent to a section of S anyway.

We will prove that S is a retract of X and that the inclusion $S \to X$ is an I-homotopy equivalence. This will show that S is fibrant and thus a minimal model of X. Let us write $i\colon S \to X$ for the inclusion map. Consider triples (T, h, p), where T is a subobject of X which contains S, $p\colon T \to S$ is a retraction (i.e. the restriction of p to S is the identity), and $h\colon I \otimes T \to X$ is a homotopy which is constant on S, and such that h_0 is the inclusion map $T \to X$, while $h_1 = ip$. Such triples are ordered in the obvious way: $(T, h, p) \leq (T', h', p')$ if $T \subset T'$, with $h'|_{I\otimes T} = h$ and $p'|_T = p$. By Zorn's lemma, we can choose a maximal triple (T, h, p). To finish the proof, it is sufficient to prove that $T = X$. In other words, it is sufficient to prove that any non-degenerate section of X belongs to T. Let $x\colon a \to X$ be a non-degenerate section which does not belong to T. Assume that the dimension of a is minimal for this property. Then ∂x must factor through T, so that, if we define T' to be the union of T and of the image of x in X, then, by Theorem 1.3.8, we have a biCartesian square of the following form.

$$\begin{array}{ccc} \partial a & \xrightarrow{\partial x} & T \\ \downarrow & & \downarrow \\ a & \xrightarrow{x} & T' \end{array}$$

We have the commutative square below.

$$\begin{array}{ccc} \{0\} \otimes \partial a & \longrightarrow & I \otimes \partial a \\ \downarrow & & \downarrow{\scriptstyle h(1\otimes \partial x)} \\ \{0\} \otimes a & \xrightarrow{x} & X \end{array}$$

If we put $u = (h(1 \otimes \partial x), x)\colon I \otimes \partial a \cup \{0\} \otimes a \to X$, we can choose a map

$H: I \otimes a \to X$ such that $H_0 = x$, while $H_{|I \otimes \partial a} = h(1 \otimes \partial x)$. If we write $y_0 = H_1$, as h_1 factors through S, we see that the boundary ∂y_0 must factor through S. Let y be the section of S which is ∂-equivalent to y_0. We choose a homotopy $K: I \otimes a \to X$ which is constant on ∂a and such that $K_0 = y_0$ and $K_1 = y$. Let

$$f: I \otimes \partial I \otimes a \cup I \otimes I \otimes \partial a \cup \{1\} \otimes I \otimes a \to X$$

be the map whose restriction to $I \otimes a = I \otimes \{\varepsilon\} \otimes a$ is H for $\varepsilon = 0$ and K for $\varepsilon = 1$, while the restriction on $I \otimes a = \{1\} \otimes I \otimes a$ is the constant homotopy with value y_0, and the restriction on $I \otimes I \otimes \partial a$ is the composition with $h(1 \otimes \partial x)$ of the projection $I \otimes I \otimes \partial a \to I \otimes \partial a$ which is constant on the second factor. Since the embedding of the source of f into $I \otimes I \otimes a$ is a trivial cofibration, there is a map

$$g: I \otimes I \otimes a \to X$$

which restricts to f. This defines a homotopy

$$L: I \otimes a \to X$$

as the restriction of g on $\{0\} \otimes I \otimes a = I \otimes a$. By construction, we have $L_0 = H_0 = x$ and $L_1 = K_1 = y$. We obtain the commutative diagram

$$\begin{array}{ccc} I \otimes \partial a & \xrightarrow{1_I \otimes \partial x} & I \otimes T \\ \downarrow & & \downarrow{\scriptstyle h} \\ I \otimes a & \xrightarrow{L} & X \end{array}$$

so that, identifying $I \otimes T'$ with $I \otimes a \amalg_{I \otimes \partial a} I \otimes T$, we define a new homotopy $h' = (L, h): I \otimes T' \to X$. Similarly, the commutative diagram

$$\begin{array}{ccc} \partial a & \xrightarrow{\partial x} & T \\ \downarrow & & \downarrow{\scriptstyle p} \\ a & \xrightarrow{y} & S \end{array}$$

defines a map $p' = (y, p): T' = a \amalg_{\partial a} T \to S$. It is clear that the triple (T', h', p') extends (T, h, p), which leads to a contradiction. □

Proposition 5.1.8 *Let X be a fibrant object and $i: S \to X$ a minimal resolution of X. Consider a map $r: X \to S$ such that $ri = 1_S$ (such a map always exists because i is a trivial cofibration with fibrant domain). Then the map r is a trivial fibration.*

Proof There exists a map $h: I \otimes X \to X$ which is constant on S and such that $h_0 = ir$ and $h_1 = 1_X$: we can see i as a trivial cofibration between cofibrant and fibrant objects in the model category of objects under S, and r is then an inverse up to homotopy in this relative situation. Consider the commutative diagram below.

$$\begin{array}{ccc} \partial a & \xrightarrow{u} & X \\ \downarrow & & \downarrow{\scriptstyle r} \\ a & \xrightarrow{v} & S \end{array}$$

We want to prove the existence of a map $w\colon a \to X$ such that $w_{|\partial a} = u$ and $rw = v$. As X is fibrant, there exists a map $k\colon I \otimes a \to X$ whose restriction to $I \otimes \partial a$ is $h(1_I \otimes u)$, while $k_0 = iv$. Let us put $w = k_1$. Then

$$\partial w = w_{|\partial a} = (h(1_I \otimes u))_1 = h_1 u = u\,.$$

It is thus sufficient to prove that $v = rw$. But k and $h(1_I \otimes w)$ coincide on $I \otimes \partial a \cup \{1\} \otimes a$ and thus, by virtue of Proposition 5.1.5, we must have $k_0 \sim (h(1_I \otimes w))_0$. In other words, since $\partial iv = iru = \partial irw$, we have $iv \sim irw$. As $ri = 1_S$, this implies that $v \sim rw$, and, by minimality of S, that $v = rw$. □

Lemma 5.1.9 *Let X be a minimal complex and $f\colon X \to X$ a map which is I-homotopic to the identity. Then f is an isomorphism.*

Proof Let us choose, once and for all, a map $h\colon I \otimes X \to X$ such that $h_0 = 1_X$ and $h_1 = f$. We will prove that the map $f_a\colon X_a \to X_a$ is bijective by induction on the dimension d of a. If a is of dimension < 0, there is nothing to prove because there is no such a. Assume that the map $f_b\colon X_b \to X_b$ is bijective for any object b of dimension $< d$. Consider two sections $x, y\colon a \to X$ such that $f(x) = f(y)$. Then, as f is injective in dimension less than d, the equations

$$f\partial x = \partial f(x) = \partial f(y) = f\partial y$$

imply that $\partial x = \partial y$. On the other hand, we can apply Proposition 5.1.5 to the maps $h(1_I \otimes x)$ and $h(1_I \otimes y)$ for $\varepsilon = 0$, and we deduce that $x \sim y$. As X is minimal, this proves that $x = y$. It remains to prove the surjectivity. Let $y\colon a \to X$ be a section. For any map $\sigma\colon b \to a$ in A such that b is of degree less than d, there is a unique section $x_\sigma\colon b \to X$ such that $f(x_\sigma) = \sigma^*(y) = y\sigma$. This implies that there is a unique map $z\colon \partial a \to X$ such that $fz = \partial y$. The map $I \otimes \partial a \xrightarrow{1_I \otimes z} I \otimes a \xrightarrow{h} X$, together with the map $\{1\} \otimes a = a \xrightarrow{y} X$, define a map $\varphi = (h(1_I \otimes z), x)$, and we can choose a filling k in the diagram below.

$$\begin{array}{ccc} I \otimes \partial a \cup \{1\} \otimes a & \xrightarrow{\varphi} & X \\ \big\downarrow & \overset{k}{\nearrow} & \\ I \otimes a & & \end{array}$$

Let us put $x = k_0$. Then $\partial x = z$, and thus $\partial f(x) = \partial y$. Applying Proposition 5.1.5 to the maps k and $h(1_I \times x)$ for $\varepsilon = 1$, we conclude that $f(x) \sim y$. The object X being a minimal complex, this proves that $f(x) = y$. □

Proposition 5.1.10 *Let X and Y be two minimal complexes. Then any weak equivalence $f\colon X \to Y$ is an isomorphism of presheaves.*

Proof If $f\colon X \to Y$ is a weak equivalence, as both X and Y are cofibrant and fibrant, there exists $g\colon Y \to X$ such that fg and gf are homotopic to the identity of Y and of X, respectively. By virtue of the preceding lemma, the maps gf and fg must be isomorphisms, which imply right away that f is an isomorphism. □

Theorem 5.1.11 *Let X be a fibrant object of $\widehat{A}$. The following conditions are equivalent.*

(i) The object X is a minimal complex.
(ii) Any trivial fibration of the form $X \to S$ is an isomorphism.
(iii) Any trivial cofibration of the form $S \to X$, with S fibrant, is an isomorphism.
(iv) Any weak equivalence $X \to S$, with S a minimal complex, is an isomorphism.
(v) Any weak equivalence $S \to X$, with S a minimal complex, is an isomorphism.

Proof It follows immediately from Proposition 5.1.10 that condition (i) is equivalent to condition (iv) as well as to condition (v). Therefore, condition (v) implies condition (iii): if $i\colon S \to X$ is a trivial cofibration with S fibrant and X minimal, then S must be minimal as well, so that i has to be an isomorphism. Let us prove that condition (iii) implies condition (ii): any trivial fibration $p\colon X \to S$ admits a section $i\colon S \to X$ which has to be a trivial cofibration with fibrant domain, and thus an isomorphism. It is now sufficient to prove that condition (ii) implies condition (i). By virtue of Theorem 5.1.7, there exists a minimal model of X, namely a trivial cofibration $S \to X$ with S a minimal complex. This cofibration has a retraction which, by virtue of Proposition 5.1.8, is a trivial fibration. Condition (ii) implies that S is isomorphic to X, and thus that X is minimal as well. □

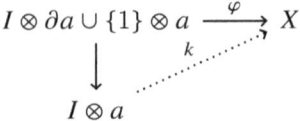

It may be convenient to work with a restricted class of fibrant objects in the following sense.

Definition 5.1.12 A class F of fibrant presheaves on A is said to be *admissible* if it has the following stability properties.

(a) It is closed under retracts: for any map $p\colon X \to Y$, which has a section, if X belongs to F, so does Y.
(b) For any trivial fibration $p\colon X \to Y$, if Y is in F, then X is in F.

The following proposition shows that we can work up to a weak equivalence, while considering such classes of presheaves.

Proposition 5.1.13 *Let F be an admissible class of fibrant presheaves on A. For any weak equivalence between fibrant presheaves $f : X \to Y$, the presheaf X if in F is and only if Y is in F.*

Proof We choose a factorisation of the weak equivalence f into a trivial cofibration $i\colon X \to T$ followed by a trivial fibration $p\colon T \to Y$. Since X is fibrant, the map i is the section of some map $r\colon T \to X$. If Y is in F, then, since p is a trivial fibration, so is T. Therefore, X belongs to F, as a retract of an element of F. For the converse, applying what precedes to the map r we see that, if X is in F, so is T. Since any trivial fibration between cofibrant objects has a section, this implies that Y is in F as well, □

Definition 5.1.14 A fibration $p\colon X \to Y$ in $\widehat{A}$ is *minimal* if it is a minimal complex as an object of $\widehat{A}/Y \simeq \widehat{A/Y}$ for the induced model category structure (whose weak equivalences, fibrations and cofibrations are the maps which have the corresponding property in $\widehat{A}$, by forgetting the base).

Proposition 5.1.15 *The class of minimal fibrations is stable by pull-back.*

Proof Consider a pull-back square

$$\begin{array}{ccc} X & \xrightarrow{u} & X' \\ {\scriptstyle p}\downarrow & & \downarrow{\scriptstyle p'} \\ Y & \xrightarrow{v} & Y' \end{array}$$

in which p' is a minimal fibration. Let $x, y : a \to X$ be two global sections which are ∂-equivalent over Y (i.e. ∂-equivalent in X, seen as a fibrant object of $\widehat{A/Y}$). Then $u(x)$ and $u(y)$ are ∂-equivalent in X' over Y', and thus $u(x) = u(y)$. As $p(x) = p(y)$, this means that $x = y$. In other words, p is a minimal fibration. □

Remark 5.1.16 The class of minimal fibrations is not stable by composition in general.

Everything we proved so far about minimal complexes has its counterpart in the language of minimal fibrations. Let us mention the properties that we will use later.

Theorem 5.1.17 *For any fibration $p\colon X \to Y$, there exists a trivial fibration $r\colon X \to S$ and a minimal fibration $q\colon S \to Y$ such that $p = qr$.*

Proof By virtue of Theorem 5.1.7 applied to p, seen as a fibrant presheaf over A/Y, there exists a trivial cofibration $i\colon S \to X$ such that $q = p_{|S}\colon S \to Y$ is a minimal fibration. As both X and S are fibrant (as presheaves over A/Y), the embedding i is a strong deformation retract, so that, by virtue of Proposition 5.1.8 (applied again in the context of presheaves over A/Y), there exists a trivial fibration $r\colon X \to S$ such that $ri = 1_S$, and such that $qr = p$. □

Remark 5.1.18 In the factorisation $p = qr$ given by the preceding theorem, q is necessarily a retract of p. Therefore, if p belongs to a class of maps which is stable under retracts, the minimal fibration q must have the same property. Similarly, as r is a trivial fibration, if p belongs to a class which is defined up to weak equivalences, then so does q. This means that this theorem can be used to study classes of fibrations which are more general than classes of fibrations of model category structures. This is where statements such as Proposition 5.1.13 might be useful.

Proposition 5.1.19 *For any minimal fibrations $p\colon X \to Y$ and $p'\colon X' \to Y$, any weak equivalence $f\colon X \to X'$ such that $p'f = p$ is an isomorphism.*

Proof This is a reformulation of Proposition 5.1.10 in the context of presheaves over A/Y. □

Lemma 5.1.20 *For any cofibration $v\colon Y \to Y'$ and any trivial fibration $p\colon X \to Y$, there exists a trivial fibration $p'\colon X' \to Y'$ and a pull-back square of the following form.*

$$\begin{array}{ccc} X & \xrightarrow{u} & X' \\ {\scriptstyle p}\downarrow & & \downarrow{\scriptstyle p'} \\ Y & \xrightarrow{v} & Y' \end{array}$$

Proof The pull-back functor $v^*\colon \widehat{A}/Y' \to \widehat{A}/Y$ has a left adjoint $v_!$ and a right adjoint v_*. We see right away that $v^*v_!$ is isomorphic to the identity (i.e. that $v_!$ is fully faithful), so that, by transposition, v^*v_* is isomorphic to the identity as well. Moreover, the functor v_* preserves trivial fibrations because its left adjoint v^* preserves monomorphisms. We define the trivial fibration p' as $v_*(p\colon X \to Y)$. □

Proposition 5.1.21 *Let $\mathcal{F}$ be a class of morphisms of presheaves over A with the following properties:*

(i) any element of $\mathcal{F}$ is a fibration;

(ii) any trivial fibration is in $\mathcal{F}$;

(iii) the class $\mathcal{F}$ is closed under retracts and under compositions.

Consider a commutative diagram of the form

$$\begin{array}{ccccc} X_0 & \xrightarrow{w} & X_1 & \xrightarrow{i_1} & X'_1 \\ & \searrow^{p_0} & \downarrow{\scriptstyle p_1} & & \downarrow{\scriptstyle p'_1} \\ & & Y & \xrightarrow{j} & Y' \end{array}$$

in which p_0, p_1 and p'_1 are in $\mathcal{F}$, w is a weak equivalence, j is a cofibration, and the square is Cartesian. Then there exists a Cartesian square

$$\begin{array}{ccc} X_0 & \xrightarrow{i_0} & X'_0 \\ \downarrow{\scriptstyle p_0} & & \downarrow{\scriptstyle p'_0} \\ Y & \xrightarrow{j} & Y' \end{array}$$

in which p'_0 is a fibration in $\mathcal{F}$, as well as a weak equivalence $w' \colon X'_0 \to X'_1$ such that $p'_1 w = p'_0$ and $i_1 w = w' i_0$.

Proof We observe that if there are two fibrations $p \colon X \to S$ and $q \colon Y \to S$ as well as a weak equivalence $u \colon X \to Y$ over S, then p is in $\mathcal{F}$ if and only if q is in $\mathcal{F}$: this is a particular instance of Proposition 5.1.13 applied to the category of presheaves on A/S for the class F of fibrations $T \to S$ which are elements of $\mathcal{F}$. This means that we may assume, without loss of generality, that the class $\mathcal{F}$ consists of all fibrations.

By virtue of Theorem 5.1.17, we can choose a trivial fibration $r'_1 \colon X'_1 \to S'$ and a minimal fibration $q' \colon S \to Y'$ such that $p'_1 = q' r'_1$. Let us write $S = Y \times_{Y'} S'$, and $k \colon S \to S'$ for the second projection. The canonical map $r_1 \colon X_1 \to S$ is a trivial fibration (being the pull-back of such a thing), and the projection $q \colon S \to Y$ is a minimal fibration by Proposition 5.1.15. We have thus a factorisation $p_1 = q r_1$. Moreover, the map $r_0 = r_1 w$ is a trivial fibration. To see this, let us choose a minimal model $u \colon T \to X_0$. Then the map $r_1 w u$ is a weak equivalence between minimal fibrations and is thus an isomorphism by Proposition 5.1.19. This means that r_0 is a trivial fibration by Proposition 5.1.8. The diagram we started from has the following form.

$$
\begin{array}{ccccc}
X_0 & \xrightarrow{w} & X_1 & \xrightarrow{i_1} & X'_1 \\
 & {\scriptstyle r_0}\searrow & \downarrow{\scriptstyle r_1} & & \downarrow{\scriptstyle r'_1} \\
 & & S & \xrightarrow{k} & S' \\
 & & \downarrow{\scriptstyle q} & & \downarrow{\scriptstyle q'} \\
 & & Y & \xrightarrow{j} & Y'
\end{array}
$$

Moreover, both squares are Cartesian. This means that we can replace j by k. In other words, without loss of generality, it is sufficient to prove the proposition in the case where p_0, p_1 and p'_1 are trivial fibrations. Under these additional assumptions, we obtain a Cartesian square

$$
\begin{array}{ccc}
X_0 & \xrightarrow{i_0} & X'_0 \\
\downarrow{\scriptstyle p_0} & & \downarrow{\scriptstyle p'_0} \\
Y & \xrightarrow{j} & Y'
\end{array}
$$

in which p'_0 is a trivial fibration by Lemma 5.1.20. The lifting problem

$$
\begin{array}{ccccc}
X_0 & \xrightarrow{w} & X_1 & \xrightarrow{i_1} & X'_1 \\
\downarrow{\scriptstyle i_0} & & {\scriptstyle w'}\nearrow & & \downarrow{\scriptstyle p'_1} \\
X'_0 & & \xrightarrow{p'_0} & & Y'
\end{array}
$$

has a solution because i_0 is a cofibration and p'_1 a trivial fibration. Moreover, any lift w' must be a weak equivalence because both p'_0 and p'_1 are trivial fibrations. □

Remark 5.1.22 Inner fibrations are not necessarily fibrations of the Joyal model structure (e.g. the nerve of any functor between connected groupoids which is not surjective on objects is not an isofibration). However, one may still consider minimal inner fibrations (this is done by Lurie in [Lur09, 2.3.3]). Indeed, we observe first that, since any invertible map in Δ^n is an identity ($n \geq 0$), any inner fibration of the form $p\colon X \to \Delta^n$ is an isofibration between ∞-categories, whence a fibration of the Joyal model category structure (Theorem 3.6.1). This implies that, as an object of *sSet*$/Y$, any inner fibration $p\colon X \to Y$ has the right lifting property with respect to inclusions of the form

$$J \times K \cup \{\varepsilon\} \times L \to J \times L, \quad \varepsilon = 0, 1,$$

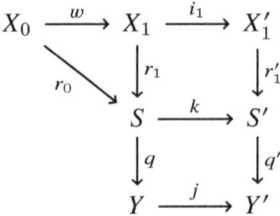

where $K \to L$ is a monomorphism of simplicial sets over Y, while $J \times L$ is considered as a simplicial set over Y with structural map given by the composition of the second projection $J \times L \to L$ with the structural map $L \to Y$: indeed, it is sufficient to prove this lifting property in the case where $L = \Delta^n$ and $K = \partial\Delta^n$, in which case we may pull back p along the given structural map $\Delta^n \to Y$. In particular, any inner fibration $X \to Y$ is a fibrant object of the model structure of Theorem 2.4.19 with the homotopical structure on the category of presheaves on $A = \mathbf{\Delta}/Y$ associated to exact cylinder $J \times (-)$ and the minimal class of J-anodyne extensions (i.e. the one obtained by performing the construction of Example 2.4.13 for $S = \varnothing$).[1] We may thus apply Theorem 5.1.7 and Proposition 5.1.8 for $A = \mathbf{\Delta}/Y$ and get a factorisation of p into a trivial fibration $r: X \to S$ followed by a minimal inner fibration $q: S \to Y$. We also observe that property (ii) of Theorem 5.1.11 is independent of the model structure we choose to work with on the category of presheaves over $A = \mathbf{\Delta}/Y$.

5.2 The Universal Left Fibration

Proposition 5.2.1 *For $0 < k < n$, the inclusion $i: \Lambda^n_k \to \Delta^n$ induces an equivalence of categories*

$$\mathbf{L}i_!: RFib(\Lambda^n_k) \xrightarrow{\sim} RFib(\Delta^n)\,.$$

Proof Since i is bijective on objects, it follows from Theorem 4.1.16 that the functor $\mathbf{R}i^*$ is conservative. Therefore, it is sufficient to prove that the functor $\mathbf{L}i_!$ is fully faithful. By virtue of Proposition 4.5.1, it is sufficient to prove that, for any object a of Δ^n, the map

$$h_a \to \mathbf{R}i^*\mathbf{L}i_!(h_a)$$

is an isomorphism in $RFib(\Lambda^n_k)$. For $a \in \{0, \ldots, n\}$, a fibrant resolution of the image of (Δ^0, a) by $i_!$ is $\Delta^a = \Delta^n/a$. The object $\mathbf{R}i^*\mathbf{L}i_!(h_a)$ is thus the intersection of Λ^n_k with Δ^a, the structural map to Λ^n_k being the inclusion. To finish the proof, it remains to check that a is final in $\Lambda^n_k \cap \Delta^a$. If $a < n$, then Δ^a is contained in the image of the face which avoids n, and therefore, since $k < n$, we have $\Delta^a \subset \Lambda^n_k$, which implies in turn that $\Lambda^n_k \cap \Delta^a = \Delta^a$. Since it is obvious that a is final in Δ^a, we only have to check the case where $a = n$. In this case, we have $\Lambda^n_k \cap \Delta^n = \Lambda^n_k$, and, as $k > 0$, Lemma 4.4.3 finishes the proof. □

[1] Despite appearances, this is not a special case of the construction of paragraph 2.5.1.

Corollary 5.2.2 *Let us consider a commutative square of the form*

$$\begin{array}{ccc} X & \xrightarrow{\tilde{\imath}} & Y \\ {\scriptstyle p}\downarrow & & \downarrow{\scriptstyle q} \\ \Lambda^n_k & \xhookrightarrow{i} & \Delta^n \end{array}$$

where $0 < k < n$. We also assume that $\tilde{\imath}$ is final (cofinal) and that both p and q are right (left) fibrations. Then the induced map $X \to \Lambda^n_k \times_{\Delta^n} Y$ is a fibrewise equivalence.

Proof In the case where $\tilde{\imath}$ is final and p and q are right fibrations, this is a reformulation of the fully faithfulness of the functor $\mathbf{L}i_!$, by Proposition 4.1.16. The dual version is obtained by applying the functor $T \mapsto T^{\mathrm{op}}$ in an appropriate way. □

Definition 5.2.3 We fix a Grothendieck universe $\mathbf{U}$. A set will be called $\mathbf{U}$-*small* if it belongs to $\mathbf{U}$. We define the simplicial set U of morphisms of simplicial sets with $\mathbf{U}$-small fibres as follows. An element of U_n is a presheaf on the slice category $\mathbf{\Delta}/\Delta^n$ with values in $\mathbf{U}$-small sets. Given a morphism $f\colon \Delta^m \to \Delta^n$, the map $f^*\colon U_n \to U_m$ is defined by composing with the induced functor $\mathbf{\Delta}/\Delta^m \to \mathbf{\Delta}/\Delta^n$.

Alternatively, if one defines $\mathbf{\Delta}$ so that its set of arrows is $\mathbf{U}$-small, an element of U_n can then be seen as map $p\colon X \to \Delta^n$, such that X takes its values in $\mathbf{U}$-small sets, together with a choice, for any map $f\colon \Delta^m \to \Delta^n$, of a Cartesian square of $\mathbf{U}$-small simplicial sets of the following form.

$$\begin{array}{ccc} f^*(X) & \xrightarrow{\tilde{f}} & X \\ \downarrow{\scriptstyle f^*p} & & \downarrow{\scriptstyle p} \\ \Delta^m & \xrightarrow{f} & \Delta^n \end{array}$$

One defines the simplicial set $\mathcal{S}$ of *left fibrations with specified* $\mathbf{U}$-*small fibres* as the subobject of U whose elements correspond to left fibrations of codomain Δ^n with specified pull-back squares of $\mathbf{U}$-small simplicial sets as above.

One checks immediately that $\mathcal{S}^{\mathrm{op}}$ can be interpreted as the simplicial set of *right fibrations with specified* $\mathbf{U}$-*small fibres*, i.e. is canonically isomorphic to the subobject of U whose elements are the right fibrations of codomain Δ^n with suitably specified pull-back squares.

There is a pointed version of U, which we denote by $U_\bullet$. A map $\Delta^n \to U_\bullet$ is a presheaf of pointed $\mathbf{U}$-small sets on $\mathbf{\Delta}/\Delta^n$. Equivalently, this can be described as a pair (X, s), where X is a presheaf of $\mathbf{U}$-small sets on $\mathbf{\Delta}/\Delta^n$, and s is a global section of X. Alternatively, if we think of X as a map $p\colon X \to \Delta^n$, one may see

s as a section of p. Forgetting the section's s defines a morphism of simplicial sets

$$\pi\colon U_\bullet \to U\,.$$

One defines similarly

$$p_{univ}\colon \mathcal{S}_\bullet \to \mathcal{S}$$

as the pull-back of $\pi\colon U_\bullet \to U$ along the inclusion $\mathcal{S} \subset U$.

The proof of the following proposition is straightforward, and the details are left to the reader.

Proposition 5.2.4 *Given a morphism of simplicial sets $f\colon X \to Y$, specifying a Cartesian square of the form*

$$\begin{array}{ccc} X & \xrightarrow{\tilde{F}} & U_\bullet \\ \downarrow{\scriptstyle f} & & \downarrow{\scriptstyle \pi} \\ Y & \xrightarrow{F} & U \end{array}$$

is equivalent to choosing, for each map $\varphi\colon \Delta^n \to Y$, a Cartesian square

$$\begin{array}{ccc} \varphi^*(X) & \xrightarrow{\tilde{\varphi}} & X \\ \downarrow{\scriptstyle \varphi^*(f)} & & \downarrow{\scriptstyle f} \\ \Delta^n & \xrightarrow{\varphi} & Y \end{array}$$

where $\varphi^*(X)$ is a **U***-small simplicial set.*

Definition 5.2.5 In the situation of the preceding proposition, we say that the morphism F *classifies* the map f.

A morphism of simplicial sets $f\colon X \to Y$ is said to have **U**-*small fibres* if there exists a map $F\colon Y \to U$ which classifies f.

We observe that a morphism of simplicial sets $f\colon X \to Y$ has **U**-small fibres if and only if, for any non-negative integer n and any map $f\colon \Delta^n \to Y$, the fibre product $Z = \Delta^n \times_Y X$ is isomorphic to a **U**-small simplicial set (i.e. the cardinal of each set Z_i, $i \geq 0$, is smaller than the cardinal of U). The excluded middle principle has the following consequence.

Corollary 5.2.6 *Let us consider a Cartesian square of simplicial sets*

$$\begin{array}{ccc} X & \longrightarrow & X' \\ \downarrow{\scriptstyle f} & & \downarrow{\scriptstyle f'} \\ A & \xrightarrow{i} & A' \end{array}$$

in which the map i is a monomorphism. If the map f is classified by a morphism of simplicial sets $F\colon A \to U$, and if f' has $\mathbf{U}$-small fibres, there exists a morphism $F'\colon A' \to U$ which classifies f', such that $F'i = F$.

Proposition 5.2.7 *A morphism of simplicial sets $p\colon X \to Y$ is a right (left) fibration if and only if, for any map $\Delta^n \to Y$, the induced map $\Delta^n \times_Y X \to \Delta^n$ is a right (left) fibration, respectively.*

Proof This follows right away from the fact that the property of being a right (left) fibration is determined by the right lifting property with respect to a set of maps with representable codomains. □

Corollary 5.2.8 *The canonical map $p_{univ}\colon \mathcal{S}_\bullet \to \mathcal{S}$ is a left fibration. Moreover, a morphism of simplicial sets $p\colon X \to Y$ with $\mathbf{U}$-small fibres is a left fibration if and only if it is classified by a map $b\colon Y \to U$ which factors through $\mathcal{S}$. In particular, any left fibration with $\mathbf{U}$-small fibres $p\colon X \to Y$ arises from a Cartesian square of the form below.*

$$\begin{array}{ccc} X & \xrightarrow{\tilde{F}} & \mathcal{S}_\bullet \\ \downarrow{\scriptstyle p} & & \downarrow{\scriptstyle p_{univ}} \\ Y & \xrightarrow{F} & \mathcal{S} \end{array}$$

Lemma 5.2.9 *Given integers $0 < k < n$, for any minimal left fibration with $\mathbf{U}$-small fibres $p\colon X \to \Lambda^n_k$, there exists a minimal left fibration with $\mathbf{U}$-small fibres $q\colon Y \to \Delta^n$, and a pull-back square of the following form.*

$$\begin{array}{ccc} X & \xrightarrow{\tilde{\imath}} & Y \\ \downarrow{\scriptstyle p} & & \downarrow{\scriptstyle q} \\ \Lambda^n_k & \xhookrightarrow{i} & \Delta^n \end{array}$$

Proof Let us factor the map ip as a left anodyne extension $i_0\colon X \to Y_0$ followed by a left fibration $p_0\colon X_0 \to \Delta^n$. By virtue of Theorem 5.1.17 (applied to the covariant model category structure over Δ^n), we can factor p_0 into a trivial fibration $q_0\colon X_0 \to Y$ followed by a minimal left fibration $q\colon Y \to \Delta^n$ (see also Remark 5.1.22). On checks that one can construct all these factorisations in such a way that all the maps have $\mathbf{U}$-small fibres. If we put $\tilde{\imath} = q_0 i_0$, we thus get the commutative square below.

$$\begin{array}{ccc} X & \xrightarrow{\tilde{\imath}} & Y \\ \downarrow{\scriptstyle p} & & \downarrow{\scriptstyle q} \\ \Lambda^n_k & \xhookrightarrow{i} & \Delta^n \end{array} \qquad (5.2.9.1)$$

The projection $\Lambda^n_k \times_{\Delta^n} Y \to \Lambda^n_k$ is a minimal left fibration (Proposition 5.1.15). On the other hand, by Corollary 5.2.2, the comparison map $X \to \Lambda^n_k \times_{\Delta^n} Y$ is a fibrewise equivalence over Λ^n_k. Therefore, Proposition 5.1.10 (applied to the covariant model structure over Λ^n_k) implies that this comparison map is an isomorphism, and thus that (5.2.9.1) is Cartesian. □

The following theorem was stated as a conjecture by Nichols-Barrer [NB07].

Theorem 5.2.10 *The simplicial set $\mathcal{S}$ is an ∞-category whose objects are the* $\mathbf{U}$*-small ∞-groupoids.*[2]

Proof By virtue of Proposition 5.2.7 and Corollary 5.2.6, to prove that $\mathcal{S}$ is an ∞-category, we only have to prove that, given integers $0 < k < n$, for any left fibration with $\mathbf{U}$-small fibres $p: X \to \Lambda^n_k$, there exists a left fibration with $\mathbf{U}$-small fibres $q: Y \to \Delta^n$ and a pull-back square of the form (5.2.9.1). By virtue of Theorem 5.1.17, there exists a factorisation of p as $p = p_0 r$ with r a trivial fibration and p_0 a minimal left fibration. We can extend p_0 and then r, using Lemmas 5.2.9 and 5.1.20 successively. The objects of $\mathcal{S}$ correspond to left fibrations with $\mathbf{U}$-small fibres whose codomain is Δ^0, which are nothing but $\mathbf{U}$-small Kan complexes, or, equivalently, $\mathbf{U}$-small ∞-groupoids. □

Corollary 5.2.11 (Joyal) *Let $i: A \to B$ be a trivial cofibration of the Joyal model category structure (e.g. an inner anodyne extension). Then, for any left (right) fibration with* $\mathbf{U}$*-small fibres $p: X \to A$, there exists a left (right) fibration with* $\mathbf{U}$*-small fibres $q: Y \to B$, and a Cartesian square of the following form.*

$$\begin{array}{ccc} X & \xrightarrow{\tilde{\imath}} & Y \\ {\scriptstyle p}\downarrow & & \downarrow{\scriptstyle q} \\ A & \xrightarrow{i} & B \end{array}$$

Proof By virtue of Corollary 5.2.6, we are only expressing the fact that the map $\mathcal{S} \to \Delta^0$ has the right lifting property with respect to i, which follows from the preceding theorem and from Theorem 3.6.1. □

Lemma 5.2.12 *Given an anodyne extension $K \to L$, for any Kan fibration with* $\mathbf{U}$*-small fibres $p: X \to K$, there exists a Kan fibration with* $\mathbf{U}$*-small fibres*

[2] The last assertion is not completely correct. The objects of $\mathcal{S}$ are $\mathbf{U}$-small ∞-groupoids X endowed with a choice, for any integer $n \geq 0$, of an abstract category-theoretic $\mathbf{U}$-small Cartesian product $\Delta^n \times X$.

$q\colon Y \to L$, and a pull-back square of the following form.

$$
\begin{array}{ccc}
X & \xrightarrow{\tilde{\imath}} & Y \\
{\scriptstyle p}\downarrow & & \downarrow{\scriptstyle q} \\
K & \xhookrightarrow{i} & L
\end{array}
\tag{5.2.12.1}
$$

Proof By virtue of Theorem 5.1.17, there exists a factorisation of p as $p = p_0 r$ with r a trivial fibration and p_0 a minimal Kan fibration. Lemma 5.1.20 thus shows that it is sufficient to consider the special case where p is a minimal Kan fibration.

Let us factor the map ip as an anodyne extension $i_0\colon X \to Y_0$ followed by a Kan fibration $p_0\colon X_0 \to L$. By virtue of Theorem 5.1.17, we can factor p_0 into a trivial fibration $q_0\colon X_0 \to Y$ followed by a minimal left fibration $q\colon Y \to L$. If we put $\tilde{\imath} = q_0 i_0$, we thus get a commutative square of the form (5.2.12.1). Since any Kan fibration is locally constant (Corollary 4.6.6), the map $X \to K \times_L Y$ is a weak homotopy equivalence over K. Propositions 5.1.15 and 5.1.19 imply that this square is Cartesian. □

Proposition 5.2.13 *Let $F\colon A \to \mathcal{S}$ be a morphism which classifies a left fibration $p\colon X \to A$. Then F factors through the maximal ∞-groupoid $k(\mathcal{S})$ if and only if the map p is a Kan fibration. In particular, for any Kan fibration $p\colon X \to A$, there exists a Cartesian square of the following form.*

$$
\begin{array}{ccc}
X & \longrightarrow & k(\mathcal{S}_\bullet) \\
{\scriptstyle p}\downarrow & & \downarrow{\scriptstyle k(p_{univ})} \\
A & \xrightarrow{F} & k(\mathcal{S})
\end{array}
\tag{5.2.13.1}
$$

Proof We have a canonical Cartesian square

$$
\begin{array}{ccc}
k(\mathcal{S}_\bullet) & \hookrightarrow & \mathcal{S}_\bullet \\
{\scriptstyle k(p_{univ})}\downarrow & & \downarrow{\scriptstyle p_{univ}} \\
k(\mathcal{S}) & \hookrightarrow & \mathcal{S}
\end{array}
$$

because the map p_{univ}, being a left fibration, is conservative. The map $k(p_{univ})$ is a left fibration whose codomain is a Kan complex, whence it is a Kan fibration. Therefore, if the classifying map $F\colon A \to \mathcal{S}$ of a left fibration factors through $k(\mathcal{S})$, we have a Cartesian square of the form (5.2.13.1) which proves that p is a Kan fibration. Conversely, if $p\colon X \to A$ is a Kan fibration, then Lemma 5.2.12

ensures that there exists a Cartesian square of the form

$$\begin{array}{ccc} X & \longrightarrow & Y \\ {\scriptstyle p}\downarrow & & \downarrow{\scriptstyle q} \\ A & \xrightarrow{\beta_A} & Ex^{\infty}(A) \end{array}$$

where β_A is the canonical anodyne extension (3.1.22.5), and q is some Kan fibration. In particular, any classifying map of p factors through the ∞-groupoid $Ex^{\infty}(A)$, hence through $k(\mathcal{S})$. □

Theorem 5.2.14 (Joyal) *For any weak categorical equivalence $f\colon A \to B$, the functor $\mathbf{L}f_!\colon RFib(A) \to RFib(B)$ is an equivalence of categories.*

Proof Let W be the class of maps f such that $\mathbf{L}f_!$ is an equivalence of categories. We want to prove that W contains the class of weak equivalences of the Joyal model category structure. By virtue of Proposition 3.6.2 it is sufficient to prove that W satisfies the following conditions.

(i) The class W has the two-out-of-three property.
(ii) Any inner anodyne extension is in W.
(iii) Any trivial fibration between ∞-categories is in W.

Property (i) follows from the facts that the class of equivalences of categories has the two-out-of-three property, and that the assignment $A \mapsto RFib(A)$ is a functor: given two composable morphisms of simplicial sets f and g, we have $\mathbf{L}g_!\mathbf{L}f_! = \mathbf{L}(gf)_!$ and $\mathbf{L}(1_A)_! = 1_{RFib(A)}$. To prove property (ii), we first remark that any inner anodyne extension is bijective on objects. Indeed, the class of morphisms of simplicial sets which induce a bijective map on objects is saturated, so that, by the small object argument, it is sufficient to check this property for the case of the inclusion of Λ^n_k into Δ^n for $0 < k < n$, in which case we obviously have an equality at the level of objects. In particular, as a consequence of Theorem 4.1.16, for any inner anodyne extension $i\colon A \to B$, the functor $\mathbf{R}i^*$ is conservative. This means that we only have to prove that $\mathbf{L}i_!$ is fully faithful. Given an object F of $RFib(A)$, we may assume that F is represented by a right fibration $p\colon X \to A$. By virtue of Corollary 5.2.11, we can choose a Cartesian square of simplicial sets of the form

$$\begin{array}{ccc} X & \xrightarrow{\tilde{\imath}} & Y \\ {\scriptstyle p}\downarrow & & \downarrow{\scriptstyle q} \\ A & \xrightarrow{i} & B \end{array}$$

in which q is a right fibration. It is now sufficient to prove that the map $\tilde{\imath}$ is a right anodyne extension. Indeed, if this is the case, the image in $RFib(A)$

of the canonical isomorphism from X to $A \times_B Y$ is then the co-unit map $F \to \mathbf{R}i^*\mathbf{L}i_!(F)$. In particular, the functor $\mathbf{L}i_!$ is fully faithful. Let us prove that $\tilde{\imath}$ is a right anodyne extension. We first consider the case where i is the canonical inclusion of Λ^n_k into Δ^n for some $0 < k < n$. Then we already know that $\mathbf{L}i_!$ is an equivalence of categories. The map $\tilde{\imath}$ is then the co-unit of this adjunction and is thus invertible in $RFib(\Delta^n)$. Therefore, it is a weak equivalence with fibrant codomain in the contravariant model category structure over Δ^n, hence a right anodyne extension (see Proposition 4.1.11). The case where i is a sum of inclusions of inner horns follows right away. For the general case, assume that i is a retract of an inner anodyne extension $i_0 : A_0 \to B_0$. Let $p_0 : X_0 \to A_0$ and $q_0 : Y_0 \to B_0$ denote the pull-backs of p and q, respectively, so that we get a new Cartesian square

$$\begin{array}{ccc} X_0 & \xrightarrow{\tilde{\imath}_0} & Y_0 \\ {\scriptstyle p_0}\downarrow & & \downarrow{\scriptstyle q_0} \\ A_0 & \xrightarrow{i_0} & B_0 \end{array}$$

of which the previous square is a retract. In particular, the map $\tilde{\imath}$ is a retract of $\tilde{\imath}_0$, and it is sufficient to prove that $\tilde{\imath}_0$ is a right anodyne extension. Using the small object argument, we can produce such a map i_0 which is a countable composition of maps

$$A_0 \xrightarrow{j_1} A_1 \xrightarrow{j_2} \cdots \xrightarrow{j_n} A_n \xrightarrow{j_{n+1}} A_{n+1} \longrightarrow \cdots,$$

each map j_n being obtained through a push-out square of the form

$$\begin{array}{ccc} \coprod_{\lambda \in L_n} \Lambda^{m_\lambda}_{k_\lambda} & \longrightarrow & A_{n-1} \\ \downarrow & & \downarrow{\scriptstyle j_n} \\ \coprod_{\lambda \in L_n} \Delta^{m_\lambda} & \longrightarrow & A_n \end{array}$$

with $0 < k_\lambda < m_\lambda$ for all $\lambda \in L_n$. Let $p_n : X_n \to A_n$ be the pull-back of q_0 along the inclusion $A_n \subset B_0$. We then have canonical pull-back squares

$$\begin{array}{ccc} X_{n-1} & \xrightarrow{\tilde{\jmath}_n} & X_n \\ {\scriptstyle p_{n-1}}\downarrow & & \downarrow{\scriptstyle p_n} \\ A_{n-1} & \xrightarrow{j_n} & A_n \end{array}$$

and it is sufficient to check that each $\tilde{\jmath}_n$ is a right anodyne extension, because $\tilde{\imath}_0$ is the countable composition $X_0 \to \varinjlim_n X_n$ of the $\tilde{\jmath}_n$. But each of these is the push-out of the pull-back of a sum of inner horn inclusions along a right fibration, which we already know to be a right anodyne extension. This achieves the proof that the class W contains all inner anodyne extensions.

It remains to prove that any trivial fibration is in W. Let $q\colon A \to B$ be a trivial fibration. Since q is surjective on objects (it even has at least a section), the functor $\mathbf{R}q^*$ is also conservative. Therefore, it is sufficient to prove that the functor $\mathbf{L}q_!$ is fully faithful. Since trivial fibrations are stable by pull-backs, the functor q^* also preserves weak equivalences. Therefore, given an object F of $RFib(A)$ represented by a morphism $p\colon X \to A$, the co-unit $F \to \mathbf{R}q^*\mathbf{L}q_!(F)$ is the image of the map $f = (p, 1_X)\colon X \to A \times_B X$. This map is a section of the trivial fibration $A \times_B X \to X$ and is thus a right anodyne extension (see Corollary 2.4.29 and Proposition 4.1.7). □

Remark 5.2.15 An inspection of the proof of Theorems 5.2.10 and 5.2.14 shows that a monomorphism of small simplicial sets $i\colon A \to B$ has the left lifting property with respect to $\mathcal{S} \to \Delta^0$ (for all universes $\mathbf{U}$) if and only if the induced functor $\mathbf{L}i_!\colon RFib(A) \to RFib(B)$ is fully faithful. In particular, the class of those monomorphisms is saturated. This is a way to prove that the class of fully faithful functors is closed under homotopy push-outs.

Corollary 5.2.16 (Joyal) *A morphism of simplicial sets $f\colon A \to B$ is a weak categorical equivalence if and only if it satisfies the following two properties.*

(a) The functor $\tau(f)\colon \tau(A) \to \tau(B)$ is essentially surjective.
(b) The functor $\mathbf{L}f_!\colon RFib(A) \to RFib(B)$ is fully faithful.

Proof Let us choose a commutative square of the form

$$\begin{array}{ccc} A & \xrightarrow{f} & B \\ {\scriptstyle a}\downarrow & & \downarrow{\scriptstyle b} \\ A' & \xrightarrow{f'} & B' \end{array}$$

in which both a and b are weak categorical equivalences, while A' and B' are ∞-categories. We know that the functor τ sends weak categorical equivalences to equivalences of categories (see Proposition 3.3.14), and it follows from the preceding theorem that the functors $\mathbf{L}a_!$ and $\mathbf{L}b_!$ are equivalences of categories. Therefore, f satisfies conditions (a) and (b) if and only if f' has the same property. Similarly, f is a weak categorical equivalence if and only if f' is a weak categorical equivalence. We may thus assume, without loss of generality, that A and B are ∞-categories. In this case, it follows from Proposition 4.5.2 that f satisfies condition (a) if and only if it is fully faithful. We also know that f is essentially surjective if and only if the functor $\tau(f)$ has this property (see Remark 3.9.4). Therefore, Theorem 3.9.7 achieves the proof. □

5.3 Homotopy Classification of Left Fibrations

Proposition 5.3.1 *Any weak categorical equivalence is final.*

Proof Let $u\colon X \to Y$ be a weak categorical equivalence. To prove that u is final, it is sufficient to prove that it induces an isomorphism in $RFib(Y)$; see Proposition 4.1.11. We may choose an inner anodyne extension $j\colon Y \to Y'$ with Y' an ∞-category. Since, by Theorem 5.2.14, the functor $\mathbf{L}j_!$ is an equivalence of categories, it is sufficient to prove that ju induces an isomorphism in $RFib(Y')$. For this, it is sufficient to prove that ju is final. In other words, it is sufficient to prove that u is final under the additional assumption that Y is an ∞-category. We recall that J' is the nerve of the contractible groupoid with sets of objects $\{0, 1\}$. We observe that inclusions of the form

$$\Lambda^n_k \to \Delta^n\,, \quad n \geq 2,\ 0 < k < n,$$

or of the form

$$J' \times \partial\Delta^n \cup \{\varepsilon\} \times \Delta^n \to J' \times \Delta^n\,, \quad n \geq 0,\ \varepsilon = 0, 1,$$

all are right anodyne extensions (Proposition 4.1.7) as well as weak categorical equivalences. Applying the small object argument to this family of inclusions, we obtain a factorisation of u of the form $u = pi$ with i both a trivial cofibration of the Joyal model structure and a right anodyne extension, and p an isofibration (whence a fibration of the Joyal model structure, by Theorem 3.6.1). Therefore, the map p is a trivial fibration, and this shows that u is final, by virtue of Corollary 4.1.9. □

Corollary 5.3.2 *Let us consider three composable maps*

$$A' \xrightarrow{i} A \xrightarrow{f} B \xrightarrow{j} B'$$

and assume that i and j are weak categorical equivalences. Then the map f is final if and only if the map jfi is final.

Proof The map f is final if and only if it defines an isomorphism in the homotopy category $RFib(B)$; see Proposition 4.1.11. The previous proposition shows that i is final, whence induces an isomorphism in $RFib(B)$. Therefore, f is final if and only if fi is final. By Theorem 5.2.14, the functor $\mathbf{L}j_!$ is an equivalence of categories, and it is clear that $\mathbf{L}j_!(f)$ is the map induced by jf. Since, again by the previous proposition, the map j is final, this shows that f is final if and only if jf has this property. □

Corollary 5.3.3 *Any right fibration is a fibration in the Joyal model category structure.*

Proof By virtue of Theorem 4.1.5, any right fibration $p\colon X \to Y$ is a fibration of the contravariant model category structure over Y. Therefore, by definition of the latter, such a map has the right lifting property with respect to right anodyne extensions over Y. Proposition 5.3.1 thus implies that p has the right lifting property with respect to any trivial cofibration of the Joyal model category structure. □

Remark 5.3.4 However, it is not true that any right fibration $X \to Y$ over a given simplicial set C is a fibration of the contravariant model category structure over C, even if C is an ∞-groupoid. Indeed, if it was true over Δ^0, this would imply that all right fibrations are Kan fibrations (see paragraph 4.1.6). But the latter is obviously not true; for instance, the inclusion of $\{0\}$ into Δ^n is always a right fibration, but it is not a Kan fibration when $n > 0$.

Proposition 5.3.5 *For any Cartesian square*

$$\begin{array}{ccc} X & \xrightarrow{\tilde{f}} & Y \\ {\scriptstyle p}\downarrow & & \downarrow{\scriptstyle q} \\ A & \xrightarrow{f} & B \end{array}$$

in which f is a weak categorical equivalence and q is a right fibration, the map $\tilde{f}$ is a weak categorical equivalence.

Proof We check first that the functor $\mathbf{L}\tilde{f}_!\colon RFib(X) \to RFib(Y)$ is fully faithful. Let $u\colon E \to X$ be a right fibration. We form a commutative square

$$\begin{array}{ccc} E & \xrightarrow{g} & F \\ {\scriptstyle u}\downarrow & & \downarrow{\scriptstyle v} \\ X & \xrightarrow{\tilde{f}} & Y \end{array}$$

in which v is a right fibration, and g is final. In other words, $(Y, q) = \mathbf{L}\tilde{f}_!(E, u)$ in $RFib(Y)$. We will prove that the map $E \to X \times_Y F$ is final as follows. Since the map $X \times_Y F \to X$ is fibrant in the contravariant model category structure over X (being a pull-back of v), and since p is a right fibration, the composed map $X \times_Y F \to A$ is a right fibration. By virtue of Proposition 4.1.11, it is sufficient to prove that this map is an isomorphism in $RFib(A)$. We now observe that $X \times_Y F$ is canonically isomorphic to $A \times_B F$. Therefore, the map $E \to X \times_Y F$ corresponds to the co-unit map $(E, pu) \to \mathbf{R}f^*\mathbf{L}f_!(E, pu)$, which is invertible, since $\mathbf{L}f_!$ is fully faithful, by Theorem 5.2.14. On the other hand, the map $E \to X \times_Y F$, which we now know to be final, can also be interpreted as the co-unit map $(E, u) \to \mathbf{R}\tilde{f}^*\mathbf{L}\tilde{f}_!(E, u)$. Therefore, the latter is an isomorphism, and this shows that $\mathbf{L}\tilde{f}_!$ is fully faithful. By Corollary 5.2.16 it suffices to check that $\tau(\tilde{f})$ is essentially surjective. Let us choose a commutative square of the form

$$\begin{array}{ccc} A & \xrightarrow{f} & B \\ {\scriptstyle a}\downarrow & & \downarrow{\scriptstyle b} \\ A' & \xrightarrow{f'} & B' \end{array}$$

in which both a and b are inner anodyne extensions, while A' and B' are ∞-categories. By Corollary 5.2.11, we can also choose a right fibration of the form $q' \colon Y' \to B'$ such that there is a pull-back square

$$\begin{array}{ccc} Y & \xrightarrow{\tilde{b}} & Y' \\ {\scriptstyle q}\downarrow & & \downarrow{\scriptstyle q'} \\ B & \xrightarrow{b} & B' \end{array}$$

and we can form the following pull-back square

$$\begin{array}{ccc} X' & \xrightarrow{\tilde{f}'} & Y' \\ {\scriptstyle p'}\downarrow & & \downarrow{\scriptstyle q'} \\ A' & \xrightarrow{f'} & B' \end{array}$$

which happens to be homotopy Cartesian. We then have a commutative square

$$\begin{array}{ccc} X & \longrightarrow & X' \\ {\scriptstyle \tilde{f}}\downarrow & & \downarrow{\scriptstyle \tilde{f}'} \\ Y & \xrightarrow{\tilde{b}} & Y' \end{array}$$

in which the right vertical map is an equivalence of ∞-categories and the two horizontal maps are bijective on objects (as pull-backs of the inner anodyne extensions a and b, respectively). To prove that $\tau(\tilde{f})$ is essentially surjective, it is thus sufficient to prove that $\tilde{b}$ is a weak categorical equivalence. For this, by virtue of Corollary 5.2.16, it is sufficient to check that $\mathbf{L}\tilde{b}_!$ is fully faithful, which we already know, since $\tilde{b}$ is the pull-back of a weak categorical equivalence along a right fibration. □

Corollary 5.3.6 *Any Cartesian square of simplicial sets of the form*

$$\begin{array}{ccc} X & \xrightarrow{g} & Y \\ {\scriptstyle p}\downarrow & & \downarrow{\scriptstyle q} \\ A & \xrightarrow{f} & B \end{array}$$

in which the map q is a right or left fibration is homotopy Cartesian in the Joyal model category structure.

Proof Since the functor $T \mapsto T^{\mathrm{op}}$ preserves and detects limits and weak categorical equivalences, and sends right fibrations to left fibrations, it is sufficient to prove the result in the case of a right fibration. Let us choose an inner anodyne extension $b\colon B \to B'$ with B' an ∞-category, as well as a pull-back square

$$\begin{array}{ccc} Y & \xrightarrow{\tilde{b}} & Y' \\ {\scriptstyle q}\downarrow & & \downarrow{\scriptstyle q'} \\ B & \xrightarrow{b} & B' \end{array}$$

in which q' is a right fibration (see Corollary 5.2.11). We factor the map bf into a trivial cofibration of the Joyal model category structure $a\colon A \to A'$, followed by a fibration $f'\colon A' \to B'$. By forming appropriate pull-backs, we also complete these data into a commutative cube

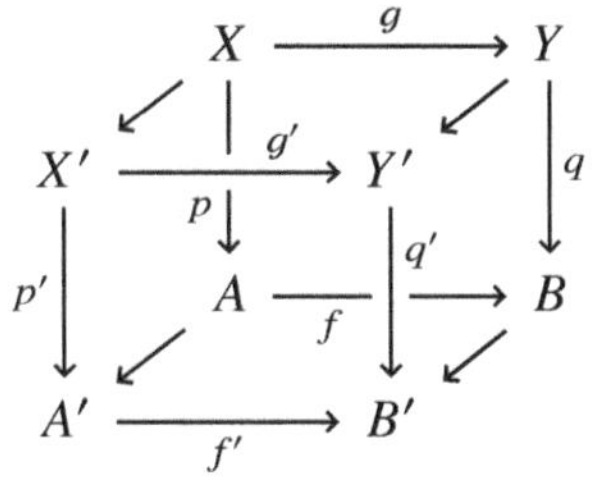

in which the non-horizontal faces are Cartesian. The slanted arrows are weak categorical equivalences: they are pull-backs of weak categorical equivalences along right fibrations, which allows one to apply the preceding proposition. Therefore, it is sufficient to prove that the front face of the cube is homotopy Cartesian. Since all the objects of this face are fibrant, this follows from the fact that f' is a fibration (see the dual version of Corollary 2.3.28). □

5.3.7 Let $p\colon X \to A$ and $q\colon Y \to A$ be two morphisms of simplicial sets. We form the map

$$(5.3.7.1) \qquad \begin{array}{c} \underline{\mathrm{Hom}}_A(X, Y) \\ \downarrow{\scriptstyle \pi_{X,Y}} \\ A \end{array}$$

corresponding to the internal Hom from (X, p) to (Y, q) in the category $sSet/A$. In other words, given a map $f\colon A' \to A$, morphisms $A' \to \underline{\mathrm{Hom}}_A(X, Y)$ over A correspond to morphisms $A' \times_A X \to A' \times_A Y$ over A'. Equivalently, we have $(\underline{\mathrm{Hom}}_A(X, Y), \pi_{X,Y}) = p_* p^*(Y, q)$ as objects of $sSet/A$.

Proposition 5.3.8 *If the morphism $q\colon Y \to A$ is a left fibration, then, for any* *map $p\colon X \to A$, the morphism $\pi_{X,Y}\colon \underline{\mathrm{Hom}}_A(X,Y) \to A$ is a fibration of the Joyal model category structure.*

Proof The functor $(X,p) \mapsto (\underline{\mathrm{Hom}}_A(X,Y), \pi_{X,Y})$ is right adjoint to the functor $(-)\times_A Y$. The latter preserves monomorphisms, and, by virtue of Proposition 5.3.5, it also preserves the class of weak equivalences of the model category structure on *sSet*/A induced by the Joyal model category structure. In particular, we have here a Quillen pair. □

Lemma 5.3.9 *Let $f\colon X \to Y$ be a morphism of simplicial sets. Assume that there is also a given map $q\colon Y \to A$. Then the following conditions are equivalent.*

(i) For any map of the form $\sigma\colon \Delta^n \to A$, for $n \geq 0$, the induced morphism $\Delta^n \times_A X \to \Delta^n \times_A Y$ is a weak categorical equivalence.
(ii) For any map $g\colon B \to A$, the induced morphism $B \times_A X \to B \times_A Y$ is a weak categorical equivalence.

In particular, condition (i) implies that the map f is a weak categorical equivalence.

Proof The class of simplicial sets B over A such that $B \times_A X \to B \times_A Y$ is a weak categorical equivalence is saturated by monomorphisms: since colimits are universal in the category of simplicial sets, this follows from Corollaries 2.3.16, 2.3.18 and 2.3.29. Corollary 1.3.10 thus implies that conditions (i) and (ii) are equivalent. □

5.3.10 A morphism of simplicial sets $X \to Y$ over A is said to be *locally a weak categorical equivalence over A* if it satisfies the equivalent conditions of Lemma 5.3.9. Under the same assumptions as in paragraph 5.3.7, we define $\underline{\mathrm{Eq}}_A(X,Y)$ as the subobject of $\underline{\mathrm{Hom}}_A(X,Y)$ whose sections over $f\colon A' \to A$ correspond to maps $A' \times_A X \to A' \times_A Y$ over A' which are locally a weak categorical equivalence over A'. We shall still write $\pi_{X,Y}\colon \underline{\mathrm{Eq}}_A(X,Y) \to A$ for the restriction of the structural map of $\underline{\mathrm{Hom}}_A(X,Y)$. We observe that, in the case where both p and q are left fibrations, the object $\underline{\mathrm{Eq}}_A(X,Y)$ classifies fibrewise equivalences; see Theorem 4.1.16 and Remark 4.1.18, keeping in mind that all the Δ^n are ∞-categories. In fact, in this case, Corollary 5.3.6 and the preceding lemma show together that a section of $\underline{\mathrm{Eq}}_A(X,Y)$ over $f\colon A' \to A$ is simply a map $A' \times_A X \to A' \times_A Y$ over A' which is a weak categorical equivalence.

Proposition 5.3.11 *If both $p\colon X \to A$ and $q\colon Y \to A$ are left fibrations, then the map $\pi_{X,Y}\colon \underline{\mathrm{Eq}}_A(X,Y) \to A$ is a fibration of the Joyal model category structure.*

Proof Let us consider a solid commutative square of the form

$$\begin{array}{ccc} K & \xrightarrow{u} & \underline{\mathrm{Eq}}_A(X,Y) \\ {\scriptstyle j}\downarrow & & \downarrow{\scriptstyle \pi_{X,Y}} \\ L & \longrightarrow & A \end{array}$$

in which j is assumed to be a trivial cofibration. The map u can be interpreted as a weak categorical equivalence from $K \times_A X$ to $K \times_A Y$ over K, and we want to find a weak categorical equivalence v from $L \times_A X$ to $L \times_A Y$ over L such that the following square of simplicial sets commutes.

$$\begin{array}{ccc} K \times_A X & \xrightarrow{u} & K \times_A Y \\ {\scriptstyle j \times_A X}\downarrow & & \downarrow{\scriptstyle j \times_A Y} \\ L \times_A X & \xrightarrow{v} & L \times_A Y \end{array}$$

Applying twice the dual version of Proposition 5.3.5 ensures that the two maps $j \times_A X$ and $j \times_A Y$ are weak categorical equivalences. Since u is also a weak categorical equivalence, we only have to find a map v from $L \times_A X$ to $L \times_A Y$ over L which extends u as in the square above. The existence of such a map v follows from Proposition 5.3.8. □

5.3.12 There is a universal morphism of left fibrations with **U**-small fibres. One considers the Cartesian product $\mathcal{S} \times \mathcal{S}$, over which there are two canonical left fibrations with **U**-small fibres $\mathcal{S}_\bullet^{(0)} \to \mathcal{S} \times \mathcal{S}$ and $\mathcal{S}_\bullet^{(1)} \to \mathcal{S} \times \mathcal{S}$ which are classified by the first and second projection to $\mathcal{S}$, respectively. The isofibration $(s,t)\colon \underline{\mathrm{Hom}}_{\mathcal{S}\times\mathcal{S}}(\mathcal{S}_\bullet^{(0)}, \mathcal{S}_\bullet^{(1)}) \to \mathcal{S}\times\mathcal{S}$ classifies morphisms between left fibrations with **U**-small fibres: for a simplicial set A, a map $(F,G)\colon A \to \mathcal{S}\times\mathcal{S}$ essentially consists of two left fibrations $p\colon X \to A$ and $q\colon Y \to A$. And a lift φ of (F,G) to $\underline{\mathrm{Hom}}_{\mathcal{S}\times\mathcal{S}}(\mathcal{S}_\bullet^{(0)}, \mathcal{S}_\bullet^{(1)})$ is equivalent to the datum of a morphism $\varphi\colon X \to Y$ over A. In particular, there is a canonical map $\mathcal{S} \to \underline{\mathrm{Hom}}_{\mathcal{S}\times\mathcal{S}}(\mathcal{S}_\bullet^{(0)}, \mathcal{S}_\bullet^{(1)})$ which corresponds to the identity of $\mathcal{S}_\bullet$ over $\mathcal{S}$. Since the identity is a fibrewise equivalence, we end up with a diagram

$$\mathcal{S} \xrightarrow{\mathrm{id}_{\mathcal{S}_\bullet}} \underline{\mathrm{Eq}}_{\mathcal{S}\times\mathcal{S}}(\mathcal{S}_\bullet^{(0)}, \mathcal{S}_\bullet^{(1)}) \xrightarrow{(s,t)} \mathcal{S}\times\mathcal{S} \tag{5.3.12.1}$$

which is a factorisation of the diagonal $\mathcal{S} \to \mathcal{S} \times \mathcal{S}$. We know that the map (s,t) is an isofibration. The nature of the map $\mathrm{id}_{\mathcal{S}_\bullet}$ is revealed by the following proposition.

Proposition 5.3.13 *The map* $\mathrm{id}_{\mathcal{S}_\bullet}\colon \mathcal{S} \to \underline{\mathrm{Eq}}_{\mathcal{S}\times\mathcal{S}}(\mathcal{S}_\bullet^{(0)}, \mathcal{S}_\bullet^{(1)})$ *is a trivial cofibration of the Joyal model category structure. In other words, the diagram*

(5.3.12.1) *is a path object of the ∞-category* $\mathcal{S}$ *in the Joyal model category structure.*

Proof This map is a section of the isofibration t, and, therefore, it is sufficient to prove that the map

$$t\colon \underline{\mathrm{Eq}}_{\mathcal{S}\times\mathcal{S}}(\mathcal{S}_\bullet^{(0)}, \mathcal{S}_\bullet^{(1)}) \to \mathcal{S}$$

is a trivial fibration. Consider a cofibration $j\colon Y \to Y'$. Then a commutative square

$$\begin{array}{ccc} Y & \xrightarrow{\xi} & \underline{\mathrm{Eq}}_{\mathcal{S}\times\mathcal{S}}(\mathcal{S}_\bullet^{(0)}, \mathcal{S}_\bullet^{(1)}) \\ {\scriptstyle j}\downarrow & & \downarrow{\scriptstyle t} \\ Y' & \xrightarrow{\xi'} & \mathcal{S} \end{array}$$

consists essentially of a commutative diagram of the form

$$\begin{array}{ccccc} X_0 & \xrightarrow{w} & X_1 & \xrightarrow{i_1} & X_1' \\ & {\scriptstyle p_0}\searrow & \downarrow{\scriptstyle p_1} & & \downarrow{\scriptstyle p_1'} \\ & & Y & \xrightarrow{j} & Y' \end{array}$$

in which p_0, p_1 and p_1' are left fibrations (with $\mathbf{U}$-small fibres), w is a weak categorical equivalence, and the square is Cartesian (where the triple (p_0, w, p_1) corresponds to ξ, the left fibration p_1' corresponds to ξ', and the Cartesian square corresponds to the equation $\xi' j = t\xi$). Since all left fibrations are fibrations of the Joyal model category structure (by the dual version of Corollary 5.3.3), Proposition 5.1.21 may be applied for $\mathcal{F}$ the class of left fibrations. Together with Corollary 5.2.6, this gives a map $\zeta\colon Y' \to \underline{\mathrm{Eq}}_{\mathcal{S}\times\mathcal{S}}(\mathcal{S}_\bullet^{(0)}, \mathcal{S}_\bullet^{(1)})$ such that $t\zeta = \xi'$ and $\zeta j = \xi$. □

5.3.14 Let $p\colon X \to A$ and $q\colon Y \to A$ be two left fibrations with $\mathbf{U}$-small fibres over a simplicial set A, classified by two morphisms F and G, respectively. A morphism from F to G in $\underline{\mathrm{Hom}}(A, \mathcal{S})$ is essentially given by a left fibration $\pi\colon W \to \Delta^1 \times A$, together with isomorphisms $X \simeq W_0$ and $Y \simeq W_1$ over A, where we denote by $\pi_i\colon W_i \to A$ the pull-back of π along $\{i\} \times A \subset \Delta^1 \times A$. We thus have the following solid commutative square

(5.3.14.1)
$$\begin{array}{ccc} \{0\} \times X & \hookrightarrow & W \\ \downarrow & \overset{\varphi}{\nearrow} & \downarrow{\scriptstyle \pi} \\ \Delta^1 \times X & \xrightarrow{1_{\Delta^1}\times p} & \Delta^1 \times A \end{array}$$

which admits a filler φ because the left-hand vertical map is a cofinal monomorphism, hence a left anodyne extension, and π is a left fibration. Taking the fibre at 1 turns φ into a morphism

$$\varphi_1 : X \simeq (\Delta^1 \times X)_1 \to W_1 \simeq Y \tag{5.3.14.2}$$

of left fibrations over A. We thus get a morphism

$$\varphi_1 : A \to \underline{\mathrm{Hom}}_{\mathcal{S}\times\mathcal{S}}(\mathcal{S}_\bullet^{(0)}, \mathcal{S}_\bullet^{(1)}) \tag{5.3.14.3}$$

which lifts the classifying map $(F, G) : A \to \mathcal{S} \times \mathcal{S}$ for p and q.

Proposition 5.3.15 *The J-homotopy class over A of the map* (5.3.14.2) *is independent of the choice of a lift in the solid commutative square* (5.3.14.1). *So is the J-equivalence class of the map* (5.3.14.3) *over $\mathcal{S} \times \mathcal{S}$.*

Proof A lift of the commutative square (5.3.14.1) is a morphism from a cofibrant object which is weakly equivalent to the initial object to a fibrant object in the category $X\backslash sSet/\Delta^1 \times A$, endowed with the model category structure induced by the covariant model category structure over $\Delta^1 \times A$. Therefore, any two such lifts are equal up to J-homotopy over $\Delta^1 \times A$. Corollary 5.3.6 implies that passing to the fibres at 1 is weakly equivalent to passing to the homotopy fibres. Therefore, two choices of lifts of the commutative square (5.3.14.1) give the same map (5.3.14.2), seen in the homotopy category of the Joyal model category structure over A. Since this is a map from a cofibrant object to a fibrant object, this identification in the homotopy category of $sSet/A$ is equivalent to an identification up to homotopy over A.

A J-homotopy between maps from X to Y over A is a map of the form $h : J \times X \to Y$, so that qh is the composition of p with the projection of $J \times X$ onto X. This defines a map $H : J \times X \to J \times Y$ over $J \times A$ defined by the projection of $J \times X$ onto J and h, where the structural maps of $J \times X$ and $J \times Y$ are $1_J \times p$ and $1_J \times q$, respectively. The map H can be seen as a map from $J \times A$ to $\underline{\mathrm{Hom}}_{\mathcal{S}\times\mathcal{S}}(\mathcal{S}_\bullet^{(0)}, \mathcal{S}_\bullet^{(1)})$. One deduces from this observation that the J-equivalence class of the map (5.3.14.3) over $\mathcal{S} \times \mathcal{S}$ only depends on the J-homotopy class over A of the map (5.3.14.2). □

Proposition 5.3.16 *Under the assumptions of paragraph 5.3.14, the following assertions are equivalent.*

(i) The morphism $\varphi_1 : X \to Y$ is a fibrewise equivalence over A.

(ii) The map $F \to G$, corresponding to W, is invertible in the ∞-category of functors $\underline{\mathrm{Hom}}(A, \mathcal{S})$.

(iii) For any object a of A, the morphism $\pi_a : W_a \to \Delta^1$, obtained by pulling back π along $\Delta^1 \times \{a\} \subset \Delta^1 \times A$, is a Kan fibration.

Proof Since a map of $\underline{\mathrm{Hom}}(A, \mathcal{S})$ is invertible if and only if it is fibrewise invertible (Corollary 3.5.12), it is sufficient to prove the case where $A = \Delta^0$ is a point. Proposition 5.2.13 ensures that conditions (ii) and (iii) are equivalent. Let us prove that conditions (iii) and (i) are equivalent. If π is a Kan fibration, the maps $W_i \to W$ are weak homotopy equivalences for $i = 0, 1$. Hence the map $\varphi\colon \Delta^1 \times X \to W$ must be a weak homotopy equivalence between Kan fibrations over Δ^1, which implies that it is a fibrewise weak homotopy equivalence. In particular, the map $\varphi_1\colon X \to Y$ is a weak homotopy equivalence. Conversely, if φ_1 is a weak homotopy equivalence, then φ is a fibrewise equivalence (recall that φ_0 is an isomorphism), hence the inclusion $X \to W$ is a weak homotopy equivalence. Lemma 4.6.9 implies that $\pi^{\mathrm{op}}\colon W^{\mathrm{op}} \to (\Delta^1)^{\mathrm{op}} \simeq \Delta^1$ is locally constant as an object of $RFib(\Delta^1)$. Theorem 4.6.10 shows that π^{op} is a Kan fibration, whence so is π. □

5.3.17 Recall that there is a subobject $h(\Delta^1, \mathcal{S}) \subset \underline{\mathrm{Hom}}(\Delta^1, \mathcal{S})$ such that morphisms $A \to h(\Delta^1, \mathcal{S})$ correspond to invertible maps in $\underline{\mathrm{Hom}}(A, \mathcal{S})$; see (3.5.7.2) and Corollary 3.5.12. The evaluation map $ev\colon \Delta^1 \times \underline{\mathrm{Hom}}(\Delta^1, \mathcal{S}) \to \mathcal{S}$ corresponds to a morphism

$$E\colon \Delta^1 \to \underline{\mathrm{Hom}}(\underline{\mathrm{Hom}}(\Delta^1, \mathcal{S}), \mathcal{S})\,.$$

As explained above, the image of E in $\underline{\mathrm{Hom}}(h(\Delta^1, \mathcal{S}), \mathcal{S})$, namely

$$E^h\colon \Delta^1 \to \underline{\mathrm{Hom}}(h(\Delta^1, \mathcal{S}), \mathcal{S})\,,$$

is an invertible map (it corresponds to the identity of $h(\Delta^1, \mathcal{S})$). One may think of E as the universal morphism in an ∞-category of the form $\underline{\mathrm{Hom}}(A, \mathcal{S})$, and of E^h as the universal invertible morphism in such an ∞-category. Indeed, these morphisms correspond to left fibrations with **U**-small fibres π and π^h, respectively, obtained by forming the following Cartesian squares.

(5.3.17.1)
$$\begin{array}{ccccc} W^h & \longrightarrow & W & \longrightarrow & \mathcal{S}_\bullet \\ {\scriptstyle \pi^h}\downarrow & & {\scriptstyle \pi}\downarrow & & \downarrow{\scriptstyle p_{univ}} \\ \Delta^1 \times h(\Delta^1, \mathcal{S}) & \hookrightarrow & \Delta^1 \times \underline{\mathrm{Hom}}(\Delta^1, \mathcal{S}) & \xrightarrow{ev} & \mathcal{S} \end{array}$$

Given a morphism $f\colon F \to G$ in $\underline{\mathrm{Hom}}(A, \mathcal{S})$, seen as a map of the form $f\colon A \to \underline{\mathrm{Hom}}(\Delta^1, \mathcal{S})$, we can form a pull-back square

(5.3.17.2)
$$\begin{array}{ccc} V & \longrightarrow & W \\ {\scriptstyle p}\downarrow & & \downarrow{\scriptstyle \pi} \\ \Delta^1 \times A & \xrightarrow{1_{\Delta^1} \times f} & \Delta^1 \times \underline{\mathrm{Hom}}(\Delta^1, \mathcal{S}) \end{array}$$

and the left fibration $p\colon V \to \Delta^1 \times A$ is the one classified by the morphism $\Delta^1 \times A \to \mathcal{S}$ corresponding to f. The property that f is invertible in the ∞-category $\underline{\mathrm{Hom}}(A, \mathcal{S})$ is equivalent to the property that the map $1_{\Delta^1} \times f$ factors through $h(\Delta^1, \mathcal{S})$. The construction of paragraph 5.3.14 applied to the left fibration π above provides a lift $\varphi_1\colon \underline{\mathrm{Hom}}(\Delta^1, \mathcal{S}) \to \underline{\mathrm{Hom}}_{\mathcal{S}\times\mathcal{S}}(\mathcal{S}_\bullet^{(0)}, \mathcal{S}_\bullet^{(1)})$ of the map $(W_0, W_1)\colon \underline{\mathrm{Hom}}(\Delta^1, \mathcal{S}) \to \mathcal{S}\times\mathcal{S}$ which classifies the source and the target of W. Furthermore, Proposition 5.3.16 ensures that there is a canonical Cartesian square of the following form

$$
\begin{array}{ccc}
h(\Delta^1, \mathcal{S}) & \longrightarrow & \underline{\mathrm{Eq}}_{\mathcal{S}\times\mathcal{S}}(\mathcal{S}_\bullet^{(0)}, \mathcal{S}_\bullet^{(1)}) \\
\downarrow & & \downarrow \\
\underline{\mathrm{Hom}}(\Delta^1, \mathcal{S}) & \xrightarrow{\varphi_1} & \underline{\mathrm{Hom}}_{\mathcal{S}\times\mathcal{S}}(\mathcal{S}_\bullet^{(0)}, \mathcal{S}_\bullet^{(1)})
\end{array}
\tag{5.3.17.3}
$$

over the ∞-category $\mathcal{S}\times\mathcal{S}$. Proposition 5.3.15 explains in which way this map is independent of the choices we made, at least up to J-homotopy over $\mathcal{S}\times\mathcal{S}$.

Lemma 5.3.18 *Let $p\colon X \to A$ and $q\colon Y \to A$ be two left fibrations classified by morphisms F and G, respectively. There is a canonical Cartesian square*

$$
\begin{array}{ccc}
\mathrm{Map}_A(X, Y) & \longrightarrow & \underline{\mathrm{Hom}}(A, \underline{\mathrm{Hom}}_{\mathcal{S}\times\mathcal{S}}(\mathcal{S}_\bullet^{(0)}, \mathcal{S}_\bullet^{(1)})) \\
\downarrow & & \downarrow{\scriptstyle (s,t)} \\
\Delta^0 & \xrightarrow{(F,G)} & \underline{\mathrm{Hom}}(A, \mathcal{S}) \times \underline{\mathrm{Hom}}(A, \mathcal{S})
\end{array}
$$

where $\mathrm{Map}_A(X, Y)$ is the mapping space of simplicial sets over A introduced in (4.1.12.1).

Proof Both the fibre of the map (s, t) over (F, G) and the simplicial set $\mathrm{Map}_A(X, Y)$ are identified with the simplicial set whose elements are the maps $\varphi\colon \Delta^n \times X \to \Delta^n \times Y$ such that the triangle

$$
\begin{array}{ccc}
\Delta^n \times X & \xrightarrow{\varphi} & \Delta^n \times Y \\
{\scriptstyle 1_{\Delta^n}\times p} \searrow & & \swarrow {\scriptstyle 1_{\Delta^n}\times q} \\
& \Delta^n \times A &
\end{array}
$$

commutes. □

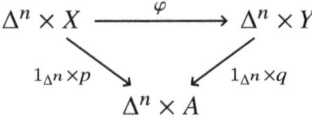

5.3.19 For two left fibrations $p\colon X \to A$ and $q\colon Y \to A$ classified by morphisms F and G as in the previous lemma, we define $\mathrm{Equiv}_A(X, Y)$ as the union of the connected components of the Kan complex $\mathrm{Map}_A(X, Y)$ corresponding to the maps $\varphi\colon X \to Y$ over A which are fibrewise equivalences (or equivalently, which become invertible in $LFib(A)$). We observe that it fits in the following Cartesian square.

$$\begin{array}{ccc} \mathrm{Equiv}_A(X,Y) & \longrightarrow & \underline{\mathrm{Hom}}(A, \underline{\mathrm{Eq}}_{\mathcal{S}\times\mathcal{S}}(\mathcal{S}^{(0)}_\bullet, \mathcal{S}^{(1)}_\bullet)) \\ \downarrow & & \downarrow \\ \mathrm{Map}_A(X,Y) & \longrightarrow & \underline{\mathrm{Hom}}(A, \underline{\mathrm{Hom}}_{\mathcal{S}\times\mathcal{S}}(\mathcal{S}^{(0)}_\bullet, \mathcal{S}^{(1)}_\bullet)) \end{array} \tag{5.3.19.1}$$

We also have a canonical Cartesian square

$$\begin{array}{ccc} k(A,\mathcal{S})(F,G) & \longrightarrow & \underline{\mathrm{Hom}}(A, h(\Delta^1,\mathcal{S})) \\ \downarrow & & \downarrow \\ \underline{\mathrm{Hom}}(A,\mathcal{S})(F,G) & \longrightarrow & \underline{\mathrm{Hom}}(A, \underline{\mathrm{Hom}}(\Delta^1,\mathcal{S})) \end{array} \tag{5.3.19.2}$$

which identifies $k(A,\mathcal{S})(F,G)$ with the union of the connected components of $\underline{\mathrm{Hom}}(A,\mathcal{S})(F,G)$ corresponding to invertible morphisms from F to G in the ∞-category $\underline{\mathrm{Hom}}(A,\mathcal{S})$ (see Corollary 3.5.12). Finally, the Cartesian square (5.3.17.3) provides a Cartesian square of the form below.

$$\begin{array}{ccc} k(A,\mathcal{S})(F,G) & \longrightarrow & \mathrm{Equiv}_A(X,Y) \\ \downarrow & & \downarrow \\ \underline{\mathrm{Hom}}(A,\mathcal{S})(F,G) & \longrightarrow & \mathrm{Map}_A(X,Y) \end{array} \tag{5.3.19.3}$$

Note that, although the construction of the horizontal maps relies on a choice, the formation of this Cartesian square is perfectly functorial in A (i.e. defines a presheaf on the category of simplicial sets over A with values in the category of Cartesian squares of simplicial sets). Furthermore, the choice we made is irrelevant up to J-homotopy: as recalled at the end of paragraph 5.3.17, the J-homotopy class over $\mathcal{S}\times\mathcal{S}$ of the map φ_1 of diagram (5.3.17.3) is independent of the choice we made, so that passing to the homotopy fibres give a map, as in the lower horizontal map of (5.3.19.3), whose J-homotopy class only depends on the one of φ_1 over $\mathcal{S}\times\mathcal{S}$. Since the square (5.3.19.3) is homotopy Cartesian in the Kan–Quillen model category structure, its upper horizontal map, which is a morphism of ∞-groupoids, only depends on the J-homotopy class of the lower horizontal map.

Proposition 5.3.20 *The morphism constructed above,*

$$k(A,\mathcal{S})(F,G) \to \mathrm{Equiv}_A(X,Y),$$

is an equivalence of ∞-groupoids.

Proof By virtue of Corollaries 3.6.7 and 3.6.4, and of Proposition 5.3.11, the diagram

$$\begin{array}{ccc} \underline{\mathrm{Hom}}(A, h(\Delta^1, \mathcal{S})) & \xrightarrow{(\varphi_1)_*} & \underline{\mathrm{Hom}}(A, \underline{\mathrm{Eq}}_{\mathcal{S}\times\mathcal{S}}(\mathcal{S}^{(0)}_\bullet, \mathcal{S}^{(1)}_\bullet)) \\ \downarrow & & \downarrow \\ \underline{\mathrm{Hom}}(A, \mathcal{S}) \times \underline{\mathrm{Hom}}(A, \mathcal{S}) & = & \underline{\mathrm{Hom}}(A, \mathcal{S}) \times \underline{\mathrm{Hom}}(A, \mathcal{S}) \end{array}$$

is a morphism of isofibrations between ∞-categories. Furthermore, projecting to the first factor, the induced maps

$$\underline{\mathrm{Hom}}(A, h(\Delta^1, \mathcal{S})) \to \underline{\mathrm{Hom}}(A, \mathcal{S})$$

and

$$\underline{\mathrm{Hom}}(A, \underline{\mathrm{Eq}}_{\mathcal{S}\times\mathcal{S}}(\mathcal{S}^{(0)}_\bullet, \mathcal{S}^{(1)}_\bullet)) \to \underline{\mathrm{Hom}}(A, \mathcal{S})$$

are trivial fibrations: by Corollary 3.6.4, it is sufficient to check this for $A = \Delta^0$, in which case this follows from Corollary 3.5.10 and Proposition 5.3.13, respectively. Therefore, since the map $(\varphi_1)_*$ is a weak equivalence between two fibrations over the ∞-category $\underline{\mathrm{Hom}}(A, \mathcal{S}) \times \underline{\mathrm{Hom}}(A, \mathcal{S})$, it must induce a weak equivalence on the fibres. □

Corollary 5.3.21 *For any simplicial set A, the operation of pulling back along the map $p_{univ}\colon \mathcal{S}_\bullet \to \mathcal{S}$ defines a bijection from the set $[A, \mathcal{S}] = \pi_0(k(A, \mathcal{S}))$ onto the set of isomorphism classes of left fibrations with* **U**-*small fibres $p\colon X \to A$ in LFib(A).*

Remark 5.3.22 Proposition 5.3.20 allows one to work up to weak categorical equivalence. For instance, assuming that we have a homotopy Cartesian square

(5.3.22.1)
$$\begin{array}{ccc} X & \longrightarrow & \mathcal{S}_\bullet \\ {\scriptstyle p}\downarrow & & \downarrow{\scriptstyle p_{univ}} \\ A & \xrightarrow{G} & \mathcal{S} \end{array}$$

in which p is a left fibration, if we denote by $q\colon Y \to A$ the left fibration classified by G, there is a canonical morphism $\xi\colon X \to Y$ over A which must be a weak categorical equivalence, hence a fibrewise equivalence: since the square (5.3.22.1) is homotopy Cartesian, this follows from Corollary 5.3.6. Let $F\colon A \to \mathcal{S}$ be a morphism which classifies the left fibration p. Proposition 5.3.20 ensures that there is an essentially unique invertible morphism $F \to G$ in $\underline{\mathrm{Hom}}(A, \mathcal{S})$ associated to the weak equivalence ξ. Conversely, if there are two functors $F, G\colon A \to \mathcal{S}$ which classify two left fibrations $p\colon X \to A$ and $q\colon Y \to A$, respectively, for any invertible morphism $F \to G$ in the ∞-category $\underline{\mathrm{Hom}}(A, \mathcal{S})$, there is an essentially unique associated weak categorical equivalence $\xi\colon X \to Y$ over A. Using ξ, one produces a commutative diagram of the form (5.3.22.1). Corollary 5.3.6 implies that this square is homotopy Cartesian because, by construction, the comparison map $X \to Y = A \times_{\mathcal{S}} \mathcal{S}_\bullet$ is the weak equivalence ξ.

5.4 Rectification of Morphisms

Lemma 5.4.1 *Let $\pi\colon W \to \Delta^n \times A$ be a left fibration. We denote by $\pi_i\colon W_i \to A$ the fibre of π at i for $0 \leq i \leq n$. The inclusion $W_n \to W$ is a weak equivalence of the covariant model category structure over A. We also choose a lift φ in the solid commutative square below.*

$$\begin{array}{ccc} \{0\} \times W_0 & \hookrightarrow & W \\ \downarrow & \overset{\varphi}{\nearrow} & \downarrow{\scriptstyle \pi} \\ \Delta^n \times W_0 & \xrightarrow{1_{\Delta^1} \times \pi_0} & \Delta^n \times A \end{array}$$

We write φ_n for the map induced by φ by passing to fibres over n. Then, if we consider this diagram as a commutative square over A through the second projection $\Delta^n \times A \to A$, for any $i \in \{0, \ldots, n\}$, the diagram

$$\begin{array}{ccc} W_0 & \xrightarrow{\varphi_n} & W_n \\ {\scriptstyle (i, 1_{W_0})}\downarrow & & \downarrow \\ \Delta^n \times W_0 & \xrightarrow{\varphi} & W \end{array}$$

commutes in $LFib(A)$. In particular, the map φ_n is equal in $LFib(A)$ to the composition of the inclusion of $W_0 \to W$ with the inverse of the invertible map $W_n \to W$.

Proof We prove first that the inclusion $W_n \to W$ is a weak equivalence of the covariant model category structure over A. We remark that this is a map between proper morphisms of codomain A: the map $W_n \to A$ is a pull-back of the left fibration π, and the map $W \to A$ is the composition of the left fibration π with the second projection $\Delta^n \times A \to A$, which is proper, by Proposition 4.4.12. Since this is a map between proper morphisms, Corollary 4.4.28 ensures that we only have to check that the inclusion of W_n into W is a fibrewise weak homotopy equivalence. But, fibrewise, it is a final map, since it is the pull-back of the final map $\{n\} \to \Delta^n$ along some left fibration.

It remains to check the commutativity of the diagram in $LFib(A)$. This is obviously true for $i = n$. Moreover, since $(0, 1_{W_0})$ is initial, the second projection $p\colon \Delta^n \times W_0 \to W_0$ is a weak equivalence of the covariant model category structure over A. Therefore, any other section of p is also a weak equivalence, and all sections of p define the same map in $LFib(A)$. This settles the case of $(i, 1_{W_0})$ for $i \neq n$. □

5.4.2 Let A be a simplicial set. We denote by $\mathrm{Arr}(sSet/A)$ the category of arrows of $sSet/A$, or, equivalently, of functors from I to $sSet/A$, where I is the category freely generated by the oriented graph $0 \to 1$.

We consider the injective model structure on $\mathrm{Arr}(sSet/A)$ associated to the covariant model category structure over A (see Proposition 2.3.11 for $\mathcal{C} = (sSet/A)^{\mathrm{op}}$). In other words, if $p_i \colon X_i \to Y_i$, $i = 0, 1$, are two morphisms of $sSet/A$, a map f from p_0 to p_1 is a commutative square of the following form.

$$\begin{array}{ccc} X_0 & \xrightarrow{f_0} & X_1 \\ {\scriptstyle p_0}\downarrow & & \downarrow{\scriptstyle p_1} \\ Y_0 & \xrightarrow{f_1} & Y_1 \end{array} \tag{5.4.2.1}$$

Such a morphism f is a weak equivalence (a cofibration, respectively) if and only if both f_0 and f_1 have this property in the covariant model category structure over A. For f to be a fibration, we require that f_1 is a left fibration and that the canonical map $X_0 \to Y_0 \times_{Y_1} X_1$ is a left fibration.

We define an adjunction

$$t_! \colon \mathrm{Arr}(sSet/A) \rightleftarrows sSet/(\Delta^1 \times A) \colon t^* \tag{5.4.2.2}$$

as follows. If $p = pr_2 \colon \Delta^1 \times A \to A$ denotes the second projection, the functor $p^* \colon sSet/A \to sSet/\Delta^1 \times A$ has a right adjoint p_*. For $\varepsilon = 0, 1$, the functor $c_\varepsilon = (\varepsilon, 1_A) \colon A \to \Delta^1 \times A$ defines a fully faithful inclusion $c_{\varepsilon,!}$ of $sSet/A$ into $sSet/\Delta^1 \times A$ which sends a simplicial set X over A to $X = \{\varepsilon\} \times X$. Furthermore, for any simplicial set X over A, there is a natural map $c_{\varepsilon,!}(X) = \{\varepsilon\} \times X \to \Delta^1 \times X = p^*(X)$ over $\Delta^1 \times A$. By transposition, this defines a natural map $p_*(W) \to c_\varepsilon^*(W) = W_\varepsilon$ for any simplicial set W over $\Delta^1 \times A$. The functor t^* corresponds to the natural transformation $p_* \to c_1^*$.

In what follows, the category $sSet/\Delta^1 \times A$ is endowed with the covariant model category structure over $\Delta^1 \times A$.

Proposition 5.4.3 *The adjunction* (5.4.2.2) *is a Quillen equivalence.*

Proof The adjunction $(c_{\varepsilon,!}, c_\varepsilon^*)$ is a Quillen pair, and, since the functor p is smooth, so is (p^*, p_*); see Propositions 4.4.17 and 4.4.18. For any monomorphism $i \colon K \to L$ over A, and for any map $p \colon X \to Y$ over $\Delta^1 \times A$, there is a correspondence between the following two lifting problems.

$$\begin{array}{ccc} K & \longrightarrow & p_*X \\ \downarrow & \nearrow & \downarrow \\ L & \longrightarrow & X_1 \times_{Y_1} p_*Y \end{array} \quad \leftrightsquigarrow \quad \begin{array}{ccc} \Delta^1 \times K \cup \{1\} \times L & \longrightarrow & X \\ \downarrow & \nearrow & \downarrow \\ \Delta^1 \times L & \longrightarrow & Y \end{array}$$

Since $\Delta^1 \times K \cup \{1\} \times L \to \Delta^1 \times L$ is a left anodyne extension whenever $K \to L$ has this property, this shows that t^* preserves both fibrations and trivial fibrations. In other words, $(t_!, t^*)$ is a Quillen pair. Similar arguments (replacing $\{1\}$ by $\{0\}$) show that the canonical morphism $p_*X \to c_0^*X = X_0$ is a trivial fibration: for any monomorphism $K \to L$, the induced embedding $\Delta^1 \times K \cup \{0\} \times L \to \Delta^1 \times L$ is always a left anodyne extension. This implies that the total derived functor $\mathbf{R}t^*$ is conservative. Therefore, it is now sufficient to prove that the total left derived functor $\mathbf{L}t_!$ is fully faithful.

For a simplicial set W over $\Delta^1 \times A$ and $\varepsilon = 0, 1$, let W/ε be the simplicial set over A obtained by pulling back along the canonical map $\Delta^1/\varepsilon \to \Delta^1$, the structural map being induced by composing with the projection of $\Delta^1/\varepsilon \times A$ to A. For $\varepsilon = 1$, this is a fancy way to look at the functor $p_!$. For $\varepsilon = 0$, this is another way to look at the functor c_0^*. The operation $W \mapsto W/\varepsilon$ is a left Quillen functor for the appropriate covariant model category structures. Furthermore, whenever W is fibrant, the map $W_\varepsilon = c_\varepsilon^*(W) \to W/\varepsilon$ is a weak equivalence (this is the first assertion of Lemma 5.4.1). Therefore, given any object F of $\mathrm{Arr}(sSet/A)$, if we choose a fibrant resolution G of $t_!F$, we obtain, for $\varepsilon = 0, 1$, the following commutative diagram of simplicial sets over A,

$$\begin{array}{ccccc} F_\varepsilon & \longrightarrow & t^*t_!(F)_\varepsilon & \longrightarrow & t^*(G)_\varepsilon \\ & \searrow & \downarrow & & \downarrow \wr \\ & & c_\varepsilon^*t_!(F) & \longrightarrow & c_\varepsilon^*(G) \\ & & \downarrow & & \downarrow \wr \\ & & t_!(F)/\varepsilon & \xrightarrow{\sim} & G/\varepsilon \end{array}$$

in which the decorated arrows $\xrightarrow{\sim}$ are weak equivalences. The map $F_\varepsilon \to t^*(G)_\varepsilon$ corresponds to the evaluation at ε of the derived unit $F \to \mathbf{R}t^*\mathbf{L}t_!(F)$. In other words, the property of fully faithfulness for the functor $\mathbf{L}t_!$ is equivalent to the property that the map $F_\varepsilon \to t_!(F)/\varepsilon$ is a weak equivalence of the covariant model category structure over A for any F and $\varepsilon = 0, 1$. Since the latter maps are natural transformations between left Quillen functors, and since the category $\mathrm{Arr}(sSet/A)$ is the category of presheaves over the Eilenberg–Zilber category $[1] \times \mathbf{\Delta}/A$, the homotopy-theoretic Corollaries 2.3.16, 2.3.18

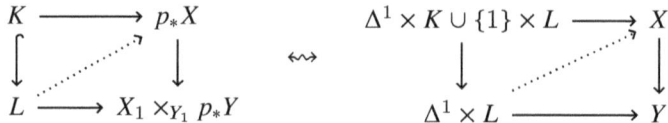

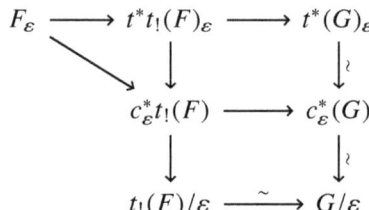

and 2.3.29, and Corollary 1.3.10, imply that we only have to check this property for F representable.

Let $a\colon \Delta^n \to A$ be a morphism of simplicial sets, defining an object (Δ^n, a) of $\mathbf{\Delta}/A$. For F the presheaf represented by $(0, (\Delta^n, a))$, $t_!(F)$ is the simplicial set $\Delta^1 \times \Delta^n$, with structural map $1_{\Delta^1} \times a\colon \Delta^1 \times \Delta^n \to \Delta^1 \times A$. Therefore, the map $F_0 \to t_!(F)/0$ is the identity, and the map $F_1 \to t_!(F)/1$ is the inclusion $\Delta^n = \{1\} \times \Delta^n \to \Delta^1 \times \Delta^n$ over A (with structural map $p(1_{\Delta^1} \times a)$ for the right-hand side), which is a homotopy equivalence. For F the presheaf represented by $(1, (\Delta^n, a))$, $t_!(F)$ is the simplicial set Δ^n over $\Delta^1 \times A$, with structural map $(1, a)\colon \Delta^n \to \Delta^1 \times A$. The map $F_0 \to t_!(F)/0$ is the identity of the empty simplicial set $\varnothing$, while the map $F_1 \to t_!(F)/1$ is the identity of Δ^n over A. □

Lemma 5.4.4 *The choice of a lift φ in the solid commutative diagram (5.3.14.1) provides, for any left fibration $\pi\colon W \to \Delta^1 \times A$, a weak equivalence $i_W\colon W_0 \to p_*(W)$ in the covariant model category structure over A, such that the following triangle commutes.*

$$\begin{array}{ccc} W_0 & \xrightarrow{\ i_W\ } & p_*(W) \\ & {\scriptstyle \varphi_1}\searrow \quad \swarrow{\scriptstyle t^*(W)} & \\ & W_1 & \end{array}$$

Proof For any map $q\colon Y \to A$, we have

$$p^*(q) = (1_{\Delta^1} \times q)\colon p^*(Y) = \Delta^1 \times Y \to \Delta^1 \times A$$

and, therefore, the composition of the unit $\eta_Y\colon Y \to p_*(\Delta^1 \times Y)$ with any of the canonical maps $p_*(\Delta^1 \times Y) \to c^*_\varepsilon(\Delta^1 \times Y) = Y$ is the identity of Y. Any choice of a lift φ in the solid commutative diagram (5.3.14.1) thus gives a commutative diagram of simplicial sets over A of the following form.

$$\begin{array}{ccccc} & & W_0 & & \\ & \diagup\!\!\diagup & \downarrow{\scriptstyle \eta_{W_0}} & \diagdown\!\!\diagdown & \\ W_0 & \xleftarrow{\ \sim\ } & p^*(\Delta^1 \times W_0) & \longrightarrow & W_0 \\ {\scriptstyle 1_{W_0} = \varphi_0}\downarrow & & \downarrow{\scriptstyle p_*(\varphi)} & & \downarrow{\scriptstyle \varphi_1} \\ W_0 & \xleftarrow{\ \sim\ } & p_*(W) & \xrightarrow{\ t^*(W)\ } & W_1 \end{array}$$

We put $i_W = p_*(\varphi)\eta_{W_0}$. We have seen in the proof of the preceding proposition that the decorated arrows of the form $\xleftarrow{\sim}$ are trivial fibrations. □

Theorem 5.4.5 *There is a canonical equivalence of categories from the homotopy category ho($\underline{\mathrm{Hom}}(A, \mathcal{S})$) onto the full subcategory of LFib(A) whose objects are the left fibrations with* **U***-small fibres of codomain A.*

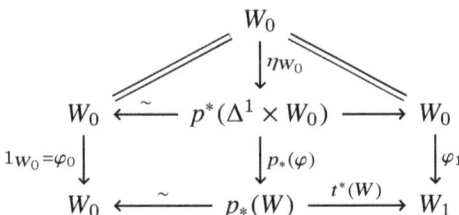

Proof We first define a functor

$$ho(\underline{\mathrm{Hom}}(A, \mathcal{S})) \to LFib(A)\,.$$

At the level of objects, this is the operation of pulling back along the left fibration $p_{univ}\colon \mathcal{S}_\bullet \to \mathcal{S}$. In other words, we associate to a morphism $F\colon A \to \mathcal{S}$ the left fibration $p\colon X \to A$ it classifies. As for maps, if $p\colon X \to A$ and $q\colon Y \to A$ are two left fibrations classified by F and G, respectively, the map

$$c\colon \mathrm{Hom}_{ho(\underline{\mathrm{Hom}}(A,\mathcal{S}))}(F, G) \to \mathrm{Hom}_{LFib(A)}(X, Y)$$

is constructed, taking into account the canonical identifications,

$$\pi_0(\underline{\mathrm{Hom}}(A, \mathcal{S})(F, G)) = \mathrm{Hom}_{ho(\underline{\mathrm{Hom}}(A,\mathcal{S}))}(F, G),$$
$$\pi_0(\mathrm{Map}_A(X, Y)) = \mathrm{Hom}_{LFib(A)}(X, Y)\,,$$

by applying the functor π_0 to the lower horizontal map of diagram (5.3.19.3). We recall that this map does not depend on the choice made to construct the map φ_1 of diagram (5.3.17.3). Let $\pi\colon W \to \Delta^2 \times A$ be a left fibration, classified by a map F. For $i = 0, 1, 2$, we denote by $\pi_i\colon W_i \to A$ the pull-back of π along the inclusion $A = \{i\} \times A \subset \Delta^2 \times A$, which is classified by the map F_i. For $i \neq j$, in $\{0, 1, 2\}$, the restriction of π to $\Delta^{\{i,j\}} \times A$ defines a map from F_i to F_j in $\underline{\mathrm{Hom}}(A, \mathcal{S})$, and we want to check that the induced triangle

(5.4.5.1)
$$\begin{array}{ccc} & W_1 & \\ \nearrow & & \searrow \\ W_0 & \longrightarrow & W_2 \end{array}$$

over A (which does not commute in *sSet*) gives a commutative triangle in $LFib(A)$. We see that $\Delta^2/i = \Delta^i$, the canonical map $\Delta^2/i \to \Delta^2$ being the obvious inclusion. For a map $q\colon Y \to \Delta^2 \times A$, we write $q/i\colon Y/i \to \Delta^i \times A$ for the pull-back of q along the inclusion $\Delta^i \times A \subset \Delta^2 \times A$. In the case where q is a left fibration, since it is proper, the inclusion $Y_i \to Y/i$ is a weak equivalence (this is the first assertion of Lemma 5.4.1). We always have $Y_0 = Y/0$ and $Y/2 = Y$. For $Y = \Delta^2 \times W_0$, we remark that we have $Y/i = \Delta^i \times W_0$. The inclusions $\Delta^0 \subset \Delta^1 \subset \Delta^2$ give an obvious commutative triangle of simplicial sets over A

(5.4.5.2)
$$\begin{array}{ccc} & W/1 & \\ \nearrow & & \searrow \\ W/0 & \longrightarrow & W/2 \end{array}$$

whose image in $LFib(A)$ can be shown to be isomorphic to the previous triangle

(5.4.5.1), using Lemma 5.4.1 (observe that the assumptions and choices, hence also their consequences, made in the statement of this lemma, are stable under base change along any map of the form $\Delta^m \times A \to \Delta^n \times A$ over A).

Now, we have a well-defined functor, and Corollary 5.2.8 ensures that, at the level of objects, its essential image consists of objects isomorphic to left fibrations with **U**-small fibres of codomain A. It remains to prove that it is fully faithful. Let us prove the property of fullness. Let $p\colon X \to A$ and $q\colon Y \to A$ be left fibrations classified by F and G, respectively, and let $\psi\colon X \to Y$ be a morphism of simplicial sets over A. Using Lemma 5.4.4, one can find a morphism $\Delta^1 \to \underline{\mathrm{Hom}}(A, \mathcal{S})$ classifying a left fibration $\pi\colon W \to \Delta^1 \times A$, out of which we can produce a commutative triangle

$$\begin{array}{ccc} W_0 & \xrightarrow{i_W} & p_*(W) \\ & {\scriptstyle\varphi_1}\searrow \quad \swarrow{\scriptstyle t^*(W)} & \\ & W_1 & \end{array}$$

with i_W a weak equivalence, and such that $t^*(W)$ and ψ are isomorphic as arrows of the homotopy category $ho(sSet/A)$. In particular, there exists a commutative diagram of the form

$$\begin{array}{ccc} W_0 & \xrightarrow{u} & X \\ {\scriptstyle\varphi_1}\downarrow & & \downarrow{\scriptstyle\psi} \\ W_1 & \xrightarrow{v} & Y \end{array}$$

in $LFib(A)$, with u and v invertible. Proposition 5.3.20 ensures that u and v are the images of isomorphisms in $ho(\underline{\mathrm{Hom}}(A, \mathcal{S}))$. Therefore, $\psi = v\varphi_1 u^{-1}$ is the image of a morphism $F \to G$. It remains to check faithfulness.

Replacing A by $\Delta^1 \times A$ in the statement of Proposition 5.3.20, and composing with the total right derived functor $\mathbf{R}t^*$ provided by Proposition 5.4.3, we get a canonical equivalence of groupoids of the form

$$k(ho(\underline{\mathrm{Hom}}(\Delta^1 \times A, \mathcal{S}))) \simeq k(ho(\mathrm{Arr}(sSet/A))) . \tag{5.4.5.3}$$

It is time for a couple of remarks.

(a) Given any maps $f\colon X \to Y$ and $g\colon X' \to Y'$ in $sSet/A$, any commutative square of the form

$$\begin{array}{ccc} X & \xrightarrow{a} & X' \\ {\scriptstyle f}\downarrow & & \downarrow{\scriptstyle g} \\ Y & \xrightarrow{b} & Y' \end{array}$$

in $ho(sSet/A)$, in which a and b are isomorphisms, can be promoted to a morphism from f to g in $k(ho(\mathrm{Arr}(sSet/A)))$.

(b) Given two maps $f: X \to Y$ and $g: X \to Y$ between fibrant objects in $sSet/A$, any homotopy from f to g provides a map $h: X' \to Y'$ in $sSet/A$ as well as a commutative diagram of the form

$$\begin{array}{ccccc} X & \xrightarrow{a} & X' & \xleftarrow{a} & X \\ {\scriptstyle f}\downarrow & & \downarrow{\scriptstyle h} & & \downarrow{\scriptstyle g} \\ Y & \xrightarrow{b} & Y' & \xleftarrow{b} & Y \end{array}$$

in $ho(sSet/A)$, in which a and b are isomorphisms (in fact, we can even impose b to be the identity).

To prove (a), we observe that we may replace f and g by weakly equivalent maps at will. In particular, we may assume that X and Y are fibrant. Therefore, both a and b are homotopy classes of maps of simplicial sets over A and the square commutes up to homotopy. There is thus a commutative diagram of the form

$$\begin{array}{ccccccc} X & \xrightarrow{(0,1_X)} & J \times X & \xleftarrow{(1,1_X)} & X & \xrightarrow{a} & X' \\ \downarrow{\scriptstyle f} & & \downarrow{\scriptstyle h} & & \downarrow{\scriptstyle ga} & & \downarrow{\scriptstyle g} \\ Y & \xrightarrow{b} & Y' & = & Y' & = & Y' \end{array}$$

in which all horizontal maps are homotopy equivalences, showing that f and g are isomorphic in the homotopy category of arrows. Property (b) is clear: we may take $X' = J \times X$ and observe that the two end-points of J are equal in the homotopy category.

Using Lemma 5.4.4 together with the equivalence of groupoids (5.4.5.3), we see that we can improve the property of fullness by asserting that any commutative diagram as in (a) above can be lifted to an analogous commutative diagram in $ho(\underline{\mathrm{Hom}}(A, \mathcal{S}))$. Therefore, relations between maps expressed as in (b) can be lifted in $ho(\underline{\mathrm{Hom}}(A, \mathcal{S}))$, whence faithfulness. □

Remark 5.4.6 The preceding theorem will be fundamental to reach the ∞-category-theoretic Yoneda lemma; see the proof of Theorem 5.8.4 below. It can be improved as follows.

Corollary 5.4.7 *Let $p: X \to A$ and $q: Y \to A$ be two left fibrations with* **U**-*small fibres classified by two morphisms F and G, respectively. The canonical map $\underline{\mathrm{Hom}}(A, \mathcal{S})(F, G) \to \mathrm{Map}_A(X, Y)$ is an equivalence of ∞-groupoids.*

Proof For any simplicial set K, if F_K denotes the composition of F with the projection $p\colon K \times A \to A$, we have

$$\underline{\mathrm{Hom}}(K, \underline{\mathrm{Hom}}(A, \mathcal{S})(F, G)) \simeq \underline{\mathrm{Hom}}(K \times A, \mathcal{S})(F_K, G_K)$$

and we also have a canonical isomorphism

$$\underline{\mathrm{Hom}}(K, \mathrm{Map}_A(X, Y)) \simeq \mathrm{Map}_{K \times A}(K \times X, K \times X)\,.$$

Hence, applying the preceding theorem for $K \times A$, we see that the map

$$\pi_0(\underline{\mathrm{Hom}}(K, \underline{\mathrm{Hom}}(A, \mathcal{S})(F, G))) \to \pi_0(\underline{\mathrm{Hom}}(K, \mathrm{Map}_A(X, Y)))$$

is bijective for any simplicial set K. Since, for any Kan complex W, the set $\pi_0(\underline{\mathrm{Hom}}(K, W))$ is the set of homotopy classes of maps from K to W in the Kan–Quillen model category structure, applying the Yoneda lemma to the homotopy category of Kan complexes, this implies the corollary. □

Corollary 5.4.8 *The category $ho(\mathcal{S})$ is equivalent to the homotopy category of* $\mathbf{U}$*-small Kan complexes.*

We can also see that changes of universes are harmless.

Proposition 5.4.9 *Let* $\mathbf{V}$ *be a Grothendieck universe which contains* $\mathbf{U}$ *as an element. Let $\mathcal{S}'$ be the ∞-category of* $\mathbf{V}$*-small ∞-groupoids (i.e. maps from $\Delta^n \to \mathcal{S}'$ correspond to left fibrations with* $\mathbf{V}$*-small fibres equipped with coherence data for base change, as in Definition 5.2.3). Then, for any simplicial set A, the inclusion map*

$$\underline{\mathrm{Hom}}(A, \mathcal{S}) \to \underline{\mathrm{Hom}}(A, \mathcal{S}')$$

is fully faithful.

Proof Let $p\colon X \to A$ and $q\colon Y \to A$ be two left fibrations with $\mathbf{U}$-small fibres classified by two morphisms F and G, respectively. We may also see F and G as maps with values in $\mathcal{S}'$, so that we get the commutative triangle below in $ho(\mathcal{S}')$.

$$\begin{array}{ccc} \underline{\mathrm{Hom}}(A, \mathcal{S})(F, G) & \longrightarrow & \underline{\mathrm{Hom}}(A, \mathcal{S}')(F, G) \\ & \searrow \quad \swarrow & \\ & \mathrm{Map}_A(F, G) & \end{array}$$

Since the two slanted maps are isomorphisms, by Corollary 5.4.7, so is the horizontal one. □

The latter proposition also means that the apparently naive notion of fibrewise $\mathbf{U}$-smallness, which consists in asking that the fibre of a left fibration at each object is equivalent to a **U**-small ∞-groupoid, is almost equivalent to the one we gave.

Corollary 5.4.10 *Let A be a simplicial set, and $p\colon X \to A$ a left fibration. We assume that, for each object a of A, the ∞-groupoid $X_a = p^{-1}(a)$ is equivalent to a **U**-small ∞-groupoid. Then there exists a left fibration with **U**-small fibres $q\colon Y \to A$ and a fibrewise equivalence $X \to Y$ over A. Equivalently, there is a homotopy Cartesian square of the following form.*

$$\begin{array}{ccc} X & \longrightarrow & \mathcal{S}_\bullet \\ {\scriptstyle p}\downarrow & & \downarrow{\scriptstyle p_{univ}} \\ A & \longrightarrow & \mathcal{S} \end{array}$$

Proof We choose a universe **V** as in the previous proposition. Let $\mathcal{S}'_{\mathbf{U}}$ be the full subcategory of $\mathcal{S}'$ whose objects are the **V**-small ∞-groupoids which are equivalent to a **U**-small ∞-groupoid. The inclusion $\mathcal{S} \to \mathcal{S}'_{\mathbf{U}}$ is essentially surjective, by definition, and fully faithful, by the previous proposition. Hence this is an equivalence of ∞-categories, which implies that the induced functor

$$\underline{\mathrm{Hom}}(A, \mathcal{S}) \to \underline{\mathrm{Hom}}(A, \mathcal{S}'_{\mathbf{U}})$$

is an equivalence of ∞-groupoids. Since we can always choose **V** so that the left fibration p has **V**-small fibres, the essential surjectivity of the latter equivalence, together with Corollary 5.4.7, proves that p is equivalent to a left fibration with **U**-small fibres $q\colon Y \to A$ in the covariant model category structure over A. □

Remark 5.4.11 In his thesis, Nichols-Barrer states Theorem 5.2.10 as a conjecture and deduces from it a version of Corollary 5.4.7; see [NB07, Conjecture 2.3.1 and Theorem 2.4.12]. He also introduces the homotopy coherent nerve of the simplicial category of Kan complexes [NB07, Proposition 2.3.5] and conjectures that it is equivalent to the ∞-category $\mathcal{S}$ above [NB07, Conjecture 2.3.10]. Since the homotopy coherent nerve functor is a Quillen equivalence relating the Joyal model category structure with the homotopy theory of simplicial categories of Dwyer and Kan (see [Lur09, Ber18]), and since the simplicial category of small Kan complexes is known to be the Dwyer–Kan localisation of the category of small simplicial sets by the class of weak homotopy equivalences, this latter conjecture essentially asserts that the ∞-category $\mathcal{S}$ is the localisation of the category of **U**-small simplicial sets by the class of weak homotopy equivalences. This reformulation of Nichols-Barrer's second conjecture will be proved below; see Theorem 7.8.9.

5.5 Bivariant Model Category Structures

5.5.1 This section addresses another approach to the theory of functors from a product of ∞-categories $A \times B$ to the ∞-category of ∞-groupoids $\mathcal{S}$. We write *bisSet* for the category of bisimplicial sets. We first recall from paragraph 3.1.15 a few basic operations in this context.

Let pr_1 and pr_2 be the first and second projection from $\mathbf{\Delta} \times \mathbf{\Delta}$ to $\mathbf{\Delta}$, respectively. We have pull-back functors

$$pr_i^* \colon sSet \to bisSet\,, \quad i = 0, 1.$$

For two simplicial sets A and B, the external product $A \boxtimes B$ is defined as

$$A \boxtimes B = pr_1^*(A) \times pr_2^*(B)\,. \tag{5.5.1.1}$$

In particular, the evaluations of $A \boxtimes B$ are of the form

$$(A \boxtimes B)_{m,n} = A_m \times B_n\,. \tag{5.5.1.2}$$

We remark that the representable presheaves on the product $\mathbf{\Delta} \times \mathbf{\Delta}$ are precisely the ones isomorphic to $\Delta^m \boxtimes \Delta^n$, for $m, n \geq 0$.

Another natural operation is induced by the diagonal functor $\delta = (1_{\mathbf{\Delta}}, 1_{\mathbf{\Delta}})\colon \mathbf{\Delta} \to \mathbf{\Delta} \times \mathbf{\Delta}$. The corresponding functor

$$\delta^* = \mathrm{diag}\colon bisSet \to sSet \tag{5.5.1.3}$$

has a left adjoint

$$\delta_! \colon sSet \to bisSet \tag{5.5.1.4}$$

as well as a right adjoint

$$\delta_* \colon sSet \to bisSet\,. \tag{5.5.1.5}$$

We recall that, for any bisimplicial set X, the evaluation of the diagonal $\mathrm{diag}(X)$ at n is

$$\delta^*(X)_n = X_{n,n}\,. \tag{5.5.1.6}$$

In particular, we have the following formula:

$$\delta^*(A \boxtimes B) = A \times B\,. \tag{5.5.1.7}$$

Given two simplicial sets A and B, we remark that we have a canonical identification:

$$(\mathbf{\Delta} \times \mathbf{\Delta})/(A \boxtimes B) = \mathbf{\Delta}/A \times \mathbf{\Delta}/B\,. \tag{5.5.1.8}$$

Using this, we can reinterpret formula (5.5.1.7) by asserting that there is a canonical Cartesian square of categories of the form

$$
\begin{array}{ccc}
\mathbf{\Delta}/(A \times B) & \xrightarrow{\delta_{A,B}} & \mathbf{\Delta}/A \times \mathbf{\Delta}/B \\
\downarrow & & \downarrow \\
\mathbf{\Delta} & \xrightarrow{\delta} & \mathbf{\Delta} \times \mathbf{\Delta}
\end{array}
\tag{5.5.1.9}
$$

where $\delta_{A,B}$ sends a triple $(\Delta^n, (a, b))$, with $a\colon \Delta^n \to A$ and $b\colon \Delta^n \to B$ two morphisms of simplicial sets, to the pair $(\Delta^n \boxtimes \Delta^n, a \boxtimes b)$, where the map $a \boxtimes b\colon \Delta^n \boxtimes \Delta^n \to A \boxtimes B$ is the one induced by functoriality of the external product.

5.5.2 For a bisimplicial set X and a simplicial set K, recall that one defines a simplicial set X^K by the formula

$$(X^K)_m = \mathrm{Hom}_{bisSet}(\Delta^m \boxtimes K, X) \simeq \varprojlim_{\Delta^n \to K} X_{m,n} \,. \tag{5.5.2.1}$$

Similarly, one defines ${}^K X$ by the formula

$$({}^K X)_n = \mathrm{Hom}_{bisSet}(K \boxtimes \Delta^n, X) \simeq \varprojlim_{\Delta^m \to K} X_{m,n} \,. \tag{5.5.2.2}$$

Definition 5.5.3 The class of *bi-anodyne extensions* is the smallest saturated class of morphisms of bisimplicial sets containing inclusions of the form

(5.5.3.1)

$$\Delta^m \boxtimes \partial\Delta^n \cup \Lambda^m_k \boxtimes \Delta^n \to \Delta^m \boxtimes \Delta^n \quad \text{and} \quad \Delta^m \boxtimes \Lambda^n_k \cup \partial\Delta^m \boxtimes \Delta^n \to \Delta^m \boxtimes \Delta^n$$

for $m \geq 1$, $n \geq 0$ and $0 \leq k \leq m$, or $m \geq 0$, $n \geq 1$ and $0 \leq k \leq n$, respectively.

We define *Kan bifibrations* as the morphisms of bisimplicial sets with the right lifting property with respect to bi-anodyne extensions.

A *Kan bicomplex* is a bisimplicial set X such that the map from X to the terminal bisimplicial set is a Kan bifibration.

We obviously have the following result.

Proposition 5.5.4 *For a morphism of bisimplicial sets $X \to Y$, the following conditions are equivalent.*

(i) The morphism $X \to L$ is a Kan bifibration.

(ii) For any anodyne extension $K \to Y$ in sSet, the induced maps

$$X^L \to X^K \times_{Y^K} Y^L \quad \textit{and} \quad {}^L X \to {}^K X \times_{{}^K Y} {}^L Y$$

are trivial fibrations.

(iii) For any monomorphism $K \to L$ in sSet, the induced maps

$$X^L \to X^K \times_{Y^K} Y^L \quad \text{and} \quad {}^L X \to {}^K X \times_{{}^K Y} {}^L Y$$

are Kan fibrations.

5.5.5 The category $\mathbf{\Delta} \times \mathbf{\Delta}$ is an Eilenberg–Zilber category, and the boundary of a representable presheaf $\Delta^m \boxtimes \Delta^n$ is $\Delta^m \boxtimes \partial\Delta^n \cup \partial\Delta^m \boxtimes \Delta^n$. Therefore, a morphism of bisimplicial sets is a trivial fibration (i.e. has the right lifting property with respect to monomorphisms) if and only if it has the right lifting property with respect to inclusions of the form

$$\Delta^m \boxtimes \partial\Delta^n \cup \partial\Delta^m \boxtimes \Delta^n \to \Delta^m \boxtimes \Delta^n \tag{5.5.5.1}$$

with $m, n \geq 0$. This gives the following statement.

Proposition 5.5.6 *For a morphism of bisimplicial sets $X \to Y$, the following conditions are equivalent.*

(i) The map p is a trivial fibration.
(ii) For any monomorphism $K \to Y$ in sSet, the induced map

$$X^L \to X^K \times_{Y^K} Y^L$$

is a trivial fibration.
(iii) For any monomorphism $K \to Y$ in sSet, the induced map

$${}^L X \to {}^K X \times_{{}^K Y} {}^L Y$$

is a trivial fibration.

Theorem 5.5.7 *There is a model category structure on the category of bisimplicial sets whose fibrant objects are the Kan bicomplexes and whose cofibrations are the monomorphisms. A morphism between Kan bicomplexes is a fibration if and only if it is a Kan bifibration. The weak equivalences are the morphisms $X \to Y$ whose associated diagonal* $\mathrm{diag}(X) \to \mathrm{diag}(Y)$ *is a weak homotopy equivalence. Moreover, both pairs $(\delta_!, \delta^*)$ and (δ^*, δ_*) are Quillen equivalences (when we endow the category of simplicial sets with the Kan–Quillen model category structure).*

Proof Let us put $I = \Delta^1 \boxtimes \Delta^0$. This is an interval, so that the Cartesian product with I defines an exact cylinder. Since, for $\varepsilon = 0, 1$, the inclusion $\{\varepsilon\} \to \Delta^1$ is an anodyne extension, for any monomorphism of simplicial sets $K \to L$, the induced map $A = \Delta^1 \times K \cup \{\varepsilon\} \times L \to \Delta^1 \times L = B$ is an anodyne extension. This implies that, for any bi-anodyne extension $X \to Y$, the induced map

$$I \times X \cup \{\varepsilon\} \times Y \to I \times Y$$

is a bi-anodyne extension for $\varepsilon = 0, 1$. Indeed, it is sufficient to prove this for $X \to Y$ a generating bi-anodyne extension of the form (5.5.3.1). We then get a map of the form $A \boxtimes V \cup B \boxtimes U \to B \boxtimes V$ associated to an anodyne extension $A \to B$ and a monomorphism $U \to V$, and Proposition 5.5.4 implies that such a map has the left lifting property with respect to Kan bifibrations. In other words, the class of bi-anodyne extensions defines a class of I-anodyne maps, and we can apply Theorem 2.4.19 to get the expected model category structure. The diagonal functor $\delta^* = \text{diag}$ preserves cofibrations, and it is clear that it sends the generating bi-anodyne extensions (5.5.3.1) to anodyne extensions. Therefore, Proposition 2.4.40 ensures that (δ^*, δ_*) is a Quillen pair. In particular, the functor δ^* preserves weak equivalences and cofibrations, and commutes with colimits. It follows from Proposition 3.1.13 that the functor $\delta_!$ preserves monomorphisms. Replacing Sd by $\delta_!$ in the proof of Proposition 3.1.18, we see that the functor $\delta_!$ sends the horn inclusions $\Lambda^n_k \to \Delta^n$ to bi-anodyne extensions. Another application of Proposition 2.4.40 gives that $(\delta_!, \delta^*)$ is a Quillen pair. Since both functors $\delta_!$ and δ^* are left Quillen functors, the class of simplicial sets X such that the unit map $X \to \delta^*\delta_!(X)$ is a weak homotopy equivalence is saturated by monomorphisms. But this class contains the representable simplicial sets Δ^n, since, in this case, the unit map is the diagonal embedding of Δ^n into its twofold product $\Delta^n \times \Delta^n$. Therefore, by virtue of Corollary 1.3.10, this class contains all simplicial sets. Similarly, the class of bisimplicial sets X such that the co-unit map $\delta_!\delta^*(X) \to X$ is a weak equivalence is saturated by monomorphisms. In the case where $X = \Delta^m \boxtimes \Delta^n$ is representable, this map simply is the morphism

$$pr_1 \boxtimes pr_2 \colon (\Delta^m \times \Delta^n) \boxtimes (\Delta^m \times \Delta^n) \to \Delta^m \boxtimes \Delta^n .$$

This map is a weak equivalence. Indeed, for any simplicial set B, the map

$$pr_1 \boxtimes 1_B \colon (\Delta^m \times \Delta^n) \boxtimes B \to \Delta^m \boxtimes B$$

is an I-homotopy equivalence, and thus a weak homotopy equivalence. By symmetry, for any simplicial set A, the morphism

$$1_A \boxtimes pr_2 \colon A \boxtimes (\Delta^m \times \Delta^n) \to A \boxtimes \Delta^n$$

is a weak equivalence. The map $pr_1 \boxtimes pr_2$ above is thus the composition of two weak equivalences. Corollary 1.3.10 implies that the co-unit map $\delta_!\delta^*(X) \to X$ is a weak equivalence for all X. In other words, the adjoint pair $(\delta_!, \delta^*)$ is a Quillen equivalence. In particular, the functor δ^* preserves weak equivalences and induces an equivalence of homotopy categories. Therefore, the Quillen pair (δ^*, δ_*) is a Quillen equivalence as well. This readily implies that the functor δ^* preserves and detects weak equivalences. □

Remark 5.5.8 If $X \to Y$ is a morphism of bisimplicial sets such that, for any integer $m \geq 0$, the induced map $X^{\Delta^m} \to Y^{\Delta^m}$ is a weak homotopy equivalence, then it is a weak equivalence of the model category structure of Theorem 5.5.7. Indeed, it is sufficient to check that its diagonal is a weak homotopy equivalence, which is precisely the conclusion of Theorem 3.1.16.

Remark 5.5.9 Although we will not use it in this book, it is a fact that the trivial cofibrations of the model category structure of Theorem 5.5.7 are precisely the bi-anodyne extensions. This can be proved in a pedestrian way, or by applying general results from Grothendieck's theory of test categories, such as [Cis06, corollaire 8.2.19]. However, this means that, if we consider the class of absolute weak equivalences associated to the homotopical structure used to construct the model category structure of Theorem 5.5.7 via Definition 2.5.2, we simply get the class of maps whose diagonal is a weak homotopy equivalence. As we did to define the covariant model category structures, we shall now consider an alternative presentation using a bivariant version of the class of left fibrations, which will provide an interesting notion of absolute weak equivalence.

Definition 5.5.10 The class of *left bi-anodyne extensions* (of *right bi-anodyne extensions*) is the smallest saturated class of morphisms of bisimplicial sets containing inclusions of the form

(5.5.10.1)

$$\Delta^m \boxtimes \partial\Delta^n \cup \Lambda^m_k \boxtimes \Delta^n \to \Delta^m \boxtimes \Delta^n \quad \text{and} \quad \Delta^m \boxtimes \Lambda^n_k \cup \partial\Delta^m \boxtimes \Delta^n \to \Delta^m \boxtimes \Delta^n$$

for $m \geq 1$, $n \geq 0$ and $0 \leq k < m$, and for $m \geq 0$, $n \geq 1$ and $0 \leq k < n$ (for $m \geq 1$, $n \geq 0$ and $0 < k \leq m$, and for $m \geq 0$, $n \geq 1$ and $0 < k \leq n$, respectively)

We define *left bifibrations* (*right bifibrations*) as the morphisms of bisimplicial sets with the right lifting property with respect to left (right) bi-anodyne extensions.

As before, we have the following.

Proposition 5.5.11 *For a morphism of bisimplicial sets $X \to Y$, the following conditions are equivalent.*

(i) The morphism $X \to Y$ is a left (right) bifibration.
(ii) For any left (right) anodyne extension $K \to L$ in sSet, the induced maps

$$X^L \to X^K \times_{Y^K} Y^L \quad \textit{and} \quad {}^L X \to {}^K X \times_{{}^K Y} {}^L Y$$

are trivial fibrations.

(iii) For any monomorphism $K \to L$ in sSet, the induced maps

$$X^L \to X^K \times_{Y^K} Y^L \quad \text{and} \quad {}^L X \to {}^K X \times_{{}^K Y} {}^L Y$$

are left (right) fibrations.

Lemma 5.5.12 *Let I be either the interval $J \boxtimes \Delta^0$ or the interval $\Delta^0 \boxtimes J$. Then the exact cylinder defined as the Cartesian product with I, together with the class of left (right) bi-anodyne extensions, form a homotopical structure in the sense of Definition 2.4.11.*

Proof We only consider the case of left bi-anodyne extensions, from which the case of right bi-anodyne extensions can be obtained by an easy duality argument. The two inclusions $\{\varepsilon\} \to J$ are left anodyne extensions because they are weak categorical equivalences (Proposition 5.3.1). Therefore, for any monomorphism of simplicial sets $K \to L$, the induced map $J \times K \cup \{\varepsilon\} \times L \to J \times L$ is a left anodyne extension. This implies, using the same arguments as in the first part of the proof of Theorem 5.5.7, that, for any left bi-anodyne extension $X \to Y$, the induced inclusion

$$I \times X \cup \{\varepsilon\} \times Y \to I \times Y$$

is a bi-anodyne extension for $\varepsilon = 0, 1$. □

Theorem 5.5.13 *Let C be a bisimplicial set. There is a model category structure on the slice category bisSet$/C$ whose cofibrations are the monomorphisms, and whose fibrant objects are the left (right) bifibrations of codomain C. For two left (right) bifibrations $p\colon X \to C$ and $q\colon Y \to C$, a map $f\colon X \to Y$ over C is a fibration if and only if it is a left (right) bifibration.*

Proof We use the construction of paragraph 2.5.1 for $A = \mathbf{\Delta} \times \mathbf{\Delta}$ and $S = C$, applied to the homotopical structure provided by the preceding lemma. □

The model category structure of the previous theorem will be called the *bicovariant model category structure over C* (the *bicontravariant model category structure over C*, respectively).

Remark 5.5.14 In the case where C is the terminal bisimplicial set, the bicovariant model category structure coincides with the model category structure of Theorem 5.5.7. To see this, it is sufficient to check that the classes of fibrant objects are the same. Let X be a bisimplicial set such that the map to the final bisimplicial set is a left bifibration. Then, for any simplicial set K, the simplicial set X^K is a Kan complex because the map $X^K \to \Delta^0$ is a left fibration. For any monomorphism of simplicial sets $K \to L$, the map $X^L \to X^K$ is a left fibration between Kan complexes, and thus a Kan fibration. By duality, the same is true for the map ${}^{L}X \to {}^{K}X$. Proposition 5.5.4 shows that X is a Kan bicomplex.

Proposition 5.5.15 *Let $u\colon C \to D$ be a morphism of bisimplicial sets. We consider the adjunction*

$$u_! : bisSet/C \rightleftarrows bisSet/D : u^* \tag{5.5.15.1}$$

where $u_!$ is the functor $(X, p) \mapsto (X, up)$. Then (5.5.15.1) is a Quillen pair for the bicovariant (bicontravariant, respectively) model category structures over C and D.

Proof The functor $u_!$ preserves monomorphisms as well as left (right) bi-anodyne extensions. Therefore, this proposition is a particular case of Proposition 2.4.40. □

Lemma 5.5.16 *Let A and B be two simplicial sets, and let $i\colon S \to T$ be a monomorphism of bisimplicial sets over $A \boxtimes B$. Assume that the map from T to $A \boxtimes B$ is a left (right) bifibration and that, for any integer $n \geq 0$, the induced map $S^{\Delta^n} \to T^{\Delta^n}$ is a left (right) anodyne extension. Then the map i is a left (right) bi-anodyne extension.*

Proof By virtue of Proposition 2.5.6, it is sufficient to prove that the map i is a weak equivalence of the bicovariant (bicontravariant) model category structure over $A \boxtimes B$. Let Λ be the set of monomorphisms of the form

$$\Delta^m \boxtimes \partial\Delta^n \cup \Lambda^m_k \boxtimes \Delta^n \to \Delta^m \boxtimes \Delta^n$$

for $m \geq 1$, $n \geq 0$ and $0 < k \leq m$ (and $0 \leq k < n$, respectively). Applying the small object argument to Λ, we factor i into a map $j\colon S \to X$ followed by a map $q\colon X \to T$, where q has the right lifting property with respect to Λ, while j belongs to the smallest saturated class of maps containing Λ. The class of morphisms of bisimplicial sets $K \to L$ such that $K^{\Delta^n} \to L^{\Delta^n}$ is a left (right) anodyne extension for any n is saturated and contains Λ. Therefore, the maps $S^{\Delta^n} \to X^{\Delta^n}$ are left (right) anodyne extensions for all $n \geq 0$. On the other hand, for all n, the map $X^{\Delta^n} \to T^{\Delta^n}$ is a left (right) fibration, and Corollary 4.1.9 ensures that it is cofinal (final). This implies that the maps $X^{\Delta^n} \to T^{\Delta^n}$ are trivial fibrations. On the other hand, for any simplicial set K, we have

$$(A \boxtimes B)^K = A \times \mathrm{Hom}_{sSet}(K, B) = \coprod_{\mathrm{Hom}_{sSet}(K,B)} A\,.$$

The projection $(A \boxtimes B)^K \to A$ is thus a Kan fibration for all K. This means that we can see the maps $X^{\Delta^n} \to T^{\Delta^n}$ as trivial fibrations between fibrant objects of the covariant (contravariant) model category structure over A. Since both functors $K \mapsto X^K$ and $K \mapsto T^K$ are continuous and send monomorphisms of simplicial sets to left (right) fibrations, we see that the class of simplicial sets K such that the map $X^K \to T^K$ is a weak equivalence of the covariant (contravariant) model category structure over A is saturated by monomorphisms. Therefore, the map $X^K \to T^K$ is a trivial fibration for all K. This implies that the map $X^L \to X^K \times_{Y^K} Y^L$ is a trivial fibration for any monomorphism $K \to L$. Therefore, the map $q: X \to Y$ is a trivial fibration of bisimplicial sets, and, since i is a monomorphism, it is a retract of j, which is in particular a left (right) bi-anodyne extension. □

Lemma 5.5.17 *The functor* $\delta^* = \operatorname{diag}: bisSet \to sSet$ *sends left (right) bi-anodyne extensions to left (right) anodyne extensions.*

Proof It is sufficient to check this on generators. Using formula (5.5.1.7), this follows right away from Proposition 3.4.3. □

Proposition 5.5.18 *Let A and B be two simplicial sets, and let* $p: X \to A \boxtimes B$ *be a left bifibration. We consider two objects a and b of A and B, respectively. We can form the fibre* $X_{a,b}$ *of p at* $a \boxtimes b: \Delta^0 \boxtimes \Delta^0 \to A \boxtimes B$. *We also choose fibrant replacements* $\tilde{A}_{/a} \to A$ *and* $\tilde{B}_{/b} \to B$ *of* $a: \Delta^0 \to A$ *and* $b: \Delta^0 \to B$ *in the contravariant model category structures over A and B, respectively. We have the pull-back* $\delta^*(X)_{/(a,b)} = (\tilde{A}_{/a} \times \tilde{B}_{/b}) \times_{(A \times B)} \delta^*(X)$. *Then the canonical map* $\delta^*(X_{a,b}) \to \delta^*(X)_{/(a,b)}$ *is a weak homotopy equivalence.*

Proof We consider the pull-back

$$X_{/a \boxtimes b} = (\tilde{A}_{/a} \boxtimes \tilde{B}_{/b}) \times_{(A \boxtimes B)} X$$

so that the map $\delta^*(X_{a,b}) \to \delta^*(X)_{/(a,b)}$ is the image by δ^* of the canonical map $X_{a,b} \to X_{/a \boxtimes b}$. By virtue of Theorem 3.1.16, it is sufficient to prove that the latter map is a weak homotopy equivalence. Replacing A and B by $\tilde{A}_{/a}$ and $\tilde{B}_{/b}$, respectively, and replacing X by $X_{/a \boxtimes b}$, we may assume that a and b are final objects of A and B, and we want to prove that $X_{a,b} \to X$ is a right bi-anodyne extension. Let us write $X_a \to \Delta^0 \boxtimes B$ for the pull-back of the map $X \to A \boxtimes B$ along the map $a \boxtimes 1_B: \Delta^0 \boxtimes B \to A \boxtimes B$. For any simplicial set K, the map $X_a^K \to X^K$ is final because it is the pull-back of the final map $\operatorname{Hom}_{sSet}(K, B) \to A \times \operatorname{Hom}_{sSet}(K, B)$ along the left fibration $X^K \to A \times \operatorname{Hom}(K, B)$. Therefore, Lemma 5.5.16 ensures that the inclusion $X_a \to X$ is a weak homotopy equivalence. Similarly, for any simplicial set K, the map ${}^K X_{a,b} \to {}^K X_a$ is final: it is the pull-back of the final map $b: \Delta^0 \to B$ along the left fibration ${}^K X_a \to B$. Therefore, applying Theorem 3.1.16 (up to a permutation of the variables), we obtain that the embedding $X_{a,b} \to X_a$ is a weak homotopy equivalence as well. The composed map $X_{a,b} \to X_a \to X$ is thus a weak homotopy equivalence. □

Corollary 5.5.19 *Under the assumptions of the preceding proposition, if we choose a fibrant replacement $q\colon Y \to A \times B$ of $\delta^*(p)\colon \delta^*(X) \to A \times B$ for the covariant model category structure over $A \times B$, the induced map on the fibre over (a, b) is an equivalence of ∞-groupoids $\delta^*(X_{a,b}) \xrightarrow{\sim} Y_{a,b}$.*

Proof It follows from Remark 5.5.14 that the fibre $X_{a,b}$ is fibrant in the model structure of Theorem 5.5.7, so that $\delta^*(X_{a,b})$ is a Kan complex. We also know that the fibres of left fibrations are Kan complexes. Therefore, it is sufficient to prove that the map $\delta^*(X) \to Y$ is a fibrewise equivalence. We have a commutative square of the form

$$\begin{array}{ccc} \delta^*(X_{a,b}) & \longrightarrow & Y_{a,b} \\ \downarrow & & \downarrow \\ \delta^*(X)_{/(a,b)} & \longrightarrow & Y_{/(a,b)} \end{array}$$

in which $Y_{/(a,b)} = (\tilde{A}_{/a} \times \tilde{B}_{/b}) \times_{(A\times B)} Y$. The vertical maps of this square are weak homotopy equivalences because of the preceding proposition for the first one, and because q is proper for the second one. The lower horizontal map is cofinal because it is the base change of a cofinal map along the smooth map $\tilde{A}_{/a} \times \tilde{B}_{/b} \to A \times B$. In conclusion, all the morphisms of this square are weak homotopy equivalences over $A \times B$. □

Lemma 5.5.20 *Let $p\colon X \to A \boxtimes B$ be a left (right) bifibration, and $n \geq 0$. Then the induced map*

$$p^{\Delta^n} : X^{\Delta^n} \to (A \times B)^{\Delta^n} = A \times B_n$$

is a left (right) fibration; in particular, the induced map $X^{\Delta^n} \to A$ is a left (right) fibration. We consider furthermore an object a of A as well as an n-simplex b of B, and we let c be the restriction of $b\colon \Delta^n \to B$ to $\{0\} \subset \Delta^n$ (to $\{n\} \subset \Delta^n$, respectively). We finally write $X^{\Delta^n}_{a,b}$ for the fibre of p^{Δ^n} at (a, b). Then there is a functorial weak equivalence of the form

$$X^{\Delta^n}_{a,b} \to \delta^*(X_{a,c})\,.$$

Proof We shall consider the case of a left fibration, the other one being deduced by duality. The first assertion is a particular case of condition (iii) of Proposition 5.5.11. Since B_n is discrete, the projection $A \times B_n \to A$ is a left fibration. Since the inclusion of $\{0\}$ into Δ^n is cofinal (hence a left anodyne extension), condition (ii) of Proposition 5.5.11 implies that the induced map

$$X^{\Delta^n} \to (A \times B_n) \times_{(A \times B_0)} X^{\Delta^0}$$

is a trivial fibration over $A \times B_n$. Passing to the fibres over the objects of $A \times B_n$, we get trivial fibrations

$$X^{\Delta^n}_{a,b} \to X^{\Delta^0}_{a,c} .$$

To finish the proof, it is sufficient to produce a functorial weak homotopy equivalence from $X^{\Delta^0}_{a,c}$ to $\delta^*(X_{a,c})$. Let us write $Y = X_{a,c}$. We shall prove that there is a canonical weak homotopy equivalence $Y^{\Delta^0} \to \delta^*(Y)$. Indeed, what precedes (for $b = c$) shows that there is a canonical anodyne extension $Y^{\Delta^0} \to Y^{\Delta^n}$ induced by the map $\Delta^n \to \Delta^0$, because the inclusion $\{0\} \to \Delta^n$ induces a trivial fibration $Y^{\Delta^n} \to Y^{\Delta^0}$ by Proposition 5.5.4. Therefore, by virtue of Theorem 3.1.16, the image of the canonical map $Y^{\Delta^0} \boxtimes \Delta^0 \to Y$ by δ^* is a weak homotopy equivalence. □

Proposition 5.5.21 *Let $p\colon X \to A \boxtimes B$ and $q\colon Y \to A \boxtimes B$ be two left (right) bifibrations, and $\varphi\colon X \to Y$ a morphism of bisimplicial sets over $A \boxtimes B$. The following conditions are equivalent.*

- *(i) The map φ is a weak equivalence of the bicovariant (bicontravariant) model category structure over $A \boxtimes B$.*
- *(ii) For any objects a and b in A and B, respectively, the map $X_{a,b} \to Y_{a,b}$ is a weak equivalence of the model category structure of Theorem 5.5.7.*
- *(iii) For any objects a and b in A and B, respectively, the map $\delta^*(X_{a,b}) \to \delta^*(Y_{a,b})$ is an equivalence of ∞-groupoids.*
- *(iv) For any integer $n \geq 0$, the map $X^{\Delta^n} \to Y^{\Delta^n}$ is a fibrewise equivalence over A.*
- *(v) For any integer $m \geq 0$, the map ${}^{\Delta^m}X \to {}^{\Delta^m}Y$ is a fibrewise equivalence over B.*

Proof The permutation of factors in $\mathbf{\Delta} \times \mathbf{\Delta}$ induces an isomorphism from $bisSet/A \boxtimes B$ onto $bisSet/B \boxtimes A$ which preserves left (right) bifibrations, so that, if conditions (i) and (iv) are equivalent, then conditions (i) and (v) are equivalent as well. Hence we may leave out condition (v). We shall focus on the case where p and q are right fibrations, since the case of left fibrations will follow by an obvious duality argument. Since the functor δ^* preserves and detects weak equivalences of Theorem 5.5.7, it is clear that conditions (ii) and (iii) are equivalent. Since right Quillen functors preserve weak equivalences between fibrant objects, the fact that condition (i) implies condition (ii) follows from Proposition 5.5.15, applied for $u = a \boxtimes b$. For any object (a, b) of $A \times B_n$, if $b_n : \Delta^0 \to B$ denotes the evaluation of b at n, we have a commutative square of the form

$$\begin{array}{ccc} X^{\Delta^n}_{a,b} & \longrightarrow & Y^{\Delta^n}_{a,b} \\ \downarrow & & \downarrow \\ \delta^*(X_{a,b_n}) & \longrightarrow & \delta^*(Y_{a,b_n}) \end{array}$$

in which the vertical maps are weak homotopy equivalences, by virtue of the preceding lemma. Since the fibre of X^{Δ^n} over a is the disjoint union of the $X^{\Delta^n}_{a,b}$, this implies that conditions (iii) and (iv) are equivalent. It is now sufficient to prove that condition (iv) implies condition (i). Theorem 4.1.16 implies that condition (iv) is equivalent to the property that the maps $X^{\Delta^n} \to Y^{\Delta^n}$ are weak equivalences of the contravariant model category structure over A (observe again that the fibre of X^{Δ^n} over a is the disjoint union of the $X^{\Delta^n}_{a,b}$). We may choose a factorisation of the map φ into a cofibration $\psi : X \to T$ followed by a trivial fibration $\pi : T \to Y$. It is clear that condition (i) for φ is equivalent to condition (i) for ψ. Similarly, since the induced maps $T^{\Delta^n} \to Y^{\Delta^n}$ are trivial fibrations as well, condition (iv) for φ is equivalent to condition (iv) for ψ. In other words, we may assume, without loss of generality, that φ is a monomorphism. But then, Lemma 5.5.16 shows that condition (iv) implies that φ is a right bi-anodyne extension, hence satisfies condition (i). □

Corollary 5.5.22 *Let $p : X \to A \boxtimes B$ and $q : Y \to A \boxtimes B$ be two bisimplicial sets over $A \boxtimes B$. A morphism $\varphi : X \to Y$ over $A \boxtimes B$ is a weak equivalence of the bicovariant (bicontravariant) model category structure over $A \boxtimes B$ if and only if the induced map $\delta^*(\varphi) : \delta^*(X) \to \delta^*(Y)$ is a weak equivalence of the covariant (contravariant) model category structure over $A \times B$.*

Proof We choose a commutative square of bisimplicial sets over $A \boxtimes B$ of the form

$$\begin{array}{ccc} X & \xrightarrow{i} & X' \\ {\scriptstyle\varphi}\downarrow & & \downarrow{\scriptstyle\varphi'} \\ Y & \xrightarrow{j} & Y' \end{array}$$

in which i and j are left bi-anodyne extensions and the structural maps of X' and Y' are left bifibrations. Lemma 5.5.17 implies that the maps $\delta^*(i)$ and $\delta^*(j)$ are left anodyne extensions. Therefore, we may assume, without loss of generality, that both p and q are left bifibrations. But then, Theorem 4.1.16 and Corollary 5.5.19 show that $\delta^*(\varphi)$ is a weak equivalence over $A \times B$ if and only if, for any objects a and b in A and B, respectively, the induced map

$\delta^*(X_{a,b}) \to \delta^*(Y_{a,b})$ is a weak homotopy equivalence. This corollary is thus a reformulation of Proposition 5.5.21. □

5.5.23 Let A and B be two simplicial sets. The functor $\delta_{A,B}\colon \mathbf{\Delta}/A \times B \to \mathbf{\Delta}/A \times \mathbf{\Delta}/B$ induces an adjunction

$$\delta_{A,B,!}\colon sSet/A \times B \rightleftarrows bisSet/A \boxtimes B : \delta^*_{A,B} \tag{5.5.23.1}$$

where the functor $\delta^*_{A,B}$ simply sends a morphism $p\colon X \to A \boxtimes B$ to $\delta^*(p)\colon \delta^*(X) \to A \times B$. This latter functor has a right adjoint $\delta_{A,B,*}$. The functor $\delta_{A,B,!}$ has an explicit description: it sends a map $p\colon X \to A \times B$ to the composed map $\varepsilon\delta_!(p)\colon \delta_!(X) \to A \boxtimes B$, where $\varepsilon\colon \delta_!(A \times B) \to A \boxtimes B$ is the co-unit map (corresponding to the identity of $A \times B = \delta^*(A \boxtimes B)$).

Theorem 5.5.24 *The pair $(\delta_{A,B,!}, \delta^*_{A,B})$ is a Quillen equivalence from the covariant (contravariant) model category structure over the Cartesian product $A \times B$ to the bicovariant (bicontravariant) model category structure over the external product $A \boxtimes B$. Furthermore, the pair $(\delta^*_{A,B}, \delta_{A,B,*})$ is also a Quillen equivalence from the bicovariant (bicontravariant) model category structure over $A \boxtimes B$ to the covariant (contravariant) model category structure over $A \times B$.*

Proof The class of maps $(a, b)\colon X \to A \times B$ such that the induced map $X \to \delta^*\delta_!(X)$ is a weak equivalence of the covariant model category structure over $A \times B$ is a class of objects of $sSet/A \times B$ which is saturated by monomorphisms. This class contains all the maps of the form $(a, b)\colon \Delta^n \to A \times B$. Indeed, the diagonal $\Delta^n \to \Delta^n \times \Delta^n$ sends the initial object 0 to the initial object $(0, 0)$, hence is a left anodyne extension. Corollary 1.3.10 thus shows that this class consists of all simplicial sets over $A \times B$. Therefore, by virtue of Corollary 5.5.22, the functor $\delta_{A,B,!}$ is a left Quillen functor, and what precedes also implies that the induced total derived functor is fully faithful, and another use of Corollary 5.5.22 shows that its right adjoint is conservative. This readily implies the first assertion of the theorem. The second one follows from the first, as in the proof of the last assertion of Theorem 5.5.7. □

Corollary 5.5.25 *The functor $\delta_!\colon sSet \to bisSet$ sends left (right) anodyne extensions to left (right) bi-anodyne extensions.*

Proof We only have to consider the case of left anodyne extensions, by the usual duality argument. The class of morphisms of simplicial sets whose image by $\delta_!$ is a left bi-anodyne extension is saturated. Therefore, it is sufficient to check that, for $n \geq 1$ and $0 \leq k < n$, the inclusion of $\delta_!(\Lambda^n_k)$ into $\delta_!(\Delta^n) = \Delta^n \boxtimes \Delta^n$ is a left bi-anodyne extension. But we know, by the previous theorem applied for $A = B = \Delta^n$, that it is a trivial cofibration of the bicovariant model structure over $\Delta^n \boxtimes \Delta^n$, whence, since the identity of $\Delta^n \boxtimes \Delta^n$ is a left bifibration, it is a left bi-anodyne extension, by Propositions 2.5.3 and 2.5.6. □

5.6 The Twisted Diagonal

5.6.1 Given a simplicial set A, we define a bisimplicial set $\mathbf{S}(A)$ by the formula

(5.6.1.1) $$\mathbf{S}(A)_{m,n} = \mathrm{Hom}_{sSet}((\Delta^m)^{\mathrm{op}} * \Delta^n, A)\,, \quad m, n \geq 0\,.$$

This defines a functor from the category of simplicial sets to the category of bisimplicial sets. There is a canonical map

(5.6.1.2) $$(s_A, t_A)\colon \mathbf{S}(A) \to A^{\mathrm{op}} \boxtimes A$$

induced by the inclusions $(\Delta^m)^{\mathrm{op}} \to (\Delta^m)^{\mathrm{op}} * \Delta^n \leftarrow \Delta^n$. We finally define $\mathrm{S}(A)$ as the diagonal of $\mathbf{S}(A)$:

(5.6.1.3) $$\mathrm{S}(A) = \delta^*(\mathbf{S}(A))\,.$$

The elements of $\mathrm{S}(A)_n$ thus correspond to morphisms $(\Delta^n)^{\mathrm{op}} * \Delta^n \to A$, for $n \geq 0$. We also have a canonical map

(5.6.1.4) $$(s_A, t_A)\colon \mathrm{S}(A) \to A^{\mathrm{op}} \times A\,.$$

Proposition 5.6.2 *If A is an ∞-category, then the induced map* (5.6.1.4) *is a left fibration, and the simplicial set* $\mathrm{S}(A)$ *is an ∞-category.*

Proof We shall first prove that the map (5.6.1.2) is a left bifibration (see Definition 5.5.10). Given two monomorphisms of simplicial sets $i\colon K \to L$ and $j\colon U \to V$, as well as two maps $f\colon L^{\mathrm{op}} \to A$ and $v\colon V \to A$, there is a one-to-one correspondence between the following lifting problems.

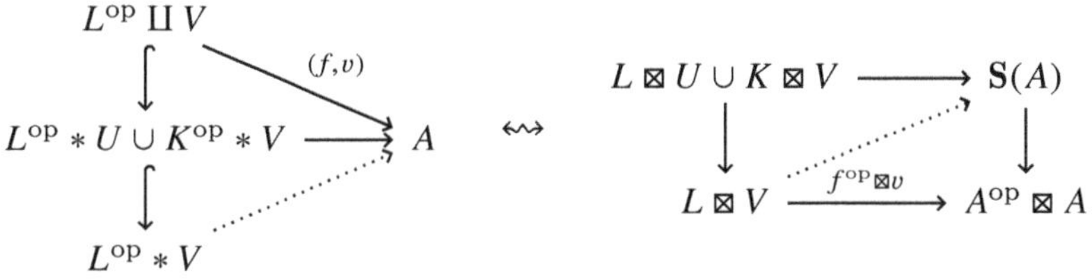

Therefore, Proposition 3.4.17 implies that (5.6.1.2) is a left bifibration. Indeed, it ensures that, for i the boundary inclusion $\partial\Delta^m \to \Delta^m$ and j the horn inclusion $\Lambda^n_k \to \Delta^n$, with $m \geq 0$, $n \geq 1$ and $0 \leq k < n$, or for i the horn inclusion $\Lambda^m_k \to \Delta^m$ and j the boundary inclusion $\partial\Delta^n \to \Delta^n$, with $m \geq 1$, $n \geq 0$ and $0 \leq k < n$, we have $L^{\mathrm{op}} * V = \Delta^{m+1+n}$ and $L^{\mathrm{op}} * U \cup K^{\mathrm{op}} * V = \Lambda^{m+1+n}_l$ for some $0 < l < m + 1 + n$. The proposition now follows from the fact that the functor δ^* sends left bifibrations to left fibrations, since its left adjoint $\delta_!$ sends left anodyne extensions to left bi-anodyne extensions (Corollary 5.5.25). □

Definition 5.6.3 The *twisted diagonal* of an ∞-category A is the ∞-category $\mathrm{S}(A)$.

Remark 5.6.4 The opposite of the twisted diagonal of the opposite of A is also known as the *twisted arrow category* of A. All the results of this chapter on the twisted diagonal have their counterpart in terms of twisted arrow categories.

If A is the nerve of a small category C, then one may describe $\mathrm{S}(A)^{\mathrm{op}}$ as the nerve of the category of elements of the presheaf $(x, y) \mapsto \mathrm{Hom}_C(x, y)$. In general, one might expect to extract the twisted diagonal of A out of the ∞-category of arrows of A from a general procedure of twisting Cartesian fibrations over a binary product. Although we shall not explain such a general procedure here, its outcome in the special case of the ∞-category of arrows is discussed in Corollary 5.6.14 below.

Proposition 5.6.5 *If A is an ∞-category, for any object a of A, we form the following Cartesian squares.*

$$\begin{array}{ccc} {}_a\mathrm{S}(A) & \longrightarrow & \mathrm{S}(A) \\ \downarrow & & \downarrow \\ A & \xrightarrow{(a,1_A)} & A^{\mathrm{op}} \times A \end{array} \qquad \begin{array}{ccc} \mathrm{S}(A)_a & \longrightarrow & \mathrm{S}(A) \\ \downarrow & & \downarrow \\ A^{\mathrm{op}} & \xrightarrow{(1_{A^{\mathrm{op}}},a)} & A^{\mathrm{op}} \times A \end{array}$$

Then there are canonical cofinal maps

$$a\backslash A \to {}_a\mathrm{S}(A) \quad \text{and} \quad (A/a)^{\mathrm{op}} \to \mathrm{S}(A)_a$$

over A and over A^{op}, respectively.

Proof We shall only consider the case of the fibre ${}_a\mathrm{S}(A)$, since the other one can be deduced from the first by appropriate duality arguments. By virtue of the dual versions of Proposition 4.1.11 and of Theorem 4.1.16, it is sufficient to prove that there is a canonical fibrewise equivalence $a\backslash A \to {}_a\mathrm{S}(A)$ over A. Let us consider the Cartesian square below.

$$\begin{array}{ccc} {}_a\mathbf{S}(A) & \longrightarrow & \mathbf{S}(A) \\ \downarrow & & \downarrow{\scriptstyle (s_A,t_A)} \\ \Delta^0 \boxtimes A & \xrightarrow{a\times 1_A} & A^{\mathrm{op}} \boxtimes A \end{array}$$

Then we have $\delta^*({}_a\mathbf{S}(A)) = {}_a\mathrm{S}(A)$. Furthermore, we see that there is a canonical identification:

$$^{\Delta^0}({}_a\mathbf{S}(A)) = a\backslash A\,.$$

Since, by the preceding proposition, the map (s_A, t_A) is a left bifibration, the property that the induced inclusion map

$$^{\Delta^0}({}_a\mathbf{S}(A)) \to \delta^*({}_a\mathbf{S}(A))$$

is a fibrewise equivalence over A is a particular case of Lemma 5.5.20. □

Corollary 5.6.6 *A functor between ∞-categories $u\colon A \to B$ is fully faithful if and only if the induced map*

$$\mathrm{S}(A) \to (A^{\mathrm{op}} \times A) \times_{(B^{\mathrm{op}} \times B)} \mathrm{S}(B)$$

is a fibrewise equivalence over $A^{\mathrm{op}} \times A$.

Proof Indeed, Corollary 4.2.10 and Proposition 5.6.5 show that the induced maps on the fibres are homotopic to the canonical maps $A(x, a) \to B(u(x), u(a))$. □

Lemma 5.6.7 *Let us consider three ∞-categories A, B and C, as well as a commutative square of the form*

$$\begin{array}{ccc} X & \xrightarrow{f} & Y \\ \downarrow{\scriptstyle p} & & \downarrow{\scriptstyle q} \\ A \times C & \xrightarrow{1_A \times g} & A \times B \end{array}$$

in which p and q are left fibrations. If the map f is fibrewise cofinal over A, then it is cofinal. In other words, if the induced map on the fibres $f_a\colon X_a \to Y_a$ is cofinal for any object a of A, then f is cofinal as well.

Proof It is sufficient to prove that f is a weak equivalence of the covariant model structure over $A \times B$, by the dual version of Proposition 4.1.11. Given any map $T \to B$ and any object b of B, we write $T/b = T \times_B B/b$. By virtue of Proposition 4.4.30, it is sufficient to prove that the induced functor f/b is a weak equivalence of the covariant model structure over A. We remark that we have a commutative square of the form

$$\begin{array}{ccc} X/b & \xrightarrow{f/b} & Y/b \\ \downarrow{\scriptstyle p/b} & & \downarrow{\scriptstyle q/b} \\ A \times C/b & \xrightarrow{1_A \times g/b} & A \times B/b \end{array}$$

which is obtained by pull-back from the original one. For any object a of A, the functor on the induced fibres $X_a \to Y_a$ is cofinal over B. Since the projection $B/b \to B$ is a right fibration, it is smooth, whence the induced functor $X_a/b \to Y_a/b$ is cofinal. This means that the functor f/b is fibrewise cofinal over A. But both functors p/b and q/b are left fibrations, since they are pull-backs of the left fibrations p and q, respectively, so that the maps $X/b \to A$ and $X/b \to A$ are proper, because any projection of the form $A \times D \to A$ is proper. Henceforth, f/b is a weak equivalence of the covariant model structure over A, by Corollary 4.4.28. □

5.6.8 Let $u : A \to B$ be a functor between ∞-categories. We have the following commutative diagrams.

$$\begin{array}{ccccc} S(A) & \longrightarrow & (A^{op} \times B) \times_{(B^{op}\times B)} S(B) & \longrightarrow & S(B) \\ {\scriptstyle (s_A,t_A)}\downarrow & & \downarrow & & \downarrow{\scriptstyle (s_B,t_B))} \\ A^{op} \times A & \xrightarrow{1_{A^{op}}\times u} & A^{op} \times B & \xrightarrow{u^{op}\times 1_B} & B^{op} \times B \end{array} \tag{5.6.8.1}$$

$$\begin{array}{ccccc} S(A) & \longrightarrow & (B^{op} \times A) \times_{(B^{op}\times B)} S(B) & \longrightarrow & S(B) \\ {\scriptstyle (s_A,t_A)}\downarrow & & \downarrow & & \downarrow{\scriptstyle (s_B,t_B))} \\ A^{op} \times A & \xrightarrow{u^{op}\times 1_A} & B^{op} \times A & \xrightarrow{1_{B^{op}}\times u} & B^{op} \times B \end{array} \tag{5.6.8.2}$$

Proposition 5.6.9 *The canonical maps*

$$S(A) \to (A^{op} \times B) \times_{(B^{op}\times B)} S(B) \quad \textit{and} \quad S(A) \to (B^{op} \times A) \times_{(B^{op}\times B)} S(B)$$

are cofinal.

Proof Proposition 5.6.5 shows that, for any object a of A, the vertical maps of the obvious commutative squares

$$\begin{array}{ccc} a\backslash A & \longrightarrow & u(a)\backslash B \\ \downarrow & & \downarrow \\ {}_aS(A) & \longrightarrow & {}_{u(a)}S(B) \end{array} \quad \text{and} \quad \begin{array}{ccc} (A/a)^{op} & \longrightarrow & (B/u(a))^{op} \\ \downarrow & & \downarrow \\ S(A)_a & \longrightarrow & S(B)_{u(a)} \end{array}$$

are cofinal. On the other hand, the upper horizontal maps are cofinal, because they preserve initial objects, so that we can apply Proposition 4.3.3. Therefore, the lower horizontal maps are cofinal as well, by Corollary 4.1.9. This proves our assertion, by the preceding lemma. □

5.6.10 There is an alternative point of view on the twisted diagonal, which consists in replacing the join operation $*$ by the diamond operation $\diamond$ of paragraph 4.2.1. For a simplicial set A, we put

$$\mathbf{S}_\diamond(A)_{m,n} = \mathrm{Hom}_{sSet}((\Delta^m)^{op} \diamond \Delta^n, A)\,. \tag{5.6.10.1}$$

This defines a bisimplicial set $\mathbf{S}_\diamond(A)$. As above, there is a canonical map

$$\mathbf{S}_\diamond(A) \to A \boxtimes A^{op} \tag{5.6.10.2}$$

induced by the inclusions $(\Delta^m)^{\mathrm{op}} \to (\Delta^m)^{\mathrm{op}} \diamond \Delta^n \leftarrow \Delta^n$, and we define $\mathrm{S}_\diamond(A)$ as the diagonal of $\mathbf{S}_\diamond(A)$:

(5.6.10.3)
$$\mathrm{S}_\diamond(A) = \delta^*(\mathbf{S}_\diamond(A))\,.$$

We also have a canonical map

(5.6.10.4)
$$\mathrm{S}_\diamond(A) \to A^{\mathrm{op}} \times A\,.$$

The canonical map $(\Delta^m)^{\mathrm{op}} \diamond \Delta^n \to (\Delta^m)^{\mathrm{op}} * \Delta^n$ provided by Proposition 4.2.2 induces the natural commutative triangle below.

(5.6.10.5)
$$\begin{array}{ccc} \mathrm{S}(A) & \longrightarrow & \mathrm{S}_\diamond(A) \\ & \searrow \quad \swarrow & \\ & A^{\mathrm{op}} \times A & \end{array}$$

Lemma 5.6.11 *For any monomorphisms of simplicial sets $i\colon U \to V$ and $j\colon S \to T$, if i is a left anodyne extension or if j is a right anodyne extension, then the induced map*

$$V \diamond S \cup U \diamond T \to V \diamond T$$

is a trivial cofibration of the Joyal model category structure.

Proof We already know this property if we replace $\diamond$ by $*$ (this is an easy consequence of Proposition 3.4.17). Proposition 4.2.3 thus implies this lemma. □

Replacing $*$ by $\diamond$ in the proof of Proposition 5.6.2 (and replacing the use of Proposition 3.4.17 by the preceding lemma), we get the following.

Proposition 5.6.12 *If A is an ∞-category, then the map* (5.6.10.2) *is a left bifibration. In particular, the induced map* (5.6.10.4) *is a left fibration, and the simplicial set $\mathrm{S}_\diamond(A)$ is an ∞-category.*

Similarly, replacing $*$ by $\diamond$ in the proof of Proposition 5.6.13 gives the following statement, where we have put

$$a\backslash\backslash A = (A^{\mathrm{op}}//a)^{\mathrm{op}}\,.$$

Proposition 5.6.13 *For any object a of an ∞-category A, we form the following Cartesian squares.*

$$\begin{array}{ccc} {}_a\mathrm{S}_\diamond(A) & \longrightarrow & \mathrm{S}_\diamond(A) \\ \downarrow & & \downarrow \\ A & \xrightarrow{(a,1_A)} & A^{\mathrm{op}} \times A \end{array} \qquad \begin{array}{ccc} \mathrm{S}_\diamond(A)_a & \longrightarrow & \mathrm{S}_\diamond(A) \\ \downarrow & & \downarrow \\ A^{\mathrm{op}} & \xrightarrow{(1_{A^{\mathrm{op}}},a)} & A^{\mathrm{op}} \times A \end{array}$$

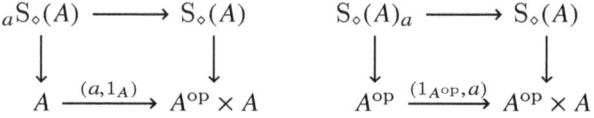

Then there are canonical fibrewise equivalences of left fibrations

$$a\backslash\backslash A \to {}_a\mathrm{S}_\diamond(A) \quad \text{and} \quad (A//a)^{\mathrm{op}} \to \mathrm{S}_\diamond(A)_a$$

over A and over A^{op}, respectively.

Corollary 5.6.14 *For any ∞-category A, the comparison map*

$$\mathrm{S}(A) \to \mathrm{S}_\diamond(A)$$

is a fibrewise equivalence of left fibrations over $A^{\mathrm{op}} \times A$ (whence an equivalence of ∞-categories). In particular, for any object a of A, there is a canonical commutative square of the form

$$\begin{array}{ccc} a\backslash A & \longrightarrow & {}_a\mathrm{S}(A) \\ \downarrow & & \downarrow \\ a\backslash\backslash A & \longrightarrow & {}_a\mathrm{S}_\diamond(A) \end{array}$$

in which all the maps are fibrewise equivalences over A.

Proof This follows right away from Propositions 4.2.9, 5.6.5 and 5.6.13. □

Remark 5.6.15 As observed in the proof of Corollary 4.2.10, the fibre of the canonical map $x\backslash\backslash A \to A$ over an object y of A is the ∞-groupoid of maps $A(x, y)$. Therefore, the preceding corollary expresses exactly how to recover the ∞-groupoid of morphisms between two objects of A as a homotopy fibre of the map $\mathrm{S}(A) \to A^{\mathrm{op}} \times A$.

5.7 Locally Small ∞-Categories

Definition 5.7.1 An ∞-category A is *locally* **U**-*small* if the twisted diagonal of A is isomorphic to a left fibration with **U**-small fibres of codomain $A^{\mathrm{op}} \times A$ in $LFib(A^{\mathrm{op}} \times A)$.

Remark 5.7.2 By virtue of Corollary 5.4.10, an ∞-category A is locally small if and only if, for any objects x and a in A, the ∞-groupoid $A(x, a)$ is equivalent to a **U**-small ∞-groupoid.

In particular, an ∞-category A is locally **U**-small if and only if its opposite A^{op} is locally **U**-small. Similarly, if $u: A \to B$ is an equivalence of ∞-categories, then A is locally **U**-small if and only if B has the same property.

Of course, a sufficient condition for an ∞-category to be locally **U**-small is that one of the left fibrations (5.6.1.4) or (5.6.10.4) has **U**-small fibres. In particular, whenever the simplicial set underlying A is **U**-small (i.e. takes its values in the chosen Grothendieck universe), the ∞-category A is locally $\mathbf{U}$-small. Such an ∞-category will be said to be $\mathbf{U}$-*small*. Another fundamental example is the following one.

Proposition 5.7.3 *For any $\mathbf{U}$-small simplicial set X, the ∞-category of functors $\underline{\mathrm{Hom}}(X, \mathcal{S})$ is locally $\mathbf{U}$-small. In particular, the ∞-category $\mathcal{S}$ of $\mathbf{U}$-small ∞-groupoids is locally $\mathbf{U}$-small.*

Proof The covariant model structure over some $\mathbf{U}$-small simplicial set X restricts to $\mathbf{U}$-small simplicial sets over X. Given two $\mathbf{U}$-small simplicial sets E and F over X, the mapping space $\mathrm{Map}_X(E, F)$ is a subcomplex of $\underline{\mathrm{Hom}}(E, F)$, which is $\mathbf{U}$-small. Therefore, Corollary 5.4.7 and the preceding remark show that $\underline{\mathrm{Hom}}(X, \mathcal{S})$ is locally $\mathbf{U}$-small. □

In order to generalise the previous proposition to abstract locally $\mathbf{U}$-small ∞-categories, we need the next two propositions.

Proposition 5.7.4 *Let $u\colon A \to B$ be a functor between locally $\mathbf{U}$-small ∞-categories. The following conditions are equivalent.*

(i) The functor $u\colon A \to B$ is fully faithful.

(ii) For any $\mathbf{U}$-small simplicial set X, the induced functor

$$u_*\colon \underline{\mathrm{Hom}}(X, A) \to \underline{\mathrm{Hom}}(X, B)$$

is fully faithful, and its essential image consists of functors $\varphi\colon X \to B$ such that, for any object x of X, there exists an object a of A, as well as an invertible map $u(a) \to \varphi(x)$ in B.

(iii) For any $\mathbf{U}$-small simplicial set X, the induced functor

$$u_*\colon ho(\underline{\mathrm{Hom}}(X, A)) \to ho(\underline{\mathrm{Hom}}(X, B))$$

is fully faithful.

(iv) There exists a factorisation of u into an equivalence of ∞-categories $i\colon A \to A'$ followed by a map $u'\colon A' \to B$ with the right lifting property with respect to inclusions of the form $\partial\Delta^n \to \Delta^n$, for $n > 0$.

Proof Given a simplicial set K and a subset $S \subset K_0$, one defines the full simplicial subset of K generated by S as the simplicial set whose maps $\Delta^n \to K_S$ correspond to maps $k\colon \Delta^n \to K$ such that, for $0 \le i \le n$, we have $k(i) \in S$. We remark that the inclusion $K_S \to K$ has the right lifting property with respect to any morphism of simplicial sets $U \to V$ such that the induced map $U_0 \to V_0$ is bijective, such as $\partial\Delta^n \to \Delta^n$ for $n > 0$ or $\Lambda^n_k \to \Delta^n$ for $n \ge 2$. In particular, if K is an ∞-category, then so is K_S, and the inclusion $K_S \to K$ is fully faithful.

If, furthermore, the set S has the property that, for any invertible map $b \to b'$ in B, we have $b \in S \Leftrightarrow b' \in S$, then the inclusion $K_S \to K$ is also an isofibration.

We choose a factorisation of u into a trivial cofibration $i: A \to A'$ followed by a fibration $u': A' \to B$ in the Joyal model category structure. We apply this construction for $K = B$ and S the essential image of u: an object b of B belongs to S if there exists an object a of A as well as an invertible map $u(a) \to b$ in B. The map $u_S: A \to B_S$ is essentially surjective. Therefore, since its composition with the fully faithful map $B_S \to B$ is fully faithful, the functor u_S is an equivalence of categories if and only if the functor u is fully faithful. Let S' be the image of $u_0: A_0 \to B_0$. Then the inclusion $B_{S'} \to B_S$ is fully faithful and essentially surjective. For a simplicial set X, let T be the set of functors $\varphi: X \to B$ such that, for any object x of X, there exists an object a of A, as well as an invertible map $u(a) \to \varphi(x)$ in B. Then we have

$$\underline{\mathrm{Hom}}(X, B_S) = \underline{\mathrm{Hom}}(X, B)_T .$$

Let T' be the set of functors $\varphi: X \to B$ such that, for any x in X_0, there exists $a \in A_0$ such that $u(a) = \varphi(x)$. Then we have

$$\underline{\mathrm{Hom}}(X, B_{S'}) = \underline{\mathrm{Hom}}(X, B)_{T'} .$$

Since the functor $\underline{\mathrm{Hom}}(X, -)$ preserves equivalences of ∞-categories, the inclusion

$$\underline{\mathrm{Hom}}(X, B)_{T'} \to \underline{\mathrm{Hom}}(X, B)_T$$

is an equivalence of ∞-categories. Moreover, condition (ii) is equivalent to the property that the functor $(u_S)_*: \underline{\mathrm{Hom}}(X, A) \to \underline{\mathrm{Hom}}(X, B_S)$ is an equivalence of ∞-categories for all X. This proves that conditions (i) and (ii) are equivalent. We also see that conditions (i) and (iv) are equivalent. One checks that condition (i) is equivalent to the property that the map $u'_{S'}: A' \to B$ is an isofibration which is an equivalence of ∞-categories, whence a trivial fibration. This proves that condition (i) is equivalent to condition (iv). If condition (iv) is verified, then the map $u_{S'}: A \to B_{S'}$ factors as an equivalence of ∞-categories followed by a trivial fibration, whence condition (i) is verified.

We already know that condition (ii) implies condition (iii) (see Remark 3.9.4). Let us assume that condition (iii) holds. Given any objects a in A, and any simplicial set X, by Proposition 4.2.12, if a also denotes the constant functor $X \to A$ with value a, then there is a canonical fibrewise equivalence of right fibrations over $\underline{\mathrm{Hom}}(X, A)$ from $\underline{\mathrm{Hom}}(X, A/a)$ to $\underline{\mathrm{Hom}}(X, A)/a$. Hence, by forming the set of connected components of the homotopy fibres over the points of A, we have canonical identifications

$$\pi_0(\underline{\mathrm{Hom}}(X, A(x, a))) \simeq \mathrm{Hom}_{ho(\underline{\mathrm{Hom}}(X,A))}(x, a) .$$

Proceeding similarly for B and $b = u(a)$, we have canonical bijections

$$\pi_0(\underline{\mathrm{Hom}}(X, A(u(x), u(a)))) \simeq \mathrm{Hom}_{ho(\underline{\mathrm{Hom}}(X,A))}(u(x), u(a)) .$$

Applying the Yoneda lemma to the homotopy category of **U**-small Kan complexes, this shows that the maps $A(x, a) \to B(u(x), u(a))$ are homotopy equivalences, hence that u is fully faithful. □

Remark 5.7.5 Since, in any model category, pulling back a weak equivalence between fibrant objects along a fibration always gives a weak equivalence, and since right lifting properties are preserved by pull-backs, condition (iv) of the preceding proposition shows that pulling back a fully faithful functor between ∞-categories along an isofibration gives a fully faithful functor.

Proposition 5.7.6 *Let A be an ∞-category. The following conditions are equivalent.*

- *(a) The ∞-category A is equivalent to a* **U***-small ∞-category.*
- *(b) There exists a* **U***-small simplicial set X as well as a weak categorical equivalence from X to A.*
- *(c) Any minimal model of A is* **U***-small.*
- *(d) There is a* **U***-small minimal model of A.*
- *(e) The set of isomorphism classes of $ho(A)$ is* **U***-small, and A is locally* **U***-small.*

Proof Conditions (c) and (d) are equivalent because all minimal models of A are isomorphic to each other. Conditions (a) and (b) are equivalent because there is a fibrant replacement functor of the Joyal model category structure which preserves **U**-smallness (e.g. the one obtained by applying the small object argument to the set of inner horn inclusions). Since any minimal model of a **U**-small ∞-category C is **U**-small, as a retract of C, it is clear that conditions (a) and (d) are equivalent. It is clear that condition (a) implies condition (e). It is thus sufficient to prove that condition (e) implies condition (d). For $n \geq 1$, the fibre $A(x_0, \dots, x_n)$ of the evaluation map

$$\underline{\mathrm{Hom}}(\Delta^n, A) \to A^{n+1}$$

at $(x_0, \dots, x_n)$ is a Kan complex which is equivalent to a finite product of Kan complexes of the form $A(x_i, x_{i+1})$ (paragraph 3.7.7). In particular, such a fibre is equivalent to a **U**-small Kan complex. We also define $\partial A(x_0, \dots, x_n)$ as the fibre at $(x_0, \dots, x_n)$ of the evaluation map

$$\underline{\mathrm{Hom}}(\partial\Delta^n, A) \to A^{n+1} .$$

We thus have a canonical Cartesian square of the following form.

$$\begin{array}{ccc}
A(x_0, \ldots, x_n) & \longrightarrow & \underline{\mathrm{Hom}}(\Delta^n, A) \\
\downarrow & & \downarrow \\
\partial A(x_0, \ldots, x_n) & \longrightarrow & \underline{\mathrm{Hom}}(\partial\Delta^n, A)
\end{array}$$

To prove condition (d), assuming condition (e), we may assume, without loss of generality, that the set of objects of A is **U**-small: we just have to choose a **U**-small set of objects E in A such that any object of A is isomorphic in $ho(A)$ to an element of E, and to replace A by the full subcategory generated by E. This means that it is sufficient to prove that the set of ∂-equivalence classes of simplices in A is **U**-small. In other words, to finish the proof, it is thus sufficient to check that the fibres of the maps

$$A(x_0, \ldots, x_n) \to \partial A(x_0, \ldots, x_n)$$

are equivalent to **U**-small Kan complexes. Note that these maps are Kan fibrations: they are isofibrations between Kan complexes. It is thus sufficient to prove that $\partial A(x_0, \ldots, x_n)$ is equivalent to a **U**-small Kan complex. We shall prove this by induction on $n \geq 1$. The case $n = 1$ is clear: Δ^0 is **U**-small. If $n > 1$, we consider $\partial\Delta^n$ as the union of Λ^n_1 and of the image of $\delta^n_1 : \Delta^{n-1} \to \Delta^n$. We then have a canonical Cartesian square over A^{n+1} of the form

$$\begin{array}{ccc}
\underline{\mathrm{Hom}}(\partial\Delta^n, A) & \longrightarrow & \underline{\mathrm{Hom}}(\Delta^{n-1}, A) \times A \\
\downarrow & & \downarrow \\
\underline{\mathrm{Hom}}(\Lambda^n_1, A) & \longrightarrow & \underline{\mathrm{Hom}}(\partial\Delta^{n-1}, A) \times A
\end{array}$$

which induces, by passing to the fibres over $(x_0, \ldots, x_n)$, a Cartesian square of the form below.

$$\begin{array}{ccc}
\partial A(x_0, \ldots, x_n) & \longrightarrow & A(x_0, x_2, \ldots, x_n) \\
\downarrow & & \downarrow \\
\Lambda^n_1 A(x_0, \ldots, x_n) & \longrightarrow & \partial A(x_0, x_2, \ldots, x_n)
\end{array}$$

Since, by induction, the vertical map of the right-hand side is a Kan fibration between Kan complexes which are equivalent to **U**-small ones, this shows that the fibres of the map $\partial A(x_0, \ldots, x_n) \to \Lambda^n_1 A(x_0, \ldots, x_n)$ are **U**-small up to homotopy. Henceforth, the latter map is equivalent to a Kan fibration with **U**-small fibres of codomain $\Lambda^n_1 A(x_0, \ldots, x_n)$, by Corollary 5.4.10. Therefore, it is sufficient to prove that its codomain is equivalent to a **U**-small Kan complex.

But the trivial fibration

$$\underline{\mathrm{Hom}}(\Delta^n, A) \to \underline{\mathrm{Hom}}(\Lambda^n_1, A)$$

induces a trivial fibration

$$A(x_0, \ldots, x_n) \to \Lambda^n_1 A(x_0, \ldots, x_n)$$

whose domain is already known to be equivalent to a **U**-small Kan complex. □

Corollary 5.7.7 *Let A be a locally* **U***-small ∞-category. For any* **U***-small simplicial set X, the ∞-category $\underline{\mathrm{Hom}}(X, A)$ is locally* **U***-small.*

Proof Let f and g be two functors from X to A. Since X is **U**-small, there is a **U**-small set S of objects of A such that both f and g factor through the full subcategory A_S spanned by S in A. By virtue of Proposition 5.7.4, we thus have an equivalence of ∞-groupoids of the form

$$\underline{\mathrm{Hom}}(X, A_S)(u, v) \simeq \underline{\mathrm{Hom}}(X, A)(u, v)\,.$$

Replacing A by A_S, we may thus assume that the set of objects of A is **U**-small. But then, the preceding proposition ensures that A is equivalent to a **U**-small ∞-category C. Therefore, the ∞-category $\underline{\mathrm{Hom}}(X, A)$ is equivalent to $\underline{\mathrm{Hom}}(X, C)$, which is **U**-small, whence locally **U**-small. □

Corollary 5.7.8 *If an ∞-category A is locally* **U***-small, so are its slices A/a for any object a.*

Proof The proof is similar to that of the previous corollary: given two objects (x, u) and (y, v) in A/a, we can find a full subcategory C of A which contains a, x and y, and which is equivalent to a **U**-small ∞-category D. We then have a canonical isomorphism

$$C/a((x, u), (y, v)) = A/a((x, u), (y, v))\,.$$

Since C/a is equivalent to a slice of D, this proves that A/a is locally **U**-small. □

Corollary 5.7.9 *An ∞-groupoid X is equivalent to a* **U***-small Kan complex if and only if $\pi_0(X)$ and $\pi_n(X, x)$ are* **U***-small for any object x in X and any positive integer n.*

Proof Since $\partial\Delta^{n+1} = Im(\delta^{n+1}_{n+1}) \cup \Lambda^{n+1}_{n+1}$, the map $\Lambda^{n+1}_{n+1} \to \Delta^0$ induces a weak homotopy equivalence $\partial\Delta^{n+1} \to \Delta^n/\partial\Delta^n$. By virtue of Proposition 3.8.10, one may interpret the elements of $\pi_n(X, x)$ as the pointed homotopy classes of maps $\Delta^n/\partial\Delta^n \to X$. The assumption of **U**-smallness on homotopy groups thus means that the set of ∂-equivalence classes in X is **U**-small, or equivalently that the minimal models of X are $\mathbf{U}$-small. This corollary is thus a direct consequence of Proposition 5.7.6. □

5.8 The Yoneda Lemma

5.8.1 Let A be a locally $\mathbf{U}$-small ∞-category. A *Hom space of* A is a morphism of ∞-categories $\mathrm{Hom}_A \colon A^{\mathrm{op}} \times A \to \mathcal{S}$ equipped with a map $\sigma_A \colon \mathrm{S}(A) \to \mathcal{S}_\bullet$ which exhibits the left fibration (5.6.1.4) as the homotopy pull-back of the universal left fibration, i.e. such that we get a homotopy Cartesian square of the following form.

(5.8.1.1)
$$\begin{array}{ccc} \mathrm{S}(A) & \xrightarrow{\sigma_A} & \mathcal{S}_\bullet \\ {\scriptstyle (s_A,t_A)}\downarrow & & \downarrow{\scriptstyle p_{univ}} \\ A^{\mathrm{op}} \times A & \xrightarrow{\mathrm{Hom}_A} & \mathcal{S} \end{array}$$

By virtue of Theorem 5.4.5, such a square determines Hom_A up to a unique isomorphism in $ho(\underline{\mathrm{Hom}}(A \times A^{\mathrm{op}}, \mathcal{S}))$. Furthermore, by Corollary 5.6.14, we may always choose the map $\mathrm{S}(A^{\mathrm{op}})^{\mathrm{op}} \to \mathcal{S}_\bullet$ so that it factors through $\mathrm{S}_\diamond(A^{\mathrm{op}})^{\mathrm{op}}$ over $A^{\mathrm{op}} \times A$, in which case, as explained in Remark 5.6.15, for any objects a and x in A, there is a canonical homotopy Cartesian square of the following form.

(5.8.1.2)
$$\begin{array}{ccc} A(x,a) & \longrightarrow & \mathcal{S}_\bullet \\ \downarrow & & \downarrow{\scriptstyle p_{univ}} \\ \Delta^0 & \xrightarrow{\mathrm{Hom}_A(x,a)} & \mathcal{S}\,. \end{array}$$

We define the *Yoneda embedding of* A as the unique functor

(5.8.1.3)
$$h_A \colon A \to \underline{\mathrm{Hom}}(A^{\mathrm{op}}, \mathcal{S})$$

which corresponds by transposition to Hom_A. In other words, we have

(5.8.1.4)
$$h_A(a)(x) = \mathrm{Hom}_A(x,a)$$

for any objects x and a of A. By construction, the functor $h_A(a) \colon A^{\mathrm{op}} \to \mathcal{S}$ classifies a left fibration which is canonically equivalent to the left fibration $a\backslash A^{\mathrm{op}} = (A/a)^{\mathrm{op}} \to A^{\mathrm{op}}$. More precisely, applying Corollary 5.6.14 to A^{op} produces a homotopy Cartesian square of the following form.

(5.8.1.5)
$$\begin{array}{ccc} (A/a)^{\mathrm{op}} & \longrightarrow & \mathcal{S}_\bullet \\ \downarrow & & \downarrow{\scriptstyle p_{univ}} \\ A^{\mathrm{op}} & \xrightarrow{h_A(a)} & \mathcal{S} \end{array}$$

A functorial way to put the identifications (5.8.1.2) and (5.8.1.4) together is the following statement.

Proposition 5.8.2 *Let a be an object of the locally* **U***-small* ∞*-category A. The left fibration classified by the Yoneda embedding composed with the functor of evaluation at a.*

(5.8.2.1)
$$A \xrightarrow{h_A} \underline{\mathrm{Hom}}(A, \mathcal{S}) \xrightarrow{a^*} \mathcal{S},$$

is fibrewise equivalent to the pull-back of the map $(s_A, t_A)\colon \mathrm{S}(A) \to A^{\mathrm{op}} \times A$ *along the embedding* $A = \{a\} \times A \subset A^{\mathrm{op}} \times A$*. Furthermore, there is a canonical homotopy Cartesian square of the following form.*

(5.8.2.2)
$$\begin{array}{ccccc} a\backslash A & \longrightarrow & & \longrightarrow & \mathcal{S}_\bullet \\ \downarrow & & & & \downarrow{\scriptstyle p_{univ}} \\ A & \xrightarrow{h_A} & \underline{\mathrm{Hom}}(A^{\mathrm{op}}, \mathcal{S}) & \xrightarrow{a^*} & \mathcal{S} \end{array}$$

Proof By definition of the ∞-category $\mathcal{S}$, if $F\colon A^{\mathrm{op}} \to \mathcal{S}$ classifies a left fibration $p\colon X \to A^{\mathrm{op}}$, the map $a^*(F) = F(a)\colon \Delta^0 \to \mathcal{S}$ classifies the fibre of p at x. This turns the first assertion into a tautology. The last assertion of the proposition is a reformulation of Corollary 5.6.14. □

Remark 5.8.3 The interest of the Hom space is that it is a *functor* (while the construction $(x, y) \mapsto A(x, y)$ is not). In particular, for any object a of A, the functor $\mathrm{Hom}_A(-, a) = h_A(a)$ takes any map $f\colon x \to y$ in A to a morphism $f^*\colon \mathrm{Hom}_A(y, a) \to \mathrm{Hom}_A(x, a)$ in $\mathcal{S}$.

Given a functor $u\colon A \to B$ between locally **U**-small ∞-categories, there is an obvious commutative square of the form

(5.8.3.1)
$$\begin{array}{ccc} \mathrm{S}(A) & \xrightarrow{\mathrm{S}(u)} & \mathrm{S}(B) \\ \downarrow & & \downarrow \\ A^{\mathrm{op}} \times A & \xrightarrow{u^{\mathrm{op}} \times u} & B^{\mathrm{op}} \times B \end{array}$$

and the induced map $\mathrm{S}(A) \to A^{\mathrm{op}} \times A \times_{B^{\mathrm{op}} \times B} \mathrm{S}(B)$, of left fibrations over the Cartesian product $A^{\mathrm{op}} \times A$, determines, by Corollaries 5.4.7 and 5.4.10, an essentially unique map $\mathrm{Hom}_A(-, -) \to \mathrm{Hom}_B(u(-), u(-))$ in $\underline{\mathrm{Hom}}(A^{\mathrm{op}} \times A, \mathcal{S})$. In other words, for any objects x and y in A, there is a functorial map

(5.8.3.2)
$$\mathrm{Hom}_A(x, y) \to \mathrm{Hom}_B(u(x), u(y))$$

in the ∞-category $\mathcal{S}$, which corresponds, up to homotopy, to the canonical map of ∞-groupoids $A(x, y) \to B(u(x), u(y))$.

Theorem 5.8.4 *Let A be a simplicial set. We consider a map $f\colon U \to A$ which we suppose is equipped with a factorisation into a right anodyne extension $i\colon U \to V$ followed by a right fibration with* **U***-small fibres $q\colon V \to A$. We also consider we are given a functor $\Phi\colon A^{op} \to \mathcal{S}$ which classifies the left fibration $q^{op}\colon V^{op} \to A^{op}$. Let*

$$\pi\colon W = \underline{\mathrm{Hom}}(A^{op}, \mathcal{S}) \times_{\underline{\mathrm{Hom}}(U^{op},\mathcal{S})} \underline{\mathrm{Hom}}(U^{op}, \mathcal{S}_\bullet) \to \underline{\mathrm{Hom}}(A^{op}, \mathcal{S})$$

be the left fibration obtained by pulling back along the functor f^* of composition with f^{op}. Then there is a canonical initial object w in W, associated to i, whose image in $\underline{\mathrm{Hom}}(A^{op}, \mathcal{S})$ is Φ. In particular, there is a homotopy Cartesian square of the following form in the Joyal model category structure.

$$\begin{array}{ccc}
\Phi\backslash \underline{\mathrm{Hom}}(A^{op}, \mathcal{S}) & \longrightarrow & \underline{\mathrm{Hom}}(U^{op}, \mathcal{S}_\bullet) \\
\downarrow & & \downarrow {\scriptstyle (p_{univ})_*} \\
\underline{\mathrm{Hom}}(A^{op}, \mathcal{S}) & \xrightarrow{f^*} & \underline{\mathrm{Hom}}(U^{op}, \mathcal{S})
\end{array}$$

Proof The last assertion about the existence of a homotopy pull-back square is a translation of the first part of the theorem. Indeed, if there is an initial object w of W whose image by π in $\underline{\mathrm{Hom}}(A^{op}, \mathcal{S})$ is Φ, the canonical functor $w\backslash W \to W$ is a trivial fibration (see the dual version of condition (iii) of Theorem 4.3.11), and, by the dual version of Proposition 4.1.2, the induced map

$$w\backslash W \to \Phi\backslash \underline{\mathrm{Hom}}(A^{op}, \mathcal{S})$$

is a trivial fibration. Choosing a section of the latter, that will define the announced homotopy Cartesian square.

The first thing to do is to understand W explicitly as follows. A map $\psi\colon E \to W$ corresponds to a pair (F, s), where F is a morphism which classifies a left fibration of the form $p\colon X \to E \times A^{op}$, and $s\colon E \times U^{op} \to X$ is a map such that ps equals the map $1_E \times f^{op}$ from $E \times U^{op}$ to $E \times A^{op}$. In other words, this is essentially determined by a commutative diagram of the form

(5.8.4.1)
$$\begin{array}{ccc}
 & & X \\
 & \overset{s}{\nearrow} & \downarrow {\scriptstyle p} \\
E \times U^{op} & \xrightarrow[1_E \times f^{op}]{} & E \times A^{op}
\end{array}$$

in which p is a left fibration with **U**-small fibres. We first remark that, if we replace s by another section s' of p over $E \times U^{op}$ which is homotopic to s in the covariant model category structure over $E \times A^{op}$, and if we denote by $\psi'\colon E \to W$ the morphism corresponding to the pair (F, s'), then ψ and ψ' are homotopic in the Joyal model category structure (i.e. ψ and ψ' are isomorphic in the homotopy category $ho(\underline{\mathrm{Hom}}(E, W))$). Indeed, if we have a homotopy $h\colon J \times E \times U^{\mathrm{op}} \to X$ over $E \times A^{\mathrm{op}}$, from s to s', this defines a map $S = (pr_1, h)\colon J \times E \times U^{\mathrm{op}} \to J \times X$ over $J \times E \times A^{\mathrm{op}}$, whence a map $\Psi\colon J \times E \to W$ corresponding to the composition of F with the projection $J \times E \to E$, endowed with the section S. The evaluations of Ψ at 0 and 1 give back ψ and ψ', respectively.

Since V^{op} has a given section i^{op} over U^{op}, one may see the pair (Φ, i^{op}) as an object w of W. It is clear that the image of w in $\underline{\mathrm{Hom}}(A^{\mathrm{op}}, \mathcal{S})$ is Φ. Given a map $\psi\colon E \to W$, corresponding to a diagram of the form (5.8.4.1), we can always choose a dotted filler λ in the solid commutative square below.

(5.8.4.2)
$$\begin{array}{ccc} E \times U^{\mathrm{op}} & \xrightarrow{\ s\ } & X \\ {\scriptstyle 1_E \times i^{\mathrm{op}}}\big\downarrow & \overset{\lambda}{\nearrow} & \big\downarrow{\scriptstyle p} \\ E \times V^{\mathrm{op}} & \longrightarrow & E \times A^{\mathrm{op}} \end{array}$$

Note that, since $1_E \times i^{\mathrm{op}}$ is cofinal, the operation of right composition with it defines a trivial fibration between Kan complexes:

$$\mathrm{Map}_{E \times A^{\mathrm{op}}}(E \times V^{\mathrm{op}}, X) \to \mathrm{Map}_{E \times A^{\mathrm{op}}}(E \times U^{\mathrm{op}}, X).$$

Therefore, the Kan complex of lifts of s are contractible. This expresses in which sense s and λ determine each other up to homotopy.

The pull-back Φ_E of Φ by the projection $E \times A^{\mathrm{op}} \to A^{\mathrm{op}}$ classifies the left fibration $1_E \times q^{\mathrm{op}}\colon E \times V^{\mathrm{op}} \to E \times A^{\mathrm{op}}$, and, if we equip it with the section $1_E \times i^{\mathrm{op}}$, this defines a map $w_E\colon E \to W$, which is nothing but the constant map with value w. By Corollary 5.4.7, the map λ determines a map $u\colon \Phi_E \to F$ in the ∞-category $\underline{\mathrm{Hom}}(E, \underline{\mathrm{Hom}}(A^{\mathrm{op}}, \mathcal{S}))$. Conversely, any such map u defines a map $\lambda\colon E \times V^{\mathrm{op}} \to X$ over $E \times A^{\mathrm{op}}$; composing λ with $1_E \times i^{\mathrm{op}}$ gives back a diagram of the form (5.8.4.1), and thus a map $\psi\colon E \to W$. The equivalence of categories of Theorem 5.4.5 means that this correspondence is compatible with composition of (invertible) maps up to homotopy, from which we deduce that the assignment $\psi \mapsto u$ defines a bijection of the form

(5.8.4.3)
$$[E, W] \simeq \coprod_{F \in [E \times A^{\mathrm{op}}, \mathcal{S}]} \pi_0(\underline{\mathrm{Hom}}(E \times A^{\mathrm{op}}, \mathcal{S})(\Phi_E, F))$$

(where $[-,-]$ is the set of maps in the homotopy category of the Joyal model category structure). For $E = \Delta^0$, the object w, seen as a map $\Delta^0 \to W$, corresponds to the identity of Φ.

For $E = W$, let us write F_1 for the map $W \times A^{\mathrm{op}} \to \mathcal{S}$ corresponding to π by transposition. It classifies a left fibration $p_1\colon X \to W \times A^{\mathrm{op}}$. The latter has a canonical section s_1 associated to the canonical projection from W to

$\underline{\mathrm{Hom}}(U^{\mathrm{op}}, \mathcal{S}_\bullet)$ (because the pull-back of p_1 along $1_W \times f^{\mathrm{op}}$ is classified by the map obtained by transposition from the canonical functor $f^*\pi$ from W to $\underline{\mathrm{Hom}}(U^{\mathrm{op}}, \mathcal{S})$). The construction above thus determines a map $u_1 : \Phi_W \to F_1$ which we may see as a functor from $\Delta^1 \times W \times A^{\mathrm{op}}$ to $\mathcal{S}$. It also corresponds through the bijection (5.8.4.3) to the identity of W. Therefore, the evaluation of u_1 at w is equivalent to the identity of Φ, because it corresponds to the map $w : \Delta^0 \to W$ through the bijection (5.8.4.3). Let

$$\tau : \Delta^1 \times \Delta^1 \to \Delta^1$$

be the unique map which sends a vertex (i, j) to 1 if and only if $i = j = 1$. Then $u_1\tau$ is a commutative square of the form

$$\begin{array}{ccc} \Phi_W & \xrightarrow{1_{\Phi_W}} & \Phi_W \\ {\scriptstyle 1_{\Phi_W}}\downarrow & & \downarrow{\scriptstyle u_1} \\ \Phi_W & \xrightarrow{u_1} & F_1 \end{array}$$

in $\underline{\mathrm{Hom}}(W \times A^{\mathrm{op}}, \mathcal{S})$. In other words, this corresponds to a map

$$\Phi_{\Delta^1 \times W} = 1_{\Phi_W} \to u_1$$

in $\underline{\mathrm{Hom}}(\Delta^1 \times W \times A^{\mathrm{op}}, \mathcal{S})$. Using correspondence (5.8.4.3), it determines a homotopy $h : \Delta^1 \times W \to W$ from the constant map with value w to the identity of W whose restriction on $\Delta^1 \times \{w\}$ is equivalent to $u_1(w)$, hence is the identity in $ho(W)$. Applying Proposition 4.3.10 for $X = W^{\mathrm{op}}$ and $\omega = w$, this shows that w is an initial object. □

Corollary 5.8.5 *The ∞-category of ∞-groupoids $\mathcal{S}$ has a final object e which classifies the Kan complex Δ^0. Furthermore, there is a fibrewise equivalence $e\backslash\mathcal{S} \to \mathcal{S}_\bullet$ over $\mathcal{S}$. In other words, for any Hom space* $\mathrm{Hom}_{\mathcal{S}} : \mathcal{S}^{\mathrm{op}} \times \mathcal{S} \to \mathcal{S}$, *there is a canonical invertible map from the functor* $\mathrm{Hom}_{\mathcal{S}}(e, -)$ *to the identity of $\mathcal{S}$ in the ∞-category* $\underline{\mathrm{Hom}}(\mathcal{S}, \mathcal{S})$.

Proof Let e be an object of $\mathcal{S}$ which classifies the identity of Δ^0. For any Kan complex X, the Kan complex $\underline{\mathrm{Hom}}(X, \Delta^0)$ is isomorphic to Δ^0, hence contractible. Therefore, Theorem 4.3.11 and Corollary 5.4.7 show that the object e is final in $\mathcal{S}$. In the case of $U = A = \Delta^0$, Theorem 5.8.4 says that there is a homotopy Cartesian square of the following form.

$$\begin{array}{ccc} e\backslash\mathcal{S} & \longrightarrow & \mathcal{S}_\bullet \\ \downarrow & & \downarrow \\ \mathcal{S} & = & \mathcal{S} \end{array}$$

The last assertion follows from the fact that $\mathrm{Hom}_{\mathcal{S}}(e,-)$ also classifies the left fibration $e\backslash\mathcal{S} \to \mathcal{S}$ up to homotopy (more precisely, from the homotopy Cartesian square (5.8.1.5) for $A = \mathcal{S}^{op}$ and $a = e$): Proposition 5.3.20 then associates an invertible map $\mathrm{Hom}_{\mathcal{S}}(e,-) \to 1_{\mathcal{S}}$ to the fibrewise equivalence $e\backslash\mathcal{S} \to \mathcal{S}_\bullet$ over $\mathcal{S}$. □

Corollary 5.8.6 *Let X be a* **U***-small simplicial set. We choose a final object e in $\mathcal{S}$, and also write e for the constant functor $X \to \mathcal{S}$ with value e. Finally, we choose a Hom space* Hom *for $\underline{\mathrm{Hom}}(X,\mathcal{S})$. Then there is a homotopy Cartesian square in the Joyal model category of the form below.*

$$\begin{array}{ccc} \underline{\mathrm{Hom}}(X,\mathcal{S}_\bullet) & \longrightarrow & \mathcal{S}_\bullet \\ {\scriptstyle (p_{univ})_*}\downarrow & & \downarrow{\scriptstyle p_{univ}} \\ \underline{\mathrm{Hom}}(X,\mathcal{S}) & \xrightarrow{\mathrm{Hom}(e,-)} & \mathcal{S} \end{array}$$

In other words, there is a fibrewise equivalence

$$\underline{\mathrm{Hom}}(X,\mathcal{S}_\bullet) \xrightarrow{\sim} e\backslash\,\underline{\mathrm{Hom}}(X,\mathcal{S})\,.$$

Proof The equivalence of ∞-categories $e\backslash\mathcal{S} \to \mathcal{S}_\bullet$ over $\mathcal{S}$ induces an equivalence of ∞-categories $\underline{\mathrm{Hom}}(X, e\backslash\mathcal{S}) \to \underline{\mathrm{Hom}}(X,\mathcal{S}_\bullet)$ over $\underline{\mathrm{Hom}}(X,\mathcal{S})$. Choosing an inverse up to homotopy over $\underline{\mathrm{Hom}}(X,\mathcal{S})$ of the latter, we will prove this corollary as follows. Proposition 4.2.12 ensures that we have a canonical equivalence of ∞-categories of the form

$$\underline{\mathrm{Hom}}(X, e\backslash\mathcal{S}) \to e\backslash\,\underline{\mathrm{Hom}}(X,\mathcal{S})$$

over $\underline{\mathrm{Hom}}(X^{op},\mathcal{S})$. Taking into consideration the homotopy Cartesian square of Proposition 5.8.2 applied to $A = \mathrm{Hom}(X,\mathcal{S})$, with $a = e$, this defines a homotopy Cartesian square of the expected form. □

Remark 5.8.7 Theorem 5.8.4 is already some version of the Yoneda lemma: given an object a in a **U**-small ∞-category A, for any functor $F\colon A^{op} \to \mathcal{S}$, this theorem, applied with $U = \Delta^0$ and $f = a$, asserts that the left fibration $h_A(a)\backslash\,\underline{\mathrm{Hom}}(A^{op},\mathcal{S}) \to \underline{\mathrm{Hom}}(A^{op},\mathcal{S})$ is fibrewise equivalent to the left fibration classified by the evaluation at a functor. In other words, the homotopy Cartesian square of the theorem provides a homotopy Cartesian square of the form

$$\begin{array}{ccc} \underline{\mathrm{Hom}}(A^{op},\mathcal{S})(h_A(a),F) & \longrightarrow & \mathcal{S}_\bullet \\ \downarrow & & \downarrow{\scriptstyle p_{univ}} \\ \Delta^0 & \xrightarrow{\quad F(a)\quad} & \mathcal{S} \end{array}$$

and Theorem 5.4.5 explains how to interpret this as an identification of the Kan complex $\underline{\mathrm{Hom}}(A^{\mathrm{op}}, \mathcal{S})(h_a, F)$ with the fibre $F(a)$. A construction of the Yoneda embedding, using the twisted diagonal, as well as this version of the Yoneda lemma, can also be found, in the setting of Rezk's complete Segal spaces, in a paper by Varshavskiĭ and Kazhdan [VK14]. However, the genuine version of the Yoneda lemma not only gives such an identification abstractly, but also it provides an explicitly given invertible map, *functorially* in a and F. This is precisely what we will do now: we shall first give an explicit description (Theorem 5.8.9), which will be functorial in F, and then provide a fully functorial version, by a cofinality argument (Theorem 5.8.13).

5.8.8 Let A be a $\mathbf{U}$-small ∞-category. Then both A and $\underline{\mathrm{Hom}}(A^{\mathrm{op}}, \mathcal{S})$ are locally $\mathbf{U}$-small. We choose a Hom space functor Hom_A for A. This defines the Yoneda functor $h_A \colon A \to \underline{\mathrm{Hom}}(A^{\mathrm{op}}, \mathcal{S})$. For each object a, we apply construction (5.8.3.2) to the functor $a \colon \Delta^0 \to A$, so that we get a canonical map

$$1_a \colon e = \mathrm{Hom}_{\Delta^0}(0,0) \to \mathrm{Hom}_A(a,a) \tag{5.8.8.1}$$

(we leave to the reader the task of checking that the object $\mathrm{Hom}_{\Delta^0}(0,0)$ is final in $\mathcal{S}$). We also choose Hom space functors for $\underline{\mathrm{Hom}}(A^{\mathrm{op}}, \mathcal{S})$ and for $\mathcal{S}$, which will be simply denoted by Hom. The evaluation functor

$$ev \colon A^{\mathrm{op}} \times \underline{\mathrm{Hom}}(A^{\mathrm{op}}, \mathcal{S}) \to \mathcal{S}\,, \quad (a, F) \mapsto F(a) \tag{5.8.8.2}$$

induces by transposition a map

$$A^{\mathrm{op}} \to \underline{\mathrm{Hom}}(\underline{\mathrm{Hom}}(A^{\mathrm{op}}, \mathcal{S}), \mathcal{S})\,, \quad a \mapsto a^*\,. \tag{5.8.8.3}$$

For each object a of A, the functor a^* is the evaluation at a. Thus construction (5.8.3.2) induces, for each functor $F \colon A^{\mathrm{op}} \to \mathcal{S}$, a map

$$\mathrm{Hom}(h_A(a), F) \to \mathrm{Hom}(h_A(a)(a), F(a)) = \mathrm{Hom}(\mathrm{Hom}_A(a,a), F(a))$$

in $\mathcal{S}$, which is functorial in F. Using the invertible map from $\mathrm{Hom}(e, F(a))$ to $F(a)$ provided by Corollary 5.8.5, there is a map

$$\mathrm{Hom}(\mathrm{Hom}_A(a,a), F(a)) \to F(a)$$

defined as the evaluation at 1_a, so that we end up with the map below:

$$\mathrm{Hom}(h_A(a), F) \to F(a)\,, \quad s \mapsto s(a)(1_a). \tag{5.8.8.4}$$

This map is functorial in F in the sense that it is the evaluation at F of a map $\mathrm{Hom}(h_A(a), -) \to a^*$ in the ∞-category $\underline{\mathrm{Hom}}(\underline{\mathrm{Hom}}(A^{\mathrm{op}}, \mathcal{S}), \mathcal{S})$.

Theorem 5.8.9 *The evaluation map* (5.8.8.4) *is invertible in the ∞-category $\mathcal{S}$ of $\mathbf{U}$-small ∞-groupoids.*

Proof Let us denote by V and W the domains of the left fibrations classified by the functors $\mathrm{Hom}(h_A(a), -)$ and a^*, respectively. We choose a functor $f: V \to W$ over $\underline{\mathrm{Hom}}(A^{\mathrm{op}}, \mathcal{S})$ corresponding to the map (5.8.8.4) via Corollary 5.4.7. It is sufficient to prove that this functor is a fibrewise equivalence over $\underline{\mathrm{Hom}}(A^{\mathrm{op}}, \mathcal{S})$; see Proposition 5.3.16. The proof of Theorem 5.8.4 asserts that the map $1_a : e \to \mathrm{Hom}_A(a, a) = h_A(a)(a)$ (interpreted via Corollary 5.4.7) defines an initial object w in W whose image in $\underline{\mathrm{Hom}}(A^{\mathrm{op}}, \mathcal{S})$ is $h_A(a)$. Similarly, the identity of $h_A(a)$ defines an object v of V. By definition, there is a fibrewise equivalence

$$h_A(a)\backslash\,\underline{\mathrm{Hom}}(A^{\mathrm{op}}, \mathcal{S}) \to V$$

which sends the canonical initial object of the left-hand side to v. Since the map (5.8.8.4) sends the identity of $h_A(a)$ to the identity of a (because a^* is a functor), Corollary 5.4.7 provides an invertible map from $f(v)$ to w. In particular, $f(v)$ is an initial object of W. Therefore, by virtue of the dual version of Proposition 4.3.3, the functor f is cofinal. Since the structural maps of V and W are left fibrations, the map f is a fibrewise equivalence over $\underline{\mathrm{Hom}}(A^{\mathrm{op}}, \mathcal{S})$, by the dual version of Theorem 4.1.16. □

5.8.10 Let A be a locally $\mathbf{U}$-small ∞-category. We choose a Hom space Hom_A for A, and we consider the commutative square

(5.8.10.1)

$$\begin{array}{ccccc}
\mathrm{S}(A) & \xrightarrow{\;v\;} & V & \longrightarrow & \mathrm{S}(\underline{\mathrm{Hom}}(A^{\mathrm{op}}, \mathcal{S})) \\
{\scriptstyle (s_A, t_A)}\downarrow & & \downarrow{\scriptstyle p} & & \downarrow \\
A^{\mathrm{op}} \times A & \xrightarrow{1_{A^{\mathrm{op}}} \times h_A} & A^{\mathrm{op}} \times \underline{\mathrm{Hom}}(A^{\mathrm{op}}, \mathcal{S}) & \xrightarrow{h_A^{\mathrm{op}} \times 1} & \underline{\mathrm{Hom}}(A^{\mathrm{op}}, \mathcal{S})^{\mathrm{op}} \times \underline{\mathrm{Hom}}(A^{\mathrm{op}}, \mathcal{S})
\end{array}$$

where the composed upper horizontal map is the one induced by functoriality of the twisted diagonal applied to the Yoneda functor h_A, while the right-hand square is defined to be Cartesian. We also consider the following commutative square

$$\begin{array}{ccccc}
\mathrm{S}(A) & \xrightarrow{\;w\;} & W & \longrightarrow & \mathcal{S}_\bullet \\
{\scriptstyle (s_A, t_A)}\downarrow & & \downarrow{\scriptstyle q} & & \downarrow{\scriptstyle p_{univ}} \\
A^{\mathrm{op}} \times A & \xrightarrow{1_{A^{\mathrm{op}}} \times h_A} & A^{\mathrm{op}} \times \underline{\mathrm{Hom}}(A^{\mathrm{op}}, \mathcal{S}) & \xrightarrow{\;ev\;} & \mathcal{S}
\end{array} \tag{5.8.10.2}$$

in which the right-hand square is defined by the property of being Cartesian as well, while the map from $\mathrm{S}(A)$ to $\mathcal{S}_\bullet$ is simply σ_A (i.e. the composed square is the very one which exhibits Hom_A as Hom space functor). We remark that the left-hand square of (5.8.10.2) is homotopy Cartesian.

To prove the functorial version of the Yoneda lemma, we need the next lemma.

Lemma 5.8.11 *The maps $v\colon \mathrm{S}(A) \to V$ and $w\colon \mathrm{S}(A) \to W$ are cofinal.*

Proof The case of the map $\mathrm{S}(A) \to V$ is a particular case of Proposition 5.6.9. For any object a of A^{op}, Theorem 5.8.4 ensures that there exists a commutative diagram of the form

$$\begin{array}{ccccc} a\backslash A & \longrightarrow & h_A(a)\backslash \underline{\mathrm{Hom}}(A^{\mathrm{op}}, \mathcal{S}) & \longrightarrow & \mathcal{S}_\bullet \\ \downarrow & & \downarrow & & \downarrow{\scriptstyle p_{univ}} \\ A & \xrightarrow{h_A} & \underline{\mathrm{Hom}}(A^{\mathrm{op}}, \mathcal{S}) & \xrightarrow{a^*} & \mathcal{S} \end{array}$$

in which the right-hand square is homotopy Cartesian, while the left-hand one is obtained by functoriality of the coslice construction. The identity of a, seen as an object of $a\backslash A$, is sent in $\mathcal{S}_\bullet$ to the object which classifies the fibre of the left fibration $S(A) \to A^{\mathrm{op}} \times A$ at (a, a), pointed by 1_a. Since the map p_{univ} is a left fibration, and since 1_a is initial in $a\backslash A$, this property determines completely the functor $a\backslash A \to \mathcal{S}_\bullet$, up to J-homotopy over $\mathcal{S}$. Similarly, the map $h_A(a)\backslash \underline{\mathrm{Hom}}(A^{\mathrm{op}}, \mathcal{S}) \to \mathcal{S}_\bullet$ is completely characterised, up to J-homotopy over $\mathcal{S}$, by the fact that it sends the identity of $h_A(a)$ to the object which classifies the fibre of $S(A) \to A^{\mathrm{op}} \times A$ at (a, a), pointed by 1_a. In particular, the composed square is equivalent to that of Proposition 5.8.2, hence is also homotopy Cartesian. Therefore, the left-hand square is homotopy Cartesian as well. In other words, fibrewise over A^{op}, there is no difference between v and w, at least up to J-homotopy over $\underline{\mathrm{Hom}}(A^{\mathrm{op}}, \mathcal{S})$ (this is a variation on (the proof of) Theorem 5.8.9). This implies that the map $\mathrm{S}(A) \to W$ is fibrewise cofinal over A^{op}, because v has this property. By Lemma 5.6.7, this proves that this map is globally cofinal. □

Remark 5.8.12 The previous lemma means that the commutative diagrams (5.8.10.1) and (5.8.10.2) exhibit both (V, p) and (W, q), as constructions of the object $\mathbf{L}(1_{A^{\mathrm{op}}} \times h_A)_!(\mathrm{S}(A))$ in the homotopy category $\mathit{LFib}(A^{\mathrm{op}} \times \underline{\mathrm{Hom}}(A^{\mathrm{op}}, \mathcal{S}))$. Moreover, the first part of the proof of the lemma also says that the Yoneda functor h_A is fully faithful. By virtue of Proposition 4.5.2, this means that the functor

$$\mathbf{L}(1_{A^{\mathrm{op}}} \times h_A)_!\colon \mathit{LFib}(A^{\mathrm{op}} \times A) \to \mathit{LFib}(A^{\mathrm{op}} \times \underline{\mathrm{Hom}}(A^{\mathrm{op}}, \mathcal{S}))$$

is fully faithful. The bijection

$$\mathrm{Hom}_{\mathit{LFib}(A^{\mathrm{op}} \times \underline{\mathrm{Hom}}(A^{\mathrm{op}}, \mathcal{S}))}((V, p), (W, q)) \simeq \mathrm{Hom}_{\mathit{LFib}(A^{\mathrm{op}} \times A)}(\mathrm{S}(A), \mathrm{S}(A))$$

induced by the functor $\mathbf{R}(1_{A^{\mathrm{op}}} \times h_A)^*$ thus implies that there is a unique map from (V, p) to (W, q) in the homotopy category $LFib(A^{\mathrm{op}} \times \underline{\mathrm{Hom}}(A^{\mathrm{op}}, \mathcal{S}))$ which restricts to the identity of $\mathrm{S}(A)$ over $A^{\mathrm{op}} \times A$. Such a map $(V, p) \to (W, q)$ is necessarily an isomorphism. The internal version of this interpretation has a homotopy-theoretic flavour, which can be summarised as follows.

Theorem 5.8.13 (Yoneda lemma) *Let A be a locally $\mathbf{U}$-small ∞-category. We choose a Hom space Hom_A for A. We also assume that there is a possibly larger universe $\mathbf{V}$ such that $\underline{\mathrm{Hom}}(A^{\mathrm{op}}, \mathcal{S})$ is locally $\mathbf{V}$-small,*[3] *and we choose a Hom space Hom for $\underline{\mathrm{Hom}}(A^{\mathrm{op}}, \mathcal{S})$. We have the following properties.*

(i) The Yoneda functor $h_A \colon A \to \underline{\mathrm{Hom}}(A^{\mathrm{op}}, \mathcal{S})$ is fully faithful. In particular, the functor h_A induces an invertible map

$$\mathrm{Hom}_A(x, y) \to \mathrm{Hom}(h_A(x), h_A(y))$$

for any objects x and y of A.

(ii) For any object x of A, and any functor $F \colon A^{\mathrm{op}} \to \mathcal{S}$, there is an invertible map

$$\mathrm{Hom}(h_A(x), F) \to F(x)$$

in the ∞-category $\mathcal{S}'$ of $\mathbf{V}$-small ∞-groupoids, which is functorial in both x and F, and which is essentially characterised by the fact that, for $F = h_A(y)$, it coincides functorially with the inverse of the map given in (i).

The functoriality of (ii) means that this map is the evaluation at (x, F) of a map in the ∞-category $\underline{\mathrm{Hom}}(A^{\mathrm{op}} \times \underline{\mathrm{Hom}}(A^{\mathrm{op}}, \mathcal{S}), \mathcal{S})$. The property of essential uniqueness means that the collection of all natural transformations as in (ii) form a contractible ∞-groupoid in a sense which will be made precise in the proof.

Proof of the Yoneda lemma We are under the hypothesis of paragraph 5.8.10, from which we take the notation and the constructions. We choose a factorisation of the map $v \colon \mathrm{S}(A) \to V$ as a left anodyne extension v' followed by a left fibration π:

$$\mathrm{S}(A) \xrightarrow{v'} V' \xrightarrow{\pi} V\,.$$

Since v is cofinal, by Lemma 5.8.11, the map π is in fact a trivial fibration. We let $p' \colon V' \to A^{\mathrm{op}} \times \underline{\mathrm{Hom}}(A^{\mathrm{op}}, \mathcal{S})$ be the composition of p and π. We thus have

[3] In the case where A is $\mathbf{U}$-small, we can take $\mathbf{V} = \mathbf{U}$, by Proposition 5.7.3.

the solid commutative square below.

$$\begin{array}{ccc} S(A) & \xrightarrow{w} & W \\ {\scriptstyle v'}\downarrow & \overset{\gamma}{\nearrow} & \downarrow{\scriptstyle q} \\ V' & \xrightarrow[p']{} & A^{op} \times \underline{\mathrm{Hom}}(A^{op}, \mathcal{S}) \end{array}$$

A filler γ of this square exists, such that $q\gamma = p'$ and $\gamma v' = w$, because v' is a left anodyne extension, while q is a left fibration (considering the Cartesian square of the right-hand side of (5.8.10.1), this follows from Proposition 5.6.2). Since the map w is cofinal (Lemma 5.8.11), the map γ is necessarily cofinal as well. Since both V' and W are left fibrant over $A^{op} \times \underline{\mathrm{Hom}}(A^{op}, \mathcal{S})$, this means that γ is in fact an equivalence of ∞-categories. We observe that the commutative square

$$\begin{array}{ccc} S(A) & \xrightarrow{v'} & V' \\ {\scriptstyle (s_A, t_A)}\downarrow & & \downarrow{\scriptstyle p'} \\ A^{op} \times A & \xrightarrow{1_{A^{op}} \times h_A} & A^{op} \times \underline{\mathrm{Hom}}(A^{op}, \mathcal{S}) \end{array}$$

is homotopy Cartesian: the trivial fibration π makes it equivalent to the left-hand square of diagram (5.8.10.2). But the map γ also makes it equivalent to the left-hand square of diagram (5.8.10.1), which proves that the latter is homotopy Cartesian as well. This already shows that the functor h_A is fully faithful, by Corollary 5.6.6 (but we already knew it from the first part of the proof of Lemma 5.8.11). On the other hand, restricting along $v'\colon S(A) \to V'$ induces a trivial fibration

$$(v')^*\colon \mathrm{Map}_C(V', W) \to \mathrm{Map}_C(S(A), W)$$

where we have put $C = A^{op} \times \underline{\mathrm{Hom}}(A^{op}, \mathcal{S})$, while the mapping spaces are those associated to the covariant model category structure over C (4.1.12.1). This shows that the space of lifts γ as above is contractible, for it is the fibre of $(v')^*$ at w, and makes precise the assertion that γ is essentially unique. The choice of a Hom space for $\underline{\mathrm{Hom}}(A^{op}, \mathcal{S})$ gives a homotopy Cartesian square of the form below.

$$\begin{array}{ccc} S(\underline{\mathrm{Hom}}(A^{op}, \mathcal{S})) & \longrightarrow & \mathcal{S}_\bullet \\ \downarrow & & \downarrow{\scriptstyle p_{univ}} \\ \underline{\mathrm{Hom}}(A^{op}, \mathcal{S})^{op} \times \underline{\mathrm{Hom}}(A^{op}, \mathcal{S}) & \xrightarrow{\mathrm{Hom}} & \mathcal{S} \end{array}$$

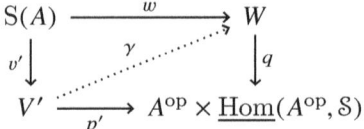

Adding the latter to the right-hand side of diagram (5.8.10.1) shows that the left fibration (s_A, t_A) is classified by the map

$$\mathrm{Hom}(h_A(-), h_A(-)) \colon A^{\mathrm{op}} \times A \to \mathcal{S}$$

up to homotopy.

Since π is a trivial fibration, it has a section σ over $A^{\mathrm{op}} \times \underline{\mathrm{Hom}}(A^{\mathrm{op}}, \mathcal{S})$. The map $\gamma\sigma$ defines, via Corollary 5.4.7, the invertible natural transformation of assertion (ii). Note that the equation $\gamma v' = w$ precisely means that this natural transformation, when restricted to representable functors, is a right inverse, up to a specified homotopy, of the natural transformation of (i) which expresses the fully faithfulness of h_A. The space of such natural transformations thus corresponds, via Theorem 5.4.5 and Corollary 5.4.7, to the space of lifts γ of the first commutative square above, which we know to be contractible. □

Corollary 5.8.14 *Let A be a locally* **U***-small ∞-category. For any simplicial set X, the Yoneda functor induces a fully faithful functor*

$$(h_A)_* \colon \underline{\mathrm{Hom}}(X, A) \to \underline{\mathrm{Hom}}(X, \underline{\mathrm{Hom}}(A^{\mathrm{op}}, \mathcal{S})) \simeq \underline{\mathrm{Hom}}(X \times A^{\mathrm{op}}, \mathcal{S})$$

which sends a functor $u \colon X \to A$ to the functor

$$X \times A^{\mathrm{op}} \to \mathcal{S}, \quad (x, a) \mapsto \mathrm{Hom}_A(a, u(x)) = h_A(u(x))(a)\,.$$

The essential image consists of those functors $X \to \underline{\mathrm{Hom}}(A^{\mathrm{op}}, \mathcal{S})$ which take their values in the class of representable presheaves.

Proof This corollary is a direct consequence of Theorem 5.8.9 and of Proposition 5.7.4. □

6

Adjoints, Limits and Kan Extensions

The title of this chapter is suggestive enough: we develop the basic concepts of category theory, namely adjoints, limits and Kan extensions.

Since we have at our disposal the Yoneda lemma, we do this in a rather classical way: an adjunction really is a functorial identification of the form[1]

$$\mathrm{Hom}(x, v(y)) \simeq \mathrm{Hom}(u(x), y)\,.$$

In the first section, we also interpret the derived functoriality of the homotopy theory of left fibrations (through the covariant model structures) as adjunctions which are compatible with the Yoneda embedding in a suitable sense (Theorem 6.1.14). All the main constructions and features of this chapter are consequences of this. A first consequence is the various characterisations of adjoint pairs of functors given by Theorem 6.1.23. This shows in particular the functoriality of the notion of adjoints: adjunctions are compatible with the formation of ∞-categories of functors. The second section is devoted to the definition of limits and colimits. For instance, small colimits exist in the ∞-category $\mathcal{S}$ of small ∞-groupoids. After checking expected properties, such as the commutation of left adjoints with colimits, a fundamental example of colimit is derived as a consequence of the Yoneda lemma: any presheaf on a small ∞-category is a canonical colimit of representable presheaves (Corollary 6.2.16). We also give an elementary proof that reflexive subcategories of (co)complete ∞-categories are (co)complete; see Proposition 6.2.17. The third section uses this to construct extensions of functors by colimits (existence is provided by Theorem 6.3.4, and uniqueness by Theorem 6.3.13). In particular, this exhibits the Yoneda embedding of a small ∞-category as the universal functor in an ∞-category with small colimits. The explicit description of these

[1] This could seem naive at first glance, when compared to the definition considered by Lurie in [Lur09], but these points of view coincide: this follows from Theorem 6.1.23 below and from [Lur09, lemma 5.2.4.1].

extensions by colimits shows that, under mild assumptions, they have right adjoints. This shows that the ∞-category of small ∞-groupoids also has limits expressed in terms of Hom (Theorem 6.3.7), and that the Yoneda embedding commutes with limits. Section 6.4 defines and studies Kan extensions as a relative version of the notions of limit and of colimit. In particular, we revisit all the computations of Section 4.4 of Chapter 4 in the language of functors with values in $\mathcal{S}$. By a transposition argument, and using compatibility of the Yoneda embedding with limits, this gives the existence and fibrewise computation of right Kan extensions in any ∞-category with small limits (Proposition 6.4.9) as well as Grothendieck's smooth base change formulas and proper base change formulas (also known as the Beck-Chevalley property) associated to appropriate pull-back squares (Theorem 6.4.13). Similarly, on our way, we check that final functors behave as expected with respect to colimits (Theorem 6.4.5).

Section 6.5 studies products: we mainly check that they correspond in $\mathcal{S}$ to ordinary products of Kan complexes. Section 6.6 takes the preceding one over an object: we check that pull-backs can be considered as binary products in sliced categories. As a byproduct, we see that ordinary pull-backs of Kan fibrations give pull-backs in $\mathcal{S}$, and that all pull-backs in $\mathcal{S}$ are obtained in this way (Remark 6.6.11). On our way to prove this, we also check that, for any right fibration $X \to A$ corresponding to a functor $F\colon A^{\mathrm{op}} \to \mathcal{S}$, presheaves on X correspond to presheaves on A over F (Proposition 6.5.7).

Finally, Section 6.7 revisits the theme of the Yoneda embedding and of extensions by colimits in a relative way (for functors with values in an arbitrary ∞-category with small colimits). This leads to an equivalence of ∞-categories which interprets the operation $A \mapsto A^{\mathrm{op}}$ as a duality operator (Theorem 6.7.2). Such a duality will be useful later to understand localisations.

6.1 Adjoints

In this section, a Grothendieck universe $\mathbf{U}$ is specified, once and for all. We assume that any locally $\mathbf{U}$-small ∞-category A comes equipped with a Hom space Hom_A.

6.1.1 Let $F\colon A^{\mathrm{op}} \to \mathcal{S}$ be a presheaf on A. The ∞-category of elements of F is the ∞-category A/F obtained by forming the Cartesian square below.

$$\begin{array}{ccc} A/F & \longrightarrow & \underline{\mathrm{Hom}}(A^{\mathrm{op}}, \mathcal{S})/F \\ \downarrow & & \downarrow \\ A & \xrightarrow{h_A} & \underline{\mathrm{Hom}}(A^{\mathrm{op}}, \mathcal{S}) \end{array} \tag{6.1.1.1}$$

The objects of A/F can be described as the pairs (a, s), where a is an object of A, and $s\colon h_A(a) \to F$ is a map in $\underline{\mathrm{Hom}}(A^{\mathrm{op}}, \mathcal{S})$, which we interpret as a morphism from $\Delta^1 = \Delta^0 * \Delta^0$ to $\underline{\mathrm{Hom}}(A^{\mathrm{op}}, \mathcal{S})$, which in turn can be seen as an object of $\underline{\mathrm{Hom}}(A^{\mathrm{op}}, \mathcal{S})/F$. We remark that, by definition, the projection $A/F \to A$ is a right fibration which is classified by the map $\mathrm{Hom}(h_A(-), F)$, where Hom is a map which classifies the twisted diagonal of $\underline{\mathrm{Hom}}(A^{\mathrm{op}}, \mathcal{S})$. The Yoneda lemma thus means that A/F is canonically fibrewise equivalent to the right fibration classified by F. In particular, for any object a of A, the canonical functor $A/a \to A/h_A(a)$ induced by the Yoneda functor is an equivalence of ∞-categories, or equivalently, a fibrewise equivalence over A.

Given an object a of A, we say that a map $s\colon h_A(a) \to F$ exhibits F as a *representable presheaf* if it is invertible in $\underline{\mathrm{Hom}}(A^{\mathrm{op}}, \mathcal{S})$. In this case, we say that F is *represented by* a. A presheaf F as above is said to be *representable* if it is represented by some object a of A.

Proposition 6.1.2 *A presheaf $F\colon A^{\mathrm{op}} \to \mathcal{S}$ is representable if and only if the ∞-category of elements A/F has final objects.*

Proof It follows from Theorem 4.3.11 that a functor $A^{\mathrm{op}} \to \mathcal{S}$ is equivalent to a functor of the form $h_A(a)$ if and only if the domain of left fibration that it classifies, $Y \to A^{\mathrm{op}}$, has an initial object whose image in $ho(A)^{\mathrm{op}}$ is isomorphic to a. □

The purpose of this chapter is to study representable presheaves in families: this is what leads to the notion of adjoint functor.

Definition 6.1.3 Let $u\colon A \to B$ and $v\colon B \to A$ be two functors between locally **U**-small ∞-categories. We say that (u, v) form an *adjoint pair*, or that u is a *left adjoint* of v, or that v is a *right adjoint* of u, if there exists a functorial invertible map of the form

$$c_{x,y}\colon \mathrm{Hom}_A(x, v(y)) \xrightarrow{\sim} \mathrm{Hom}_B(u(x), y)$$

in the ∞-category of **U**-small ∞-groupoids $\mathcal{S}$ (where the word functorial means that this map is the evaluation at (x, y) of a specified morphism c in the ∞-category of functors $\underline{\mathrm{Hom}}(A^{\mathrm{op}} \times B, \mathcal{S})$).

An *adjunction* from A to B is a triple (u, v, c), where u and v are functors as above, while c is an invertible map from $\mathrm{Hom}_A(-, v(-))$ to $\mathrm{Hom}_B(u(-), -)$ which exhibits v as a right adjoint of u.

Remark 6.1.4 The canonical isomorphism $A^{\mathrm{op}} \times B \simeq (B^{\mathrm{op}})^{\mathrm{op}} \times A^{\mathrm{op}}$ shows that $u\colon A \to B$ is a left adjoint of $v\colon B \to A$ if and only if $v^{\mathrm{op}}\colon B^{\mathrm{op}} \to A^{\mathrm{op}}$ is a left adjoint of $u^{\mathrm{op}}\colon A^{\mathrm{op}} \to B^{\mathrm{op}}$.

Remark 6.1.5 If $u\colon A \to B$ and $v\colon B \to A$ form a pair of adjoint functors between locally $\mathbf{U}$-small ∞-categories, then

$$ho(u)\colon ho(A) \to ho(B) \quad \text{and} \quad ho(v)\colon ho(A) \to ho(B)$$

form a pair of adjoint functors between locally $\mathbf{U}$-small categories in the usual sense. Indeed, we have the following natural bijections:

$$\begin{aligned}
\mathrm{Hom}_{ho(A)}(x, v(y)) &= \pi_0(A(x, v(y))) \\
&\simeq \pi_0(\mathcal{S}(e, \mathrm{Hom}_A(x, v(y)))) \\
&\simeq \pi_0(\mathcal{S}(e, \mathrm{Hom}_B(u(x), y))) \\
&\simeq \pi_0(B(u(x), y)) \\
&= \mathrm{Hom}_{ho(B)}(u(x), y).
\end{aligned}$$

Proposition 6.1.6 *Let $u\colon A \to B$ be an equivalence of locally $\mathbf{U}$-small ∞-categories. Any quasi-inverse of u can be promoted to a right adjoint of u. Conversely, any right adjoint of u is a quasi-inverse of u.*

Proof Let $v\colon B \to A$ be a quasi-inverse of u (i.e. v is an inverse of u up to homotopy with respect to the Joyal model structure). In particular, there is an invertible natural transformation from uv to the identity of B, whence an invertible natural transformation of the form

$$\mathrm{Hom}_B(x, u(v(b))) \xrightarrow{\sim} \mathrm{Hom}_B(x, b)\,.$$

We can furthermore replace the variable x by $u(a)$, where a runs in the ∞-category A^{op}. On the other hand, since u is in particular fully faithful, it defines an invertible functorial map in $\mathcal{S}$ of the form

$$\mathrm{Hom}_A(a, v(b)) \xrightarrow{\sim} \mathrm{Hom}_B(u(a), u(v(b)))\,.$$

Since we can compose invertible maps in the ∞-category $\underline{\mathrm{Hom}}(A^{\mathrm{op}} \times B, \mathcal{S})$, this exhibits v as a right adjoint of u.

Conversely, if v is a right adjoint of u, using that u is fully faithful, we have invertible natural transformations of the form

$$\mathrm{Hom}_B(u(a), b) \xleftarrow{\sim} \mathrm{Hom}_A(a, v(b)) \xrightarrow{\sim} \mathrm{Hom}_B(u(a), u(v(b)))\,.$$

Since the functor $u^*\colon \underline{\mathrm{Hom}}(B^{\mathrm{op}}, \mathcal{S}) \to \underline{\mathrm{Hom}}(A^{\mathrm{op}}, \mathcal{S})$ is an equivalence of ∞-categories, this means that there is an invertible map from $h_B uv$ to h_B in the ∞-category $\underline{\mathrm{Hom}}(B^{\mathrm{op}}, \mathcal{S})$, hence an invertible map from uv to the identity of B, by Corollary 5.8.14. Therefore, the functor u being an equivalence of ∞-categories, the functor v must have the same property and is a quasi-inverse of u. □

Proposition 6.1.7 *Let $u\colon A \to B$ and $v\colon B \to A$ be a pair of adjoint functors between locally* **U***-small ∞-categories. Given any functor $w\colon B \to A$, any invertible natural transformation from w to v exhibits canonically w as a right adjoint of u.*

Proof If we have an invertible functorial map $w(b) \to v(b)$, we have an invertible functorial map

$$\mathrm{Hom}_A(a, w(b)) \xrightarrow{\sim} \mathrm{Hom}_A(a, v(b))$$

which we can compose with the invertible functorial map

$$\mathrm{Hom}_A(x, v(y)) \xrightarrow{\sim} \mathrm{Hom}_B(u(x), y)$$

which equips w with the structure of a right adjoint of u. □

Proposition 6.1.8 *Let $u\colon A \to B$ and $p\colon B \to C$ be two functors between locally* **U***-small ∞-categories. Assume that they both have right adjoints v and q, respectively. Then vq is canonically a right adjoint of pu.*

Proof Let us choose two functorial invertible maps

$$c_{x,y}\colon \mathrm{Hom}_A(x, v(y)) \xrightarrow{\sim} \mathrm{Hom}_B(u(x), y),$$
$$d_{x,y}\colon \mathrm{Hom}_B(x, q(y)) \xrightarrow{\sim} \mathrm{Hom}_C(p(x), y).$$

For any object x of A and any object y of C, the maps $d_{u(x),y}$ and $c_{x,q(y)}$ are invertible:

$$\mathrm{Hom}_A(x, v(q(y))) \simeq \mathrm{Hom}_B(u(x), q(y)) \simeq \mathrm{Hom}_C(p(u(x)), y).$$

Since $\underline{\mathrm{Hom}}(A^{\mathrm{op}} \times C, \mathcal{S})$ is an ∞-category, one can compose these maps so that they are part of an invertible natural transformation which exhibits vq as a right adjoint of pu. □

Proposition 6.1.9 *Let $u\colon A \to B$ be a functor between locally* **U***-small ∞-categories. For any two adjunctions of the form (u, v, c) and (u, v', c'), the functorial maps $v(y) \xrightarrow{\sim} v'(y)$ which are compatible with c and c' form a contractible ∞-groupoid (i.e. there is a unique way to identitfy v and v' as right adjoints of u).*

Proof By virtue of Corollary 5.8.14, the Yoneda functor induces a fully faithful embedding of the form

$$\underline{\mathrm{Hom}}(B, A) \to \underline{\mathrm{Hom}}(B, \underline{\mathrm{Hom}}(A^{\mathrm{op}}, \mathcal{S})) = \underline{\mathrm{Hom}}(B \times A^{\mathrm{op}}, \mathcal{S}).$$

On the other hand, c and c' specify identifications of $\mathrm{Hom}_B(u(-), -)$ with $F = \mathrm{Hom}_A(-, v(-))$ and $F' = \mathrm{Hom}_A(-, v'(-))$, respectively. Let γ be a composition of an inverse of c with c'. The natural transformations $v \to v'$ which are compatible with c and c' are thus the objects in the homotopy fibre of the equivalence of ∞-groupoids

$$\underline{\mathrm{Hom}}(B, A)(v, v') \to \underline{\mathrm{Hom}}(B \times A^{\mathrm{op}}, \mathcal{S})(F, F')$$

at the point γ. □

6.1.10 Let $u\colon A \to B$ be a functor between locally **U**-small ∞-categories. It induces a functor

$$u^*\colon \underline{\mathrm{Hom}}(B^{\mathrm{op}}, \mathcal{S}) \to \underline{\mathrm{Hom}}(A^{\mathrm{op}}, \mathcal{S}) \tag{6.1.10.1}$$

defined by precomposing with u^{op}. For an object b of B, we define $A/b = B/b \times_B A$.

Proposition 6.1.11 *Under the assumptions of the previous paragraph, the following conditions are equivalent.*

(i) The functor u has a right adjoint.
(ii) For any object b of B, the presheaf $u^*(h_B(b))$ is representable.
(iii) For any object b of B, the ∞-category A/b has a final object.

Proof The canonical functor $B/b \to B/h_B(b)$ is a weak equivalence between fibrant objects of the contravariant model category structure over B, whence so is the pull-back $A/b \to B/h_B(b) \times_B A$. This shows that the left fibration

$$(B/h_B(b) \times_B A)^{\mathrm{op}} \to A^{\mathrm{op}}$$

is classified, up to homotopy, by the functor $a \mapsto \mathrm{Hom}(h_B(u(a)), h_B(b))$. By the Yoneda lemma, the latter functor is equivalent to the functor $a \mapsto \mathrm{Hom}_A(u(a), b)$. In other words, A/b is fibrewise equivalent to $A/h_B(b)$ over A. Proposition 6.1.2 thus implies that $u^*(h_B(b))$ is representable if and only if A/b has a final object. If v is a right adjoint of u, for any object b of B, there is a functorial invertible map of the form

$$\mathrm{Hom}_B(u(a), b) \xrightarrow{\sim} \mathrm{Hom}_A(a, v(b)) \,.$$

In other words, there is an invertible map from $u^*(h_B(b))$ to $h_A(v(b))$, whence a proof that the presheaf $u^*(h_B(b))$ is representable. Conversely, let us assume that $u^*(h_B(b))$ is representable for any object b of B. Let A' be the full subcategory of representable presheaves in $\underline{\mathrm{Hom}}(A^{\mathrm{op}}, \mathcal{S})$. The Yoneda functor defines an equivalence of ∞-categories $i\colon A \to A'$. Let us choose a quasi-inverse $r\colon A' \to A$ of i. There exists an invertible natural transformation $\gamma_F\colon F \to h_A(r(F))$ for any representable presheaf F on A. We define

$v = ru^*h_B\colon B \to A$. Restricting γ to u^*h_B defines an invertible natural transformation

$$c_{x,y}\colon \mathrm{Hom}_B(u(x), y) \xrightarrow{\sim} \mathrm{Hom}_A(x, v(y))$$

which exhibits v as a right adjoint of u. □

Remark 6.1.12 Condition (iii) of the preceding proposition shows that the property of having a right adjoint does not depend on the chosen ambient universe. The latter was only chosen in order to express adjunctions in terms of Hom space functors, as we usually do in basic category theory.

Corollary 6.1.13 *If a functor $A \to B$ has a right adjoint, then it is cofinal.*

Proof We know that, if an ∞-category C has a final object, then it is weakly contractible: seen as a map $\Delta^0 \to C$, the final object is a right anodyne extension, whence an anodyne extension, and thus a trivial cofibration of the Kan–Quillen model category structure. Our assertion is thus a consequence of the previous proposition and of Corollary 4.4.31. □

Proposition 6.1.14 *Let $u\colon A \to B$ be a morphism of* $\mathbf{U}$*-small simplicial sets. The functor*

$$u^*\colon \underline{\mathrm{Hom}}(B^{\mathrm{op}}, \mathcal{S}) \to \underline{\mathrm{Hom}}(A^{\mathrm{op}}, \mathcal{S})$$

has a left adjoint

$$u_!\colon \underline{\mathrm{Hom}}(A^{\mathrm{op}}, \mathcal{S}) \to \underline{\mathrm{Hom}}(B^{\mathrm{op}}, \mathcal{S})\,.$$

More precisely, if $F\colon A^{\mathrm{op}} \to \mathcal{S}$ classifies a right fibration with $\mathbf{U}$*-small fibres of the form $p\colon X \to A$, one may choose a commutative square*

(6.1.14.1)
$$\begin{array}{ccc} X & \xrightarrow{i} & Y \\ {\scriptstyle p}\downarrow & & \downarrow{\scriptstyle q} \\ A & \xrightarrow{u} & B \end{array}$$

in which i is final and q is a right fibration with $\mathbf{U}$*-small fibres, and define $u_!(F)\colon B^{\mathrm{op}} \to \mathcal{S}$ as any choice of functor which classifies the left fibration q^{op}. The unit map $F \to u^*u_!(F)$, which exhibits the functor* $\mathrm{Hom}(F, u^*(-))$ *as represented by $u_!(F)$, is the map corresponding to the morphism $X \to A \times_B Y$, induced by i, by Corollary 5.4.7.*

Furthermore, there is a canonical invertible map

$$h_B u \xrightarrow{\sim} u_! h_A\,.$$

Proof Let us consider the case where $X = A$ and p is the identity of A. In other words, F is then the constant functor with value the final object e in $\mathcal{S}$. If $\Phi\colon B^{op} \to \mathcal{S}$ is a functor which classifies q^{op}, then Theorem 5.8.4 and Corollary 5.8.6 ensure that we have two canonical homotopy Cartesian squares of the form below.

$$\begin{array}{ccccc} \Phi\backslash \underline{\mathrm{Hom}}(B^{op}, \mathcal{S}) & \longrightarrow & \underline{\mathrm{Hom}}(A^{op}, \mathcal{S}_\bullet) & \longrightarrow & \mathcal{S}_\bullet \\ \downarrow & & \downarrow{\scriptstyle (p_{univ})_*} & & \downarrow{\scriptstyle p_{univ}} \\ \underline{\mathrm{Hom}}(B^{op}, \mathcal{S}) & \xrightarrow{u^*} & \underline{\mathrm{Hom}}(A^{op}, \mathcal{S}) & \xrightarrow{\mathrm{Hom}(e,-)} & \mathcal{S} \end{array}$$

In other words, we have a canonical invertible natural transformation

$$\mathrm{Hom}(\Phi, G) \xrightarrow{\sim} \mathrm{Hom}(e, u^*(G))\,.$$

In the general case, if we still denote by Φ a functor which classifies q^{op}, applying twice the computation above (once for p, and once for up), we have invertible natural transformations of the form

$$\mathrm{Hom}(F, u^*(G)) \xrightarrow{\sim} \mathrm{Hom}(e, p^*u^*(G)) = \mathrm{Hom}(e, (up)^*(G)) \xleftarrow{\sim} \mathrm{Hom}(\Phi, G)\,.$$

In other words, the functor $\mathrm{Hom}(F, u^*(-))$ is representable by Φ. One may then apply (the proof of) Proposition 6.1.11.

Let us choose a universe $\mathbf{V}$ such that $C = \underline{\mathrm{Hom}}(B^{op}, \mathcal{S})^{op}$ is $\mathbf{V}$-small. For any presheaf $F\colon B^{op} \to \mathcal{S}$ and any object a of A, there are functorial invertible maps:

$$\begin{aligned} \mathrm{Hom}(h_B(u(a)), F) &\xrightarrow{\sim} F(u(a)) \\ &= u^*(F)(a) \\ &\xleftarrow{\sim} \mathrm{Hom}(h_A(a), u^*(F)) \\ &\xrightarrow{\sim} \mathrm{Hom}(u_! h_A(a), F)\,. \end{aligned}$$

Since, by virtue of Corollary 5.8.14, the Yoneda embedding of C induces a fully faithful functor from $\underline{\mathrm{Hom}}(A^{op}, C)$ into the ∞-category of presheaves of $\mathbf{V}$-small ∞-groupoids on $A \times C$, this defines a canonical invertible map $h_B u \xrightarrow{\sim} u_! h_A$. □

Proposition 6.1.15 *Let $u\colon A \to B$ be a functor between $\mathbf{U}$-small ∞-categories. The following conditions are equivalent.*

(i) The functor u is fully faithful.

(ii) For any objects x and y in A, the induced map

$$\mathrm{Hom}_A(x, y) \to \mathrm{Hom}_B(u(x), u(y))$$

is invertible in $\mathcal{S}$.

(iii) *The functor* $u_!\colon \underline{\mathrm{Hom}}(A^{\mathrm{op}}, \mathcal{S}) \to \underline{\mathrm{Hom}}(B^{\mathrm{op}}, \mathcal{S})$ *is fully faithful.*

Proof The equivalence between conditions (i) and (ii) is obvious, by definition of the canonical map (5.8.3.2). Since, by the preceding proposition, we have a canonical invertible natural transformation from $h_B u$ to $u_! h_A$, we can reinterpret condition (ii) by saying that, for any object y of A, the unit map

$$h_A(y) \to u^* u_!(h_A(y))$$

is invertible. But, since the unit map of the adjunction between $u_!$ and u^* coincides with the unit map of the derived adjunction between $\mathbf{L}u_!$ and $\mathbf{R}u^*$, the latter property also means that the map $F \to \mathbf{R}u^*\mathbf{L}u_!(F)$ is invertible in $RFib(A)$ for any F representing a right fibration $X \to A$ such that X has a final object. Therefore, Proposition 4.5.1 shows that conditions (ii) and (iii) are equivalent. □

Proposition 6.1.16 *Let* $u\colon A \to B$ *and* $f\colon K \to L$ *be two morphisms of simplicial sets. For any object* F *in* $LFib(L \times A)$*, the base change map*

$$\mathbf{L}(1_K \times u)_!\mathbf{R}(f \times 1_A)^*(F) \to \mathbf{R}(f \times 1_B)^*\mathbf{L}(1_L \times u)_!(F)$$

associated to the Cartesian square

$$\begin{array}{ccc} K \times A & \xrightarrow{f\times 1_A} & L \times A \\ {\scriptstyle 1_K\times u}\downarrow & & \downarrow{\scriptstyle 1_L\times u} \\ K \times B & \xrightarrow{f\times 1_B} & L \times B \end{array}$$

is an isomorphism in $LFib(K \times B)$.

Proof We consider first the case where $K = \Delta^0$, so that f is simply an object of L. We choose a factorisation of f into a final map $\tilde{f}\colon \Delta^0 \to L_{/f}$ followed by a right fibration $\pi\colon L_{/f} \to L$. For any map $W \to L$, we define $W_{/f} = L_{/f} \times_L W$ and $W_f = \{f\} \times_L W$. Let $p\colon X \to L \times A$ be a left fibration, and let us choose a factorisation of $(1_L \times u)p$ into a cofinal map $j\colon X \to Y$ followed by a left fibration $q\colon Y \to L \times B$. We can form the following Cartesian squares of simplicial sets over A.

$$\begin{array}{ccccc} X_f & \longrightarrow & X_{/f} & \longrightarrow & X \\ {\scriptstyle p_f}\downarrow & & {\scriptstyle p_{/f}}\downarrow & & \downarrow{\scriptstyle p} \\ A & \xrightarrow{(\tilde{f}, 1_A)} & L_{/f} \times A & \xrightarrow{\pi\times 1_A} & L \times A \end{array}$$

Since all the structural maps to A are proper (being compositions of left fibrations and of Cartesian projection of the form $E \times A \to A$), the horizontal maps of the left-hand square are weak equivalences of the covariant model category structure over A (because they are fibrewise equivalences, so that we can apply Corollary 4.4.28). Similarly, in the diagram of simplicial sets over B,

$$\begin{array}{ccccc} Y_f & \longrightarrow & Y_{/f} & \longrightarrow & Y \\ {\scriptstyle q_f}\downarrow & & {\scriptstyle q_{/f}}\downarrow & & \downarrow{\scriptstyle q} \\ B & \xrightarrow{(\tilde{f},1_B)} & L_{/f}\times B & \xrightarrow{\pi\times 1_B} & L\times B \end{array}$$

the horizontal maps of the left-hand Cartesian square are weak equivalences of the covariant model category structure over B. On the other hand, the map $j_{/f}\colon X_{/f}\to Y_{/f}$ is cofinal, since this is a pull-back of the cofinal map $j\colon X\to Y$ along the smooth map $L_{/f}\to L$. We thus have a commutative square

$$\begin{array}{ccc} X_f & \longrightarrow & X_{/f} \\ {\scriptstyle j_f}\downarrow & & \downarrow{\scriptstyle j_{/f}} \\ Y_f & \longrightarrow & Y_{/f} \end{array}$$

in which the two horizontal maps as well as the right vertical map are weak equivalences of the covariant model category structure over B. This proves that the map $X_f\to Y_f$ is cofinal: this is a weak equivalence of the covariant model category structure over B with fibrant codomain, so that we may apply Proposition 2.5.6. In conclusion, if we put $F=(X,p)$, we have $\mathbf{R}(f\times 1_A)^*(F)=(X_f,p_f)$, and we also have $\mathbf{L}(1_L\times u)_!(F)=(Y,q)$ and $\mathbf{R}(f\times 1_B)^*\mathbf{L}(1_L\times u)_!(F)=(Y_f,q_f)$. The fact that the map $X_f\to Y_f$ is cofinal with fibrant codomain over B means that $\mathbf{L}u_!(X_f,p_f)=(Y_f,q_f)$. With these conventions, the base change map

$$\mathbf{L}u_!\mathbf{R}(f\times 1_A)^*(F)\to\mathbf{R}(f\times 1_B)^*\mathbf{L}(1_L\times u)_!(F)$$

is nothing but an equality.

To prove the case where K is an arbitrary simplicial set, it is sufficient to prove that, for any point $x\colon\Delta^0\to K$, the induced map

$$\mathbf{R}(x\times 1_B)^*\mathbf{L}(1_K\times u)_!\mathbf{R}(f\times 1_A)^*(F)\to\mathbf{R}(f(x)\times 1_B)^*\mathbf{L}(1_L\times u)_!(F)$$

is an isomorphism in $LFib(B)$. We are thus reduced to the case where $K=\Delta^0$ which was treated in detail above. □

Corollary 6.1.17 *Let $u\colon A\to B$ be a morphism of* $\mathbf{U}$*-small simplicial sets. For any simplicial set X, the functor*

$$u^*=\underline{\mathrm{Hom}}(X^{\mathrm{op}},u^*)\colon\underline{\mathrm{Hom}}(X^{\mathrm{op}},\underline{\mathrm{Hom}}(B^{\mathrm{op}},\mathcal{S}))\to\underline{\mathrm{Hom}}(X^{\mathrm{op}},\underline{\mathrm{Hom}}(A^{\mathrm{op}},\mathcal{S}))$$

has a left adjoint of the form

$$u_! = \underline{\mathrm{Hom}}(X^{\mathrm{op}}, u_!) \colon \underline{\mathrm{Hom}}(X^{\mathrm{op}}, \underline{\mathrm{Hom}}(A^{\mathrm{op}}, \mathcal{S})) \to \underline{\mathrm{Hom}}(X^{\mathrm{op}}, \underline{\mathrm{Hom}}(B^{\mathrm{op}}, \mathcal{S}))\,.$$

Proof Let $\mathbf{V}$ be a universe such that X is $\mathbf{V}$-small. Replacing $\mathcal{S}$ by the ∞-category of ∞-groupoids $\mathcal{S}'$, Proposition 6.1.14 gives a left adjoint

$$(1_X \times u)_! \colon \underline{\mathrm{Hom}}(X^{\mathrm{op}} \times A^{\mathrm{op}}, \mathcal{S}') \to \underline{\mathrm{Hom}}(X^{\mathrm{op}} \times B^{\mathrm{op}}, \mathcal{S}')$$

of the pull-back functor $(1_X \times u)^*$. It is sufficient to check that $(1_X \times u)_!$ preserves presheaves with values in $\mathbf{U}$-small ∞-groupoids, which follows from the fact that it can be computed pointwise over X, as shown by Proposition 6.1.16. □

Definition 6.1.18 Let A be a simplicial set. A $\mathbf{U}$-*small presheaf* over A is a functor $F \colon A^{\mathrm{op}} \to \mathcal{S}$ such that there exists a commutative square of the form

$$\begin{array}{ccc} X & \longrightarrow & \mathcal{S}_\bullet \\ \big\downarrow & & \big\downarrow{\scriptstyle p_{univ}} \\ A^{\mathrm{op}} & \xrightarrow{F} & \mathcal{S} \end{array}$$

with X a $\mathbf{U}$-small simplicial set, such that the induced map $X \to A^{\mathrm{op}} \times_{\mathcal{S}} \mathcal{S}_\bullet$ is cofinal.

A simplicial set A is *locally* $\mathbf{U}$-*small* if there exists a weak categorical equivalence $A \to B$ with B a locally $\mathbf{U}$-small ∞-category.

Example 6.1.19 If A is a $\mathbf{U}$-small simplicial set, any presheaf $A^{\mathrm{op}} \to \mathcal{S}$ is $\mathbf{U}$-small.

Example 6.1.20 Let A be a locally $\mathbf{U}$-small ∞-category. Any representable presheaf on A is $\mathbf{U}$-small: for $F = \mathrm{Hom}_A(-, a)$, one may take $X = \Delta^0$.

Proposition 6.1.21 *Let A be a locally* $\mathbf{U}$*-small simplicial set. The ∞-category of* $\mathbf{U}$*-small presheaves over A is a locally* $\mathbf{U}$*-small ∞-category.*

Proof Let F be a $\mathbf{U}$-small presheaf and $G \colon A^{\mathrm{op}} \to \mathcal{S}$ any presheaf. We choose a $\mathbf{U}$-small simplicial set X, and a map $p \colon X \to A$, such that there exists an invertible map $p_!(e) \xrightarrow{\sim} F$. We then have an invertible map

$$\mathrm{Hom}(e, p^*(G)) \xrightarrow{\sim} \mathrm{Hom}(p_!(e), G) \simeq \mathrm{Hom}(F, G)$$

in a large enough ∞-category of ∞-groupoids. Since, by virtue of Proposition 5.7.3, $\underline{\mathrm{Hom}}(X^{\mathrm{op}}, \mathcal{S})$ is locally $\mathbf{U}$-small, this shows that $\mathrm{Hom}(F, G)$ is equivalent to a $\mathbf{U}$-small ∞-groupoid. □

Theorem 6.1.22 *Let $u \colon A \to B$ and $v \colon B \to A$ be a pair of adjoint functors* *between locally* $\mathbf{U}$*-small* ∞*-categories. Then, for any* $\mathbf{U}$*-small simplicial set* X*, the functors*

$$u = \underline{\mathrm{Hom}}(X, u) : \underline{\mathrm{Hom}}(X, A) \to \underline{\mathrm{Hom}}(X, B)$$

and

$$v = \underline{\mathrm{Hom}}(X, v) : \underline{\mathrm{Hom}}(X, B) \to \underline{\mathrm{Hom}}(X, A)$$

form a pair of adjoint functors between locally $\mathbf{U}$*-small* ∞*-categories. Furthermore, the formation of this adjunction is functorial in* X*: for any map* $p\colon Y \to X$*, the image by* p^* *(the functor of composition with* p*) of the unit map* $F \to v(u(F))$ *and of the co-unit map* $u(v(G)) \to G$ *are the unit map* $p^*F \to v(u(p^*F))$ *and the co-unit map* $u(v(p^*G)) \to p^*G$*.*

Proof Corollary 5.7.7 ensures that the property of local $\mathbf{U}$-smallness is preserved by the functor $\underline{\mathrm{Hom}}(X, -)$. The natural invertible map

$$\mathrm{Hom}_B(u(x), y) \xrightarrow{\sim} \mathrm{Hom}_A(x, v(y))$$

produces an invertible map from $u^* h_B$ to $h_A v$. On the other hand, the last assertion of Proposition 6.1.14 gives us an invertible natural transformation from $h_B u$ to $u_! h_A$. This means that we can see the adjunction between u and v as a restriction (along the Yoneda embedding) of the canonical adjunction between $u_!$ and u^*. Corollaries 5.8.14 and 6.1.17 thus imply that there is a functorial invertible map

$$\mathrm{Hom}(u(F), G) \xrightarrow{\sim} \mathrm{Hom}(F, v(G))$$

for any functors $F\colon X \to A$ and $G\colon X \to B$. The assertion about the compatibility of unit maps comes from the explicit description of the unit maps of the adjunction between $u_!$ and u^* provided by Proposition 6.1.14. Applying what precedes to the pair $(v^{\mathrm{op}}, u^{\mathrm{op}})$ also gives the assertion about co-unit maps. □

Theorem 6.1.23 *Let* $u\colon A \to B$ *and* $v\colon B \to A$ *be two functors between* $\mathbf{U}$*-small* ∞*-categories. The following conditions are equivalent.*

(i) There exists an adjunction of the form (u, v, c)*.*
(ii) There exists an invertible map $h_A v \to u^*(h_B)$ *in* $\underline{\mathrm{Hom}}(B, \underline{\mathrm{Hom}}(A^{\mathrm{op}}, \mathcal{S}))$*.*
(iii) There exists a functorial map $\varepsilon_b\colon u(v(b)) \to b$ *such that* (b, ε_b) *defines a final object of* A/b *for any object* b *of* B*.*
(iv) For any simplicial set X*, we have an adjunction of the form*

$$ho(u)\colon ho(\underline{\mathrm{Hom}}(X, A)) \rightleftarrows ho(\underline{\mathrm{Hom}}(X, B)) : ho(v),$$

functorially in X*.*

(v) *There exist two natural transformations* $\varepsilon : uv \to 1_B$ *and* $\eta : 1_A \to vu$, *as well as two commutative triangles*

$$\begin{array}{ccc} u & \xrightarrow{1_u} & u \\ & {\scriptstyle u\eta}\searrow \quad \nearrow{\scriptstyle \varepsilon_u} & \\ & uvu & \end{array} \qquad \textit{and} \qquad \begin{array}{ccc} v & \xrightarrow{1_v} & v \\ & {\scriptstyle \eta_v}\searrow \quad \nearrow{\scriptstyle v\varepsilon} & \\ & vuv & \end{array}$$

in $\underline{\mathrm{Hom}}(A, B)$ *and in* $\underline{\mathrm{Hom}}(B, A)$, *respectively.*

(vi) *The functor* $\mathbf{R}(u\times 1_C)^* : RFib(B\times C) \to RFib(A\times C)$ *is a left adjoint of the functor* $\mathbf{R}(v\times 1_C)^* : RFib(A\times C) \to RFib(B\times C)$, *for any* ∞*-category* C, *functorially in* C.

(vii) *There is an invertible natural transformation from* $v_!$ *to* u^* *in the* ∞*-category* $\underline{\mathrm{Hom}}(\underline{\mathrm{Hom}}(A^{\mathrm{op}}, \mathcal{S}), \underline{\mathrm{Hom}}(B^{\mathrm{op}}, \mathcal{S}))$.

Proof The fact that conditions (i) and (ii) are equivalent to each other is essentially a reformulation of the definition. The Yoneda lemma implies that a map $\varepsilon_b : u(v(b)) \to b$ can be seen as map from $h_A(v(b))$ to $u^*(h_B(b))$. Saying that (b, ε_b) defines a final object of A/b means that the corresponding map $h_A(v(b)) \to u^*(h_B(b))$ is invertible. Therefore, the functoriality of the Yoneda lemma shows that conditions (ii) and (iii) are equivalent.

The preceding theorem and Remark 6.1.5 show that condition (i) implies condition (iv). If condition (iv) holds, for any functors $F : X \to A$ and $G : X \to B$, we have commutative triangles of the form

$$\begin{array}{ccc} u(F) & \xrightarrow{1_{u(F)}} & u(F) \\ & {\scriptstyle u\eta_F}\searrow \quad \nearrow{\scriptstyle \varepsilon_{u(F)}} & \\ & uvu(F) & \end{array} \qquad \text{and} \qquad \begin{array}{ccc} v(G) & \xrightarrow{1_{v(G)}} & v(G) \\ & {\scriptstyle \eta_{v(G)}}\searrow \quad \nearrow{\scriptstyle v\varepsilon_G} & \\ & vuv(G) & \end{array}$$

in $\underline{\mathrm{Hom}}(X, B)$ and in $\underline{\mathrm{Hom}}(X, A)$, respectively. In the case where $X = A$ and $F = 1_A$ or where $X = B$ and $F = 1_B$, this proves condition (v).

Let us prove that condition (v) implies condition (vi). Condition (v) translates into analogous conditions for the functors

$$u^* : \underline{\mathrm{Hom}}(B^{\mathrm{op}}, \mathcal{S}) \to \underline{\mathrm{Hom}}(A^{\mathrm{op}}, \mathcal{S}) \quad \text{and} \quad v^* : \underline{\mathrm{Hom}}(A^{\mathrm{op}}, \mathcal{S}) \to \underline{\mathrm{Hom}}(B^{\mathrm{op}}, \mathcal{S}) .$$

Indeed, the map η induces a map $\eta^* : u^* v^* \to 1$ and the map ε induces a map $\varepsilon^* : 1 \to v^* u^*$. Therefore, after applying the Boardman–Vogt construction, this gives rise to an adjunction

$$u^* : ho(\underline{\mathrm{Hom}}(B^{\mathrm{op}}, \mathcal{S})) \rightleftarrows ho(\underline{\mathrm{Hom}}(A^{\mathrm{op}}, \mathcal{S})) : v^*$$

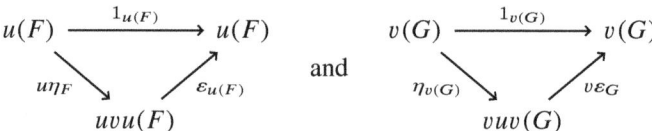

in the usual sense. Since we may enlarge the universe at will, using the equivalences $RFib(X) \simeq LFib(X^{op})$, Theorem 5.4.5 ensures that we obtain an adjunction of the following form:

$$\mathbf{R}u^* : RFib(B) \rightleftarrows RFib(A) : \mathbf{R}v^* .$$

Since condition (iv) for the pair (u, v) implies condition (iv) for the pair $(u \times 1_C, v \times 1_C)$ for any ∞-category C, this shows that condition (v) implies condition (vi); note that, in the formulation of (v), no universe is fixed, so that we may always enlarge the universe and assume that C is $\mathbf{U}$-small.

Let us prove that condition (vi) implies condition (vii). By virtue of Theorem 5.4.5, since pulling back preserves the size of fibres, condition (vi) implies that, for any ∞-category C, the functor

$$ho(u \times 1_C)^* : ho(\underline{\mathrm{Hom}}(B^{op} \times C^{op}, \mathcal{S})) \to ho(\underline{\mathrm{Hom}}(A^{op} \times C^{op}, \mathcal{S}))$$

is a left adjoint of the functor

$$ho(v \times 1_C)^* : ho(\underline{\mathrm{Hom}}(A^{op} \times C^{op}, \mathcal{S})) \to ho(\underline{\mathrm{Hom}}(B^{op} \times C^{op}, \mathcal{S})),$$

functorially in C. In particular, the functor u^* preserves representable presheaves. This means that u has a right adjoint w, by Proposition 6.1.11. Since we already know that condition (i) implies condition (vi), we see that the functor $(u \times 1_C)^*$ is left adjoint to $(w \times 1_C)^*$. Hence, the functors w^* and v^* are J-homotopic: they induce the same map from the set of isomorphism classes of functors from C^{op} to $\underline{\mathrm{Hom}}(A^{op}, \mathcal{S})$ to the set of isomorphism classes of functors from C^{op} to $\underline{\mathrm{Hom}}(B^{op}, \mathcal{S})$, for any ∞-category C. Therefore, there are invertible maps $v_! \simeq w_! \simeq u^*$, by Propositions 6.1.14, 6.1.7 and 6.1.9.

Finally, if condition (vii) holds, since $v_! h_B$ is canonically equivalent to $h_A v$, we get condition (ii), whence condition (i). □

Remark 6.1.24 Let $u \colon X \to A$ be a right fibration between $\mathbf{U}$-small simplicial sets. We define a functor

$$u_\sharp \colon \underline{\mathrm{Hom}}(X^{op}, \mathcal{S}) \to \underline{\mathrm{Hom}}(A^{op}, \mathcal{S}) \tag{6.1.24.1}$$

as follows. We choose a functor $\Phi \colon A^{op} \to \mathcal{S}$ which classifies the left fibration $u^{op} \colon X^{op} \to A^{op}$; in particular, we now have specified pull-backs of u along any map of codomain A. Given a simplicial set K and a functor $F \colon K \to \underline{\mathrm{Hom}}(X^{op}, \mathcal{S})$ classifying a right fibration $p \colon Y \to K^{op} \times X$, we define $u_\sharp(F)$ as the functor which classifies the composed right fibration

$$q \colon Y \xrightarrow{p} K^{op} \times X \xrightarrow{1_{K^{op}} \times u} K^{op} \times A \tag{6.1.24.2}$$

(we observe that this is well defined, since the specified pull-backs of p and

$1_{K^{\mathrm{op}}} \times u$ determine pull-backs of q). There is a canonical map

(6.1.24.3) $$F \to u^*(u_\sharp(F))$$

induced by the canonical map $Y \to X \times_A Y$ via Theorem 5.4.5. Similarly, for any functor $G\colon K^{\mathrm{op}} \times A \to \mathcal{S}$ classifying a right fibration $q\colon Z \to K^{\mathrm{op}} \times A$, there is a canonical map

(6.1.24.4) $$u_\sharp(u^*(G)) \to G$$

corresponding to the projection $X \times_A Z \to Z$. Note that both maps (6.1.24.3) and (6.1.24.4) are natural transformations, because they are well defined for every simplicial set K. Using Theorem 5.4.5, we see right away that these natural transformations satisfy condition (v) of the previous theorem. Therefore, the functor $u_\sharp$ introduced above is canonically the left adjoint of the functor u^*. Hence $u_\sharp \simeq u_!$.

6.2 Limits and Colimits

6.2.1 Let A be an ∞-category, and I a simplicial set. We write

(6.2.1.1) $$c_I\colon A \to \underline{\mathrm{Hom}}(I, A).$$

The functor associates to an object a of A the constant functor from I to A with value a (i.e. c_I is the functor defined by right composition with the map $I \to \Delta^0$). Given a functor $F\colon I \to A$, we can form the slice category $\underline{\mathrm{Hom}}(I, A)/F$, and we define the ∞-category A/F by forming the pull-back below.

(6.2.1.2) $$\begin{array}{ccc} A/F & \longrightarrow & \underline{\mathrm{Hom}}(I, A)/F \\ \downarrow & & \downarrow \\ A & \xrightarrow{c_I} & \underline{\mathrm{Hom}}(I, A) \end{array}$$

An object of A/F is characterised by a pair (a, λ), where a is an object of A, while $\lambda\colon c_I(a) \to F$ is a natural transformation (i.e. a map in $\underline{\mathrm{Hom}}(I, A)$). Indeed, such a λ is a morphism of simplicial sets $\Delta^1 = \Delta^0 * \Delta^0 \to \underline{\mathrm{Hom}}(I, A)$ which sends the final object of Δ^1 to F, and thus determines an object of the slice $\underline{\mathrm{Hom}}(I, A)/F$ whose image in $\underline{\mathrm{Hom}}(I, A)$ is the image by c_I of the source of λ, namely a. Such a pair (a, λ) will be called a *cone* of F.

Definition 6.2.2 Let A be an ∞-category, and I a simplicial set.

A *limit* of a functor $F\colon I \to A$ is a cone $(\varprojlim F, \lambda)$ whose corresponding object in A/F is final. In this case, we shall also say that the natural transformation λ exhibits the object $\varprojlim F$ as a limit of F.

A *colimit* of a functor $F\colon I \to A$ is a pair $(\varinjlim F, \lambda)$ such that $(\varinjlim F, \lambda^{\mathrm{op}})$ is a limit of the functor $F^{\mathrm{op}}\colon I^{\mathrm{op}} \to A^{\mathrm{op}}$. As above, we shall say that λ exhibits the object $\varinjlim F$ as a colimit of F.

Notation 6.2.3 We shall sometimes write

$$\varprojlim_{i\in I} F_i \quad (\text{or} \quad \varinjlim_{i\in I} F_i)$$

for the projective limit (or the injective limit, respectively) of a functor $F\colon I \to A$.

Remark 6.2.4 Since the final objects of any ∞-category C form either the empty ∞-category or a contractible ∞-groupoid, given two limits (L, λ) and (L', λ') of F, the ∞-groupoid of maps from (L, λ) to (L', λ') in A/F form a contractible ∞-groupoid. In other words (L, λ) and (L', λ') are canonically equivalent, as in ordinary category theory.

Remark 6.2.5 If we choose a universe $\mathbf{U}$ such that both A and $\underline{\mathrm{Hom}}(I, A)$ are locally small, then giving a limit of a functor $F\colon I \to A$ is equivalent to producing an invertible map

$$\mathrm{Hom}(c_I(a), F) \xrightarrow{\sim} \mathrm{Hom}(a, \varprojlim F)$$

functorially in a.

Definition 6.2.6 Let I be a simplicial set. An ∞-category A has *limits of type I* (*colimits of type I*) if any functor $F\colon I \to A$ has a limit (a colimit, respectively).

Given a universe $\mathbf{U}$, an ∞-category is $\mathbf{U}$-*complete* ($\mathbf{U}$-*cocomplete*) if it has limits (colimits, respectively) of type I for any $\mathbf{U}$-small simplicial set I. If there is no ambiguity about the universe $\mathbf{U}$, we will simply say that such an ∞-category is complete (cocomplete).

Remark 6.2.7 Using Propositions 6.1.6 and 6.1.8, one can see that, if $I \to J$ is a weak categorical equivalence, then an ∞-category A has limits (or colimits) of type I if and only if it has limits (or colimits) of type J. In particular, the notion of complete (or cocomplete) ∞-category can be defined in terms of ∞-categories only.

Example 6.2.8 Limits of type $\varnothing$ in an ∞-category A simply are the final objects of A. Indeed, there is only one functor $F\colon \varnothing \to A$, so that we have

$A/F = A$ (because $\underline{\mathrm{Hom}}(\varnothing, A) \simeq \Delta^0$). Hence the final objects of A/F are the final objects of A. Similarly, colimits of type $\varnothing$ are the initial objects.

Proposition 6.2.9 *Let A be an ∞-category and I a simplicial set. The constant functor*

$$c_I : A \to \underline{\mathrm{Hom}}(I, A)$$

has a right adjoint if and only if A has limits of type I. If this is the case this right adjoint sends each functor $F : I \to A$ to its limit, and we denote it by

$$\varprojlim : \underline{\mathrm{Hom}}(I, A) \to A\,.$$

Proof This is a particular case of Proposition 6.1.11 in the case where $B = \underline{\mathrm{Hom}}(I, A)$ and $u = c_I$. □

Corollary 6.2.10 *Let A be an ∞-category, and I a simplicial set. If A has limits of type I, then so does $\underline{\mathrm{Hom}}(X, A)$ for any simplicial set X. Furthermore, limits are computed fibrewise: for any functor $F : I \to \underline{\mathrm{Hom}}(X, A)$, the limit of I evaluated at an object x of X is canonically equivalent to the limit of $F_x : I \to A$.*

Proof This is a direct consequence of the formula

$$\underline{\mathrm{Hom}}(I, \underline{\mathrm{Hom}}(X, A)) \simeq \underline{\mathrm{Hom}}(X, \underline{\mathrm{Hom}}(I, A))$$

and of the preceding proposition, by Theorem 6.1.22. □

Definition 6.2.11 Let $\mathbf{U}$ be a universe, and A a $\mathbf{U}$-small simplicial set. We write $\widehat{A} = \underline{\mathrm{Hom}}(A^{\mathrm{op}}, \mathcal{S})$ for the category of *presheaves on A* (with values in the ∞-category $\mathcal{S}$ of ∞-groupoids).

Proposition 6.2.12 *Let $\mathbf{U}$ be a universe, and A a $\mathbf{U}$-small simplicial set. Then the ∞-category $\widehat{A}$ is cocomplete.*

Proof The corollary above implies that it is sufficient to check that $\mathcal{S}$ is cocomplete. And by virtue of the preceding proposition, this is a direct consequence of Proposition 6.1.14. □

An important example of colimit is the following one.

Proposition 6.2.13 *Let A be a $\mathbf{U}$-small ∞-category. The colimit of the Yoneda embedding $h_A : A \to \widehat{A}$ is the final object of $\widehat{A}$ (i.e. the constant presheaf with value the one-point ∞-groupoid).*

Proof The Yoneda embedding corresponds to the functor Hom_A, which classifies the left fibration $S(A) \to A^{op} \times A$. On the other hand, there is a canonical Cartesian square of the form

$$\begin{array}{ccc} S(A) & \xrightarrow{w} & W \\ {\scriptstyle (s_A, t_A)}\downarrow & & \downarrow{\scriptstyle q} \\ A^{op} \times A & \xrightarrow{1_{A^{op}} \times h_A} & A^{op} \times \widehat{A} \end{array}$$

in which q is the left fibration classified by the evaluation functor from $A^{op} \times \widehat{A}$ to $\mathcal{S}$, and w is cofinal; see Lemma 5.8.11. Let $p\colon A^{op} \times A \to A^{op}$ denote the first projection. Then we have an essentially commutative diagram of the form

$$\begin{array}{ccccc} ho(\underline{\mathrm{Hom}}(A^{op} \times A, \mathcal{S})) & \xrightarrow{\sim} & ho(\underline{\mathrm{Hom}}(A, \widehat{A})) & \xrightarrow{ho(\underrightarrow{\lim})} & ho(\widehat{A}) \\ \downarrow & & & & \downarrow \\ LFib(A^{op} \times A) & & \xrightarrow{\mathbf{L}p_!} & & LFib(A^{op}) \end{array}$$

in which the vertical maps are fully faithful functors obtained by applying Theorem 5.4.5 a couple of times. In this picture, the colimit of h_A in $\widehat{A}$ is identified with the image of $S(A)$ by the functor $\mathbf{L}p_!$. On the other hand, since w is cofinal, this coincides with the image of W by $\mathbf{L}\hat{p}_!$, where $\hat{p}$ is the projecion from $A^{op} \times \widehat{A}$ to A^{op}. But the functor corresponding to the evaluation functor is the identity of $\widehat{A}$. In other words, the colimit of h_A is canonically isomorphic to the colimit of the identity of $\widehat{A}$, which is thus the evaluation of the identity at the final presheaf, by Lemma 4.4.22. □

Definition 6.2.14 Let $u\colon A \to B$ be a functor between ∞-categories, and I a simplicial set. We say that u *commutes with limits of type* I if, for any functor $F\colon I \to A$, the induced functor $A/F \to B/u(F)$ preserves final objects (where $u(F)$ is the functor $i \mapsto u(F(i))$).

We say that u *commutes with colimits of type* I if u^{op} commutes with limits of type I^{op}.

Proposition 6.2.15 *If a functor has a left adjoint, it commutes with all limits.*

Proof Let $v\colon B \to A$ be a functor between ∞-categories, and I a simplicial set. If $u\colon A \to B$ is a left adjoint of v, then the functor $u\colon \underline{\mathrm{Hom}}(I, A) \to \underline{\mathrm{Hom}}(I, B)$ is a left adjoint of the functor $v\colon \underline{\mathrm{Hom}}(I, B) \to \underline{\mathrm{Hom}}(I, A)$ functorially in I; see Theorem 6.1.22. We may choose a universe $\mathbf{U}$ such that A, B and I are $\mathbf{U}$-small. In particular, for any functor $F\colon I \to B$ equipped with a limit $\underleftarrow{\lim} F$

in B, there are functorial equivalences

$$\begin{aligned}\mathrm{Hom}(a, v(\varprojlim F)) &\xleftarrow{\sim} \mathrm{Hom}(u(a), \varprojlim F)\\ &\xleftarrow{\sim} \mathrm{Hom}(c_I(u(a)), F)\\ &= \mathrm{Hom}(u(c_I(a)), F)\\ &\xrightarrow{\sim} \mathrm{Hom}(c_I(a), v(F))\end{aligned}$$

which exhibit $v(\varprojlim F)$ as a limit of $v(F)$. □

Corollary 6.2.16 *Let A be a $\mathbf{U}$-small simplicial set and $F\colon A^{\mathrm{op}} \to \mathcal{S}$ a functor which classifies a right fibration with $\mathbf{U}$-small fibres $p\colon X \to A$. There is a canonical invertible map in $\widehat{A}$ of the form*

$$\varinjlim_{x \in X} h_A(p(x)) = \varinjlim h_A p \xrightarrow{\sim} F\,.$$

Proof By virtue of Proposition 6.1.14, the functor

$$p^*\colon \widehat{A} \to \widehat{X}$$

has a left adjoint $p_!$, whose restriction along the Yoneda embedding of X coincides with the composed functor $h_A p$. Moreover, the presheaf F is canonically identified with the image of the final object e by $p_!$. On the other hand, Proposition 6.2.13 tells us that the colimit of h_X in $\widehat{X}$ is the final object e. We thus have canonical maps:

$$\varinjlim_{x \in X} h_A(p(x)) \xrightarrow{\sim} \varinjlim p_! h_X \to p_!(\varinjlim h_X) \xrightarrow{\sim} p_!(e) = F\,.$$

To finish the proof, it is thus sufficient to justify that the middle map above is invertible, which follows from the fact that $p_!$ commutes with colimits, since it has a right adjoint p^*. □

A very useful source of (co)limits is provided by the following principle.

Proposition 6.2.17 *Let $u\colon A \to B$ be a functor between ∞-categories equipped with a fully faithful right adjoint $v\colon B \to A$, and $F\colon I \to B$ be a functor. If $v(F)$ has a limit (or a colimit) in A, then F has a limit (or a colimit) in B, which is nothing other than the image by u of the limit (or of the colimit, respectively) of $v(F)$. In particular, if A is complete (or cocomplete), so is B.*

Proof If $v(F)\colon I \to A$ has a colimit, then the dual version of the preceding proposition tells us that $u(v(F))\colon I \to B$ has a colimit, which is nothing other than the image by u of the colimit of $v(F)$. But, by virtue of Proposition 5.7.4, the functor

$$v\colon \underline{\mathrm{Hom}}(I, B) \to \underline{\mathrm{Hom}}(I, A)$$

is fully faithful, so that we have functorial invertible maps of the form

$$\mathrm{Hom}(X, F) \xrightarrow{\sim} \mathrm{Hom}(v(X), v(F)) \xleftarrow{\sim} \mathrm{Hom}(X, u(v(F)))\,.$$

Applying the Yoneda lemma to the ∞-category $\underline{\mathrm{Hom}}(I, B)$, this shows that the canonical map $u(v(F)) \to F$ is invertible. In particular, any colimit of $u(v(F))$ provides a colimit of F.

If $v(F)$ has a limit in A, then $u(\varprojlim v(F))$ is a limit of F in B. The proof will require a few preliminary steps, though. Let $x \to y$ be a map of A such that $u(x) \to u(y)$ is invertible in B. Then the induced map

$$c_I(u(x)) = u(c_I(x)) \to u(c_I(y)) = c_I(u(y))$$

is invertible in $\underline{\mathrm{Hom}}(I, B)$. Therefore, we have the following commutative diagram in $\mathcal{S}$ (up to a choice of a suitable universe, so that it makes sense):

$$\begin{array}{ccc}
\mathrm{Hom}(y, \varprojlim v(F)) & \longrightarrow & \mathrm{Hom}(x, \varprojlim v(F)) \\
\wr\uparrow & & \wr\uparrow \\
\mathrm{Hom}(c_I(y), v(F)) & \longrightarrow & \mathrm{Hom}(c_I(x), v(F)) \\
\wr\uparrow & & \wr\uparrow \\
\mathrm{Hom}(u(c_I(y)), F) & \xrightarrow{\sim} & \mathrm{Hom}(u(c_I(x)), F)
\end{array}$$

which shows, by the Yoneda lemma, that the upper horizontal map above is invertible in $\mathcal{S}$. Now, if an object a of A has the property that, for any map $x \to y$ in A whose image by u is invertible, the induced map

$$\mathrm{Hom}(y, a) \to \mathrm{Hom}(x, a)$$

is invertible, then the canonical unit map $\eta_a : a \to v(u(a))$ is invertible. Indeed, this implies that there exists a map $r : vu(a) \to a$ which is a retraction of η_a in $ho(A)$. By naturality of the unit map η, this means that there is a commutative diagram of the form

$$\begin{array}{ccccc}
a & \xrightarrow{\eta_a} & v(u(a)) & \xrightarrow{r} & a \\
\downarrow{\scriptstyle \eta_a} & & \downarrow{\scriptstyle \eta_{v(u(a))}} & & \downarrow{\scriptstyle \eta_a} \\
v(u(a)) & \xrightarrow{v(u(\eta_a))} & v(u(v(u(a)))) & \xrightarrow{v(u(r))} & v(u(a))
\end{array}$$

which turns η_a into a retract of $\eta_{v(u(a))}$ in $ho(A)$. Since v is fully faithful, the map $\eta_{v(u(a))}$ is an isomorphism, and this shows that η_a is invertible.

The image of the canonical map $\varprojlim v(F) \to v(u(\varprojlim v(F)))$ by u is invertible: indeed, the image of this map by u is a section of the canonical map

$$u(v(u(\varprojlim v(F)))) \to u(\varprojlim v(F)),$$

and the latter is invertible because $ho(v)\colon ho(B) \to ho(A)$ is fully faithful. Therefore, this map is invertible. Finally, we have the following invertible maps, functorially in b:

$$\begin{aligned}
\operatorname{Hom}(c_I(b), F) &\xrightarrow{\sim} \operatorname{Hom}(v(c_I(b)), v(F)) \\
&= \operatorname{Hom}(c_I(v(b)), v(F)) \\
&\xrightarrow{\sim} \operatorname{Hom}(v(b), \varprojlim v(F)) \\
&\xrightarrow{\sim} \operatorname{Hom}(v(b), v(u(\varprojlim v(F)))) \\
&\xleftarrow{\sim} \operatorname{Hom}(b, u(\varprojlim v(F))).
\end{aligned}$$

This proves that $u(\varprojlim v(F))$ is a limit of F in B, as required. □

Lemma 6.2.18 *Let $p\colon X \to Y$ be an isofibration between ∞-categories. For any object x of X and $y = p(x)$, the induced functor $X/x \to Y/y$ is an isofibration.*

Proof Since the functor $Y/y \to Y$ is a right fibration, hence is conservative, a lifting problem of the form

$$\begin{array}{ccc}
\{1\} & \longrightarrow & X/x \\
\big\downarrow & \overset{h}{\nearrow} & \big\downarrow \\
\Delta^1 & \xrightarrow{g} & Y/y
\end{array}$$

with g invertible in Y (and h invertible in X/x) corresponds to a lifting problem of the form

$$\begin{array}{ccc}
\Delta^1 & \xrightarrow{f} & X \\
\big\downarrow{\scriptstyle \delta^2_0} & \overset{H}{\nearrow} & \big\downarrow{\scriptstyle p} \\
\Delta^2 & \xrightarrow{G} & Y
\end{array}$$

with $g = G|_{\Delta^{\{0,1\}}}$ (and $h = H|_{\Delta^{\{0,1\}}}$) invertible. Since p is an isofibration, we may find an invertible map $h'\colon x_0 \to x_1$, where x_1 is the domain of f, such that $p(h') = g$. Since p is also an inner fibration, the induced map $(h', f)\colon \Lambda^2_1 \to X$ is then the restriction of some map $H\colon \Delta^2 \to X$ such that $pH = G$. □

Proposition 6.2.19 *We consider a Cartesian square in the category of ∞-categories of the form*

$$\begin{array}{ccc}
X' & \xrightarrow{u} & X \\
\big\downarrow{\scriptstyle p'} & & \big\downarrow{\scriptstyle p} \\
Y' & \xrightarrow{v} & Y
\end{array}$$

in which p is an isofibration, as well as a functor $F\colon A \to X'$. We assume that uF has a limit in X whose image by p is a limit of puF, and that $p'F$ has a limit in Y whose image by v is a limit of $vp'F$. Then F has a limit in X' whose image by u is a limit of uF and whose image by p' is a limit of $p'F$.

Proof With the slice construction of (6.2.1.2), we see that we have an induced Cartesian square of the form

$$\begin{array}{ccc} X'/F & \longrightarrow & X/uF \\ \downarrow & & \downarrow \\ Y'/p'F & \longrightarrow & Y/puF \end{array}$$

(simply because right adjoints commute with pull-backs). Furthermore, the functor $X/uF \to Y/puF$ is still an isofibration: by construction, we have a Cartesian square of the form

$$\begin{array}{ccc} X/uF & \longrightarrow & \underline{\mathrm{Hom}}(A,X)/uF \\ \downarrow & & \downarrow \\ Y/puF & \longrightarrow & \underline{\mathrm{Hom}}(A,Y)/puF \end{array}$$

in which the vertical map of the right-hand side is an isofibration, by the previous lemma. Therefore, replacing X' by X'/F, X by X/uF and so forth, we see that we may assume $A = \varnothing$. In other words, we have to prove that, if Y' has a final object y' and X has a final object x, such that both $v(y')$ and $p(x)$ are final objects of Y, then there is a final object x' of X' such that $u(x')$ is a final object of X and $p'(x')$ is a final object of Y'. Since $v(y')$ and $p(x)$ are final objects, there exists an invertible map $g\colon v(y') \to p(x)$. The functor p being an isofibration, there exists an invertible map $f\colon x_0 \to x$ such that $p(f) = g$. In particular, x_0 is also a final object of X, and, replacing x by x_0, we may assume, without loss of generality, that $p(x) = v(y')$. There is then a unique object x' in X' such that $u(x') = x$ and $p'(x') = y'$. It remains to check that x' is a final object of X'. Using condition (iv) of Theorem 4.3.11, we see that it is sufficient to prove that the canonical functor $X'/x' \to X'$ is an equivalence of ∞-categories. But we have a canonical Cartesian square

$$\begin{array}{ccc} X'/x' & \longrightarrow & X/x \\ \downarrow & & \downarrow \\ Y'/y' & \longrightarrow & Y/y \end{array}$$

in which the vertical map of the right-hand side is an isofibration between ∞-categories (by the previous lemma), whence a fibration of the Joyal model category structure. Therefore, this is a homotopy Cartesian square. Since X' is also the homotopy pull-back of X and Y' over Y, it is sufficient to observe that each of the functors $Y'/y' \to Y'$, $X/x \to X$ and $Y/y \to Y$ are trivial fibrations (by Theorem 4.3.11). □

Corollary 6.2.20 *Let*

$$\begin{array}{ccc} X' & \xrightarrow{u} & X \\ \downarrow{\scriptstyle p'} & & \downarrow{\scriptstyle p} \\ Y' & \xrightarrow{v} & Y \end{array}$$

be a Cartesian square in the category of ∞-categories, with p an isofibration. We also assume that there is a simplicial set I such that X, Y and Y′ have limits of type I and that both functors p and v commute with limits of type I. Then X′ has limits of type I and both u and p′ commute with limits of type I.

6.3 Extensions of Functors by Colimits

We fix a universe of sets **U**.

6.3.1 We consider a **U**-small ∞-category A as well as an ∞-category C with **U**-small colimits (but we do not require C to be locally **U**-small). For convenience, we choose a universe **V**, containing **U**, such that both $\widehat{A}$ and C are **V**-small. As usual, we let $\mathcal{S}$ and $\mathcal{S}'$ denote the ∞-categories of **U**-small ∞-groupoids and of **V**-small ∞-groupoids, respectively, so that we have a full embedding $\mathcal{S} \subset \mathcal{S}'$. We consider a functor

$$u \colon A \to C \tag{6.3.1.1}$$

and we want to extend it by colimits to a functor defined on the category of presheaves on A.

Proposition 6.3.2 *Let $\widehat{C}$ be the full subcategory of $\underline{\mathrm{Hom}}(C^{\mathrm{op}}, \mathcal{S}')$ whose objects are* **U***-small presheaves on C. The Yoneda embedding $h_C \colon C \to \widehat{C}$ has a left adjoint*

$$L \colon \widehat{C} \to C\,.$$

Proof Let $F \colon C^{\mathrm{op}} \to \mathcal{S}'$ be a **U**-small presheaf. It classifies a left fibration with **V**-small fibres $p^{\mathrm{op}} \colon Y^{\mathrm{op}} \to C^{\mathrm{op}}$, and there exist a **U**-small simplicial set X as well as a final map $i \colon X \to Y$. If we put $q = pi$, we know that F is the image by $q_!$ of the final object e. In other words, for any presheaf $G \colon C^{\mathrm{op}} \to \mathcal{S}'$, there is a functorial invertible map of the following form:

$$\mathrm{Hom}(F, G) \xrightarrow{\sim} \mathrm{Hom}(q_!(e), G)\,.$$

We may assume that X is an ∞-category. Indeed, we can choose an inner anodyne extension $j \colon X \to X'$ with $\mathbf{U}$-small codomain X', and since Y is an ∞-category, there exists a map $i' \colon X' \to Y$ such that $i'j = i$. The map i' is also final, by Corollary 5.3.2. Therefore, we may replace X by X', and assume, without loss of generality, that X is an ∞-category. In this case, e is the colimit of h_X in $\widehat{X}$. Since the functor $q_!$ commutes with colimits, we thus have functorial invertible maps:

$$\mathrm{Hom}(F, G) \xrightarrow{\sim} \mathrm{Hom}(\varinjlim_{x \in X} q_!(h_X(x)), G) \xrightarrow{\sim} \mathrm{Hom}(\varinjlim_{x \in X} h_C(q(x)), G)\,.$$

In the case where $G = h_C(y)$ is represented by an object y of C, we have

$$\mathrm{Hom}(F, h_C(y)) \xrightarrow{\sim} \mathrm{Hom}(\varinjlim_{x \in X} h_C(q(x)), h_C(y)) \xrightarrow{\sim} \mathrm{Hom}(h_C q, h_C(y)_X)\,,$$

where, given an object i in an ∞-category I, $i_X = c_X(i)$ is the constant diagram $X \to I$ with value i. On the other hand, since the Yoneda embedding h_C is fully faithful, we have a canonical invertible map

$$\mathrm{Hom}(q, y_X) \xrightarrow{\sim} \mathrm{Hom}(h_C q, h_C(y)_X)\,.$$

Since C has $\mathbf{U}$-small colimits, if we put $L(F) = \varinjlim q$, we see that we produce an invertible map

$$\mathrm{Hom}(L(F), y) \xrightarrow{\sim} \mathrm{Hom}(F, h_C(y))$$

functorially in y. Proposition 6.1.11 thus shows that there is a way to extend the assignment $F \mapsto L(F)$ to a left adjoint of $h_C \colon C \to \widehat{C}$. □

Proposition 6.3.3 *The functor $u_!$ restricts to a functor*

$$u_! : \widehat{A} \to \widehat{C}$$

(i.e. $u_!$ sends presheaves on A with values in $\mathbf{U}$-small ∞-groupoids to $\mathbf{U}$-small presheaves on C) which commutes with $\mathbf{U}$-small colimits. If we assume furthermore that C is locally $\mathbf{U}$-small, then the functor u^* restricts to a functor

$$u^* : \widehat{C} \to \widehat{A}$$

(where $\widehat{A} = \underline{\mathrm{Hom}}(A^{\mathrm{op}}, \mathcal{S})$ is the ∞-category of presheaves on A with $\mathbf{U}$-small fibres).

Proof The functor

$$u_! \colon \underline{\mathrm{Hom}}(A^{\mathrm{op}}, \mathcal{S}') \to \underline{\mathrm{Hom}}(C^{\mathrm{op}}, \mathcal{S}')$$

sends $\widehat{A}$ into $\widehat{C}$: if $X \to A$ is the right fibration classified by a presheaf $F \colon A^{\mathrm{op}} \to \mathcal{S}$, and if $Y \to C$ is the one classified by $u_!(F)$, then the associated functor $X \to Y$ is final, while X is $\mathbf{U}$-small. The fact that $u_! \colon \widehat{A} \to \widehat{C}$ commutes with $\mathbf{U}$-small colimits comes from the fact that the inclusion $\mathcal{S} \to \mathcal{S}'$ commutes with $\mathbf{U}$-small colimits (because of the explicit description of colimits given by Proposition 6.1.14), which implies that the functor

$$\underline{\mathrm{Hom}}(A^{\mathrm{op}}, \mathcal{S}) \subset \underline{\mathrm{Hom}}(A^{\mathrm{op}}, \mathcal{S}') \xrightarrow{u_!} \underline{\mathrm{Hom}}(C^{\mathrm{op}}, \mathcal{S}')$$

commutes with $\mathbf{U}$-small colimits.

In the case where C is locally $\mathbf{U}$-small, by virtue of Proposition 6.1.21 and by the Yoneda lemma, we know that $\mathbf{U}$-small presheaves have $\mathbf{U}$-small fibres. On the other hand, by Proposition 5.4.9 and Corollary 5.4.10, the full subcategory of $\underline{\mathrm{Hom}}(A^{\mathrm{op}}, \mathcal{S}')$ which consists of functors with $\mathbf{U}$-small fibres is equivalent to $\widehat{A}$. Therefore, the functor u^* sends $\widehat{C}$ to $\widehat{A}$. □

We finally reach the point where Kan's theorem (Theorem 1.1.10) holds for ∞-categories as well: we can now extend functors by colimits as follows.

Theorem 6.3.4 *The composed functor*

$$Lu_! \colon \widehat{A} \to C$$

commutes with $\mathbf{U}$*-small colimits, and there is a canonical invertible natural transformation*

$$u \xrightarrow{\sim} Lu_! h_A \,.$$

Furthermore, if C is locally $\mathbf{U}$*-small, then the functor $Lu_!$ has a right adjoint, namely the functor*

$$\begin{aligned} u^* h_C \colon C &\to \widehat{A} \\ x &\mapsto \mathrm{Hom}_C(u(-), x) \,. \end{aligned}$$

Proof The functor $u_!$ commutes with $\mathbf{U}$-small colimits, by the previous proposition, and so does the functor L, since it is a left adjoint. Therefore, the composed functor $Lu_!$ commutes with $\mathbf{U}$-small colimits. In the case where C is locally $\mathbf{U}$-small, the fact that $Lu_!$ is a left adjoint of $u^* h_C$ comes from the previous two propositions and from the compatibility of adjunctions with composition of functors; see Proposition 6.1.8. Finally, we have an invertible natural transformation from $h_C u$ to $u_! h_A$, and, since h_C is fully faithful, the canonical map from Lh_C to the identity of C is invertible, which provides an invertible map $Lh_C u \xrightarrow{\sim} u$. Choosing an inverse to the latter ends the proof. □

Corollary 6.3.5 *For any locally* **U***-small ∞-category C, for any object X of C, the functor*

$$\mathrm{Hom}_C(X, -)\colon C \to \mathcal{S}$$

commutes with limits.

Proof We apply Theorem 6.3.4 for $A = \Delta^0$ and $u = X\colon \Delta^0 \to C$. The functor $u^* h_C = \mathrm{Hom}_C(X, -)$ thus has a left adjoint $Lu_!$. Henceforth, it commutes with limits, by Proposition 6.2.15. □

Remark 6.3.6 Here is a practical consequence of the preceding corollary. Let A and B be two cocomplete locally **U**-small ∞-categories, and let $u, r\colon A \to B$ be two functors equipped with right adjoints v and s, respectively. We assume that there are natural transformations $b\colon u \to r$ and $a\colon s \to v$, such that, for any objects x and y in A and B respectively, the square

$$\begin{array}{ccc} \mathrm{Hom}_{ho(A)}(r(x), y) & \xrightarrow{\sim} & \mathrm{Hom}_{ho(B)}(x, s(y)) \\ {\scriptstyle b^*}\downarrow & & \downarrow{\scriptstyle a_*} \\ \mathrm{Hom}_{ho(A)}(u(x), y) & \xrightarrow{\sim} & \mathrm{Hom}_{ho(B)}(x, v(y)) \end{array} \tag{6.3.6.1}$$

commutes. Then, the squares

$$\begin{array}{ccc} \mathrm{Hom}_A(r(x), y) & \xrightarrow{\sim} & \mathrm{Hom}_B(x, s(y)) \\ {\scriptstyle b^*}\downarrow & & \downarrow{\scriptstyle a_*} \\ \mathrm{Hom}_A(u(x), y) & \xrightarrow{\sim} & \mathrm{Hom}_B(x, v(y)) \end{array} \tag{6.3.6.2}$$

commute in $ho(S)$. Indeed, it is sufficient to check that the squares

(6.3.6.3)

$$\begin{array}{ccc} \mathrm{Hom}_{ho(\mathcal{S})}(s, \mathrm{Hom}_A(r(x), y)) & \xrightarrow{\sim} & \mathrm{Hom}_{ho(\mathcal{S})}(s, \mathrm{Hom}_B(x, s(y))) \\ {\scriptstyle b^*}\downarrow & & \downarrow{\scriptstyle a_*} \\ \mathrm{Hom}_{ho(\mathcal{S})}(s, \mathrm{Hom}_A(u(x), y)) & \xrightarrow{\sim} & \mathrm{Hom}_{ho(\mathcal{S})}(s, \mathrm{Hom}_B(x, v(y))) \end{array}$$

commute for any object s in $\mathcal{S}$ classifying a Kan fibration of the form $K \to \Delta^0$. For an object t of a cocomplete ∞-category C, let us define

$$s \otimes t = \varinjlim_{k \in K^{op}} t, \tag{6.3.6.4}$$

as the colimit indexed by K of the constant diagram with value t. We then have natural invertible maps

$$s \otimes \varphi(t) \xrightarrow{\sim} \varphi(s \otimes t) \tag{6.3.6.5}$$

for any cocontinuous functor φ. In particular, since $s \otimes e = s$, by Corollary 6.2.16, the preceding theorem tells us that we have canonical bijections:

$$\mathrm{Hom}_{ho(C)}(s \otimes t, z) \simeq \mathrm{Hom}_{ho(\mathcal{S})}(s, \mathrm{Hom}_C(t, z))\,. \tag{6.3.6.6}$$

This shows that the squares (6.3.6.3) commute and thus that the squares (6.3.6.2) commute.

Theorem 6.3.7 *The ∞-category of* **U***-small ∞-groupoids is complete. More precisely, for any* **U***-small ∞-category I and any functor $F\colon I \to \mathcal{S}$, the map to the final functor $h_{I^{op}} \to e$ and the Yoneda lemma induce a cone*

$$c_I \,\mathrm{Hom}(e, F) \to F$$

which exhibits $\mathrm{Hom}(e, F)$ *as the limit of F in $\mathcal{S}$.*

Proof We know that e is the colimit of $h_{I^{op}}$ in $\underline{\mathrm{Hom}}(I, \mathcal{S})$, by Proposition 6.2.13. Therefore, the preceding corollary exhibits $\mathrm{Hom}(e, F)$ as a limit of the functor $i \mapsto \mathrm{Hom}(h_{I^{op}}(i), F)$. We conclude with the observation that the Yoneda lemma identifies the latter with the functor F itself. □

Corollary 6.3.8 *For any simplicial set X, the ∞-category $\underline{\mathrm{Hom}}(X, \mathcal{S})$ is complete (with respect to* **U***).*

Proof Since $\mathcal{S}$ is complete, we can apply Corollary 6.2.10 to $C = \mathcal{S}^{op}$. □

Proposition 6.3.9 *Let A be a* **U***-small ∞-category, let C be an ∞-category with* **U***-small colimits, and let $\varphi\colon \widehat{A} \to C$ be a functor. We put $u = \varphi h_A\colon A \to C$. There is a canonical natural transformation $Lu_! \to \varphi$ such that the induced composed map $u \xrightarrow{\sim} Lu_! h_A \to \varphi h_A = u$ is the identity. Furthermore, the following conditions are equivalent.*

(i) The functor φ commutes with **U***-small colimits.*
(ii) The canonical natural transformation $Lu_! \to \varphi$ is invertible.

If, furthermore, the ∞-category C is locally **U***-small, then the above conditions are also equivalent to each of the following three conditions.*

(iii) The functor φ commutes with all colimits.
(iv) The functor φ has a right adjoint.
(v) The functor φ is a left adjoint of the functor $u^* h_C$.

Proof Since the adjunctions are functorial, by Theorem 6.1.22, to determine a map

$$Lu_! = Lu_!(1_{\widehat{A}}) \to \varphi$$

in $\underline{\mathrm{Hom}}(\widehat{A}, C)$, it is sufficient to define a map $1_{\widehat{A}} \to u^* h_C(\varphi)$. Functors $\widehat{A} \to \widehat{A}$ correspond to functors $A^{\mathrm{op}} \times \widehat{A} \to \mathcal{S}$. From this point of view, the functors $1_{\widehat{A}}$ and $u^* h_C(\varphi)$ correspond to the functors

$$(a, F) \mapsto F(a) \quad \text{and} \quad (a, F) \mapsto \mathrm{Hom}_C(u(a), \varphi(F)),$$

respectively. By the Yoneda lemma, we have a functorial invertible map

$$\mathrm{Hom}_{\widehat{A}}(h_A(a), F) \xrightarrow{\sim} F(a)\,.$$

Therefore, since $\varphi h_A = u$, it is sufficient to produce a functorial map of the form

$$\mathrm{Hom}_{\widehat{A}}(X, F) \to \mathrm{Hom}_C(\varphi(X), \varphi(F))\,.$$

Such a map is provided out of φ by a functoriality argument; see Remark 5.8.3.

It remains to prove the second part of the proposition. Since, by virtue of Theorem 6.3.4, the functor $Lu_!$ commutes with **U**-small colimits, it is clear that condition (ii) implies condition (i). Let us check that condition (i) implies condition (ii). Let $F: A^{\mathrm{op}} \to \mathcal{S}$ be a presheaf on A, classifying a right fibration $p: X \to A$. By virtue of Corollary 6.2.16, there is a canonical invertible map

$$\varinjlim_{x \in X} h_A(p(x)) \to F\,.$$

Assuming that φ commutes with **U**-small colimits, we obtain a commutative diagram of the form

$$\begin{array}{ccc}
\varinjlim_{x\in X} Lu_!(h_A(p(x))) & \longrightarrow & \varinjlim_{x\in X} u(p(x)) \\
\downarrow \wr & & \downarrow \wr \\
Lu_!(F) & \longrightarrow & \varphi(F)
\end{array}$$

in $ho(C)$, in which the vertical maps are invertible, because they exhibit $Lu_!(F)$ and $\varphi(F)$ as colimits of the functors $Lu_! h_A p$ and up, respectively. Since the map $Lu_! h_A(a) \to \varphi(h_A(a)) = u(a)$ is invertible for every object a of A, the upper horizontal map of the square above is invertible as well, which shows that condition (ii) holds true.

Finally, let us assume that C is locally **U**-small. Then, by Theorem 6.3.4, the functor $Lu_!$ is left adjoint to $u^* h_C$. Therefore, Proposition 6.1.7 shows that condition (ii) implies condition (v), which, in turn, obviously gives condition (iv). We know that condition (iv) implies condition (iii), by Proposition 6.2.15. It is clear that condition (iii) implies condition (i). □

Corollary 6.3.10 *For any functor between* **U**-*small* ∞-*categories* $f\colon A \to B$, *the functor* $f^*\colon \underline{\mathrm{Hom}}(B^{\mathrm{op}}, \mathcal{S}) \to \underline{\mathrm{Hom}}(A^{\mathrm{op}}, \mathcal{S})$ *has a right adjoint*

$$f_*\colon \underline{\mathrm{Hom}}(A^{\mathrm{op}}, \mathcal{S}) \to \underline{\mathrm{Hom}}(B^{\mathrm{op}}, \mathcal{S})$$

which associates to a presheaf F *on* A *the presheaf* $b \mapsto \mathrm{Hom}(f^* h_B(b) F)$.

Proof One may interpret Proposition 6.1.16 as a proof that the functor u^* commutes with **U**-small colimits. We conclude by applying the preceding proposition for $\varphi = f^*$. □

Definition 6.3.11 Given two ∞-categories A and B, we write $\underline{\mathrm{Hom}}_!(A, B)$ for the full subcategory of $\underline{\mathrm{Hom}}(A, B)$ whose objects are the functors $A \to B$ which commute with **U**-small colimits.

Proposition 6.3.12 *Let* I *be a* **U**-*small simplicial set. If an* ∞-*category* B *has colimits of type* I, *then so does* $\underline{\mathrm{Hom}}_!(A, B)$ *for any* ∞-*category* A. *Furthermore, the inclusion functor from* $\underline{\mathrm{Hom}}_!(A, B)$ *into* $\underline{\mathrm{Hom}}(A, B)$ *commutes with colimits of type* I.

Proof By Corollary 6.2.10, we know that $\underline{\mathrm{Hom}}(A, B)$ has colimits of type I and that these are computed fibrewise. Moreover the colimit functor

$$\varinjlim\colon \underline{\mathrm{Hom}}(I, \underline{\mathrm{Hom}}(A, B)) \to \underline{\mathrm{Hom}}(A, B)$$

commutes with colimits, since it is a left adjoint. Therefore, it preserves the property of commuting with **U**-small colimits. □

Theorem 6.3.13 *Let* A *be a* **U**-*small* ∞-*category. The Yoneda embedding* $h_A\colon A \to \underline{\mathrm{Hom}}(A^{\mathrm{op}}, \mathcal{S}) = \widehat{A}$ *is the universal functor into an* ∞-*category with* **U**-*small colimits. In other words, given an* ∞-*category* C *with* **U**-*small colimits, the restriction functor*

$$h_A^*\colon \underline{\mathrm{Hom}}_!(\widehat{A}, C) \to \underline{\mathrm{Hom}}(A, C)$$
$$\varphi \mapsto \varphi h_A$$

is an equivalence of ∞-*categories. In particular, it is possible to turn the assignment* $u \mapsto Lu_!$ *into a functor: as a quasi-inverse of the equivalence of* ∞-*categories above.*

Proof The preceding proposition gives the essential surjectivity of the functor h_A^*. It remains to check the property of full faithfulness. Let us consider two functors u and v from A to C. Let $f\colon u \to v$ be a natural transformation. We can see f as a functor

$$f\colon A \to \underline{\mathrm{Hom}}(\Delta^1, C)$$

and Theorem 6.3.4 tells us how to extend the latter into a cocontinuous functor

$$Lf_!\colon \widehat{A} \to \underline{\mathrm{Hom}}(\Delta^1, C).$$

Since the evaluation at ε functor

$$ev_\varepsilon\colon \underline{\mathrm{Hom}}(\Delta^1, C) \to C$$

commutes with colimits for $\varepsilon = 0, 1$, we have natural identifications

$$Lu_! \xrightarrow{\sim} ev_0 Lf_! \quad \text{and} \quad Lv_! \xrightarrow{\sim} ev_1 Lf_!\,.$$

Therefore, the functor $Lf_!$ provides a natural transformation $Lu_! \to Lv_!$ whose restriction to A gives back f (because $Lf_! h_A$ and f are canonically equivalent). Conversely, any natural transformation $Lu_! \to Lv_!$ defines a functor

$$\varphi\colon \widehat{A} \to \underline{\mathrm{Hom}}(\Delta^1, C)$$

such that $ev_0\varphi = Lu_!$ and $ev_1\varphi = Lv_!$. Let $f\colon u \to v$ be the map induced by φh_A. Then, by virtue of the preceding proposition, we have a canonical equivalence from $Lf_!$ to φ. This means that the map

$$\mathrm{Hom}_{ho(\underline{\mathrm{Hom}}_!(\widehat{A},C))}(Lu_!, Lv_!) \to \mathrm{Hom}_{ho(\underline{\mathrm{Hom}}(A,C))}(u, v)$$

is a bijection for all u and v. Given an object s of $\mathcal{S}$ which classifies an ∞-groupoid K, we may consider this bijection replacing u by $s \otimes u$; see (6.3.6.4). Then formula (6.3.6.5) and Propositions 6.3.9 and 6.3.12 mean that $L(s \otimes u)_! \simeq s \otimes Lu_!$, and we deduce from formula (6.3.6.6) that the natural map

$$\mathrm{Hom}_{\underline{\mathrm{Hom}}_!(\widehat{A},C)}(Lu_!, Lv_!) \to \mathrm{Hom}_{\underline{\mathrm{Hom}}(A,C)}(u, v)$$

is invertible in $\mathcal{S}$ as follows. For any object s of $\mathcal{S}$, the map

$$\mathrm{Hom}_{ho(\mathcal{S})}(s, \mathrm{Hom}_{\underline{\mathrm{Hom}}_!(\widehat{A},C)}(Lu_!, Lv_!)) \to \mathrm{Hom}_{ho(\mathcal{S})}(s, \mathrm{Hom}_{\underline{\mathrm{Hom}}(A,C)}(u, v))$$

is isomorphic to the bijective map

$$\mathrm{Hom}_{ho(\underline{\mathrm{Hom}}_!(\widehat{A},C))}(L(s \otimes u)_!, Lv_!) \to \mathrm{Hom}_{ho(\underline{\mathrm{Hom}}(A,C))}(s \otimes u, v)\,.$$

We conclude by applying the Yoneda lemma to $ho(\mathcal{S})$. □

6.4 Kan Extensions

We still fix a universe **U**.

6.4.1 Let $u\colon A \to B$ be a morphism of simplicial sets, and C an ∞-category. We are interested in the construction and computation of a left adjoint $u_!$ or a right adjoint u_* of the pull-back functor $u^*\colon \underline{\mathrm{Hom}}(B, C) \to \underline{\mathrm{Hom}}(A, C)$, which are called the *left Kan extension* and the *right Kan extension* of u in C, respectively. These operators thus generalise the notions of colimit and of limit, which correspond to the case where $B = \Delta^0$ is the final simplicial set. In his work on homotopy theory, Grothendieck calls $u_!$ the *homological push-forward* (or direct image) functor, and u_* the *cohomological push-forward* functor. We will also consider the pull-back functor $u^*\colon \underline{\mathrm{Hom}}(B^{\mathrm{op}}, C) \to \underline{\mathrm{Hom}}(A^{\mathrm{op}}, C)$ (obtained by precomposing with u^{op}); its left adjoint will also be denoted by $u_!$ and its right adjoint by u_*, despite the ambiguity it might cause.[2] If we do this for an abstract ∞-category C, we see that there is no difference: since $\underline{\mathrm{Hom}}(A, C)^{\mathrm{op}} = \underline{\mathrm{Hom}}(A^{\mathrm{op}}, C^{\mathrm{op}})$, one can switch from the point of view of functors $A \to C$ to that of functors $A^{\mathrm{op}} \to D$ (with $D = C^{\mathrm{op}}$), and this will exchange the roles of $u_!$ and u_*. For the reader who wonders why we change the variance of the functors we consider, we can only say that we have never found it satisfactory to emphasise one side more than the other. Covariant functors are extremely useful, of course, and we use them continually, but contravariant functors also appear naturally, and their proximity to topoi allows us to apply many topological intuitions which are also useful in practice (e.g. smooth functors, proper functors, and the associated base change formulas).

In this chapter, we will be interested in the case where both A and B are **U**-small, while C is (co)complete and locally **U**-small. The general strategy will consist in looking at the case where $C = \mathcal{S}$ is the ∞-category of **U**-small ∞-groupoids, and to use (variations on) the Yoneda lemma to extend to the general case. We remark that, for a morphism of **U**-small simplicial sets $u\colon A \to B$, the functor $u^*\colon \underline{\mathrm{Hom}}(B, \mathcal{S}) \to \underline{\mathrm{Hom}}(A, \mathcal{S})$ has both a left adjoint and a right adjoint, by Proposition 6.1.14 and Corollary 6.3.10, respectively. That is a good starting point.

[2] We remark that, in the case where $C = \mathcal{S}$, these conventions are compatible with the fact that the homotopy categories of $\underline{\mathrm{Hom}}(A, \mathcal{S})$ and of $\underline{\mathrm{Hom}}(A^{\mathrm{op}}, \mathcal{S})$ correspond to $LFib(A)$ and $RFib(A)$, respectively.

6.4.2 Let us consider the following commutative square of simplicial sets.

$$\begin{array}{ccc} A' & \xrightarrow{u} & A \\ {\scriptstyle q}\downarrow & & \downarrow{\scriptstyle p} \\ B' & \xrightarrow{v} & B \end{array} \tag{6.4.2.1}$$

We also assume that there is an ∞-category C such that the left adjoint $\varphi_!$ of the pull-back functor

$$\varphi^*\colon \underline{\mathrm{Hom}}(Y, C) \to \underline{\mathrm{Hom}}(X, C)$$

associated to $\varphi\colon X \to Y$ exists for $\varphi = p$ and $\varphi = q$. Then, for any functor $F\colon A \to C$, there is a canonical base change map of the form

$$q_!\, u^*(F) \to v^*\, p_!(F) \tag{6.4.2.2}$$

in $\underline{\mathrm{Hom}}(B', C)$, which is constructed as follows. Applying the functor u^* to the unit map $F \to p^* p_!(F)$ gives a functor

$$u^*(F) \to u^* p^* p_!(F) = q^* v^* p_!(F)$$

which determines the map (6.4.2.2). This map is a natural transformation. To see this, we remark that we may apply the preceding construction replacing C by $\underline{\mathrm{Hom}}(T, C)$ for any simplicial set T (the fact that our hypothesis still applies follows from Theorem 6.1.22). For $T = \underline{\mathrm{Hom}}(A, C)$ and $F\colon A \to \underline{\mathrm{Hom}}(T, C)$ the functor obtained by transposition of the evaluation functor $ev\colon A \times \underline{\mathrm{Hom}}(A, C) \to C$, construction (6.4.2.2) above provides a map in the ∞-category of functors

$$\underline{\mathrm{Hom}}(B', \underline{\mathrm{Hom}}(\underline{\mathrm{Hom}}(A, C), C)) \simeq \underline{\mathrm{Hom}}(\underline{\mathrm{Hom}}(A, C), \underline{\mathrm{Hom}}(B', C))\,.$$

Dually, assume that the pull-back functor

$$\varphi^*\colon \underline{\mathrm{Hom}}(Y^{\mathrm{op}}, C) \to \underline{\mathrm{Hom}}(X^{\mathrm{op}}, C)$$

associated to $\varphi\colon X \to Y$ has a right adjoint φ_* for $\varphi = u$ and $\varphi = v$. Then, for any functor $F\colon A^{\mathrm{op}} \to C$, there is a functorial base change map of the form

$$v^*\, p_*(F) \to q_*\, u^*(F) \tag{6.4.2.3}$$

obtained by applying the previous construction to C^{op}.

Proposition 6.4.3 *Let us consider a Cartesian square of* **U**-*small simplicial sets of the form* (6.4.2.1). *If v is smooth, or if p is proper, then, for any functor $F\colon A \to \mathcal{S}$, the base change map $q_!\, u^*(F) \to v^*\, p_!(F)$ is invertible.*

Proof It is sufficient to check that this map is an isomorphism in the homotopy category $ho(\underline{\mathrm{Hom}}(B', \mathcal{S}))$. Let $\pi\colon X \to A$ be the left fibration classified by F. By virtue of Theorem 5.4.5, and of Proposition 6.1.14, this map corresponds to the base change map

$$\mathbf{L}q_! \,\mathbf{R}u^*(X) \to \mathbf{R}v^* \,\mathbf{L}p_!(X)$$

in $LFib(B')$. This proposition is thus a reformulation of Corollary 4.4.20 and of Theorem 4.4.24. □

Corollary 6.4.4 *We consider a Cartesian square of* **U**-*small simplicial sets of the form* (6.4.2.1). *If v is smooth, or if p is proper, then, for any functor $F\colon A^{\mathrm{op}} \to \mathcal{S}$, the base change map $v^*\, p_*(F) \to q_*\, u^*(F)$ is invertible.*

Proof For any functor $X\colon B'^{\mathrm{op}} \to \mathcal{S}$, the induced map $u_!q^*(X) \to p^*v_!(X)$ is invertible, because v^{op} is proper or p^{op} is smooth, so that we may apply the preceding proposition. We thus have a commutative square in $ho(\mathcal{S})$ of the form

$$\begin{array}{ccc} \mathrm{Hom}(p^*r_!(X), F) & \xrightarrow{\sim} & \mathrm{Hom}(X, v^*p_*(F)) \\ \wr\downarrow & & \downarrow \\ \mathrm{Hom}(u_!q^*(X), F) & \xrightarrow{\sim} & \mathrm{Hom}(X, q_*u^*(F)) \end{array}$$

(it is sufficient to check this in a naive way, by Remark 6.3.6). The Yoneda lemma applied to $\underline{\mathrm{Hom}}(B'^{\mathrm{op}}, \mathcal{S})$ thus proves the corollary. □

Theorem 6.4.5 *Let $u\colon A \to B$ be a morphism of* **U**-*small simplicial sets. The following conditions are equivalent.*

(a) *The morphism u is final.*
(b) *For any ∞-category C, a functor $F\colon B \to C$ has a colimit in C if and only if the functor $u^*(F) = Fu$ has a colimit in C. If this is the case, then the canonical map $\varinjlim u^*(F) \to \varinjlim F$ is invertible.*
(c) *For any ∞-category D, a functor $F\colon B^{\mathrm{op}} \to D$ has a limit in D if and only if the functor $u^*(F) = Fu^{\mathrm{op}}$ has a limit in D. If this is the case, the canonical map $\varprojlim F \to \varprojlim u^*(F)$ is invertible.*
(d) *For any functor $F\colon B^{\mathrm{op}} \to \mathcal{S}$ the canonical map $\varprojlim F \to \varprojlim u^*(F)$ is invertible.*

Proof Conditions (b) and (c) are essentially the same, via the identification $C = D^{\mathrm{op}}$. Condition (d) means that, for any functor $F\colon B^{\mathrm{op}} \to \mathcal{S}$, if e denotes the final functor, the map

$$\mathrm{Hom}(e, F) \to \mathrm{Hom}(u^*(e), u^*(F)) = \mathrm{Hom}(e, u^*(F)) \simeq \mathrm{Hom}(u_!(u^*(e)), F)$$

is invertible. This means the canonical map $u_!(u^*(e) \to e$ is invertible. It corresponds through the equivalence of Theorem 5.4.5 to the map $u^{\mathrm{op}} : A^{\mathrm{op}} \to B^{\mathrm{op}}$ in $LFib(B^{\mathrm{op}})$. The latter is invertible if and only if u is final, by Proposition 4.1.11. Hence conditions (a) and (d) are equivalent. Note that condition (a) is independent of the chosen universe. In particular, condition (d) for **U** implies its analogue for any larger universe. This means that, to prove the equivalence between conditions (c) and (d), one may always assume that C is locally **U**-small. We observe that, by Corollary 6.3.5, a functor $F : I \to C$ has a limit in C if and only if the limit $\varprojlim h_C(F)$ exists in $\underline{\mathrm{Hom}}(C^{\mathrm{op}}, \mathcal{S})$ and is representable, in which case we always have

$$\mathrm{Hom}_C(X, \varprojlim F) \simeq \varprojlim \mathrm{Hom}_C(X, F)$$

for all objects X of C, or in other words, a canonical invertible map

$$h_C(\varprojlim F) \xrightarrow{\sim} \varprojlim h_C(F)\,.$$

If condition (d) holds true, for any functor $F : B^{\mathrm{op}} \to C$ and any object X of C, we have canonical isomorphisms

$$\varprojlim \mathrm{Hom}_C(X, F) \simeq \varprojlim u^* \mathrm{Hom}_C(X, F) = \varprojlim \mathrm{Hom}(X, u^* F)$$

in $ho(\mathcal{S})$. This shows that $\varprojlim h_C(F)$ is representable if and only if $\varprojlim h_C(u^* F)$ is representable, hence condition (c). □

Corollary 6.4.6 *For any simplicial set A with a final object a, the colimit (the limit) of any functor $F : A \to C$ (of any functor $F : A^{\mathrm{op}} \to C$, respectively) is canonically equivalent to the evaluation $F(a)$.*

6.4.7 Let $u : A \to B$ be a functor between **U**-small ∞-categories. Given an object b of B, we form the following Cartesian square

(6.4.7.1)
$$\begin{array}{ccc} A/b & \xrightarrow{p} & A \\ {\scriptstyle u/b}\downarrow & & \downarrow{\scriptstyle u} \\ B/b & \xrightarrow{q} & B \end{array}$$

in which q is the canonical map $B/b \to B$. Given any functor $F : A^{\mathrm{op}} \to C$, where C is an ∞-category, we put

(6.4.7.2)
$$F/b = p^*(F) : (A/b)^{\mathrm{op}} \to C\,.$$

Similarly, for any functor $G : B^{\mathrm{op}} \to C$, we define

(6.4.7.3)
$$G/b = q^*(G) : (B/b)^{\mathrm{op}} \to C.$$

Therefore, if the functors u_* and $(u/b)_*$ exist in C, we have a canonical base change map

$$u_*(F)/b \to (u/b)_*(F/b)\,. \tag{6.4.7.4}$$

We observe that the limit of $(u/b)_*(F/b)$ is the limit of F/b and that the evaluation of $u_*(F)/b$ at the object $(b, 1_b)$ is the evaluation of $u_*(F)$ at b. Since $(b, 1_b)$ is a final object of B/b, the preceding corollary thus means that evaluating the map (6.4.7.4) at $(b, 1_b)$ gives a functorial map of the form

$$u_*(F)_b \to \varprojlim F/b, \tag{6.4.7.5}$$

where we write $H_b = b^*(H)$ for the fibre at b of any functor $H\colon B^{\mathrm{op}} \to C$.

Proposition 6.4.8 *Let $u\colon A \to B$ be a functor between* **U***-small ∞-categories, and $F\colon A^{\mathrm{op}} \to \mathcal{S}$ a presheaf on A. Then, for any object b of B, the canonical map* (6.4.7.4) *is invertible. Henceforth, the canonical map* (6.4.7.5) *is invertible as well.*

Proof Since q is a right fibration, it is smooth. Therefore, this is a particular instance of the base change formula provided by Corollary 6.4.4. □

Proposition 6.4.9 *Let C be a locally* **U***-small ∞-category, and $u\colon A \to B$ a functor between* **U***-small ∞-categories. If limits of type $(A/b)^{\mathrm{op}}$ exist in C for all objects b of B, then the functor $u^*\colon \underline{\mathrm{Hom}}(B^{\mathrm{op}}, C) \to \underline{\mathrm{Hom}}(A^{\mathrm{op}}, C)$ has a right adjoint u_*. Moreover, for any functor $F\colon B^{\mathrm{op}} \to C$ and any object b of B, the maps* (6.4.7.4) *and* (6.4.7.5) *are invertible.*

Proof Let $F\colon A^{\mathrm{op}} \to C$ be a functor. Applying the Yoneda embedding of C, we get a functor $h_C(F)\colon A^{\mathrm{op}} \to \underline{\mathrm{Hom}}(C^{\mathrm{op}}, \mathcal{S})$. The functor

$$u^*\colon \underline{\mathrm{Hom}}(B^{\mathrm{op}}, \mathcal{S}) \to \underline{\mathrm{Hom}}(A^{\mathrm{op}}, \mathcal{S})$$

has a right adjoint u_*, by Corollary 6.3.10. Therefore, for any simplicial set X, the functor

$$u^* = \underline{\mathrm{Hom}}(X, u^*)\colon \underline{\mathrm{Hom}}(X, \underline{\mathrm{Hom}}(B^{\mathrm{op}}, \mathcal{S})) \to \underline{\mathrm{Hom}}(X, \underline{\mathrm{Hom}}(A^{\mathrm{op}}, \mathcal{S}))$$

has a right adjoint u_*, which is defined fibrewise over X, by Theorem 6.1.22. This means that, for any functor $\Phi\colon X \to \underline{\mathrm{Hom}}(B^{\mathrm{op}}, \mathcal{S})$, the functor $u_*(\Phi)$ evaluated at b simply is $u_*(\Phi_b)$. In the case where $X = C^{\mathrm{op}}$ and $\Phi = h_C(F)$, this gives a functor

$$u_*(h_C(F))\colon B^{\mathrm{op}} \to \underline{\mathrm{Hom}}(C^{\mathrm{op}}, \mathcal{S})\,.$$

Using the preceding proposition and Corollary 6.3.5, we see that, for any objects

x and b of C and B, respectively, we have functorial identifications of the form below:

$$\begin{aligned}(u_*(h_C(F))_b)_x &\simeq u_*(\mathrm{Hom}_C(x, F))_b \\ &\simeq \varprojlim \mathrm{Hom}_C(x, F/b) \\ &\simeq \mathrm{Hom}_C(x, \varprojlim F/b).\end{aligned}$$

In other words, $u_*(h_C(F))_b$ is representable for all b. This means that $u_*(h_C(F))$ is in the essential image of the fully faithful functor

$$\underline{\mathrm{Hom}}(B^{\mathrm{op}}, h_C) : \underline{\mathrm{Hom}}(B^{\mathrm{op}}, C) \to \underline{\mathrm{Hom}}(B^{\mathrm{op}}, \underline{\mathrm{Hom}}(C^{\mathrm{op}}, \mathcal{S}))$$

provided by Corollary 5.8.14. □

Corollary 6.4.10 *Let C be a complete locally $\mathbf{U}$-small ∞-category. For any map between $\mathbf{U}$-small simplicial sets $u : A \to B$, the pull-back functor $u^* : \underline{\mathrm{Hom}}(B^{\mathrm{op}}, C) \to \underline{\mathrm{Hom}}(A^{\mathrm{op}}, C)$ has a right adjoint.*

Proof In the case where A and B are ∞-categories, this is a particular case of the preceding proposition. Otherwise, one chooses a commutative square of the form

$$\begin{array}{ccc} A & \xrightarrow{f} & A' \\ {\scriptstyle u}\downarrow & & \downarrow{\scriptstyle v} \\ B & \xrightarrow{g} & B' \end{array}$$

in which f and g are inner anodyne maps, and A' and B' are ∞-categories. For any weak categorical equivalence $\varphi : I \to J$, the pull-back functor

$$\varphi^* : \underline{\mathrm{Hom}}(J^{\mathrm{op}}, C) \to \underline{\mathrm{Hom}}(I^{\mathrm{op}}, C)$$

is an equivalence of ∞-categories, hence has a right adjoint φ_*, which is also a left adjoint, by Proposition 6.1.6. Hence Proposition 6.1.8 implies that $g^* v_* f_*$ is a right adjoint of u^*. □

6.4.11 Let $\Phi : C \to D$ be a functor between complete locally $\mathbf{U}$-small categories, and $u : A \to B$ a morphism of $\mathbf{U}$-small simplicial sets. By virtue of Theorem 6.1.22, for any simplicial set X and any functor $R : X \to \underline{\mathrm{Hom}}(A, C)$, there is a canonical map

(6.4.11.1) $$\Phi(u_*(R)) \to u_*(\Phi(R))$$

in $\underline{\mathrm{Hom}}(X, \underline{\mathrm{Hom}}(B, D))$ which corresponds, by transposition, to the map

$$u^*(\Phi(u_*(R))) = \Phi(u^* u_*(F)) \to \Phi(R),$$

obtained as the image by Φ of the co-unit map $u^*u_*(R) \to R$. In the case where $X = \underline{\mathrm{Hom}}(A, C)$ and R is the identity of X, this provides, for any functor $F\colon A \to C$, a canonical map

$$\Phi(u_*(F)) \to u_*\Phi((F)), \tag{6.4.11.2}$$

functorially in F.

Proposition 6.4.12 *If the functor Φ above commutes with* **U***-small limits, then, for any functor $F\colon A \to C$, the canonical map $\Phi(u_*(F)) \to u_*(\Phi(F))$ is invertible.*

Proof As in the proof of the preceding corollary, one can reduce to the case where A and B are ∞-categories. The last assertion of Proposition 6.4.9 means that we may even reduce to the case where $B = \Delta^0$. This latter case is precisely the property of commuting with **U**-small limits. □

Theorem 6.4.13 *Let us consider a pull-back square of* **U***-small simplicial sets of the following form.*

$$\begin{array}{ccc} A' & \xrightarrow{u} & A \\ {\scriptstyle q}\downarrow & & \downarrow{\scriptstyle p} \\ B' & \xrightarrow{v} & B \end{array}$$

If p is proper, or if v is smooth, then, for any ∞-category C with **U***-small limits, and for any functor $F\colon A^{\mathrm{op}} \to C$, the base change map*

$$v^*\, p_*(F) \to q_*\, u^*(F)$$

is invertible.

Proof Let us choose a universe **V**, containing **U**, such that C is **V**-small. The Yoneda embedding of C commutes with limits, whence with u_* for any morphism of **U**-small simplicial sets u, by the preceding proposition. As in the proof of Theorem 6.4.5, one reduces the assertion to the case where $C = \mathcal{S}'$ is the ∞-category of **V**-small ∞-groupoids. The latter case is already known, by Corollary 6.4.4. □

6.5 The Cartesian Product

A universe **U** is given once and for all, hence, the ∞-category of **U**-small ∞-groupoids $\mathcal{S}$.

6.5.1 One defines a functor

(6.5.1.1) $$\times\colon \mathcal{S}\times\mathcal{S}\to\mathcal{S}\,,\quad (x,y)\mapsto x\times y$$

as follows. Given a simplicial set A and two functors $F, G\colon A\to\mathcal{S}$ which classify two left fibrations with **U**-small fibres $p\colon X\to A$ and $q\colon Y\to A$, respectively, one defines $F\times G$ as the functor which classifies the left fibration $r\colon Z\to A$ appearing in the Cartesian square below.

(6.5.1.2) $$\begin{array}{ccc} Z & \longrightarrow & X\times Y \\ {\scriptstyle r}\downarrow & & \downarrow{\scriptstyle p\times q} \\ A & \xrightarrow{(1_A,1_A)} & A\times A \end{array}$$

Note that there is a canonical way to specify pull-backs of $p\times q$ out of the specified pull-backs of p and q: one writes the map $p\times q$ as a composition of $p\times 1_A$ and of $1_A\times q$, and we observe that pull-backs of $p\times 1_A$ and of $1_A\times q$ are also pull-backs of p and of q, respectively. Therefore, the left fibration r is well defined (as opposed to defined up to a unique isomorphism), as well as the map $F\times G\colon A\to\mathcal{S}$ which classifies r. Furthermore, if ever one specifies a section of p and a section of q, this determines a section of $p\times q$, hence a section of r itself. In other words, there is a canonical way to promote the map (6.5.1.1) to a Cartesian square of the following form.

(6.5.1.3) $$\begin{array}{ccc} \mathcal{S}_\bullet\times\mathcal{S}_\bullet & \xrightarrow{\times} & \mathcal{S}_\bullet \\ {\scriptstyle p_{univ}\times p_{univ}}\downarrow & & \downarrow{\scriptstyle p_{univ}} \\ \mathcal{S}\times\mathcal{S} & \xrightarrow{\times} & \mathcal{S} \end{array}$$

This means that the map (6.5.1.2) classifies the product of two copies of the universal left fibration with **U**-small fibres.

We can do this with many variables: given a **U**-small set I, a family of functors $F_i\colon A\to\mathcal{S}$, $i\in I$, which classifies a family of left fibrations with **U**-small fibres $(p_i\colon X_i\to A)_i$, the functor $\prod_{i\in I}F_i\colon A\to\mathcal{S}$ classifies the left fibration $r\colon Z\to A$ appearing in the Cartesian square below.

(6.5.1.4) $$\begin{array}{ccc} Z & \longrightarrow & \prod_{i\in I}X_i \\ {\scriptstyle r}\downarrow & & \downarrow{\scriptstyle \prod_{i\in I}p_i} \\ A & \xrightarrow{(1_A)_{i\in I}} & \prod_{i\in I}A \end{array}$$

We choose the Cartesian square below.

(6.5.1.5)
$$\begin{array}{ccc} \mathcal{S}^I_\bullet & \xrightarrow{\prod_{i\in I}} & \mathcal{S}_\bullet \\ {\scriptstyle p^I_{univ}}\downarrow & & \downarrow{\scriptstyle p_{univ}} \\ \mathcal{S}^I & \xrightarrow{\prod_{i\in I}} & \mathcal{S} \end{array}$$

In other words, the map $\prod_{i\in I} \colon \mathcal{S}^I \to \mathcal{S}$ classifies the left fibration with U-small fibres p^I_{univ}. Combining diagrams (6.5.1.4) and (6.5.1.5) we then have the commutative diagram below, in which all the squares are Cartesian.

(6.5.1.6)
$$\begin{array}{ccccccc} Z & \longrightarrow & \prod_{i\in I} X_i & \longrightarrow & \mathcal{S}^I_\bullet & \xrightarrow{\prod_{i\in I}} & \mathcal{S}_\bullet \\ {\scriptstyle r}\downarrow & & \downarrow{\scriptstyle \prod_{i\in I} p_i} & & \downarrow{\scriptstyle p^I_{univ}} & & \downarrow{\scriptstyle p_{univ}} \\ A & \xrightarrow{(1_A)_{i\in I}} & \prod_{i\in I} A & \xrightarrow{\prod_{i\in I} F_i} & \mathcal{S}^I & \xrightarrow{\prod_{i\in I}} & \mathcal{S} \end{array}$$

We observe that, for any object x of $\mathcal{S}$, and $i \in I$, if we have a family $(x_i)_{i\in I}$ such that one of the x_i is equal to x and all the other factors are equal to the final object e, then $\prod_{i\in I} x_i = x$. Indeed, this comes from the fact that all the squares of the following diagram are Cartesian, in which the lower left horizontal map is the functor $(\prod_{i\in I} x_i)_!$ (i.e. the functor which is defined by e on the jth factors for $j \neq i$ and by the identity of $\mathcal{S}$ on the ith factor).

(6.5.1.7)
$$\begin{array}{ccccc} \mathcal{S}_\bullet & \longrightarrow & \mathcal{S}^I_\bullet & \xrightarrow{\prod_{i\in I}} & \mathcal{S}_\bullet \\ {\scriptstyle p_{univ}}\downarrow & & {\scriptstyle p^I_{univ}}\downarrow & & \downarrow{\scriptstyle p_{univ}} \\ \mathcal{S} & \longrightarrow & \mathcal{S}^I & \xrightarrow{\prod_{i\in I}} & \mathcal{S} \end{array}$$

Since we always have a canonical map to the final object e, this means that, for any finite family $(\prod_{i\in I} x_i)$ of objects of $\mathcal{S}$, there are canonical maps

(6.5.1.8)
$$\prod_{i\in I} x_i \to x_i$$

called the *ith projection* for $1 \leq i \leq n$.

Proposition 6.5.2 *The family of maps* (6.5.1.8) *exhibits $\prod_{i\in I} x_i$ as the limit of the diagram $(x_i)_{i\in I}$ (seen as a functor $I \to \mathcal{S}$). This identification is functorial.*

Proof Let us consider a family $(x_i)_{i\in I}$ of endofunctors of $\mathcal{S}$. To check that the projections (6.5.1.8) exhibit $\prod_{i\in I} x_i$ as the limit of the diagram $(x_i)_{i\in I}$ in $\underline{\mathrm{Hom}}(\mathcal{S}, \mathcal{S})$, since limits are computed fibrewise in categories of functors, it is sufficient to check that, for any object y of $\mathcal{S}$, the projections

$$\prod_{i\in I} x_i(y) \to x_i(y)$$

exhibit $\prod_{i \in I} x_i(y)$ as the limit of the diagram $(x_i(y))_{i \in I}$ in $\mathcal{S}$. This shows that the last assertion, about functoriality, is automatically true. It thus remains to prove the first assertion. Let C be a locally $\mathbf{U}$-small category which has limits of type I. Let us consider an I-indexed family of objects x_i in C. We denote by x the limit of the corresponding functor $F \colon I \to C$ (with $F(i) = x_i$). Since we have

$$\underline{\mathrm{Hom}}(I, C) \simeq C^I = \prod_{i \in I} C,$$

we get the canonical isomorphism

$$\underline{\mathrm{Hom}}(I, C)/F \simeq \prod_{i \in I} C/x_i \,.$$

Therefore, we then have the following commutative diagram, in which all the squares are Cartesian, where, for each i, the functor $h_i \colon C^{\mathrm{op}} \to \mathcal{S}$ classifies the left fibration $(C/x_i)^{\mathrm{op}} \to C^{\mathrm{op}}$.

$$\begin{array}{ccccccc} C/F & \longrightarrow & \prod_{i \in I} C/x_i & \longrightarrow & \prod_{i \in I} \mathcal{S}_\bullet^{\mathrm{op}} & \longrightarrow & \mathcal{S}_\bullet^{\mathrm{op}} \\ \downarrow & & \downarrow & & \downarrow & & \downarrow \\ C & \longrightarrow & C^I & \xrightarrow{\prod_{i \in I} h_i^{\mathrm{op}}} & \prod_{i \in I} \mathcal{S}^{\mathrm{op}} & \xrightarrow{(\prod_{i \in I})^{\mathrm{op}}} & \mathcal{S}^{\mathrm{op}} \end{array}$$

In other words, since there is a canonical fibrewise equivalence $C/x \to C/F$ over C, we have proved that x also represents the presheaf

$$y \mapsto \prod_{i \in I} \mathrm{Hom}_C(y, x_i) \,.$$

For $C = \mathcal{S}$, since the functor $\mathrm{Hom}_{\mathcal{S}}(e, -)$ is canonically isomorphic to the identity, this shows that $\prod_{i \in I} x_i$ is the limit of the family $(x_i)_{i \in I}$ in $\mathcal{S}$. □

Proposition 6.5.3 *Let y be an object of $\mathcal{S}$, seen as a functor $\Delta^0 \to \mathcal{S}$. There is a canonical equivalence $Ly_!(x) \xrightarrow{\sim} x \times y$ for all objects x of $\mathcal{S}$, functorially in both x and y. In particular, there is a natural equivalence*

$$\mathrm{Hom}_{\mathcal{S}}(x \times y, z) \simeq \mathrm{Hom}_{\mathcal{S}}(x, \mathrm{Hom}_{\mathcal{S}}(y, z)) \,.$$

Proof Applying Theorem 6.3.13 for $A = \Delta^0$ and $C = \mathcal{S}$, we obtain the diagram

$$\mathcal{S} \xleftarrow{\sim} \underline{\mathrm{Hom}}_!(\mathcal{S}, \mathcal{S}) \subset \underline{\mathrm{Hom}}(\mathcal{S}, \mathcal{S})$$

which, in turn, defines the functor

$$\begin{aligned} \mathcal{S} \times \mathcal{S} &\to \mathcal{S} \\ (x, y) &\mapsto Ly_!(x) \,. \end{aligned}$$

Since, for the terminal object e of $\mathcal{S}$, we have $Ly_!(e) \simeq y$ (functorially in y), there is a natural map $Ly_!(x) \to y$. Since $Le_! \simeq 1_{\mathcal{S}}$, the map $y \to e$ induces a natural transformation $Ly_!(x) \to x$. Since, by the previous proposition, $x \times y$ is the limit of (x, y) functorially in both variables, this defines a functorial map

$$Ly_!(x) \to x \times y\,.$$

We shall prove that, for each fixed y, this map is invertible. The functoriality of this natural transformation means that it is defined for x an arbitrary functor $x\colon A \to \mathcal{S}$ (with A any simplicial set). As such, one may identify $x \times y$ with $q_!q^*(x)$, functorially in x, where $q\colon L \to \Delta^0$ is the Kan fibration classified by y, by Remark 6.1.24. In particular, the functor $(-)\times y$ commutes with **U**-small colimits. If an object x of $\mathcal{S}$ classifies the Kan fibration $p\colon X \to \Delta^0$, then there is a canonical invertible map

$$\varinjlim_{k\in K} e \simeq x\,.$$

Therefore, it is sufficient to check that the canonical map $Ly_!(e) \to e\times y = y$ is an equivalence, which is true by construction. The last assertion is a particular case of Theorem 6.3.4 for $u = y$. □

Definition 6.5.4 Let C be an ∞-category, and I a small set. The *product* (*coproduct*) of an I-indexed family of objects $(x_i)_{i\in I}$ of C is its limit (colimit), seen as a functor $I \to C$; one writes such a product (coproduct) as

$$\prod_{i\in I} x_i \quad \Big(\coprod_{i\in I} x_i, \text{ respectively}\Big)\,.$$

In the case where $I = \{1,\dots,n\}$ for some non-negative integer n, one also writes

$$\prod_{i\in I} x_i = x_1 \times \cdots \times x_n \quad \Big(\coprod_{i\in I} x_i = x_1 \amalg \cdots \amalg x_n, \text{ respectively}\Big)\,.$$

Proposition 6.5.5 *Let I be a small set. If an ∞-category C has I-indexed products, then the category $ho(C)$ has I-indexed products. Furthermore, the canonical functor $C \to N(ho(C))$ commutes with products.*

Proof It follows right away from Theorem 1.6.6 that the canonical functor $ho(C^I) \to ho(C)^I$ is an isomorphism of categories, functorially in I and C. Therefore, Remark 6.1.5 shows that $ho(C)$ has I-indexed products whenever it is so for C, and that the functor $C \to N(ho(C))$ commutes with I-indexed products. □

Proposition 6.5.6 *Let s be an object of an ∞-category C such that the product with s always exists in C. Then the canonical functor $C/s \to C$ has a right adjoint. The latter assigns to each object t of C the product $s\times t$, equipped with* *the canonical map $s \times t \to s$, and the co-unit is given by the other canonical map $s \times t \to t$.*

Proof After enlarging the ambient universe, we may assume that C is $\mathbf{U}$-small. Let t be an object of C; we choose a product y of s and t, and write $g\colon y \to s$ for the canonical map. By Proposition 6.5.2 there is a functorial equivalence

$$\mathrm{Hom}_C(x, s) \times \mathrm{Hom}_C(x, t) \simeq \mathrm{Hom}_C(x, y)\,.$$

In other words, up to a fibrewise equivalence over C, the right fibration $C/y \to C$ is the pull-back of the product right fibration $C/s \times C/t \to C \times C$ along the diagonal $C \to C \times C$. Hence there is a homotopy Cartesian square of the form

$$\begin{array}{ccc} C/y & \longrightarrow & C/t \\ \downarrow & & \downarrow \\ C/s & \longrightarrow & C \end{array}$$

(in which the map $C/y \to C$ is the one induced by g, up to homotopy). Given a map $f\colon x \to s$ in C, seen as an object (x, f) of C/s (through the equality $\Delta^0 * \Delta^0 = \Delta^1$), this means that the homotopy fibre of the map $C/y \to C/s$ at (x, f) is canonically equivalent to the homotopy fibre of $C/t \to C$ at x, which is the ∞-groupoid $C(x, t)$. On the other hand, there is the canonical commutative square below, in which the upper horizontal map is a trivial fibration, because $C/s \to C$ is a right fibration.

$$\begin{array}{ccc} (C/s)/(y, g) & \longrightarrow & C/y \\ \downarrow & & \downarrow \\ C/s & \longrightarrow & C \end{array}$$

This means that the homotopy fibre of the map $C/y \to C/s$ at (x, f) is the ∞-groupoid $C/s((x, f), (y, g))$. In other words, by Propositions 5.3.20 and 5.8.2, there is an equivalence

$$\mathrm{Hom}_{C/s}((x, f), (y, g)) \simeq \mathrm{Hom}_C(x, t)$$

in $\mathcal{S}$, functorially in x. We conclude with Proposition 6.1.11. □

Proposition 6.5.7 *Let $p\colon X \to A$ be a right fibration between $\mathbf{U}$-small simplicial sets, classified by a functor $F\colon A^{\mathrm{op}} \to \mathcal{S}$. There is a canonical equivalence of ∞-categories*

$$\underline{\mathrm{Hom}}(X^{\mathrm{op}}, \mathcal{S}) \simeq \underline{\mathrm{Hom}}(A^{\mathrm{op}}, \mathcal{S})/F$$

such that the canonical functor $\underline{\mathrm{Hom}}(A^{\mathrm{op}}, \mathcal{S})/F \to \underline{\mathrm{Hom}}(A^{\mathrm{op}}, \mathcal{S})$ corresponds to the functor $p_!\colon \underline{\mathrm{Hom}}(X^{\mathrm{op}}, \mathcal{S}) \to \underline{\mathrm{Hom}}(A^{\mathrm{op}}, \mathcal{S})$.

Proof We have $p_!(e) = F$, so that there is a canonical commutative square of the form

$$\begin{array}{ccc} \underline{\mathrm{Hom}}(X^{\mathrm{op}}, \mathcal{S}) & \xrightarrow{p_!} & \underline{\mathrm{Hom}}(A^{\mathrm{op}}, \mathcal{S}) \\ \uparrow & & \uparrow \\ \underline{\mathrm{Hom}}(X^{\mathrm{op}}, \mathcal{S})/e & \longrightarrow & \underline{\mathrm{Hom}}(A^{\mathrm{op}}, \mathcal{S})/F \end{array}$$

in which the left vertical map is a trivial fibration, because e is a final object. Choosing a section of this trivial fibration and composing with the lower horizontal map above defines a comparison functor

$$\underline{\mathrm{Hom}}(X^{\mathrm{op}}, \mathcal{S}) \to \underline{\mathrm{Hom}}(A^{\mathrm{op}}, \mathcal{S})/F .$$

To prove that the latter is an equivalence of ∞-categories, it is sufficient to prove that, for any simplicial set K, the induced map

$$\pi_0(k(\underline{\mathrm{Hom}}(K^{\mathrm{op}}, \underline{\mathrm{Hom}}(X^{\mathrm{op}}, \mathcal{S})))) \to \pi_0(k(\underline{\mathrm{Hom}}(K^{\mathrm{op}}, \underline{\mathrm{Hom}}(A^{\mathrm{op}}, \mathcal{S})/F)))$$

is bijective. By Proposition 4.2.12, there is an equivalence

$$\underline{\mathrm{Hom}}(K^{\mathrm{op}}, \underline{\mathrm{Hom}}(A^{\mathrm{op}}, \mathcal{S})/F) \simeq \underline{\mathrm{Hom}}(K^{\mathrm{op}}, \underline{\mathrm{Hom}}(A^{\mathrm{op}}, \mathcal{S})/F_K)$$

where F_K denotes the functor obtained by composing F with the Cartesian projection $A^{\mathrm{op}} \times K^{\mathrm{op}} \to A^{\mathrm{op}}$. We are thus reduced to proving that the induced functor

$$ho(\underline{\mathrm{Hom}}(K^{\mathrm{op}} \times X^{\mathrm{op}}, \mathcal{S})) \to ho(\underline{\mathrm{Hom}}(K^{\mathrm{op}} \times A^{\mathrm{op}}, \mathcal{S})/F_K)$$

induces a bijection when one passes to the sets of isomorphism classes of objects. For this, it is sufficient to prove that this functor is essentially surjective, full and conservative. To prove the essential surjectivity, let us consider a functor $\Phi\colon K^{\mathrm{op}} \times A^{\mathrm{op}} \to \mathcal{S}$ together with a map $f\colon \Phi \to F_K$. If Φ classifies a right fibration with **U**-small fibres $\pi\colon E \to K \times A$, by virtue of Theorem 5.4.5, the map f may be represented by a map $\tilde{f}\colon E \to K \times X$ over $K \times A$. We then choose a factorisation of $\tilde{f}$ into a final map $i\colon E \to E'$, followed by a right fibration $\pi'\colon E' \to K \times X$, and observe that π' is fibrewise equivalent to a right fibration with **U**-small fibres: for any **U**-small simplicial set L and any map $L \to K$, it follows from Proposition 6.1.16 that the pull-back $E'_L = L \times_K E'$ is equivalent to a right fibration with **U**-small fibres. Therefore, by virtue of Corollary 5.4.10, there exists a functor $\Psi\colon K^{\mathrm{op}} \times X^{\mathrm{op}} \to \mathcal{S}$ which classifies π', at least up to a fibrewise equivalence over $K \times X$. The final map i induces an isomorphism from the image of Ψ with the pair (Φ, f). It remains to prove the properties of fullness and of conservativity. To prove the property of fullness, we first observe that, if two right fibrations $f, g\colon E \to K \times X$ are J-homotopic, then the corresponding objects of $ho(\underline{\mathrm{Hom}}(K^{op} \times X^{op}, \mathcal{S}))$ are isomorphic. Indeed, if $h\colon J \times E \to K \times X$ is a homotopy from f to g, we may factor h into a final map followed by a right fibration $h'\colon E' \to K \times X$, and both (E, f) and (E, g) will be weakly equivalent to (E', h') in the contravariant model category structure over $K \times X$. We also observe that any map in $ho(\underline{\mathrm{Hom}}(K^{op} \times A^{op}, \mathcal{S})/F_K)$ is induced by a functor $\Phi\colon \Delta^2 \to \underline{\mathrm{Hom}}(K^{op} \times A^{op}, \mathcal{S})$ such that $\Phi(2) = F_K$. Applying Theorem 5.4.5 once again, for the contravariant model structure over $(\Delta^2)^{op} \times K \times A$, these two observations imply the property of fullness. Since, by Proposition 3.4.8, the functor

$$\underline{\mathrm{Hom}}(K^{op} \times A^{op}, \mathcal{S})/F_K \to \underline{\mathrm{Hom}}(K^{op} \times A^{op}, \mathcal{S})$$

is conservative, it is now sufficient to prove that the functor

$$\mathbf{L}(1_K \times p)_!\colon RFib(K \times X) \to RFib(K \times A)$$

is conservative. Theorem 4.1.16 and Proposition 6.1.16 show that it is sufficient to prove this for $K = \Delta^0$ (just to simplify the notation). Let $u\colon E \to F$ and $r\colon F \to X$ be two morphisms of simplicial sets such that r and $q = ru$ are right fibrations. By virtue of Proposition 4.1.11, u is a final map if and only if it is a weak equivalence of the contravariant model category structure over X. But pr and pq are also right fibrations, whence u is a final map if and only if it is a weak equivalence of the contravariant model category structure over A. This implies the property of conservativity, and thus achieves the proof. □

The special case where $X = A/a$ gives the following corollary.

Corollary 6.5.8 *Let a be an object of a* **U***-small ∞-category A. There is a canonical equivalence of ∞-categories*

$$\underline{\mathrm{Hom}}((A/a)^{op}, \mathcal{S}) \simeq \underline{\mathrm{Hom}}(A^{op}, \mathcal{S})/h_A(a)\,.$$

The special case where $A = \Delta^0$ is also of interest.

Corollary 6.5.9 *Let x be an object of $\mathcal{S}$, corresponding to a* **U***-small ∞-groupoid X. There is a canonical equivalence of ∞-categories of the form*

$$\underline{\mathrm{Hom}}(X^{op}, \mathcal{S}) \simeq \mathcal{S}/x\,.$$

Proposition 6.5.10 *Let s be an object of an ∞-category such that the canonical functor $C/s \to C$ has a right adjoint. Then, for any object t of C, the product of s and t exists in C.*

Proof We may assume that C is **U**-small. Let $p\colon C/s \to C$ be the canonical functor, and let us choose a right adjoint q of p. Then, by virtue of Theorem 6.1.23, The functor $p_!\colon \underline{\mathrm{Hom}}((C/s)^{op}, \mathcal{S}) \to \underline{\mathrm{Hom}}(C^{op}, \mathcal{S})$ is a left adjoint of the functor $q_! \simeq p^*$. Since $\underline{\mathrm{Hom}}(C^{\mathrm{op}}, \mathcal{S})$ has **U**-small limits, it has finite products, hence by virtue of Propositions 6.5.7 and 6.5.6, the functor $p_!q_! \simeq (pq)_!$ takes a presheaf $F\colon C^{\mathrm{op}} \to \mathcal{S}$ to the presheaf $h_C(s) \times F$. With the identification $h_C(p(q(t))) \simeq (pq)_!(h_C(t))$, this means that $p(q(t))$ represents the presheaf

$$x \mapsto \mathrm{Hom}(x, s) \times \mathrm{Hom}(x, t)$$

on C, whence that the product of s and t is representable. □

6.6 Fibre Products

6.6.1 Let us consider $\Delta^1 \times \Delta^1$. We may identify the horn Λ^2_2 with the full subcategory of $\Delta^1 \times \Delta^1$ whose objects are the pairs (i, j) such that $1 \in \{i, j\}$. We define

$$\varphi\colon \Delta^1 \times \Lambda^2_2 \to \Delta^1 \times \Delta^1$$

as the unique functor such that

$$\varphi(\varepsilon, (i, j)) = \begin{cases} (0, 0) & \text{if } \varepsilon = 0, \\ (i, j) & \text{if } \varepsilon = 1. \end{cases}$$

Lemma 6.6.2 *The obvious commutative square*

$$\begin{array}{ccc} \{0\} \times \Lambda^2_2 & \hookrightarrow & \Delta^1 \times \Lambda^2_2 \\ \downarrow & & \downarrow{\scriptstyle\varphi} \\ \Delta^0 & \xrightarrow{(0,0)} & \Delta^1 \times \Delta^1 \end{array}$$

is homotopy coCartesian in the Joyal model category structure.

Proof One checks that we have coCartesian squares of the following form.

$$\begin{array}{ccccccc} \Delta^0 \times \Lambda^2_2 & \longrightarrow & (\Delta^0 \times \Lambda^2_2) \amalg \Lambda^2_2 & \xrightarrow{\sim} & \Delta^0 \times \partial\Delta^1 \times \Lambda^2_2 & \longrightarrow & \Delta^0 \times \Delta^1 \times \Lambda^2_2 \\ \downarrow & & \downarrow & & \downarrow & & \downarrow \\ \Delta^0 & \longrightarrow & \Delta^0 \amalg \Lambda^2_2 & = & \Delta^0 \amalg \Lambda^2_2 & \longrightarrow & \Delta^0 \diamond \Lambda^2_2 \end{array}$$

Furthermore, we observe that $\Delta^0 * \Lambda^2_2 = \Delta^1 \times \Delta^1$. In other words, the commutative square of the lemma induces the canonical map

$$\Delta^0 \diamond \Lambda^2_2 \to \Delta^0 * \Lambda^2_2 = \Delta^1 \times \Delta^1$$

which is known to be a weak categorical equivalence, by Proposition 4.2.3. □

6.6.3 Given an ∞-category C, specifying a map $\Lambda^2_2 \to C$ is the same as providing a diagram of the following form in C.

(6.6.3.1)
$$\begin{array}{ccc} & & x \\ & & \big\downarrow{\scriptstyle f} \\ y' & \xrightarrow{v} & y \end{array}$$

Such a datum will be called a *lower corner in C*.

A *commutative square in C* is a map $\Delta^1 \times \Delta^1 \to C$. Hence a commutative square in C consists of a diagram of the form

(6.6.3.2)
$$\begin{array}{ccc} x' & \xrightarrow{u} & x \\ \big\downarrow{\scriptstyle f'} & & \big\downarrow{\scriptstyle f} \\ y' & \xrightarrow{v} & y \end{array}$$

together with a map $h\colon x' \to y$, as well as with two morphisms $\Delta^2 \to C$ which express that the triangles

$$\begin{array}{ccc} x' & \xrightarrow{u} & X \\ & \searrow^{h} & \big\downarrow{\scriptstyle f} \\ & & y \end{array} \quad \text{and} \quad \begin{array}{ccc} x' & & \\ \big\downarrow{\scriptstyle f'} & \searrow^{h} & \\ y' & \xrightarrow{v} & y \end{array}$$

commute (to see this, recall that, as explained in the proof of Lemma 3.1.25, the nerve of any partially ordered set E is isomorphic to the colimit of the nerve of its non-empty finite totally ordered subsets, and apply this principle to the product $E = [1] \times [1]$).

Finally, we observe that giving a commutative square of the form (6.6.3.2) is equivalent to giving a morphism of lower corners of the following form.

(6.6.3.3)
$$\begin{array}{ccc} & & x' \\ & & \big\downarrow{\scriptstyle 1_{x'}} \\ x' & \xrightarrow{1_{x'}} & x' \end{array} \quad \to \quad \begin{array}{ccc} & & x \\ & & \big\downarrow{\scriptstyle f} \\ y' & \xrightarrow{v} & y \end{array}$$

This is made precise by the next proposition, taking into account that two parallel arrows in a homotopy Cartesian square have equivalent homotopy fibres.

Proposition 6.6.4 *Let C be an ∞-category. There is a canonical homotopy Cartesian square of the following form, in which the lower horizontal map is* *the constant diagram functor.*

$$\begin{array}{ccc} \underline{\mathrm{Hom}}(\Delta^1 \times \Delta^1, C) & \xrightarrow{\varphi^*} & \underline{\mathrm{Hom}}(\Delta^1 \times \Lambda^2_2, C) \\ {\scriptstyle ev_{0,0}}\downarrow & & \downarrow{\scriptstyle ev_0} \\ C & \longrightarrow & \underline{\mathrm{Hom}}(\Lambda^2_2, C) \end{array}$$

Proof Since the functor $\underline{\mathrm{Hom}}(-, C)$ sends homotopy coCartesian squares of ∞-categories to homotopy Cartesian ones, this follows right away from Lemma 6.6.2. □

Definition 6.6.5 A *Cartesian square* in an ∞-category C is a commutative square of the form (6.6.3.2) which exhibits, via the induced map (6.6.3.3), the object x' as the limit of the induced lower corner (6.6.3.1).

A *coCartesian square* in an ∞-category C is a commutative square which can be interpreted as a Cartesian square in C^{op} via the unique isomorphism $(\Delta^1)^{\mathrm{op}} \simeq \Delta^1$.

Remark 6.6.6 Given a diagram of the form (6.6.3.1) in an ∞-category, its limit, whenever it exists, is usually denoted by $y' \times_y x$ and called the *fibre product* of y' and x over y. In this case, a commutative square of the form (6.6.3.2) defines a map $x' \to y' \times_y x$, and the latter is invertible if and only if the given square is Cartesian.

Proposition 6.6.7 *Let us a consider a commutative square of the form*

(6.6.7.1)
$$\begin{array}{ccc} x' & \xrightarrow{u} & x \\ {\scriptstyle f'}\downarrow & & \downarrow{\scriptstyle f} \\ y' & \xrightarrow{v} & y \end{array}$$

in an ∞-category C (there is in particular a specified composition h both of f and u, and of v and f'). Since $(1,1)$ is a final object of $\Delta^1 \times \Delta^1$, it corresponds to a commutative square of the form

(6.6.7.2)
$$\begin{array}{ccc} (x', h) & \xrightarrow{u} & (x, f) \\ {\scriptstyle f'}\downarrow & & \downarrow{\scriptstyle f} \\ (y', v) & \xrightarrow{v} & (y, 1_y) \end{array}$$

in C/y. The square (6.6.7.1) *is Cartesian in C if and only if the square* (6.6.7.2) *is Cartesian in C/y.*

Proof We may assume that C is $\mathbf{U}$-small for some universe $\mathbf{U}$. Since the

Yoneda embedding is fully faithful, commutes with limits, and is compatible with slicing over an object, by Corollary 6.5.8, we may replace C by $\underline{\mathrm{Hom}}(C^{\mathrm{op}}, \mathcal{S})$. In particular, by Proposition 6.5.7, we may assume that C as well as all its slices are locally $\mathbf{U}$-small and are complete and cocomplete. Proving this proposition then amounts to showing that the canonical functor $C/y \to C$ preserves Cartesian squares (it will then detect Cartesian squares because it is conservative, since it is a right fibration). Let us assume that the square (6.6.7.2) is Cartesian. Let us form the limit z of the lower corner of diagram (6.6.7.1),

(6.6.7.3)
$$\begin{array}{ccc} z & \xrightarrow{u} & x \\ {\scriptstyle g'}\downarrow & & \downarrow{\scriptstyle g} \\ y' & \xrightarrow{v} & y \end{array}$$

and let us call $k\colon z \to y$ the canonical map. We get a commutative square of C/y of the following form.

(6.6.7.4)
$$\begin{array}{ccc} (z,k) & \xrightarrow{u} & (x,g) \\ {\scriptstyle g'}\downarrow & & \downarrow{\scriptstyle f} \\ (y',v) & \xrightarrow{v} & (y,1_y) \end{array}$$

We want to prove that the comparison map $x' \to z$, obtained from (6.6.7.1), is invertible. But this comparison map comes from a map of C/y, because it is compatible, by construction with the whole diagram (6.6.7.1). On the other hand, the commutative square (6.6.7.4) defines a map $(z,k) \to (x',h)$ in C/y. The composition $(x',h) \to (z,h) \to (x',h)$ is the identity, because it can be seen as a map in $\underline{\mathrm{Hom}}(\Lambda^2_2, C/y)/(v,f)$, where (v,f) is the lower corner defined by the maps v and f in diagram (6.6.7.2). Hence the composition $x' \to z \to x'$ is the identity in C. Similarly, the composition $z \to x' \to z$ is the identity because it can be promoted to a map in $\underline{\mathrm{Hom}}(\Lambda^2_2, C)/(v,f)$, where, this time, (v,f) is the lower corner defined by the maps v and f in diagram (6.6.7.1). □

Proposition 6.6.8 *Let C be an ∞-category with finite products. For any objects x and y of C and any maps $x \to e$ and $y \to e$, where e is a final object of C, there is a canonical Cartesian square of the form*

(6.6.8.1)
$$\begin{array}{ccc} x \times y & \xrightarrow{p} & y \\ {\scriptstyle q}\downarrow & & \downarrow \\ x & \longrightarrow & e \end{array}$$

where p and q are the projections which exhibit $x \times y$ as the product of x and y.

Proof Applying Proposition 6.6.4 to C^{op}, we see that there is an isomorphism between the ∞-category of diagrams

$$\begin{array}{ccc} z & \xrightarrow{f} & y \\ {\scriptstyle g}\downarrow & & \downarrow \\ x & \longrightarrow & e \end{array}$$

and that of the following maps.

$$\begin{array}{ccc} z & \xrightarrow{f} & y \\ {\scriptstyle g}\downarrow & & \\ x & & \end{array} \quad \to \quad \begin{array}{ccc} e & \xrightarrow{1_e} & e \\ {\scriptstyle 1_e}\downarrow & & \\ e & & \end{array}$$

Since the constant upper corner defined by e is a final object of $\underline{\mathrm{Hom}}(\Lambda^2_0, C)$, there is an essentially unique way to extend any diagram $x \leftarrow z \rightarrow y$ to a commutative square whose lower right vertex is equal to e, such as diagram (6.6.8.1). As in the proof of the previous proposition, we reduce to the case where C has **U**-small limits. We see right away that the comparison map from $x \times y$ to the limit of the diagram $x \rightarrow e \leftarrow y$ induced by diagram (6.6.8.1) is invertible. □

Theorem 6.6.9 *Let C be an ∞-category. Limits of shape Λ^2_2 exist in C if and only if, for any object y of C, the slice category C/y has finite products. Furthermore, for a commutative square*

$$\begin{array}{ccc} x' & \xrightarrow{u} & x \\ \downarrow{\scriptstyle f'} & & \downarrow{\scriptstyle f} \\ y' & \xrightarrow{v} & y \end{array}$$

in C, with canonical map $h\colon x' \rightarrow y$, the following conditions are equivalent.

(i) It is Cartesian.
(ii) The map u exhibits the pair (x', f'), seen as an object of C/y', as a representation of the presheaf

$$\begin{aligned} (C/y')^{\mathrm{op}} &\rightarrow \mathcal{S} \\ (t, g\colon t \rightarrow y') &\mapsto \mathrm{Hom}_{C/y}((t, vg), (x, f)), \end{aligned}$$

where the functor $(t, g) \mapsto (t, vg)$ is obtained by composing the canonical functor $(C/y)/(y', v) \rightarrow C/y$ with a choice of section of the canonical trivial fibration $(C/y)/(y', v) \rightarrow C/y'$.
(iii) The maps f' and u exhibit (x', h) as the product of (x, f) and of (y', v) in C/y.

Proof The first assertion is a direct consequence of Propositions 6.5.6 and 6.6.7. This also proves the rest of the theorem in the case where, furthermore, the ∞-category C has limits of type Λ^2_2. The general case follows from the fact that, since the Yoneda embedding is fully faithful, is compatible with slices and commutes with limits, we may choose a universe $\mathbf{U}$ such that C is locally $\mathbf{U}$-small, and replace C by the ∞-category $\underline{\mathrm{Hom}}(C^{\mathrm{op}}, \mathcal{S})$. □

Corollary 6.6.10 *Let C be an ∞-category. We consider a functor $\Delta^1 \times \Delta^2 \to C$, seen as a commutative diagram whose restriction to $\Delta^1 \times \Lambda^2_1$ is of the form below.*

$$\begin{array}{ccccc} x'' & \xrightarrow{u'} & x' & \xrightarrow{u} & x \\ {\scriptstyle p''}\downarrow & & {\scriptstyle p'}\downarrow & & \downarrow{\scriptstyle p} \\ y'' & \xrightarrow{v'} & y' & \xrightarrow{v} & y \end{array}$$

We assume that the square

$$\begin{array}{ccc} x' & \xrightarrow{u} & x \\ {\scriptstyle p'}\downarrow & & \downarrow{\scriptstyle p} \\ y' & \xrightarrow{v} & y \end{array}$$

is Cartesian. Then the square

$$\begin{array}{ccc} x'' & \xrightarrow{u'} & x' \\ {\scriptstyle p''}\downarrow & & \downarrow{\scriptstyle p'} \\ y'' & \xrightarrow{v'} & y' \end{array}$$

is Cartesian if and only if the composed square

$$\begin{array}{ccc} x'' & \xrightarrow{u''} & x \\ {\scriptstyle p''}\downarrow & & \downarrow{\scriptstyle p} \\ y'' & \xrightarrow{v''} & y \end{array}$$

is Cartesian.

Proof We easily reduce to the case where $C = \mathcal{S}$ using the Yoneda embedding (and choosing a universe $\mathbf{U}$ so that C is $\mathbf{U}$-small). In other words, without loss of generality, we may assume that C has pull-backs. For any map $w : s \to t$ in C the functor of composition by w

$$w_! : C/s \to C/t$$

then has a right adjoint w^*, obtained by pulling back along w; see Theorem 6.6.9.

Furthermore, interpreting the Yoneda embedding through the dual version of Theorem 5.4.5, we see that the composition of the functors

$$v'_!: C/y'' \to C/y' \quad \text{and} \quad v_!: C/y' \to C/y$$

is J-homotopic to the functor

$$v''_!: C/y'' \to C/y\,.$$

Therefore, the composition of v^* and v'^* is a right adjoint to $v''_!$, by Propositions 6.1.7 and 6.1.8. In other words, whenever the comparison map

$$(x', p') \to v^*(x, p)$$

is invertible, the comparison map

$$(x'', p'') \to v''^*(x, p)$$

is isomorphic to the comparison map

$$(x'', p'') \to v'^*(x', p')\,.$$

Hence one is invertible if and only if the other is invertible. □

Remark 6.6.11 Given any model category $\mathcal{C}$, we may consider the injective model category structure on the category of arrows $\mathrm{Arr}(\mathcal{C})$. This operation is functorial in the sense that any Quillen adjunction (or Quillen equivalence) from $\mathcal{C}$ to $\mathcal{D}$ induces a Quillen adjunction (or equivalence) from $\mathrm{Arr}(\mathcal{C})$ to $\mathrm{Arr}(\mathcal{D})$. In particular, since (a variant of) Proposition 5.4.3 provides, for any simplicial set A, a Quillen equivalence of the form

$$t_!: \mathrm{Arr}(sSet/A) \rightleftarrows sSet/(\Delta^1 \times A) : t^*\,,$$

where $sSet/C$ is equipped with the contravariant model category structure over C for $C = A$ or $C = \Delta^1 \times A$, we get a Quillen equivalence

$$\mathrm{Arr}(\mathrm{Arr}(sSet/A)) \rightleftarrows \mathrm{Arr}(sSet/(\Delta^1 \times A))$$

as well as a Quillen equivalence

$$\mathrm{Arr}(sSet/(\Delta^1 \times A)) \rightleftarrows sSet/(\Delta^1 \times \Delta^1 \times A)\,.$$

Therefore, any commutative square of the form

$$\begin{array}{ccc} X' & \xrightarrow{u} & X \\ {\scriptstyle q}\downarrow & & \downarrow{\scriptstyle p} \\ Y' & \xrightarrow{v} & Y \end{array} \qquad (6.6.11.1)$$

may be interpreted as an object of $RFib(\Delta^1 \times \Delta^1 \times Y)$. In the case when all the simplicial sets of diagram (6.6.11.1) are **U**-small for a universe **U**, for any map with **U**-small codomain $\pi: Y \to A$, we get in fact an object in $ho(\underline{\mathrm{Hom}}(\Delta^1 \times \Delta^1 \times A, \mathcal{S}))$, by Theorem 5.4.5. If both p and q are right fibrations with **U**-small fibres, this can be refined as follows. We may choose two functors

$$F: Y^{\mathrm{op}} \to \mathcal{S} \quad \text{and} \quad F': Y'^{\mathrm{op}} \to \mathcal{S}$$

classifying p and q, respectively. Then the commutative square above corresponds to a map

$$F' \to v^*(F) \tag{6.6.11.2}$$

in $\underline{\mathrm{Hom}}(Y'^{\mathrm{op}}, \mathcal{S})$ via the correspondence provided by Corollary 5.4.7. The map (6.6.11.2) is invertible if and only if the square (6.6.11.1) is homotopy Cartesian. In the case when all the simplicial sets in (6.6.11.1) are **U**-small, for any left fibration $\pi: Y \to A$, the square defines a commutative diagram of the form

$$\begin{array}{ccc} \pi_! v_!(F') & \longrightarrow & \pi_!(F) \\ \big\downarrow & & \big\downarrow \\ \pi_! v_!(e) & \longrightarrow & \pi_!(e) \end{array} \tag{6.6.11.3}$$

which is Cartesian if and only if the square (6.6.11.1) is homotopy Cartesian, by Proposition 6.5.7 and Theorem 6.6.9. Moreover, all Cartesian squares of the ∞-category $\underline{\mathrm{Hom}}(A^{\mathrm{op}}, \mathcal{S})$ are obtained in this way. In particular, Cartesian squares of $\mathcal{S}$ correspond to homotopy Cartesian squares of the form (6.6.11.1) in the Kan–Quillen model category structure, restricted to **U**-small simplicial sets.

Theorem 6.6.12 *Let* **U** *be a universe, and A a* **U***-small ∞-category. For any morphism $p: F \to G$ in $\widehat{A} = \underline{\mathrm{Hom}}(A^{\mathrm{op}}, \mathcal{S})$, the pull-back functor*

$$p^*: \widehat{A}/G \to \widehat{A}/F, \quad (Y \to G) \mapsto (Y \times_G F \to F)$$

has a right adjoint (and thus commutes with colimits).

Proof The presheaf G classifies a right fibration $A' \to A$. Replacing A by A', we see that, by virtue of Proposition 6.5.7, we may assume, without loss of generality, that $G = e$ is the final object. By Proposition 6.3.9, it is sufficient to prove that the functor p^* commutes with colimits. Since the canonical functor $\widehat{A}/F \to \widehat{A}$ is conservative (being a right fibration) and commutes with colimits (having a right adjoint), it is sufficient to prove that the functor $X \mapsto X \times F$ commutes with colimits. Since the evaluation functors $X \mapsto X(a)$, $a \in A_0$, form a conservative family of functors which commute with colimits as well as with limits (hence with products), it is sufficient to prove that the functor

$x \mapsto x \times y$ commutes with colimits in $\mathcal{S}$ for all y. But, in this case, we already know that such a functor has a right adjoint, by Proposition 6.5.3. □

6.7 Duality

We fix a universe **U**.

6.7.1 Let B be an ∞-category with **U**-small colimits, and A a **U**-small ∞-category. The *relative Yoneda embedding*

$$h_{A/B}\colon A \times B \to \underline{\mathrm{Hom}}(A^{\mathrm{op}}, B) \tag{6.7.1.1}$$

is the functor defined by $h_{A/B}(a, x) = a \otimes x$, where

$$a \otimes x = a_!(x) \tag{6.7.1.2}$$

(we recall that, by virtue of Theorem 6.3.13, the assignment $a \mapsto a_!$ is indeed a functor). In other words, we have functorial identifications

$$\mathrm{Hom}(a \otimes x, f) \simeq \mathrm{Hom}(x, f(a)) \tag{6.7.1.3}$$

(in the ∞-category of **V**-small ∞-groupoids for **V** large enough). Given another ∞-category C with **U**-small colimits, any functor

$$\Phi\colon \underline{\mathrm{Hom}}(A^{\mathrm{op}}, B) \to C \tag{6.7.1.4}$$

restricts to a functor $\Phi h_{A/B}\colon A \times B \to C$, which, by transposition, defines a functor

$${}^{\mathrm{t}}\Phi\colon B \to \underline{\mathrm{Hom}}(A, C)\,. \tag{6.7.1.5}$$

If Φ commutes with **U**-small colimits, then so does ${}^{\mathrm{t}}\Phi$. Indeed, we have

$${}^{\mathrm{t}}\Phi(x)(a) = \Phi(a_!(x))$$

and, since Φ commutes with **U**-small colimits, and since colimits are computed fibrewise in $\underline{\mathrm{Hom}}(A, C)$, this proves our claim. The assignment $\Phi \mapsto {}^{\mathrm{t}}\Phi$ thus restricts to a functor

$$\underline{\mathrm{Hom}}_!(\underline{\mathrm{Hom}}(A^{\mathrm{op}}, B), C) \to \underline{\mathrm{Hom}}_!(B, \underline{\mathrm{Hom}}(A, C))\,. \tag{6.7.1.6}$$

The aim of this section is to prove the following theorem.

Theorem 6.7.2 *The functor* (6.7.1.6) *is an equivalence of ∞-categories.*

Remark 6.7.3 In the case where $B = \mathcal{S}$ is the ∞-category of **U**-small ∞-groupoids, Theorem 6.3.13 provides a canonical equivalence of ∞-categories

$$\underline{\mathrm{Hom}}_!(B, \underline{\mathrm{Hom}}(A, C)) \simeq \underline{\mathrm{Hom}}(A, C)$$

so that the functor (6.7.1.6) is an equivalence of ∞-categories, by a second application of Theorem 6.3.13. In general, Theorem 6.7.2 is thus a generalisation of Theorem 6.3.13, and the ingredients of the proof have a lot in common, as we shall see below.

6.7.4 Let $\Psi\colon B \to \underline{\mathrm{Hom}}(A, C)$ be a functor which commutes with **U**-small colimits. The associated functor $\bar{\Psi}\colon A \times B \to C$ extends by **U**-small colimits to a functor

$$\bar{\Psi}_!\colon \widehat{A \times B} \to \widehat{C} \tag{6.7.4.1}$$

where, for any ∞-category D, $\widehat{D}$ is the full subcategory of **U**-small presheaves on D; indeed, since we have a canonical invertible map of the form

$$\bar{\Psi}_! h_{A\times B} \simeq h_C \bar{\Psi}\,, \tag{6.7.4.2}$$

and since $\bar{\Psi}_!$ commutes with colimits, it preserves the property of being a **U**-small colimit of representable presheaves. On the other hand, for any object a in A and any object x in B, there is a canonical identification of the form

$$a \otimes h_B(x) \xrightarrow{\sim} h_{A\times B}(a, x)\,. \tag{6.7.4.3}$$

To see this, using that, by virtue of Proposition 6.5.3, the Cartesian product commutes with colimits in each variable in $\mathcal{S}$, we observe that, for any object b of A and any object y of B, we have

$$(a \otimes h_B(x))(b, y) = \mathrm{Hom}_A(b, a) \times \mathrm{Hom}_B(y, x)\,. \tag{6.7.4.4}$$

In other words, formulas (6.7.4.2) and (6.7.4.3) give a functorial invertible map

$$\bar{\Psi}_!(a \otimes h_B(x)) \simeq h_C(\Psi(x)(a))\,. \tag{6.7.4.5}$$

Lemma 6.7.5 *Let* **V** *be any universe such that B is* **V***-small, and let $\mathcal{S}'$ be the ∞-category of* **V***-small ∞-groupoids. For any functor $F\colon I \to B$, the presheaf on $I^{\mathrm{op}} \times B$ defined by $(i, x) \mapsto \mathrm{Hom}_B(x, F(i))$ is* **U***-small.*

Proof The presheaf $(i, x) \mapsto \mathrm{Hom}_B(x, F(i))$ corresponds to the left fibration obtained as the pull-back along $1_{B^{\mathrm{op}}} \times F\colon B^{\mathrm{op}} \times I \to B^{\mathrm{op}} \times B$ of the left fibration $(s_B, t_B)\colon \mathrm{S}(B) \to B^{\mathrm{op}} \times B$ (5.6.1.4). Proposition 5.6.9 provides a canonical cofinal functor of the form

$$\mathrm{S}(I) \to (B^{\mathrm{op}} \times I) \times_{B^{\mathrm{op}}\times B} \mathrm{S}(B)\,.$$

Since $\mathcal{S}(I)$ is $\mathbf{U}$-small, this proves the lemma. □

6.7.6 Let $L\colon \widehat{C} \to C$ be the functor provided by Proposition 6.3.2. We now have the composed functor

$$\Psi'\colon \underline{\mathrm{Hom}}(A^{\mathrm{op}}, B) \xrightarrow{\underline{\mathrm{Hom}}(A^{\mathrm{op}}, h_B)} \widehat{A \times B} \xrightarrow{\bar{\Psi}_!} \widehat{C} \xrightarrow{L} C\,. \tag{6.7.6.1}$$

Lemma 6.7.7 *If $\Psi = {}^t\Phi$ for some functor $\Phi\colon \underline{\mathrm{Hom}}(A^{\mathrm{op}}, B) \to C$, then there is a canonical natural transformation $\Psi' \to \Phi$. The latter is invertible whenever the functor Φ commutes with $\mathbf{U}$-small colimits.*

Proof Since Φ is a functor, we get a functorial map of the form

$$\mathrm{Hom}_B(x, f(a)) \simeq \mathrm{Hom}_{\underline{\mathrm{Hom}}(A^{\mathrm{op}}, B)}(a \otimes x, f) \to \mathrm{Hom}_C(\Phi(a \otimes x), \Phi(f))$$

for any objects a in A and x in B, and for any functor $f\colon A^{\mathrm{op}} \to B$. By transposition, this defines a map

$$L\bar{\Psi}_!(h_B(f)) \to \Phi(f)$$

functorially in f. By construction, when $f = a \otimes x$, this map is the isomorphism

$$L\bar{\Psi}_!(a \otimes h_B(x)) \simeq \Phi(a \otimes x)$$

provided by formula (6.7.4.5) for $\Psi = {}^t\Phi$. Let $\mathbf{V}$ be a universe containing $\mathbf{U}$ and such that B and C are $\mathbf{V}$-small. We denote by $\mathcal{S}'$ the ∞-category of $\mathbf{V}$-small ∞-groupoids. The adjunction given by Proposition 6.3.2

$$L\colon \widehat{B} \rightleftarrows B\colon h_B$$

induces, by Theorem 6.1.22, an adjunction

$$L\colon \underline{\mathrm{Hom}}(A^{\mathrm{op}}, \widehat{B}) \rightleftarrows \underline{\mathrm{Hom}}(A^{\mathrm{op}}, B)\colon h_B\,.$$

For a functor $f\colon A^{\mathrm{op}} \to B$, since $F = h_B(f)$ is a $\mathbf{U}$-small presheaf on the product $A \times B$, by Corollary 6.2.16 and Lemma 6.7.5, interpreted via formula (6.7.1.3) and the Yoneda lemma, there is a canonical isomorphism of the form

$$\varinjlim_{x \to f(a)} a \otimes h_B(x) \simeq h_B(f)$$

in $\underline{\mathrm{Hom}}(A^{\mathrm{op}}, \widehat{B})$. Since the functor L commutes with all colimits (Proposition 6.2.15), and since Lh_B is isomorphic to the identity, this induces a canonical isomorphism

$$\varinjlim_{x \to f(a)} a \otimes x \simeq f$$

in $\underline{\mathrm{Hom}}(A^{\mathrm{op}}, B)$. This shows that the map $L\bar{\Psi}_!(h_B(f)) \to \Phi(f)$ is invertible whenever Φ commutes with **U**-small colimits, because it is then sufficient to prove this property for $f = a \otimes x$. $\square$

Proof of Theorem 6.7.2 We deduce from the previous lemma that the functor (6.7.1.6) induces a bijection

$$\pi_0(k(\underline{\mathrm{Hom}}_!(\underline{\mathrm{Hom}}(A^{\mathrm{op}}, B), C))) \to \pi_0(k(\underline{\mathrm{Hom}}_!(B, \underline{\mathrm{Hom}}(A, C)))) .$$

For any simplicial set X and any ∞-category D with **U**-small colimits, the ∞-category $\underline{\mathrm{Hom}}(X, D)$ has **U**-small colimits. The bijection above thus holds for $C = \underline{\mathrm{Hom}}(X, D)$. On the other hand, we know that a functor between ∞-categories $u \colon E \to F$ is an equivalence of ∞-categories if and only if the induced map

$$\pi_0(k(\underline{\mathrm{Hom}}(X, E))) \to \pi_0(k(\underline{\mathrm{Hom}}(X, F)))$$

is bijective for all X. Using the identifications

$$\underline{\mathrm{Hom}}(X, \underline{\mathrm{Hom}}_!(\underline{\mathrm{Hom}}(A^{\mathrm{op}}, B), D)) \simeq \underline{\mathrm{Hom}}_!(\underline{\mathrm{Hom}}(A^{\mathrm{op}}, B), C)$$

and

$$\underline{\mathrm{Hom}}(X, \underline{\mathrm{Hom}}_!(B, \underline{\mathrm{Hom}}(A, D))) \simeq \underline{\mathrm{Hom}}_!(B, \underline{\mathrm{Hom}}(A, C)) ,$$

this achieves the proof. $\square$

7

Homotopical Algebra

This chapter aims at providing tools to describe localisations of ∞-categories. In other words, it is all about the process of inverting maps. This is fundamental, because inverting maps in (the nerve of) an ordinary category is the source of many examples of ∞-categories.

The first section defines localisations through the appropriate universal property and gives a general construction of these. It also explores the basic features that can be derived from the universal property: existence of final objects or of finite products in a localisation, compatibility with adjunctions, and fully faithfulness of adjoints of localisation functors. We also study a class of examples of localisations given by proper functors with weakly contractible fibres (although it looks like a pleasant exercise about proper base change formulas, it will provide quite concrete computational tools, mainly in Section 7.3 below); see Proposition 7.1.12. Section 7.2 studies (right) calculus of fractions. For instance, a sufficient condition for such a calculus to hold is provided by the possibility of approximating weak equivalences by trivial fibrations in a suitable sense (Corollary 7.2.18). If ever all weak equivalences can be approximated by trivial fibrations, we deduce from our understanding of locally constant presheaves (i.e. from an ∞-categorical analogue of Quillen's theorem B) that pull-backs along suitable fibrations give Cartesian squares in the localisation (Theorem 7.2.25). Section 7.3 is about providing formulas to decompose limits as 'simple' limits of smaller diagrams, where 'simple' limits are limits of sequences of maps, pull-backs and products (Theorem 7.3.22). In the process, we see that all ∞-categories are localisations of ordinary categories (Proposition 7.3.15).

In Section 7.4 we study functors indexed by finite direct categories (i.e. by categories whose nerve is a finite simplicial set). We introduce a notion of ∞-category with weak equivalences and fibrations. For instance, any (nerve of a) Quillen model category, or any (nerve of a) category of fibrant objects in the sense of Brown, gives rise to such a thing. We explain how diagrams indexed by a finite direct category with values in an ∞-category with weak equivalences and fibrations is an ∞-category with weak equivalences and fibrations in two ways: although we define weak equivalences fibrewise, we can define fibrations fibrewise as well, or as Reedy fibrations. In particular, for any finite direct category I, I-indexed presheaves can be approximated by Reedy fibrant presheaves. Furthermore, limits of Reedy fibrant presheaves always exist, and their formation is compatible with fibrewise weak equivalences. This will play an important role in the construction and analysis of derived functors.

The goal of Section 7.5 is to explain why the localisation of any ∞-category with weak equivalences and fibrations has finite limits, and to construct derived functors. This is achieved by studying first the special case where all objects are fibrant: we observe that the localisation of an ∞-category of fibrant objects always has finite limits in a canonical way (Proposition 7.5.6). As a byproduct, we obtain a systematic method to invert maps in a compatible way with respect to finite limits (Proposition 7.5.11). Then comes the problem of studying fibrant approximation from an ∞-category-theoretic point of view: we prove a local finality theory property (Proposition 7.5.16) which explains how to compute the left Kan extensions along the localisation functor restricted to fibrant objects (Corollary 7.5.17). As rather direct consequences, we derive two important features: the fact that the localisation of its subcategory of fibrant objects is equivalent to the localisation of any ∞-category with weak equivalences and fibrations (Theorem 7.5.18), and the construction of right derived functors through the usual method: by picking a fibrant resolution of each object (7.5.25). The compatibility of the formation of derived functors with composition (Proposition 7.5.29) and the ∞-categorical analogue of Quillen's derived adjunction theorem (7.5.30) follow rather immediately.

Section 7.6 studies necessary and sufficient conditions for a left exact functor to induce an equivalence of finitely complete ∞-categories (this is quite meaningful, since the right derived functors of the previous section are left exact). It appears that such a condition is simply that the induced functor on the homotopy categories is an equivalence of categories in the very ordinary sense. Another equivalent condition is an ∞-categorical analogue of Waldhausen's approximation property (Theorem 7.6.10). As a consequence, we obtain nice coherence results of the following form. Given an ∞-category with weak equivalences and fibrations C and a fibrant object x in C, there is a canonical equivalence from the localisation of C/x to the slice of the localisation of C over x (Corollary 7.6.13). Similarly, the formation of the localisation of an ∞-category with weak equivalences and fibrations commutes with the functor $\underline{\mathrm{Hom}}(N(I)^{\mathrm{op}}, -)$ for any finite direct category I (Theorem 7.6.17). In Section 7.7, we see how to ensure the existence of small products, hence of small limits, in the localisation of an ∞-category with weak equivalences and fibrations. This is why we introduce homotopy complete ∞-categories with weak equivalences and fibrations. We then characterise the localisation with a universal property among ∞-categories with small limits and continuous functors (Theorem 7.7.7). Section 7.8 relies on all the preceding ones to prove that the localisation of the covariant model category over a simplicial set X is canonically equivalent to the ∞-category of functors from X to $\mathcal{S}$; see Theorem 7.8.9. As we sketch in Remark 7.8.11, this may be seen as a variation of the axiomata of Eilenberg and Steenrod to characterise homology theories.

In Section 7.9, we come back to the problem of computing localisations of diagram categories. An important special case of an ∞-category with weak equivalences and fibrations is when all maps are fibrations. This is what happens when we want to invert a class of maps which is closed under finite limits in an ∞-category with finite limits C. In this case, the property of homotopy completeness simply means that C has small products and that the class of weak equivalences is closed under small products. Furthermore, in this situation, using the duality theorem from Section 6.7 of Chapter 6, we show that inverting weak equivalences commutes with the functor $\underline{\mathrm{Hom}}(X^{\mathrm{op}}, -)$ for any small ∞-category X (Proposition 7.9.2). We then study the possibility of extending this kind of result to homotopy complete ∞-categories with weak equivalences and fibrations. We observe that we cannot expect this for X arbitrary, but rather for X the nerve of a small category (Remark 7.9.3). Under a mild additional assumption which is always satisfied in practice, we show that this is the only obstruction (Theorem 7.9.8). As a corollary, we see that, for any small category A, the localisation of the category of small simplicial presheaves on A is canonically equivalent to the ∞-category of presheaves of small ∞-groupoids on $N(A)$ (Corollary 7.9.9).

In Section 7.10, we explain how to compute mapping spaces. For this we need to work with locally small ∞-categories. Therefore, we first give sufficient conditions for (localisations of) ∞-categories to be locally small (Proposition 7.10.1 and Corollary 7.10.5). We then characterise mapping spaces of complete locally small ∞-categories (Proposition 7.10.6), out of which we see that the main methods to construct mapping spaces in Quillen model structures give the right answer.

Finally Section 7.11 gives a brief introduction to presentable ∞-categories. We define them as the ∞-categories which are equivalent to cocontinuous localisations of ∞-categories of presheaves on small ∞-categories by a small set of maps. We show various stability properties (such as the stability by left Bousfield localisations), and prove that cocontinuous functors between presentable ∞-categories always have a right adjoint. Finally, we interpret a result of Dugger in order to characterise presentable ∞-categories as localisations of combinatorial model structures.

7.1 Localisation

7.1.1 Let C be a simplicial set, and $W \subset C$ be a simplicial subset. Given an ∞-category D, we write $\underline{\mathrm{Hom}}_W(C, D)$ for the full subcategory of $\underline{\mathrm{Hom}}(C, D)$ which consists of functor $f: C \to D$ such that, for any map $u: x \to y$ in W, the induced map $f(u): f(x) \to f(y)$ is invertible in D. In other words, there is a pull-back square of the following form, induced by the operation of restriction along $W \subset C$.

(7.1.1.1)
$$\begin{array}{ccc} \underline{\mathrm{Hom}}_W(C, D) & \longrightarrow & \underline{\mathrm{Hom}}(C, D) \\ \downarrow & & \downarrow \\ \underline{\mathrm{Hom}}(W, k(D)) & \longrightarrow & \underline{\mathrm{Hom}}(W, D) \end{array}$$

Definition 7.1.2 A *localisation of C by W* is a functor $\gamma: C \to W^{-1}C$ such that:

(i) $W^{-1}C$ is an ∞-category;
(ii) the functor γ sends the morphisms of W to invertible morphisms in $W^{-1}C$;
(iii) for any ∞-category D, composing with γ induces an equivalence of ∞-categories of the form
$$\underline{\mathrm{Hom}}(W^{-1}C, D) \xrightarrow{\sim} \underline{\mathrm{Hom}}_W(C, D)\,.$$

Proposition 7.1.3 *The localisation of C by W always exists and is essentially unique.*

Proof For any Kan complex K, the functor $\underline{\mathrm{Hom}}(-, K)$ takes anodyne extensions to trivial fibrations. Therefore, if we put $W' = Ex^{\infty}(W)$, we have a trivial fibration
$$\underline{\mathrm{Hom}}(W', k(D)) \to \underline{\mathrm{Hom}}(W, k(D)) = \underline{\mathrm{Hom}}_W(W, D)\,.$$

We define C' by forming the following push-out square.

(7.1.3.1)
$$\begin{array}{ccc} W & \longrightarrow & C \\ \downarrow & & \downarrow \\ W' & \longrightarrow & C' \end{array}$$

Since $\underline{\mathrm{Hom}}(W', k(D)) = \underline{\mathrm{Hom}}(W', D)$, we have $\underline{\mathrm{Hom}}_{W'}(C', D) = \underline{\mathrm{Hom}}(C', D)$, and therefore, the Cartesian square obtained by applying the functor $\underline{\mathrm{Hom}}(-, D)$ to the coCartesian square (7.1.3.1), together with the Cartesian square (7.1.1.1), give a Cartesian square of the following form.

$$\begin{array}{ccc} \underline{\mathrm{Hom}}(C', D) & \longrightarrow & \underline{\mathrm{Hom}}_W(C, D) \\ \downarrow & & \downarrow \\ \underline{\mathrm{Hom}}(W', D) & \longrightarrow & \underline{\mathrm{Hom}}(W, k(D)) \end{array} \tag{7.1.3.2}$$

If we choose a fibrant resolution $W^{-1}C$ of C' in the Joyal model category structure, we get a trivial fibration of the form

$$\underline{\mathrm{Hom}}(W^{-1}C, D) \to \underline{\mathrm{Hom}}(C', D)\,.$$

On the other hand, since the lower horizontal map of diagram (7.1.3.2) is a trivial fibration, so is the upper one, hence the inclusion $\gamma\colon C \to W^{-1}C$ is a localisation of C by W. By definition, the map γ exhibits $W^{-1}C$ as a representation of the functor $\pi_0(k(\underline{\mathrm{Hom}}_W(C, -)))$ in the homotopy category of the Joyal model category structure. The Yoneda lemma thus implies that the pair $(W^{-1}C, \gamma)$ is unique up to a unique isomorphism in the homotopy category of the Joyal model category structure. □

Remark 7.1.4 One may always choose the localisation of C by W so that γ is the identity at the level of objects. Indeed, in the construction given in the above, the map $W \to W'$ is the identity on objects, and one may choose the map $C' \to W^{-1}C$ to be an inner anodyne extension, whence a retract of a countable composition of sums of push-outs of maps which are the identity on objects (namely, the inner horn inclusions).

Remark 7.1.5 One defines $\overline{W} \subset C$, the *saturation of* W *in* C, by forming the following Cartesian square.

$$\begin{array}{ccc} \overline{W} & \hookrightarrow & C \\ \downarrow & & \downarrow{\scriptstyle \gamma} \\ k(W^{-1}C) & \hookrightarrow & W^{-1}C \end{array}$$

We thus have inclusions

$$Sk_1(W) \subset W \subset \overline{W}\,.$$

One checks that, for any ∞-category D, these inclusions induce equalities

$$\underline{\mathrm{Hom}}_{Sk_1(W)}(C, D) = \underline{\mathrm{Hom}}_W(C, D) = \underline{\mathrm{Hom}}_{\overline{W}}(C, D)\,.$$

Therefore, $W^{-1}C$ is also the localisation of C by $Sk_1(W)$, as well as the localisation of C by $\overline{W}$. We observe that the inclusion $\overline{W} \to C$ is a fibration of the Joyal model category structure (because it is a pull-back of such a thing). Therefore, if ever C is an ∞-category, so is $\overline{W}$. We will say that W is *saturated* if $W = \overline{W}$.

Remark 7.1.6 The functor $\tau(C) \to ho(W^{-1}C)$ exhibits $ho(W^{-1}C)$ as the localisation of the ordinary category $\tau(C)$ by the class of maps W_1 as considered in Definition 2.2.8 (this is seen directly, by inspection of the universal properties). In particular, given a small category C and a set of maps W in C, seen as a simplicial subset of $Sk_1(N(C))$, there is a canonical equivalence of categories

$$ho(W^{-1}N(C)) \simeq W^{-1}C,$$

where the right-hand side is the ordinary localisation of C by W. However, even in this case, there is no reason why the canonical functor

$$W^{-1}N(C) \to ho(W^{-1}N(C)) \simeq N(W^{-1}C)$$

would be an equivalence of ∞-categories. In fact, the localisation $W^{-1}N(C)$ often has much better properties than the ordinary one.

Proposition 7.1.7 *The functor $\gamma^{op}\colon C^{op} \to (W^{-1}C)^{op}$ is the localisation of C^{op} by W^{op}.*

Proof This is for the equality $\underline{\mathrm{Hom}}_{W^{op}}(C^{op}, D) = \underline{\mathrm{Hom}}_{W}(C, D^{op})^{op}$. □

Proposition 7.1.8 *Let $f\colon C \to C'$ be a weak categorical equivalence. For any simplicial subset $W \subset C$, the localisation of C' by $f(W)$ is the localisation of C by W.*

Proof The map f induces an equivalence of ∞-categories

$$\underline{\mathrm{Hom}}(C', D) \simeq \underline{\mathrm{Hom}}(C, D)$$

which restricts to an equivalence of ∞-categories

$$\underline{\mathrm{Hom}}_{f(W)}(C', D) \simeq \underline{\mathrm{Hom}}_{W}(C, D)$$

for any ∞-category D. □

Remark 7.1.9 Assume that a universe $\mathbf{U}$ is given. Let $W \subset C$ be an inclusion of $\mathbf{U}$-small simplicial sets, and let $\gamma\colon C \to W^{-1}C$ be the associated localisation. Then the functor

$$\gamma^*\colon \underline{\mathrm{Hom}}(W^{-1}C^{op}, \mathcal{S}) \to \underline{\mathrm{Hom}}(C^{op}, \mathcal{S})$$

is fully faithful. Furthermore, its essential image consists of those presheaves

$F: C^{op} \to \mathcal{S}$ such that, for any map $u: x \to y$ in W, the induced map $u^*: F(y) \to F(x)$ is invertible in $\mathcal{S}$.

Indeed, by definition, the functor γ^* defines an equivalence of ∞-categories

$$\underline{\mathrm{Hom}}(W^{-1}C^{op}, \mathcal{S}) \simeq \underline{\mathrm{Hom}}_{W^{op}}(C^{op}, \mathcal{S}) .$$

Since the functor γ^* has adjoints, these properties may be translated as follows. For any presheaf G on $W^{-1}C$, the unit map $G \to \gamma_*\gamma^*(G)$ is invertible, and so is the co-unit map $\gamma_!\gamma^*(G) \to G$. For a presheaf F on C which sends the maps of W to invertible maps, the co-unit map $\gamma^*\gamma_*(F) \to F$ and the unit map $F \to \gamma^*\gamma_!(F)$ are invertible (simply because the restrictions of $\gamma_!$ and of γ_* to $\underline{\mathrm{Hom}}_{W^{op}}(C^{op}, \mathcal{S})$ are adjoints of the equivalence of ∞-categories induced by functor γ^*).

Proposition 7.1.10 *For any inclusion of simplicial sets $W \subset C$, the localisation functor $\gamma: C \to W^{-1}C$ is both final and cofinal.*

Proof We may assume that C is $\mathbf{U}$-small for some universe $\mathbf{U}$. For any functor $F: W^{-1}C^{op} \to \mathcal{S}$, since γ^* is fully faithful, by Remark 7.1.9, we have an isomorphism $F \simeq \gamma_*\gamma^*(F)$, hence an invertible map

$$\varprojlim F \simeq \varprojlim \gamma_*\gamma^*(F) \simeq \varprojlim \gamma^*(F)$$

in $\mathcal{S}$. By virtue of Theorem 6.4.5, this shows that γ is final. Proposition 7.1.7 ensures that we can apply the above to γ^{op}, whence γ is cofinal as well. □

Proposition 7.1.11 *We consider a universe $\mathbf{U}$. Let $W \subset C$ be an inclusion of $\mathbf{U}$-small simplicial sets. We consider a functor $f: C \to X$, where X is a small ∞-category. Then f exhibits X as the localisation of C by W if and only if the following three conditions are verified.*

(a) *The functor f sends the maps of W to invertible maps of X.*
(b) *The functor f is essentially surjective.*
(c) *The functor f^* induces an equivalence of ∞-categories*

$$f^*: \underline{\mathrm{Hom}}(X^{op}, \mathcal{S}) \xrightarrow{\sim} \underline{\mathrm{Hom}}_{W^{op}}(C^{op}, \mathcal{S}) .$$

Proof It is clear that conditions (a), (b) and (c) are verified whenever f is the localisation of C by W (see Remark 7.1.4 for condition (b)). To prove the converse, we consider the localisation functor $\gamma: C \to W^{-1}C$. Condition (a) implies that there exists a functor $g: W^{-1}C \to X$ such that $g\gamma$ is isomorphic to f. We thus have the commutative diagram (up to J-homotopy)

$$\begin{array}{ccc} \underline{\mathrm{Hom}}(X^{\mathrm{op}}, \mathcal{S}) & \xrightarrow{g^*} & \underline{\mathrm{Hom}}(W^{-1}C^{\mathrm{op}}, \mathcal{S}) \\ & {\scriptstyle f^*}\searrow \quad \swarrow{\scriptstyle \gamma^*} & \\ & \underline{\mathrm{Hom}}_{W^{\mathrm{op}}}(C^{\mathrm{op}}, \mathcal{S}) & \end{array}$$

in which both γ^* and f^* are equivalences of ∞-categories, by condition (c). Therefore, the functor g^* is an equivalence of ∞-categories. This implies that its left adjoint $g_!$ is an equivalence of ∞-categories as well. In particular, the functor $g_!$ is fully faithful, whence the functor g is fully faithful, by Proposition 6.1.15. Since f is essentially surjective, by condition (b), so is g. Therefore, the functor g is an equivalence of ∞-categories. Hence, in the commutative triangle (up to J-homotopy)

$$\begin{array}{ccc} \underline{\mathrm{Hom}}(X, D) & \xrightarrow{g^*} & \underline{\mathrm{Hom}}(W^{-1}C, D) \\ & {\scriptstyle f^*}\searrow \quad \swarrow{\scriptstyle \gamma^*} & \\ & \underline{\mathrm{Hom}}_{W}(C, D) & \end{array}$$

the functors g^* and γ^* are equivalences of ∞-categories for any ∞-category D. □

Proposition 7.1.12 *Let $p\colon A \to B$ be an inner fibration between ∞-categories. We assume that p is proper or smooth, and that, for any object b of B, the fibre $A_b = p^{-1}(b)$ is weakly contractible. Then, for any Cartesian square of ∞-categories*

$$\begin{array}{ccc} A' & \xrightarrow{u} & A \\ {\scriptstyle p'}\downarrow & & \downarrow{\scriptstyle p} \\ B' & \xrightarrow{v} & B \end{array}$$

the functor p' exhibits B' as the localisation of A' by the set of maps whose image in B' are identities.

Proof We shall prove the case where p is proper, the case where p is smooth being deduced by duality. We may assume that A and B are $\mathbf{U}$-small for some universe $\mathbf{U}$. Since the properties of being proper and of having weakly contractible fibres is closed under pull-backs, it is sufficient to prove that p exhibits B as a localisation of A. We first prove that the functor

$$p^*\colon \underline{\mathrm{Hom}}(B^{\mathrm{op}}, \mathcal{S}) \to \underline{\mathrm{Hom}}(A^{\mathrm{op}}, \mathcal{S})$$

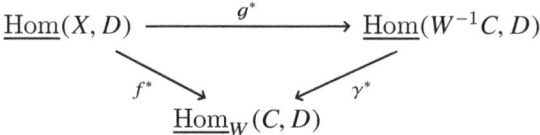

is fully faithful. By virtue of Theorem 6.4.13, for any presheaf F on B, and for any object b of B, the evaluation of the unit map

$$F \to p_* p^*(F)$$

at b is isomorphic to the unit map

$$F(b) \to \varprojlim_{a \in A_b^{\mathrm{op}}} F(b) .$$

To prove that the latter is invertible, by Theorem 6.4.5, it is sufficient to prove that the unique map $A_b \to \Delta^0$ is final, which may be seen as a very special case of Corollary 4.4.31, for instance.

Let $W = \bigcup_{b \in B_0} A_b \subset A$. If a presheaf F on A sends the maps of W to invertible maps in $\mathcal{S}$, then, for each object b of B, the restriction of F to A_b is locally constant (i.e. sends all maps of A_b to invertible maps of $\mathcal{S}$). But, since the weak homotopy type of A_b is the point, the map $A_b \to \Delta^0$ is a localisation. Let us put $F_{|A_b} = j^*(F)$, with $j \colon A_b \to A$ the inclusion. The above implies that, for any choice of an object x in A_b, $F_{|A_b}$ is equivalent to the constant presheaf with value $F(x)$. Furthermore, the diagram

$$\Delta^0 \xrightarrow{x} A_b \to \Delta^0$$

induces a diagram

$$F(x) \to \varprojlim_{a \in A_b^{\mathrm{op}}} F_{|A_b} \to F(x)$$

whose composition is the identity of $F(x)$ and in which the first map is invertible, by applying condition (d) of Theorem 6.4.5 to the final map $A_b \to \Delta^0$. This implies that the evaluation at an object a of A of the co-unit map

$$p^* p_*(F) \to F$$

is invertible whenever F belongs to $\underline{\mathrm{Hom}}_{W^{\mathrm{op}}}(A^{\mathrm{op}}, \mathcal{S})$: using Theorem 6.4.13, we see that it is isomorphic to the invertible map

$$(p^* p_*(F))(a) = p_*(F)(p(a)) \simeq \varprojlim_{x \in A_{p(a)}^{\mathrm{op}}} F_{|A_{p(a)}} \to F(a)$$

induced by $a \colon \Delta^0 \to A_{p(a)}$ (applying what precedes for $x = a$ and $b = p(a)$). This shows that the essential image of the functor p^* is exactly the subcategory $\underline{\mathrm{Hom}}_{W^{\mathrm{op}}}(A^{\mathrm{op}}, \mathcal{S})$. The functor p is also essentially surjective: each fibre A_b is weakly contractible, whence non-empty. Proposition 7.1.11 thus ends the proof. □

Proposition 7.1.13 *Let I be a finite set, and $W_i \subset C_i$, $i \in I$, a collection of inclusions of simplicial sets. Let us assume that, for any object x in some C_i, the identity of x belongs to W_i. We thus have the localisation*

$$\gamma_i : C_i \to W_i^{-1} C_i$$

of C_i by W_i for each i. Then the product map

$$\prod_{i \in I} \gamma_i : \prod_{i \in I} C_i \to \prod_{i \in I} W_i^{-1} C_i$$

is the localisation of $\prod_{i \in I} C_i$ by $\prod_{i \in I} W_i$.

Proof By an obvious induction, it is sufficient to consider the case where $I = \{1, 2\}$. Let us put $C = C_1 \times C_2$ and $W = W_1 \times W_2$. We have the canonical isomorphism

$$\underline{\mathrm{Hom}}(W_1^{-1}C_1 \times W_2^{-1}C_2, D) \simeq \underline{\mathrm{Hom}}(W_1^{-1}C_1, \underline{\mathrm{Hom}}(W_2^{-1}C_2, D)) .$$

Since the functor $\underline{\mathrm{Hom}}(W_1^{-1}C_1, -)$ preserves ∞-categories, we have an equivalence of ∞-categories

$$\underline{\mathrm{Hom}}(W_1^{-1}C_1, \underline{\mathrm{Hom}}(W_2^{-1}C_2, D)) \to \underline{\mathrm{Hom}}(W_1^{-1}C_1, \underline{\mathrm{Hom}}_{W_2}(C_2, D)).$$

Composing with the equivalence of ∞-categories

$$\underline{\mathrm{Hom}}(W_1^{-1}C_1, \underline{\mathrm{Hom}}_{W_2}(C_2, D)) \to \underline{\mathrm{Hom}}_{W_1}(C_1, \underline{\mathrm{Hom}}_{W_2}(C_2, D)) ,$$

we see that $\gamma = \gamma_1 \times \gamma_2$ induces an equivalence of ∞-categories

$$\underline{\mathrm{Hom}}(W_1^{-1}C_1, \underline{\mathrm{Hom}}(W_2^{-1}C_2, D)) \to \underline{\mathrm{Hom}}_{W_1}(C_1, \underline{\mathrm{Hom}}_{W_2}(C_2, D)) .$$

On the other hand, since W_1 and W_2 contain all identities, all maps in W can be obtained as the diagonal of a commutative square of the following form.

$$\begin{array}{ccc} (x_1, x_2) & \xrightarrow{(f_1, 1_{x_2})} & (y_1, x_2) \\ {\scriptstyle (1_{x_1}, f_2)} \downarrow & & \downarrow {\scriptstyle (1_{y_1}, f_2)} \\ (x_1, y_2) & \xrightarrow{(f_1, 1_{y_2})} & (y_1, y_2) \end{array}$$

As a consequence, the canonical isomorphism

$$\underline{\mathrm{Hom}}(C_1 \times C_2, D) \simeq \underline{\mathrm{Hom}}(C_1, \underline{\mathrm{Hom}}(C_2, D))$$

induces a canonical isomorphism

$$\underline{\mathrm{Hom}}_W(C, D) \simeq \underline{\mathrm{Hom}}_{W_1}(C_1, \underline{\mathrm{Hom}}_{W_2}(C_2, D))$$

for any ∞-category D. □

Proposition 7.1.14 *Let $f\colon A \to B$ be a functor between ∞-categories, equipped with a right adjoint $g\colon B \to A$. We suppose that there are simplicial subset $V \subset A$ and $W \subset B$ such that $f(V) \subset W$ and $g(W) \subset V$. Then there exist commutative diagrams up to J-equivalence of the form*

$$\begin{array}{ccc} A & \xrightarrow{f} & B \\ \downarrow & & \downarrow \\ V^{-1}A & \xrightarrow{\overline{f}} & W^{-1}B \end{array} \qquad \text{and} \qquad \begin{array}{ccc} B & \xrightarrow{g} & A \\ \downarrow & & \downarrow \\ W^{-1}B & \xrightarrow{\overline{g}} & V^{-1}A \end{array}$$

in which the vertical maps are the localisation functors, and such that $\overline{f}$ and $\overline{g}$ canonically form an adjoint pair.

Proof Let $\underline{\mathrm{Hom}}_V^W(A, B)$ be the full subcategory of $\underline{\mathrm{Hom}}(A, B)$ which consists of functors $\varphi\colon A \to B$ such that $\varphi(V) \subset W$. There is a canonical functor of the form

$$\underline{\mathrm{Hom}}_V^W(A, B) \to \underline{\mathrm{Hom}}_V(A, W^{-1}B) \xleftarrow{\sim} \underline{\mathrm{Hom}}(V^{-1}A, W^{-1}B).$$

Therefore, given any functors $f_0, f_1\colon A \to B$ such that $f_i(V) \subset W$ for $i = 0, 1$, we have canonical commutative squares up to J-homotopy of the form

$$\begin{array}{ccc} A & \xrightarrow{f_i} & B \\ \downarrow & & \downarrow \\ V^{-1}A & \xrightarrow{\overline{f}_i} & W^{-1}B \end{array}$$

for $i = 0, 1$. Moreover, any natural transformation $h\colon f_0 \to f_1$ determines canonically a natural transformation $\overline{h}\colon \overline{f}_0 \to \overline{f}_1$ which is compatible with h via the commutative square above. Moreover, the assignment $h \mapsto \overline{h}$ is functorial; in particular, it is compatible with composition. This proposition thus follows from the characterisation of adjunctions given by condition (v) of Theorem 6.1.23. □

Remark 7.1.15 Let $W \subset C$ be an inclusion of simplicial sets, equipped with the associated localisation functor $\gamma\colon C \to W^{-1}C$. If e is a final (or initial) object of C, then $\gamma(e)$ is a final (or initial) object of $W^{-1}C$. Indeed, $\gamma(e)\colon \Delta^0 \to W^{-1}C$ is the composition of the final (or cofinal) map $e\colon \Delta^0 \to C$ and of the map γ, which is both final and cofinal, by Proposition 7.1.10.

Corollary 7.1.16 *Let C be an ∞-category with finite products, equipped with a simplicial subset $W \subset C$. We assume that W contains all the identities of C, and that the product functor*

$$\times\colon C \times C \to C$$

sends $W \times W$ into W. Then the localisation of C by W has finite products and the localisation functor $\gamma\colon C \to W^{-1}C$ commutes with finite products.

Proof As explained in the above remark, we already know that the final object of C is sent to a final object of $W^{-1}C$. By an easy induction, we see that it is thus sufficient to prove that C has binary products. The product functor is the right adjoint of the diagonal functor $C \to C \times C$. Therefore, this corollary follows straight away from Propositions 7.1.13 and 7.1.14. □

Proposition 7.1.17 *Let C be an ∞-category equipped with a simplicial subset $W \subset C$. Any right adjoint (or left adjoint) of the localisation functor $\gamma\colon C \to W^{-1}C$ is fully faithful.*

Proof We may assume that C is $\mathbf{U}$-small for some universe $\mathbf{U}$. If γ has a right adjoint v, then, by virtue of Theorem 6.1.23, the functor

$$\gamma^*\colon \underline{\mathrm{Hom}}(W^{-1}C^{\mathrm{op}}, \mathcal{S}) \to \underline{\mathrm{Hom}}(C^{\mathrm{op}}, \mathcal{S})$$

is isomorphic to $v_!$. Therefore, since γ^* is fully faithful, so is $v_!$. By virtue of Proposition 6.1.15, this shows that v is fully faithful. The case of a left adjoint of γ follows from the case of a right adjoint, by Proposition 7.1.7. □

Proposition 7.1.18 *Let $u\colon A \to B$ be a functor between ∞-categories which has a fully faithful right adjoint $v\colon B \to A$. Let $W = k(B) \times_B A$ be the subcategory of maps of A which become invertible in B. Then the functor u exhibits B as the localisation of A by W.*

Proof We may assume that A and B are $\mathbf{U}$-small for some universe $\mathbf{U}$. Let $\gamma\colon A \to W^{-1}A$ be the localisation of A by W. For any presheaf $F\colon B^{\mathrm{op}} \to \mathcal{S}$, the functor $u^*(F)$ belongs to $\underline{\mathrm{Hom}}_{W^{\mathrm{op}}}(A^{\mathrm{op}}, \mathcal{S})$. Furthermore, since u is a left adjoint of v, condition (vii) of Theorem 6.1.23 tells us that we have a functorial identification $u^*(F) \simeq v_!(F)$. In other words, by Remark 7.1.9, the following map is invertible:

$$v_!(F) \to \gamma^*\gamma_! v_!(F)\,.$$

Since, by virtue of Proposition 6.1.15, the unit map $F \to v^* v_!(F)$ is invertible, and since $v^*\gamma^* = (\gamma v)^*$, this shows that the unit map

$$F \to (\gamma v)^*(\gamma v)_!(F)$$

is invertible. In other words, the functor γv is fully faithful, by Proposition 6.1.15. On the other hand, the functor γv is essentially surjective: for any object a of A, the unit map $a \to vu(a)$ belongs to W, for its image by u is a section of the invertible co-unit map $uvu(a) \to u(a)$. Therefore, the functor

γv is an equivalence of categories. Since there is an invertible map $uv \simeq 1_A$, we deduce that the functors $\gamma v u$ and γ are isomorphic. Therefore, for any ∞-category D, the triangle

$$\begin{array}{ccc} \underline{\mathrm{Hom}}(W^{-1}A, D) & \xrightarrow{\;v^*\gamma^*\;} & \underline{\mathrm{Hom}}(B, D) \\ & \searrow \quad \swarrow {\scriptstyle u^*} & \\ & \underline{\mathrm{Hom}}_W(A, D) & \end{array}$$

commutes up to J-homotopy, and the horizontal maps as well as the slanted map of the left-hand side are equivalences of ∞-categories. Therefore, the functor u exhibits B as the localisation of A by W. □

7.2 Calculus of Fractions

The aim of this chapter is to provide tools to compute localisations of ∞-categories. We fix a universe **U**.

7.2.1 Let $W \subset C$ be an inclusion of **U**-small simplicial sets, and

$$\gamma \colon C \to W^{-1}C$$

the corresponding localisation functor. For convenience, we assume that C is an ∞-category, and that the embedding $W \to C$ is an isofibration (in particular, W is an ∞-category as well).

Definition 7.2.2 Let x be an object of C. A *putative right calculus of fractions at the object* x is a functor $\pi(x) \colon W(x) \to C$ with the following properties:

(a) $W(x)$ is a **U**-small simplicial set with a final object x_0 whose image by $\pi(x)$ is x;
(b) the image by $\pi(x)$ of any map $z_0 \to x_0$ in $W(x)$ lies in W.

Example 7.2.3 There is always a minimal putative right calculus of fractions at an object x, which consists in putting $W(x) = \Delta^0$ and $\pi(x) = x$. There is a maximal one which consists in picking the full subcategory $W(x)$ of $C//x$ whose objects correspond to all possible maps of the form $s \colon z \to x$ in W; the functor $\pi(x)$ is then the composition of the inclusion $W(x) \subset C//x$ with the canonical projection from $C//x$ to C.

Definition 7.2.4 A functor $F \colon C \to \mathcal{S}$ is *W-local* if it sends the maps of W to invertible maps of $\mathcal{S}$.

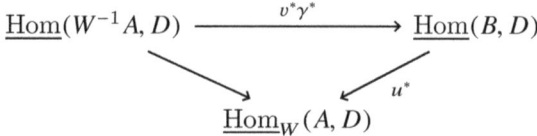

Let $p \colon X \to C$ be a morphism of simplicial sets. We say that p is *W-local*

if there exists a W-local functor $F: C \to \mathcal{S}$ classifying a left fibration with $\mathbf{U}$-small fibres $q: Y \to C$, as well as a cofinal map $i: X \to Y$ such that $qi = p$.

Remark 7.2.5 Let $\overline{W} = \gamma^{-1}(k(W^{-1}C))$ be the saturation of W in the sense of Remark 7.1.5. If a functor $F: C \to \mathcal{S}$ is W-local, then it is $\overline{W}$-local: indeed, such a functor F is equivalent to the composition of the localisation functor γ with a functor from $W^{-1}C$ to $\mathcal{S}$. Therefore, a morphism of simplicial sets $p: X \to C$ is W-local if and only if it is $\overline{W}$-local.

Definition 7.2.6 A *right calculus of fractions at an object x of C* is a putative right calculus of fractions at x, with the property that the map $\pi(x): W(x) \to C$ is W-local.

A *right calculus of fractions of W in C* is the data of right calculus of fractions $\pi(x): W(x) \to C$ at x, for each object x of C.

A *left calculus of fractions of W in C* is a right calculus of fractions of W^{op} in C^{op}.

7.2.7 From now on, we fix a right calculus of fractions $W(x)$ at a given object x of C. We thus have a functor $\pi(x): W(x) \to C$, which induces a functor

$$\pi(x)_!: \underline{\mathrm{Hom}}(W(x), \mathcal{S}) \to \underline{\mathrm{Hom}}(C, \mathcal{S}). \tag{7.2.7.1}$$

Since this functor commutes with colimits, while the map $W(x)/x_0 \to W(x)$ is a trivial fibration, applying Proposition 6.2.13 for $A = W(x)^{\mathrm{op}}$, we see that $\pi(x)_!$ sends the final object e to the following functor (where the colimit is indexed by $(W(x)/x_0)^{\mathrm{op}}$, the objects of which we consider as maps of the form $s_0: z_0 \to x_0$, whose images by $\pi(x)$ are denoted by $s: z \to x$):

$$\pi(x)_!(e) = \Big(y \mapsto \varinjlim_{s_0 : z_0 \to x_0} \mathrm{Hom}_C(z, y)\Big). \tag{7.2.7.2}$$

We observe that the property that $\pi(x)$ is W-local precisely means that the functor $\pi(x)_!(e)$ is W-local.

Theorem 7.2.8 *If there is a right calculus of fractions at x, then there is a canonical equivalence, functorially in y:*

$$\varinjlim_{s_0 : z_0 \to x_0} \mathrm{Hom}_C(z, y) \simeq \mathrm{Hom}_{W^{-1}C}(\gamma(x), \gamma(y)).$$

Proof By virtue of Remark 7.2.5, we may assume, without loss of generality, that W is saturated in the sense of Remark 7.1.5 (without changing $W(x)$). We shall say that a morphism $u: F \to G$ of $\underline{\mathrm{Hom}}(C, \mathcal{S})$ is a *W-equivalence* if its image $\gamma_!(u): \gamma_!(F) \to \gamma_!(G)$ is invertible in $\underline{\mathrm{Hom}}(W^{-1}C, \mathcal{S})$. Since the W-local functors form the essential image of the functor γ^*, applying the

Yoneda lemma to $\underline{\mathrm{Hom}}(W^{-1}C, \mathcal{S})$, we see that a morphism u as above is a W-equivalence if and only if, for any W-local functor $\Phi\colon C \to \mathcal{S}$, the induced map

$$u^*\colon \mathrm{Hom}(G, \Phi) \to \mathrm{Hom}(F, \Phi)$$

is invertible. By Proposition 7.1.18, the functor $\gamma_!$ exhibits the ∞-category of functors $\underline{\mathrm{Hom}}(W^{-1}C, \mathcal{S})$ as the localisation of $\underline{\mathrm{Hom}}(C, \mathcal{S})$ by the subcategory of W-equivalences.

The above also applies to the localisation $W^{-1}W = Ex^{\infty}(W)$. In particular, one may see the ∞-category $\underline{\mathrm{Hom}}(W^{-1}W, \mathcal{S})$ as the localisation of $\underline{\mathrm{Hom}}(W, \mathcal{S})$ by the W-equivalences. Furthermore, if $i\colon W \to C$ is the inclusion map, the induced restriction functor

$$i^*\colon \underline{\mathrm{Hom}}(C, \mathcal{S}) \to \underline{\mathrm{Hom}}(W, \mathcal{S})$$

preserves W-local objects. Therefore, its left adjoint

$$i_!\colon \underline{\mathrm{Hom}}(W, \mathcal{S}) \to \underline{\mathrm{Hom}}(C, \mathcal{S})$$

preserves W-equivalences. Furthermore, since W is saturated, we observe that the image of $\pi(x)$ in C lies in W. There is thus a unique map $q\colon W(x) \to W$ such that $iq = \pi(x)$. Therefore, the functor $x_0\colon \Delta^0 \to W(x)$ defines a map

$$h_{W^{\mathrm{op}}}(x) \to q_!(e)$$

in $\underline{\mathrm{Hom}}(W, \mathcal{S})$. One shows that this is a W-equivalence as follows. We have to show that its image in $\underline{\mathrm{Hom}}(W^{-1}W, \mathcal{S})$ is invertible. But, since right fibrations are conservative, it follows from Corollary 6.5.9 that the colimit functor

$$\varinjlim\colon \underline{\mathrm{Hom}}(W^{-1}W, \mathcal{S}) \to \mathcal{S}$$

is conservative. Our claim thus follows from the fact that the map $x_0\colon \Delta^0 \to W(x)$ is final, using condition (b) of Theorem 6.4.5. Applying the functor $i_!$, this provides a canonical W-equivalence $h_{C^{\mathrm{op}}}(x) \to \pi(x)_!(e)$. In other words, applying the functor $\gamma_!$ gives an invertible map

$$h_{W^{-1}C^{\mathrm{op}}}(\gamma(x)) = \gamma_!(h_{C^{\mathrm{op}}}(x)) \simeq \gamma_!\pi(x)_!(e)\,.$$

On the other hand, since $\pi(x)_!(e)$ is W-local, the unit map

$$\pi(x)_!(e) \to \gamma^*\gamma_!\pi(x)_!(e)$$

is invertible. We thus have

$$\pi(x)_!(e) \simeq \gamma^* h_{W^{-1}C^{\mathrm{op}}}(\gamma(x))$$

which really is a reformulation of the theorem, by formula (7.2.7.2). □

Corollary 7.2.9 *If there is a right calculus of fractions $W(x)$ at x, then there is a canonical equivalence of the form*

$$\varinjlim_{s_0\colon z_0 \to x_0} F(z) \simeq \gamma_!(F)(\gamma(x))$$

for all presheaves $F\colon C^{\mathrm{op}} \to \mathcal{S}$ (where the colimit is indexed by $W(x)^{\mathrm{op}}$).

Proof Since the functors

$$F \mapsto \varinjlim_{s_0\colon z_0 \to x_0} F(z) \quad \text{and} \quad F \mapsto F(x)$$

commute with colimits, and since any presheaf is a colimit of representable presheaves (Corollary 6.2.16), it is sufficient to restrict ourselves to the case where F is representable. Since $\gamma_!(h_C(y)) \simeq h_{W^{-1}C}(\gamma(y))$, this is then a reformulation of Theorem 7.2.8. □

Remark 7.2.10 The weak homotopy type corresponding to the colimit object $\varinjlim_{s_0\colon z_0 \to x_0} \mathrm{Hom}_C(z, y)$ of $\mathcal{S}$ may be described as follows. Assuming that a putative right calculus of fractions $\pi(x)\colon W(x) \to C$ is chosen at x, we put

$$\mathit{Span}_W(x, y) = C//y \times_C W(x) \tag{7.2.10.1}$$

(we observe that this is in fact a homotopy pull-back, by Corollary 5.3.6, and that we could have chosen C/y instead of $C//y$, by Proposition 4.2.9). Since x_0 is a final object in $W(x)$, and since the projection of $C//x$ to C is a right fibration, there is a functor $w\colon W(x) \to C//x$ which sends x_0 to 1_x such that $\pi(x)$ is the composition of w and the projection from $C//x \to C$. This defines a functor from $\mathit{Span}_W(x, y)$ to $C//y \times_C C//x$ whose image may be described as the subcategory of the ∞-category $\underline{\mathrm{Hom}}(\Lambda^2_0, C)$ whose objects correspond to the functors $\Lambda^2_0 \to C$ of the form

$$x \xleftarrow{\;s\;} z \xrightarrow{\;f\;} y \tag{7.2.10.2}$$

in which (z, s) belongs to the image of w, so that, in particular, the map s is in W (in the case of the maximal putative right calculus of fractions at x, this fully determines $\mathit{Span}_W(x, y)$, but not in general). By virtue of the dual version of Proposition 6.4.8, the evaluation of $\pi(x)_!(e)\colon C \to \mathcal{S}$ at y is the $\mathit{Span}_W(x, y)$-indexed colimit of the constant functor with value the terminal object e. In other words, by Proposition 6.1.14, the evaluation of $\pi(x)_!(e)\colon C \to \mathcal{S}$ at y corresponds to the weak homotopy type of $\mathit{Span}_W(x, y)$.

By the preceding theorem, the right calculus of fractions at x thus provides a canonical isomorphism

$$Ex^\infty(Span_W(x, y)) \simeq W^{-1}C(\gamma(x), \gamma(y)) \tag{7.2.10.3}$$

in the homotopy category $ho(sSet)$. In particular, by Propositions 3.1.31 and 3.7.2, the right calculus of fractions at x implies that, for any object y of C, we have a canonical bijection of the form

$$\pi_0(Span_W(x, y)) \simeq \mathrm{Hom}_{ho(W^{-1}C)}(\gamma(x), \gamma(y)) \tag{7.2.10.4}$$

which assigns to a span of the form (7.2.10.2) the map $\gamma(f)\gamma(s)^{-1}$.

7.2.11 When we have a right calculus of fractions, one may deduce from the preceding remark a description of $ho(W^{-1}C)$ by a right calculus of fractions in the ordinary sense as follows.

We define two parallel maps $f, g\colon x \to y$ to be *W-homotopic* if they become equal in $ho(W^{-1}C)$. The relation of being W-homotopic is an equivalence relation on $\mathrm{Hom}_{ho(C)}(x, y)$ which is compatible with composition, whence there is a unique functor

$$q\colon ho(C) \to \pi(C) \tag{7.2.11.1}$$

which is the identity on objects such that, for each pair of objects x and y in $ho(C)$, the induced map

$$\mathrm{Hom}_{ho(C)}(x, y) \to \mathrm{Hom}_{\pi(C)}(x, y)$$

is the quotient of $\mathrm{Hom}_{ho(C)}(x, y)$ by the relation of W-homotopy. The canonical functor from $ho(C)$ to $ho(W^{-1}C)$ factors canonically through a faithful functor

$$\overline{\gamma}\colon \pi(C) \to ho(W^{-1}C). \tag{7.2.11.2}$$

We define $\pi(W)$ to be the set of maps obtained as the maps which are images of elements of W in $\pi(C)$.

Corollary 7.2.12 *If there is a right calculus of fractions of W in C, then the set of maps $\pi(W)$ satisfies the axioms for a right calculus of fractions in the usual sense of Gabriel and Zisman.*[1] *Furthermore, the functor* (7.2.11.2) *exhibits $ho(W^{-1}C)$ as the localisation of $\pi(C)$ by $\pi(W)$.*

Proof The last assertion is immediate, and is only stated for the record. By definition, the first assertion means that the following conditions are satisfied.

(1) All identities are in W.

[1] The axioms are recalled in the proof.

(2) For any diagram $y' \xrightarrow{g} y \xleftarrow{t} x$ in $\pi(C)$, with $t \in \pi(W)$, there is a commutative square in $\pi(C)$ of the form

$$\begin{array}{ccc} x' & \xrightarrow{f} & x \\ {\scriptstyle s}\downarrow & & \downarrow{\scriptstyle t} \\ y' & \xrightarrow{g} & y \end{array}$$

with $s \in \pi(W)$.

(3) For any parallel morphisms $f, g \colon x \to y$ in $\pi(C)$, if $t \colon y \to y'$ is a map in $\pi(W)$ such that $tf = fg$, there exists a map $s \colon x' \to x$ such that $fs = gs$.

Property (1) follows from the existence of a right calculus of fractions for W. Since the functor $\overline{\gamma}$ is faithful, properties (2) and (3) follow straight away from the fact that, by virtue of the bijection (7.2.10.4), any map in $ho(W^{-1}C)$ may be written as $q(f)q(s)^{-1}$, where $f \colon z \to y$ is any map in $\pi(C)$, and $s \colon z \to x$ is an element of W. □

Remark 7.2.13 In practice, one can describe the relation of W-homotopy above as the more concrete path-homotopy relation. See Proposition 7.6.14 below.

Definition 7.2.14 Let F be a set of maps in C.

We say that F is *closed under compositions* if all identities of C are in F and if, for composable maps $u \colon x \to y$ and $v \colon y \to z$ in F, any composition of u and v is in F.

We say that F is *closed under pull-backs* if the following properties are verified.

(a) For any map $f \colon x \to y$ in F, and for any map $v \colon y' \to y$ in C, there exists a Cartesian square of the form

$$\begin{array}{ccc} x' & \xrightarrow{u} & x \\ {\scriptstyle f'}\downarrow & & \downarrow{\scriptstyle f} \\ y' & \xrightarrow{v} & y \end{array}$$

such that f' belongs to F.

(b) Any map isomorphic to an element of F in $\underline{\mathrm{Hom}}(\Delta^1, C)$ belongs to F.

Lemma 7.2.15 *Let $v \colon y' \to y$ be a map in C. We denote by $v_! \colon C/y' \to C/y$ any choice of functor encoding composition by v (e.g. one may construct $v_!$ by composing the functor $C/v \to C/y$ with some section of the trivial fibration $C/v \to C/y'$). If the pull-back of f along any map of codomain y exists in C,* *then, for any map $v\colon y' \to y$ in F, and any object x of C, the induced functor*

$$v_! \times_C 1_{C/x}\colon C/y' \times_C C/x \to C/y \times_C C/x$$

has a right adjoint, which we can describe as follows. It sends an object (z, s, f), of $C/y \times_C C/x$, seen as a diagram

$$\begin{array}{ccc} & z & \\ {}^{s}\swarrow & & \searrow^{f} \\ x & & y \end{array}$$

to the object (z', s', f') obtained by forming the Cartesian square

$$\begin{array}{ccc} z' & \xrightarrow{w} & z \\ {\scriptstyle f'}\downarrow & & \downarrow{\scriptstyle f} \\ y' & \xrightarrow{v} & y \end{array}$$

and by choosing a composition s' of s and w.

Proof Since the Yoneda embedding is fully faithful, continuous and compatible with slices, we see that proving the lemma for $\underline{\mathrm{Hom}}(C^{op}, \mathcal{S})$ implies that it holds for C itself. In other words, we may assume, without loss of generality, that C has finite products and pull-backs. We observe then that there is a canonical equivalence of ∞-categories

$$C/y \times_C C/x \simeq C/x \times y$$

over $C \times C$. Replacing v by $1_x \times v\colon x \times y' \to x \times y$, we see that we may assume that x is the final object. We are thus reduced to proving that the functor $v_!\colon C/y' \to C/y$ has a right adjoint, which follows from Theorem 6.6.9. □

Theorem 7.2.16 *If W is closed under compositions and under pull-backs, then there is a right calculus of fractions of W in C, defined by the maximal putative right calculus of fractions (see Example 7.2.3).*

Proof We fix an object x of C. We let $W(x)$ be the maximal putative right calculus of fractions at x: the full subcategory of C/x whose objects are the couples $(z, s\colon z \to x)$ with s in W.

Let $v\colon y' \to y$ be a map in W. With the notation of the previous lemma, we have a functor

$$v_!\colon C/y' \to C/y$$

which encodes the operation of composing with v, and we know that the induced functor

$$v_! \times_C 1_{C/x}\colon C/y' \times_C C/x \to C/y \times_C C/x$$

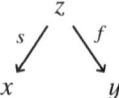

has a right adjoint. The explicit description of the latter shows that, when we restrict ourselves to the subcategories in the diagram

$$v_! \times_C 1_{W(x)} : C/y' \times_C W(x) \to C/y \times_C W(x),$$

we still get a functor with a right adjoint (it might help to replace C/z by $C//z$ though). But functors with right adjoints are cofinal, by Corollary 6.1.13, whence are weak homotopy equivalences.This means that the functor $\pi(x)_!(e)$ sends v to an invertible map of $\mathcal{S}$. Therefore, the map $\pi(x)\colon W(x) \to C$ is W-local for any object x. □

Remark 7.2.17 Under the form given by Corollary 7.2.9, the previous theorem can be found in one of Hoyois' papers [Hoy17, proposition 3.4]. Under the form of formula (7.2.10.3), this can also be found in the work of Nuiten [Nui16, corollary 3.14].

Corollary 7.2.18 *Let us assume that there is a subcategory W' of W which is closed under compositions and under pull-backs, and such that the saturation of W' in C contains W (see Remark 7.1.5). Then there is a right calculus of fractions of W in C*

Proof Since W and W' have the same saturation in C, the localisation by W' coincides with the localisation by W, so that any calculus of fractions of W' in C is also a calculus of fractions of W in C. We may thus apply the preceding theorem to W'. □

Definition 7.2.19 We say that W has the *two-out-of-three property*, if, for any commutative triangle of C,

$$\begin{array}{ccc} & y & \\ {}^{f}\nearrow & & \searrow^{g} \\ x & \xrightarrow{h} & z \end{array}$$

in which two maps among f, g and h are in W, then so is the third.

Definition 7.2.20 A *class of trivial fibrations* (with respect to W) is a set of maps F in C such that:

(a) $F \subset W$;
(b) F is closed under compositions and under pull-backs;
(c) for any map $f\colon x \to y$ in W, there exist commutative triangles of the form

$$\begin{array}{ccccc} & & x & & \\ & // & \downarrow{\scriptstyle s} & \searrow^{f} & \\ x & \xleftarrow{p} & z & \xrightarrow{q} & y \end{array}$$

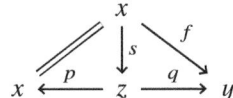

with p and q in F.

The elements of F will often be called *trivial fibrations* (although this notion depends on the choice of F).

Remark 7.2.21 The existence of a class of trivial fibrations implies that there is a right calculus of fractions, by Corollary 7.2.18.

Definition 7.2.22 A map $p\colon y' \to y$ in C is *W-local* if, for any map $f\colon t \to y$ and any element $v\colon t' \to t$ of W, one can form the following Cartesian squares in C

$$\begin{array}{ccccc} s' & \xrightarrow{u} & s & \xrightarrow{f'} & y' \\ \downarrow{\scriptstyle q'} & & \downarrow{\scriptstyle q} & & \downarrow{\scriptstyle p} \\ t' & \xrightarrow{v} & t & \xrightarrow{f} & y \end{array}$$

in such a way that the map u belongs to W.

Remark 7.2.23 In the case where W contains all the identities (which is implied by the existence of a (putative) right calculus of fractions), one can form the pull-back of a W-local map along any map: this is because, in the above definition, we are then allowed to consider the case where v is the identity of t. One then checks immediately that the set of W-local maps is closed under compositions and under pull-backs.

Remark 7.2.24 In the case where C is the nerve of a small category, the notion of category with weak equivalences and a class of trivial fibrations has been introduced by Horel [Hor16] under the name of partial Brown categories. Similarly, the following theorem as well as its proof are extensions of statements and proofs that Weiss made in [Wei99] in the setting of Waldhausen categories with functorial cylinder objects.

Theorem 7.2.25 *Let us assume that W has the two-out-of-three property and that there is a class of trivial fibrations with respect to W. For any Cartesian square in C of the form*

$$\begin{array}{ccc} s & \xrightarrow{f'} & y' \\ \downarrow{\scriptstyle q} & & \downarrow{\scriptstyle p} \\ t & \xrightarrow{f} & y \end{array} \tag{7.2.25.1}$$

in which the map p is W-local, the square

(7.2.25.2)
$$\begin{array}{ccc} \gamma(s) & \xrightarrow{\gamma(f')} & \gamma(y') \\ \big\downarrow{\scriptstyle \gamma(q)} & & \big\downarrow{\scriptstyle \gamma(p)} \\ \gamma(t) & \xrightarrow{\gamma(f)} & \gamma(y) \end{array}$$

is Cartesian in $W^{-1}C$.

Proof By Remark 6.6.11, the image of the Cartesian square (7.2.25.1) by the Yoneda embedding of C may be realised as a Cartesian square of the contravariant model category structure over C of the form

(7.2.25.3)
$$\begin{array}{ccc} C_{/s} & \xrightarrow{f'_!} & C_{/y'} \\ \big\downarrow{\scriptstyle q_!} & & \big\downarrow{\scriptstyle p_!} \\ C_{/t} & \xrightarrow{f_!} & C_{/y} \end{array}$$

in which all maps are right fibrations (and each structural map of the form $C_{/c} \to C$ is a right fibration whose domain has a final object which is sent to c in C). Each object x of C will be equipped with the right calculus of fractions $W(x)$ associated to some choice of a class of trivial fibrations, i.e. which consists of the full subcategory of C/x whose objects are the pairs (z, s), with $s\colon z \to x$ a trivial fibration. Pulling back along the map $\pi(x)\colon W(x) \to C$, we thus get a Cartesian square over $W(x)$ of the form

(7.2.25.4)
$$\begin{array}{ccc} W(x)_{/s} & \xrightarrow{f'_!} & W(x)_{/y'} \\ \big\downarrow{\scriptstyle q_!} & & \big\downarrow{\scriptstyle p_!} \\ W(x)_{/t} & \xrightarrow{f_!} & W(x)_{/y} \end{array}$$

(in which we have put $W(x)_{/c} = C_{/c} \times_C W(x)$). By virtue of the interpretation of Theorem 7.2.8 made in Remark 7.2.10, it is sufficient to prove that diagram (7.2.25.4) is homotopy Cartesian in the Kan–Quillen model category structure. For this purpose, we shall apply the dual version of Proposition 4.6.11 to the functor $p_!\colon W(x)_{/y'} \to W(x)_{/y}$. In other words, we have to give ourselves a map $w\colon a_0 \to a_1$ in $W(x)_{/y}$ and see that pulling back along the induced functor

(7.2.25.5)
$$w_!\colon (W(x)_{/y})/a_0 \to (W(x)_{/y})/a_1$$

gives a weak homotopy equivalence of the form

(7.2.25.6)
$$(W(x)_{/y'})/a_0 \to (W(x)_{/y'})/a_1 .$$

Another way to formulate this is that composing the Yoneda embedding of $W(x)_{/y}$ with the pull-back functor along $p_!$, and then with the colimit functor,

$$W(x)_{/y} \xrightarrow{h_{W(x)_{/y}}} \underline{\mathrm{Hom}}((W(x)_{/y})^{\mathrm{op}}, \mathcal{S}) \xrightarrow{(p_!)^*} \underline{\mathrm{Hom}}((W(x)_{/y'})^{\mathrm{op}}, \mathcal{S}) \xrightarrow{\varinjlim} \mathcal{S}$$

gives a locally constant functor (i.e. sends all maps of $W(x)_{/y}$ to invertible maps). In particular, the set of maps w of $W(x)_{/y}$ such that (7.2.25.6) is a weak homotopy equivalence has the two-out-of-three property. Up to equivalence, such a map w corresponds essentially to a commutative diagram in C, of the form

$$\begin{array}{ccc} z_0 & \xrightarrow{g_0} & y \\ {\scriptstyle s_0}\big\downarrow & \searrow^{w} & \big\uparrow{\scriptstyle g_1} \\ x & \xleftarrow{s_1} & z_1 \end{array}$$

such that s_0 and s_1 are trivial fibrations. Choosing a commutative diagram of the form

$$\begin{array}{ccccc} & & z_0 & & \\ & \swarrow\!\!= & \big\downarrow & \searrow^{w} & \\ z_0 & \xleftarrow{p} & z_2 & \xrightarrow{q} & z_1 \end{array}$$

such that p and q are trivial fibrations, we see that it is sufficient to prove that the map (7.2.25.6) is a weak homotopy equivalence when the map $w\colon z_0 \to z_1$ is a trivial fibration. We then form the following Cartesian squares.

$$\begin{array}{ccccc} z'_0 & \xrightarrow{w'} & z'_1 & \xrightarrow{g'_1} & y' \\ \big\downarrow{\scriptstyle p_0} & & \big\downarrow{\scriptstyle p_1} & & \big\downarrow{\scriptstyle p} \\ z_0 & \xrightarrow{w} & z_1 & \xrightarrow{g_1} & y \end{array}$$

We let $q_i\colon z'_i \to x$ be a composition of s_i and p_i for $i = 0, 1$. We also write $g'_0\colon z'_0 \to y'$ for the pull-back of g_0 along p (so that g'_0 is a composition of g'_1 and w'). One may see (z_i, q_i, g'_i) as an object a'_i of $C/x \times_C C/y'$. Using Lemma 7.2.15, we see that $(W(x)_{/y})/a_i$ is the full subcategory of $(C/x \times_C C/y')/a'_i$ whose objects correspond to commutative diagrams of the form

$$\begin{array}{ccc} z & \xrightarrow{\varphi} & y' \\ {\scriptstyle \sigma}\big\downarrow & \searrow^{\psi} & \big\uparrow{\scriptstyle g_i} \\ x & \xleftarrow{q_i} & z_i \end{array}$$

such that σ is a trivial fibration. Since the canonical map $C/x \times_C C/y' \to C$ is a right fibration, there are canonical trivial fibrations

$$(C/x \times_C C/y')/a'_i \to C/z'_i$$

for $i = 0, 1$. By virtue of Theorem 6.6.9, the functor

$$w'_! : C/z'_0 \to C/z'_1$$

has a right adjoint. This means that the induced functor

$$w'_! : (C/x \times_C C/y')/a'_0 \to (C/x \times_C C/y')/a'_1$$

also has a right adjoint. We finally observe that the latter restricts to a right adjoint of the induced functor (7.2.25.6). In particular, by Corollary 6.1.13, the functor (7.2.25.6) is cofinal, whence a weak homotopy equivalence. □

7.3 Constructions of Limits

7.3.1 Given a small category I and a functor $F : I \to \mathcal{C}at$, one can form the *Grothendieck construction* $p_F : \int F \to I$ as follows. The objects of the category $\int F$ are pairs (i, x) where i is an object of I, and x is an object of $F(i)$. Morphisms $(i, x) \to (j, y)$ are pairs (u, v), where $u : i \to j$ is a map in I, and $v : F(u)(x) \to y$ is a map in $F(j)$. The composition of two maps $(u_0, v_0) : (i_0, x_0) \to (i_1, x_1)$ and $(u_1, v_1) : (i_1, x_1) \to (i_2, x_2)$ is given by

$$(u_1, v_1) \circ (u_0, v_0) = (u_1 u_0, v_1 F(u_1)(v_0)) .$$

The functor $p : \int F \to I$ is defined by $p(i, x) = i$ on objects and by $p(u, v) = u$ on morphisms. We observe that, for any functor $\varphi : J \to I$, we have a canonical Cartesian square of the following form, where $\varphi^*(F) = F\varphi$.

(7.3.1.1)
$$\begin{array}{ccc} \int \varphi^*(F) & \longrightarrow & \int F \\ {\scriptstyle p_{\varphi^*(F)}}\downarrow & & \downarrow{\scriptstyle p_F} \\ J & \xrightarrow{\varphi} & I \end{array}$$

In particular, for any object i of I, there is the canonical Cartesian square below, where e denotes the final category.

(7.3.1.2)
$$\begin{array}{ccc} F(i) & \longrightarrow & \int F \\ \downarrow & & \downarrow{\scriptstyle p_F} \\ e & \xrightarrow{i} & I \end{array}$$

Proposition 7.3.2 *The morphism $N(p_F) : N(\int F) \to N(I)$ is proper.*

Proof Since this is the nerve of a functor, this is an inner fibration between ∞-categories. We shall prove that condition (ii) of Theorem 4.4.36 is satisfied. Let i be an object of I. We write F/i for the composition of F with $I/i \to I$.

Since the nerve functor commutes with the formation of slices and preserves adjunctions, it is sufficient to prove that the canonical functor

$$F(i) \to \int F/i$$

is final. By virtue of the dual version of Corollary 6.1.13, it is sufficient to prove that it has a left adjoint. An object of $\int F/i$ is a triple (j, α, x), where $\alpha\colon j \to i$ is a map of I, and x is an object of $F(j)$. We define a functor $\int F/i \to F(i)$ by sending (j, α, x) to $F(\alpha)(x)$. The unit map is

$$(\alpha, 1_{F(\alpha)(x)})\colon (j, \alpha, x) \to (i, 1_i, F(\alpha)(x))$$

and the co-unit map is the identity. □

7.3.3 For a small category A and a presheaf of sets F on A, one associates the category of elements A/F; see Definition 1.1.7. Since this construction is functorial, for any functor $F\colon I \to \underline{\mathrm{Hom}}(A^{\mathrm{op}}, \mathit{Set})$, we obtain a functor

$$A/F\colon I \to \mathit{Cat}\,, \quad i \mapsto A/F(i)\,.$$

There is a canonical functor

$$\pi_F\colon \int A/F \to A/\varinjlim F \tag{7.3.3.1}$$

defined as follows. We let $\ell_i\colon F(i) \to \varinjlim F$ be the canonical map, for each object i of I. The objects of $\int A/F$ are triples (i, a, s), where i is an object of I, a is an object of A, and $s \in F(a)$. Morphisms $(i, a, s) \to (j, b, t)$ are couples (u, v), where $u\colon i \to j$ is a map in I, while $v\colon a \to b$ is a map in A such that $v^*(t) = s$. The functor π_F is defined by $\pi_F(i, a, s) = (a, \ell_i(s))$ on objects and by $\pi_F(u, v) = v$ on maps.

Proposition 7.3.4 *The functor $N(\pi_F)\colon N(\int A/F) \to N(A/\varinjlim F)$ is smooth.*

Proof We may assume that $A/\varinjlim F = A$: since $(A/\varinjlim F)/(F(i), \ell_i) \simeq A/F(i)$ for all i, we may replace A by $A/\varinjlim F$. This mainly simplifies notation. Let a_0 be an object of A. As in the proof of the preceding proposition, we observe that it is sufficient to prove that the fully faithful functor

$$\pi_F^{-1}(a_0) \to a_0\backslash\Big(\int A/F\Big)\,, \quad (i, a, s) \mapsto (i, a, s, 1_{a_0})$$

has a right adjoint. The fibre $\pi_F^{-1}(a_0)$ may be identified with the Grothendieck construction associated with the functor $i \mapsto F(i)(a_0)$ (where we consider sets as categories in which all maps are identities). The objects of $a_0\backslash(\int A/F)$ may be described as 4-tuples (i, a, s, f), where (i, a, s) is an object of $\int A/F$, while $f\colon a_0 \to a$ is a morphism in A. The right adjoint sends such an object (i, a, s, f)

to $(i, a_0, f^*(s))$. The co-unit is given by the map $(1_i, f)\colon (i, a_0, f^*(s), 1_{a_0}) \to (i, a, s, f)$. □

Proposition 7.3.5 *We fix a universe* **U**. *Let A and I be* **U***-small categories, and* $F\colon I \to \underline{\mathrm{Hom}}(A^{\mathrm{op}}, Set)$ *a functor taking its values in* **U***-small presheaves of sets. We assume that one of the following three conditions is satisfied.*

(a) The category I is discrete.
(b) We have $N(I) = \Lambda^2_0$*, and the map* $F(0) \to F(1)$ *is a monomorphism.*
(c) The category I is filtered.

We consider an ∞-category C with **U***-small limits, and a functor*

$$\Phi\colon N(A/\varinjlim F)^{\mathrm{op}} \to C\,.$$

For each object i of I, we write $\Phi_i = \ell_i^*(\Phi) = \Phi_{|A/F(i)^{\mathrm{op}}}$*, where* $\ell_i\colon A/F(i) \to A/\varinjlim F$ *is the canonical functor. Then the assignment* $i \mapsto \varprojlim \Phi_i$ *extends naturally to a functor* $N(I)^{\mathrm{op}} \to C$ *such that there is a canonical invertible map of the form*

$$\varprojlim_{x\in A/\varinjlim F^{\mathrm{op}}} \Phi(x) \simeq \varprojlim_{i\in N(I)^{\mathrm{op}}} \varprojlim_{x\in N(A/F(i))^{\mathrm{op}}} \Phi_i(x)\,.$$

Proof Let $\Psi = N(p_{A/F})_*(N(\pi_F)_*(\Phi))$. Since, by Proposition 7.3.2, the map $N(p_{A/F})$ is proper, we can apply Theorem 6.4.13 to the Cartesian square

$$\begin{array}{ccc} N(A/F(i)) & \longrightarrow & N(\int A/F) \\ \downarrow & & \downarrow{\scriptstyle N(p_{A/F})} \\ \Delta^0 & \longrightarrow & N(I) \end{array}$$

and deduce that there are canonical isomorphisms of the form

$$\Psi(i) \simeq \varprojlim_{x\in N(A/F(i))^{\mathrm{op}}} \Phi_i(x)\,.$$

There is a canonical map

$$\varprojlim \Phi \to \varprojlim N(\pi_F)^*(\Phi) \simeq \varprojlim N(p_{A/F})_* N(\pi_F)^*(\Phi) = \varprojlim \Psi\,.$$

By Theorem 6.4.5, to finish the proof, it is sufficient to check that the map $N(\pi_F)$ is final. Since, by virtue of Proposition 7.3.4, it is smooth, Propositions 7.1.12 and 7.1.10 show that it is sufficient to prove that the fibres of $N(\pi_F)$ are weakly contractible. Let $x = (a, s)$ be an object of $A/\varinjlim F$. We shall describe the fibre $\pi_F^{-1}(x)$. This is where we have to go case by case. However, whenever x is fixed, we observe that $\pi_F^{-1}(x)$ is isomorphic to the fibre at x of the map

$\pi_{F(a)}\colon \int F(a) \to \varinjlim_I F(a)$ (where we see sets as categories in which the only maps are identities). In other words, to do these computations, we may assume that A is the final category, and thus that F takes its values in the category of **U**-small sets. In particular, we shall drop A and a from the notation below, and thus identify $x = s$.

If I is discrete, then the functor π_F is an isomorphism. In particular, the nerve of its fibre at x is isomorphic to Δ^0, which is obviously contractible.

If $N(I) = \Lambda^2_0$, and if $F(0) \to F(1)$ is a monomorphism, we have two cases to consider. We have a coCartesian square in the category of presheaves of sets on A of the following form, where the horizontal maps will be considered as inclusions.

$$\begin{array}{ccc} F(0) & \hookrightarrow & F(1) \\ \downarrow & & \downarrow \\ F(2) & \hookrightarrow & \varinjlim F \end{array}$$

If x belongs to the set $F(2)$, then $(2, x)$ is a final object of $\pi_F^{-1}(x)$: one may identify the nerve of $\pi_F^{-1}(x)$ with $(\amalg_E \Delta^0) * \Delta^0$, where E is the fibre of the map $F(0) \to F(2)$ at x. If x does not belong to the set $F(2)$, then, since the induced map

$$F(1) \setminus F(0) \to \varinjlim F \setminus F(2)$$

is bijective, the only object of $\pi_F^{-1}(x)$ is $(1, x)$, and this identifies the nerve of $\pi_F^{-1}(x)$ with Δ^0.

It remains to consider the case where I is filtered. One then observes by direct inspection that the fibre $\pi_F^{-1}(x)$ itself is filtered. Finally, we remark that the nerve of any small filtered category J is weakly contractible: since J is the J-indexed colimit of its slices, for any finite partially ordered set E, any functor $E \to J$ factors through some slice J/j, so that Lemma 4.3.15 proves that the nerve of J is weakly contractible. □

Remark 7.3.6 The previous proposition may be applied in situations where the functor F is not given explicitly. In fact, in practice, this is the functor A/F that will be given. To see that this will be harmless, we first observe that there is a canonical isomorphism

$$\varinjlim A/F \simeq A/\varinjlim F$$

(i.e. the functor $A/(-)$ commutes with colimits). In other words, the proposition above is about computing limits indexed by certain colimits of small categories. Recall that the nerve of a functor between small categories $u\colon A \to B$ is a right fibration if and only if, for any object a of A, the induced map $A/a \to B/u(a)$ is an isomorphism of categories (this follows right away from Proposition 4.1.2). In this case there is a unique presheaf of sets F on B such that $A \simeq B/F$ over B: one defines $F(b)$ as the fibre of u at b, which is then always a discrete category, whence a set. In this case, we say that u is a *discrete fibration*, or a *Grothendieck fibration with discrete fibres*. We may consider functors $G\colon I \to Cat/A$ such that, for each object i of I, the structural map $G(i) \to A$ is a discrete fibration. These are isomorphic to functors of the form $A/F(i)$ as above. We let the reader reformulate the preceding proposition directly in terms of the functor G.

7.3.7 Let A be a small category. We consider a *décalage* on A, that is an object ω, a functor $D\colon A \to A$, equipped with two natural transformations (where ω also stands for the constant functor with value ω)

(7.3.7.1)
$$1_A \xrightarrow{\eta} D \xleftarrow{\pi} \omega\,.$$

We also consider a *categorical realisation* of this décalage, by which we mean a functor $i\colon A \to Cat$ with the following properties.

(i) For any object a of A, the category $i(a)$ comes equipped with a specified final object e_a.
(ii) For any object a of A, the map $i(\eta_a)\colon i(a) \to i(D(a))$ is a sieve, i.e. it is a discrete fibration which is injective on objects.[2]
(iii) For any map $f\colon a \to b$ in A, the commutative square

(7.3.7.2)
$$\begin{array}{ccc} a & \xrightarrow{\eta_a} & D(a) \\ {\scriptstyle f}\downarrow & & \downarrow{\scriptstyle D(f)} \\ b & \xrightarrow{\eta_b} & D(b) \end{array}$$

is Cartesian.
(iv) For any object a of A, the commutative square

(7.3.7.3)
$$\begin{array}{ccc} \varnothing & \longrightarrow & \omega \\ \downarrow & & \downarrow{\scriptstyle \pi_a} \\ a & \xrightarrow{\eta_a} & D(a) \end{array}$$

is Cartesian.

Given any small category C, one may form the presheaf $i^*C = \mathrm{Hom}_{Cat}(i(-), C)$, and there is a canonical functor

(7.3.7.4)
$$\tau_C\colon A/i^*C \to C$$

[2] Alternatively, one may also see sieves as those functors $j\colon U \to V$ which are injective on objects and fully faithful, with the additional property that a map $f\colon w \to v$ in V is in the image of j if and only if there exists an object u in U such that $j(u) = v$.

defined on objects by $\tau_C(a,u) = u(e_a)$. Given a map $f\colon (a,u) \to (b,v)$ in A/i^*C, there is a map $i(f)(e_b) \to e_a$ whose image by u is the definition of $\tau_C(f)$.

Proposition 7.3.8 *Under the hypothesis of the above, for any small category C, the nerve of the functor $\tau_C\colon A/i^*C \to C$ is proper with contractible fibres. In particular, it is always final and cofinal.*

Proof The last assertion follows from the first, by Propositions 7.1.12 and 7.1.10. We shall thus focus on the proof that τ_C is proper with contractible fibres. For any object c of the category C, we observe that the obvious commutative square

$$\begin{array}{ccc} A/i^*(C/c) & \longrightarrow & A/i^*(C) \\ \big\downarrow{\scriptstyle \tau_{C/c}} & & \big\downarrow{\scriptstyle \tau_C} \\ C/c & \longrightarrow & C \end{array}$$

is Cartesian. Therefore, by virtue of Theorem 4.4.36, it is sufficient to prove that, for any small category C equipped with a final object e, the fibre $N(\tau_C^{-1}(e))$ is contractible and the nerve of the inclusion $\tau_C^{-1}(e) \to A/i^*C$ is final.

For this, we shall extend the décalage on A to a décalage on A/i^*C. Given an object a of A and a functor $u\colon i(a) \to C$, we define

$$D(u)\colon i(D(a)) \to C$$

as the unique functor such that $D(u)i(\eta_a) = u$, and such that $D(u)(x) = e$ for any object x which does not belong to the image of $i(\eta_a)$. This makes sense because of property (ii) above. The following diagram commutes (where we also write e for any constant functor with value e):

$$\begin{array}{ccccc} i(a) & \xrightarrow{i(\eta_a)} & i(D(a)) & \xleftarrow{i(\pi_a)} & i(\omega) \\ & {\scriptstyle u}\searrow & \big\downarrow{\scriptstyle D(u)} & \swarrow{\scriptstyle e} & \\ & & C & & \end{array}$$

because of property (iv). Property (iii) ensures that, for any map $f\colon (a,u) \to (b,v)$ in A/i^*C, the induced diagram

$$\begin{array}{ccc} i(D(a)) & \xrightarrow{i(D(f))} & i(D(b)) \\ & {\scriptstyle D(u)}\searrow \quad \swarrow{\scriptstyle D(v)} & \\ & C & \end{array}$$

commutes. In other words, we have defined a functor

$$\begin{aligned} D_C\colon A/i^*C &\to A/i^*C \\ (a,u) &\mapsto (D(a), D(u)) \end{aligned}$$

as well as natural transformations

$$1_{A/i^*C} \xrightarrow{\eta} D_C \xleftarrow{\pi} (\omega, e)\,.$$

We see that the full subcategory $\tau_C^{-1}(e)$ of A/i^*C is stable under the operator D_C, which proves that $N(\tau_C^{-1}(e))$ is contractible.

Finally, let us consider an object $\xi = (a_0, u_0)$ of A/i^*C, and form the following Cartesian square.

$$\begin{array}{ccc} \xi\backslash\tau_C^{-1}(e) & \longrightarrow & \xi\backslash(A/i^*C) \\ \downarrow & & \downarrow \\ \tau_C^{-1}(e) & \longrightarrow & A/i^*C \end{array}$$

It is sufficient to prove that the nerve of $\xi\backslash\tau_C^{-1}(e)$ is contractible. Given a triple (a,u,f) corresponding to an object (a,u) of $\tau_C^{-1}(e)$, i.e. a functor of the form $u\colon i(a) \to C$ such that $u(e_a) = e$, equipped with a map $f\colon a_0 \to a$ such that $u \circ i(f) = u_0$, one defines

$$D_\xi(a,u,g) = (D(a), D(u), \eta_a f)\,.$$

Defining $D_\xi(g) = D(g)$ on morphisms, this defines a functor

$$D_\xi\colon \xi\backslash\tau_C^{-1}(e) \to \xi\backslash\tau_C^{-1}(e)$$

which comes equipped with natural transformations

$$1_{\xi\backslash\tau_C^{-1}(e)} \xrightarrow{\eta} D_\xi \xleftarrow{D(f)} (D(a_0), D(u_0), \eta_{a_0})\,.$$

This shows that the nerve of $\xi\backslash\tau_C^{-1}(e)$ is contractible. □

Corollary 7.3.9 *For any small category C, there is a canonical functor $\tau_C\colon \mathbf{\Delta}/N(C) \to C$ whose nerve is proper with contractible fibres, whence is both final and cofinal.*

Proof We observe that the functor $D([n]) = [n] * [0] = [n+1]$ can be naturally extended to a décalage on $\mathbf{\Delta}$ with $\omega = [0]$. We conclude by applying the preceding proposition in the case where $A = \mathbf{\Delta}$ and $i\colon A \to \mathit{Cat}$ is the inclusion (so that $i^* = N$ is the nerve functor). □

Let $\mathbf{\Delta}_+$ be the subcategory of $\mathbf{\Delta}$ with the same objects, but with the strictly increasing maps as morphisms. The preceding corollary has a version without degeneracies.

Corollary 7.3.10 *Let $i\colon \mathbf{\Delta}_+ \to Cat$ be the inclusion functor. Evaluating at the maximal elements defines, for any small category C, a functor $\tau_C : \mathbf{\Delta}_+/i^*C \to C$ whose nerve is proper with contractible fibres, whence is both final and cofinal.*

Proof The same décalage as in the previous corollary has the same consequences for the same reasons. □

Proposition 7.3.11 *For any small category A, the functor*

$$\underline{\mathrm{Hom}}(A^{\mathrm{op}}, Set) \to sSet$$
$$F \mapsto N(A/F)$$

commutes with colimits.

Proof Let $\tau_A : \mathbf{\Delta}/N(A) \to A$ be the functor defined by evaluating at maximal elements. The induced functor

$$\tau_A^*\colon \underline{\mathrm{Hom}}(A^{\mathrm{op}}, Set) \to \underline{\mathrm{Hom}}((\mathbf{\Delta}/N(A))^{\mathrm{op}}, Set) \simeq sSet/N(A)$$

has a right adjoint and thus commutes with colimits. One checks that its composition with the obvious colimit preserving functor

$$sSet/N(A) \to sSet$$

is isomorphic to the functor $F \mapsto N(A/F)$, whence the assertion. □

7.3.12 Let C be a simplicial set, and $W \subset C$. A *weak localisation of C by W* is a map $C \to L$ such that, for any inner anodyne extension $L \to D$ such that D is an ∞-category, the composed map $C \to D$ is the localisation of C by W. For instance, if we form the push-out

$$\begin{array}{ccc} W & \hookrightarrow & C \\ \downarrow & & \downarrow \\ Ex^{\infty}(W) & \longrightarrow & Ex^{\infty}(W) \cup C \end{array} \tag{7.3.12.1}$$

the map $f\colon C \to L = Ex^{\infty}(W) \cup C$ is a weak localisation of C by W; see the proof of Proposition 7.1.3. The uniqueness of localisations implies that, if $f'\colon C \to L'$ is another weak localisation by W, any map $l\colon L \to L'$ such that $lf = f'$ is a weak categorical equivalence. We also observe that, if $C \to L$ is a weak localisation by W, and if V denotes the image of W in L, then the map $L \to Ex^{\infty}(V) \cup L$ is a weak categorical equivalence. Indeed, for any inner fibration $L \to D$ such that D is an ∞-category, we have $V \subset k(D)$, so that the localisation of D by V is equivalent to D. In other words, we obtain a commutative diagram

$$\begin{array}{ccccc} C & \longrightarrow & L & \longrightarrow & D \\ \downarrow & & \downarrow & & \downarrow \\ Ex^{\infty}(W) \cup C & \longrightarrow & Ex^{\infty}(V) \cup L & \longrightarrow & Ex^{\infty}(V) \cup D \end{array} \tag{7.3.12.2}$$

in which the maps on the bottom line, as well as the middle vertical map and the right vertical map, are weak categorical equivalences, because the map from C to any vertex of this diagram which is distinct from C is a weak localisation by W.

Lemma 7.3.13 *Let I be a small category, $F, G \colon I \to sSet$ be functors, and $\alpha \colon F \to G$ be a natural transformation. We suppose that there is a subfunctor $W \subset F$ such that, for any object i of I, the map $F(i) \to G(i)$ is a weak localisation of $F(i)$ by $W(i)$. We assume furthermore that one of the following conditions is satisfied.*

(a) The category I is discrete.
(b) We have $N(I) = \Lambda^2_0$, the maps $F(0) \to F(1)$ and $G(0) \to G(1)$ are inclusions, and $W(0) = F(0) \cap W(1)$.
(c) The category I is filtered.

Then the map $\varinjlim F \to \varinjlim G$ is a weak localisation of $\varinjlim F$ by the image of $\varinjlim W$ in $\varinjlim F$.

Proof Since the case where I is empty is obvious, case (a) follows straight away from (b) and (c). Assuming that (c) holds, we proceed as follows. We remark that, for any functor $X \colon I \to sSet$, using the natural trivial cofibration (3.1.22.5), we have a commutative diagram

$$\begin{array}{ccc} & \varinjlim X & \\ \swarrow & & \searrow \\ \varinjlim Ex^{\infty}(X) & \longrightarrow & Ex^{\infty}(\varinjlim X) \end{array}$$

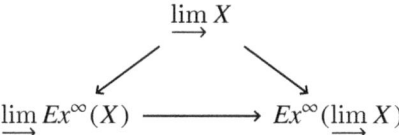

in which the slanted maps are weak homotopy equivalences, by Corollary 4.1.17. The horizontal map is thus an equivalence of ∞-groupoids. Let V be the image of W in G. For each index i, the identity of $G(i)$ exhibits $G(i)$ as a weak localisation by $V(i)$ (because the 1-simplices of $V(i)$ all are invertible in $\tau(G(i))$). By virtue of Corollary 3.9.8, we now have the commutative diagram below, in which the horizonal maps are weak categorical equivalences.

$$\begin{array}{ccccc}
\varinjlim(Ex^\infty(W)\cup F) & = & \varinjlim(Ex^\infty(W))\cup(\varinjlim F) & \longrightarrow & Ex^\infty(\varinjlim W)\cup(\varinjlim F)\\
\downarrow & & \downarrow & & \downarrow\\
\varinjlim(Ex^\infty(V)\cup G) & = & \varinjlim(Ex^\infty(V))\cup(\varinjlim G) & \longrightarrow & Ex^\infty(\varinjlim V)\cup(\varinjlim G)
\end{array}$$

Furthermore, as seen in paragraph 7.3.12, the maps

$$G(i)\to Ex^\infty(V(i))\cup G(i)\leftarrow Ex^\infty(W(i))\cup F(i)$$

are weak categorical equivalences for all i, whence so are the maps

$$\varinjlim G\to\varinjlim(Ex^\infty(V)\cup G)\leftarrow\varinjlim(Ex^\infty(W)\cup F)\,.$$

Since the map $\varinjlim F\to Ex^\infty(\varinjlim W)\cup(\varinjlim F)$ is a weak localisation, this proves the assertion of the lemma under assumption (c).

Under assumption (b), we proceed as follows. Given an ∞-category C, a simplicial set A and a simplicial subset $V\subset A$, recall that we write $\underline{\mathrm{Hom}}_V(A,C)$ for the full subcategory of $\underline{\mathrm{Hom}}(A,C)$ which sends the morphisms of V to invertible morphisms of C. The restriction functor along $F(0)\subset F(1)$ induces an isofibration

$$\underline{\mathrm{Hom}}(F(1),C)\to\underline{\mathrm{Hom}}(F(0),C)\,.$$

Therefore, since inner horn inclusions are bijective on objects, we get an inner fibration

$$\underline{\mathrm{Hom}}_{W(1)}(F(1),C)\to\underline{\mathrm{Hom}}_{W(0)}(F(0),C)\,.$$

We also observe that if two maps $f,g\colon A\to C$ are isomorphic in $\underline{\mathrm{Hom}}(A,C)$, then f belongs to $\underline{\mathrm{Hom}}_V(A,C)$ if and only if g has the same property. This implies that the inner fibration above is in fact an isofibration as well. The assumption that $W(0)=F(0)\cap W(1)$ ensures that the induced map $\varinjlim W\to\varinjlim F$ is a monomorphism. Furthermore the set of 1-simplices of $\varinjlim W$ consists of the disjoint union of $W(2)_1$ with the complement of $W(0)_1$ in $W(1)_1$. The pull-back square

$$\begin{array}{ccc}
\underline{\mathrm{Hom}}(\varinjlim F,C) & \longrightarrow & \underline{\mathrm{Hom}}(F(1),C)\\
\downarrow & & \downarrow\\
\underline{\mathrm{Hom}}(F(2),C) & \longrightarrow & \underline{\mathrm{Hom}}(F(0),C)
\end{array}$$

thus induces the pull-back square below.

$$\begin{array}{ccc}
\underline{\mathrm{Hom}}_{\varinjlim W}(\varinjlim F, C) & \longrightarrow & \underline{\mathrm{Hom}}_{W(1)}(F(1), C) \\
\downarrow & & \downarrow \\
\underline{\mathrm{Hom}}_{W(2)}(F(2), C) & \longrightarrow & \underline{\mathrm{Hom}}_{W(0)}(F(0), C)
\end{array}$$

Hence the latter is homotopy Cartesian in the Joyal model structure. On the other hand, we have a canonical homotopy Cartesian square of the following form.

$$\begin{array}{ccc}
\underline{\mathrm{Hom}}(\varinjlim G, C) & \longrightarrow & \underline{\mathrm{Hom}}(G(1), C) \\
\downarrow & & \downarrow \\
\underline{\mathrm{Hom}}(G(2), C) & \longrightarrow & \underline{\mathrm{Hom}}(G(0), C)
\end{array}$$

Therefore, the equivalences of ∞-categories

$$\underline{\mathrm{Hom}}(G(i), C) \to \underline{\mathrm{Hom}}_{W(i)}(F(i), C)$$

induce an equivalence of ∞-categories

$$\underline{\mathrm{Hom}}(\varinjlim G, C) \to \underline{\mathrm{Hom}}_{\varinjlim W}(\varinjlim F, C)\,.$$

This shows that $\varinjlim G$ is the weak localisation of $\varinjlim F$ by $\varinjlim W$ under assumption (b). □

7.3.14 The maps

$$\tau_{\Delta^n} = N(\tau_{[n]})\colon N(\mathbf{\Delta}/\Delta^n) = N(\mathbf{\Delta}/N([n])) \to N([n]) = \Delta^n \tag{7.3.14.1}$$

are functorial in Δ^n, and thus, by Proposition 7.3.11, extend uniquely to a natural map

$$\tau_X\colon N(\mathbf{\Delta}/X) \to X\,. \tag{7.3.14.2}$$

Let $j\colon \mathbf{\Delta}_+ \to \mathbf{\Delta}$ be the inclusion functor. Then, for any simplicial set X, there is a canonical inclusion $\mathbf{\Delta}_+/j^*(X) \to \mathbf{\Delta}/X$, whence a canonical map

$$\tau_{+,X}\colon N(\mathbf{\Delta}_+/j^*X) \to X\,. \tag{7.3.14.3}$$

We define

(7.3.14.4)

$$W_X = \bigcup_{x\in X_0} \tau_X^{-1}(x) \quad \text{and} \quad W_{+,X} = \bigcup_{x\in X_0} \tau_{+,X}^{-1}(x) = W_X \cap N(\mathbf{\Delta}_+/j^*X)\,.$$

Proposition 7.3.15 *The functor $\tau_X\colon N(\Delta/X) \to X$ exhibits X as the weak localisation of $N(\Delta/X)$ by W_X. Similarly, the functor $\tau_{+,X}\colon N(\Delta_+/j^*X) \to X$ exhibits X as a weak localisation of $N(\Delta_+/j^*X)$ by $W_{+,X}$.*

Proof Since τ is a natural transformation between functors which preserve colimits as well as monomorphisms, the class of simplicial sets X such that τ_X and $\tau_{+,X}$ are a weak localisation by W_X and $W_{+,X}$, respectively, is saturated by monomorphisms, by the previous lemma. By virtue of Corollary 1.3.10, it is sufficient to prove that this class contains Δ^n for all $n \geq 0$, which follows right away from Corollaries 7.3.9 and 7.3.10 for $C = [n]$, by Proposition 7.1.12. □

Proposition 7.3.16 *We fix a universe **U**. Let I be a **U**-small category, and $X\colon I \to sSet$ a functor taking its values in **U**-small simplicial sets, with colimit $Y = \varinjlim X$. We assume that one condition among the three following ones is satisfied.*

- *(a) The category I is discrete. The simplicial set Y is then the sum of the family $(X(i))_{i\in I}$.*
- *(b) We have $N(I) = \Lambda^2_0$, and the map $X(0) \to X(1)$ is a monomorphism. The simplicial set Y then fits in the following coCartesian square.*

$$\begin{array}{ccc} X(0) & \hookrightarrow & X(1) \\ \downarrow & & \downarrow \\ X(2) & \hookrightarrow & Y \end{array}$$

- *(c) The category I is filtered.*

*We consider an ∞-category C with **U**-small limits, and a functor*

$$\Phi\colon Y^{\mathrm{op}} \to C\,.$$

For each object i of I, we write $\Phi_i = \ell_i^*(\Phi) = \Phi_{|X(i)^{\mathrm{op}}}$, where we denote by $\ell_i\colon X(i) \to \varinjlim X = Y$ the canonical map. Then the assignment $i \mapsto \varprojlim \Phi_i$ extends naturally to a functor $N(I)^{\mathrm{op}} \to C$ such that there is a canonical invertible map of the form

$$\varprojlim_{y\in Y^{\mathrm{op}}} \Phi(y) \simeq \varprojlim_{i\in N(I)^{\mathrm{op}}} \varprojlim_{x\in X(i)^{\mathrm{op}}} \Phi_i(x)\,.$$

Proof For any **U**-small simplicial set K, we have a final functor

$$\tau_K\colon N(\Delta/K) \to K\,,$$

by Proposition 7.3.15. Therefore, for any functor $\Psi\colon K^{\mathrm{op}} \to C$, there is a canonical equivalence $\varprojlim \Psi \simeq \varprojlim \tau_K^*\Psi$, by Theorem 6.4.5. Therefore Proposition

7.3.5 for $A = \mathbf{\Delta}$ and $F = X$ tells us that the assignment

$$i \mapsto \varprojlim \Phi_i \simeq \varprojlim \tau^*_{X(i)} \Phi_i$$

can be promoted to a functor from $N(I)^{\mathrm{op}}$ to C such that we have

$$\varprojlim_{y \in Y^{\mathrm{op}}} \Phi(y) \simeq \varprojlim \tau_Y^* \Phi \simeq \varprojlim_{i \in N(I)^{\mathrm{op}}} \varprojlim \tau^*_{X(i)} \Phi_i \simeq \varprojlim_{i \in N(I)^{\mathrm{op}}} \varprojlim_{x \in X(i)^{\mathrm{op}}} \Phi_i(x) .$$

Hence the assertion. □

Corollary 7.3.17 *Let C be an ∞-category which has countable products as well as pull-backs. Then it has limits of type $N(\mathbf{Z}_{\geq 0})$ (where $\mathbf{Z}_{\geq 0}$ is the set of non-negative integers, totally ordered in the usual sense). More precisely, for any functor $X : N(\mathbf{Z}_{\geq 0})^{\mathrm{op}} \to C$, which one may see informally as a diagram of the form*

(7.3.17.1) $$\cdots \to X_{n+1} \xrightarrow{p_n} X_n \to \cdots \to X_1 \xrightarrow{p_0} X_0 ,$$

there is a canonical Cartesian square of the following form in C,

(7.3.17.2) $$\begin{array}{ccc} \varprojlim X & \longrightarrow & \prod_{n \geq 0} X_n \\ \downarrow & & \downarrow {\scriptstyle (pr_{n+1}, pr_n)_{n \geq 0}} \\ \prod_{n \geq 0} X_{n+1} & \xrightarrow{\prod_{n \geq 0}(1_{X_{n+1}}, p_n)} & \prod_{n \geq 0} X_{n+1} \times X_n \end{array}$$

where the map $pr_i : \prod_{n \geq 0} X_n \to X_i$ is the ith projection.

Proof Let Y be the union of the images of the maps $u_i : \Delta^1 \to N(\mathbf{Z}_{\geq 0})$ defined by $u_i(t) = i + t$ for $t = 0, 1$, with $i \geq 0$. The inclusion map

$$Y \to N(\mathbf{Z}_{\geq 0})$$

is a weak categorical equivalence:[3] indeed, it is a filtered colimit of the weak categorical equivalences provided by Proposition 3.7.4, so that we may apply Corollary 3.9.8. Since any weak categorical equivalence is final, it is thus sufficient to prove that C has limits of type Y. By virtue of Theorem 1.3.8, there is a canonical coCartesian square of the form

$$\begin{array}{ccc} \coprod_{y \in \Sigma} \partial \Delta^1 & \longrightarrow & Sk_0(Y) \\ \downarrow & & \downarrow \\ \coprod_{y \in \Sigma} \Delta^1 & \longrightarrow & Y \end{array}$$

where $\Sigma \simeq \mathbf{Z}_{\geq 0}$ is the set of non-degenerate maps of the form $f : \Delta^1 \to$

[3] Hence the informal way of considering functors indexed by $N(\mathbf{Z}_{\geq 0})^{\mathrm{op}}$ as sequences of maps as above is not that informal after all.

Y. Therefore, in the case where C has $\mathbf{U}$-small limits for some universe $\mathbf{U}$, Proposition 7.3.16 shows that we have a Cartesian square of the form (7.3.17.2). In the general case, we may assume that C is $\mathbf{U}$-small. Then, for any tower of maps of the form (7.3.17.1) in $\underline{\mathrm{Hom}}(C^{\mathrm{op}}, \mathcal{S})$, the limit fits in a Cartesian square of the form (7.3.17.2). If, furthermore, all the presheaves X_n are representable, since C is assumed to have countable products as well as pull-backs, and since the Yoneda embedding preserves limits, this shows that $\varprojlim_{n \geq 0} X_n$ exists in C, and that the latter is constructed with the pull-back square (7.3.17.2) in C. □

7.3.18 Let I be a small category. We say that I is *cycle-free* if, for any object i of I, any map of the form $i \to i$ is an identity.

An object i of a cycle-free category I has *finite length* if there is an integer n such that, for any finite sequence of maps of the form

$$i_0 \xrightarrow{f_1} i_1 \xrightarrow{f_2} \cdots \xrightarrow{f_m} i_m = i\,,$$

in which none of the maps f_i is an identity, we have $m \leq n$. If this is the case, the smallest such integer n is called the *length of i* and is denoted by $\ell(i)$.

Definition 7.3.19 A *direct category* is a small category I which is cycle-free and whose objects all have finite length.

Example 7.3.20 The nerve of a small category I is a finite simplicial set (i.e. has finitely many non-degenerate simplices) if and only if I is finite and direct. In particular, any finite partially ordered set I is a direct category.

Example 7.3.21 The property of being direct is local in the following sense: a small category I is direct if and only if the slice I/i is direct for any object i of I. In particular, if I is direct, for any discrete fibration $J \to I$, the category J is direct. Therefore, since the category $\mathbf{\Delta}_+$ is direct (the length of $[n]$ is n), for any simplicial set X, the category $\mathbf{\Delta}_+/j^*X$ is direct as well.

Theorem 7.3.22 *For an ∞-category C and a universe $\mathbf{U}$, the following conditions are equivalent.*

(i) The ∞-category C has $\mathbf{U}$-small limits.
(ii) The ∞-category C has limits of type $N(I)^{\mathrm{op}}$ for any $\mathbf{U}$-small direct category I.
(iii) The ∞-category C has limits of type $N(I)$ for any $\mathbf{U}$-small direct category I.
(iv) The ∞-category C has $\mathbf{U}$-small products as well as pull-backs.

Proof Since the category $\mathbf{\Delta}_+/j^*X$ is a $\mathbf{U}$-small direct category for any $\mathbf{U}$-small simplicial set X, Theorem 6.4.5 and Proposition 7.3.15 show that conditions (i),

(ii) and (iii) are equivalent. It is clear that condition (i) implies condition (iv). Therefore, it only remains to prove that condition (iv) implies condition (i). By virtue of Corollary 7.3.17, condition (iv) implies that C has limits indexed by $N(\mathbf{Z}_{\geq 0})^{\mathrm{op}}$. Proceeding as at the end of the proof of Corollary 7.3.17 (possibly after having chosen a universe $\mathbf{V}$ containing $\mathbf{U}$ and such that C is $\mathbf{V}$-small), we see that Proposition 7.3.16 implies that the class of simplicial sets X such that C has limits of type X is saturated by monomorphisms within the subcategory of $\mathbf{U}$-small simplicial sets. Applying the $\mathbf{U}$-small version of Corollary 1.3.10, this implies that condition (i) is fulfilled. □

A similar proof gives the following.

Proposition 7.3.23 *Let $f: C \to D$ be a functor between ∞-categories. We assume that we are given a universe $\mathbf{U}$ such that C has $\mathbf{U}$-small limits. Then the following conditions are equivalent.*

(i) The functor f commutes with $\mathbf{U}$-small limits.
(ii) The functor f commutes with limits of type $N(I)^{\mathrm{op}}$ for any $\mathbf{U}$-small direct category I.
(iii) The functor f commutes with limits of type $N(I)$ for any $\mathbf{U}$-small direct category I.
(iv) The functor f commutes with $\mathbf{U}$-small products as well as with pull-backs.

There is a variant with finite limits.

Definition 7.3.24 *Finite limits* are limits in ∞-categories which are indexed by finite simplicial sets (e.g. by nerves of finite direct categories).

Example 7.3.25 Finite products are finite limits. Pull-backs are finite limits.

Lemma 7.3.26 *Let $\mathcal{C}$ be a class of simplicial sets. We assume that $\mathbf{C}$ is closed under finite coproducts and that, for any push-out square of simplicial sets*

$$\begin{array}{ccc} X & \longrightarrow & X' \\ \downarrow & & \downarrow \\ Y & \longrightarrow & Y' \end{array}$$

in which the vertical maps are monomorphisms, if X, X' and Y all are in $\mathcal{C}$, so is Y'. If $\mathcal{C}$ contains Δ^n for all $n \geq 0$, then $\mathcal{C}$ contains all finite simplicial sets (in particular, the nerve of any finite direct category is in $\mathcal{C}$).

The proof is similar to the proof of Corollary 1.3.10 and is left to the reader.

Theorem 7.3.27 *For an ∞-category C, the following conditions are equivalent.*

(i) The ∞-category C has finite limits.
(ii) The ∞-category C has limits of type $N(I)^{\mathrm{op}}$ for any finite direct category I.
(iii) The ∞-category C has a final object as well as pull-backs.

Proof Final objects and pull-backs are limits indexed by $\varnothing$ and Λ^2_2, respectively, which both are nerves of finite partially ordered sets. Therefore, condition (iii) is implied by condition (ii). Condition (i) obviously implies condition (ii). Using Theorem 6.6.9, we see that the existence of a final object and of pull-backs implies the existence of finite products. Hence, using Proposition 7.3.16 and the preceding lemma, we see that condition (iii) implies condition (i). □

Similarly, we get the following result.

Proposition 7.3.28 *Let $f\colon C \to D$ be a functor between ∞-categories, C having finite limits. The following conditions are equivalent.*

(i) The functor f commutes with finite limits.
(ii) The functor f commutes with limits of type $N(I)^{\mathrm{op}}$ for any finite direct category I.
(iii) The functor f preserves final objects and commutes with pull-backs.

Corollary 7.3.29 *Let C be an ∞-category with finite limits. For any object x of C, the slice category C/x has finite limits. Moreover, for any functor $f\colon C \to D$ which commutes with finite limits, and for any object x of C, the induced functor $C/x \to D/f(x)$ commutes with finite limits.*

Proof By virtue of Theorem 6.6.9, all limits of type Λ^2_2 exist in C/x, and the functor $C/x \to C$ commutes with them. Since C/x has a final object, Theorem 7.3.27 shows that it has finite limits. The last assertion follows similarly from Theorem 6.6.9 and from Proposition 7.3.28. □

7.4 Finite Direct Diagrams

7.4.1 Let I be a direct category. There is a filtration of I given by the full subcategories $I^{(n)}$ which are made of objects i of length $\ell(i) \leq n$. The inclusion functors

$$I^{(n)} \to I^{(n+1)}$$

all are sieves (in particular, discrete fibrations), so that the categories $I^{(n)}$ are also direct categories.

Given a set $S \subset \mathrm{Ob}(I)$, we define $I \setminus S$ as the full subcategory of I whose objects are those which do not belong to S. We define the boundary of $I^{(n)}$ as

$$\partial I^{(n)} = I^{(n)} \setminus \{i \in \mathrm{Ob}(I) \mid \ell(i) = n\}\,.$$

Again $\partial I^{(n)} \subset I^{(n)}$ is a sieve, whence $\partial I^{(n)}$ is a direct category. We say that a direct category I has *finite length* if $I^{(n)} = I$ for n large enough. The smallest such n is called the *length of I* and is denoted by $\ell(I)$. If I is of finite length, its *boundary* is defined as

$$\partial I = \partial I^{(\ell(I))} = I^{(\ell(I)-1)}\,.$$

Given an object i of I, the slice category I/i is a direct category as well. Moreover, I/i has finite length, and $\ell(i) = \ell(I/i)$. In particular, we have the *boundary category at i*: $\partial I/i = (I/i)^{\ell(i)-1}$.

Proposition 7.4.2 *Let I be a direct category of finite length. We let S be any subset of the set of objects i of I such that $\ell(i) = \ell(I)$. Then there is the following canonical biCartesian square of categories in which all maps are discrete fibrations.*

$$\begin{array}{ccc} \coprod_{i \in S} \partial I/i & \hookrightarrow & \coprod_{i \in S} I/i \\ \downarrow & & \downarrow \\ I \setminus S & \hookrightarrow & I \end{array} \tag{7.4.2.1}$$

Proof It is clear that the obvious commutative square of the form (7.4.2.1) is Cartesian and that all its maps are discrete fibrations. We recall that the forgetful functor from discrete fibrations over I to categories over I commutes with colimits (being a left adjoint), and that there is an equivalence relating discrete fibrations $X \to I$ and presheaves of sets on I. Therefore, since colimits of presheaves are computed fibrewise, to prove that (7.4.2.1) is coCartesian, it is sufficient to prove that, for any object j of I, the pull-back of the square (7.4.2.1) along $\{j\} \to I$ is a coCartesian square of sets (or, more precisely, of discrete categories). If j is in S, then we get

$$\begin{array}{ccc} \varnothing & \longrightarrow & \{(j, 1_j)\} \\ \downarrow & & \downarrow \\ \varnothing & \longrightarrow & \{j\} \end{array}$$

and, otherwise, we obtain

$$\begin{array}{ccc} \coprod_{i \in S} \coprod_{u \in \mathrm{Hom}_I(j,i)} \{(j,u)\} & = & \coprod_{i \in S} \coprod_{u \in \mathrm{Hom}_I(j,i)} \{(j,u)\} \\ \downarrow & & \downarrow \\ \{j\} & \longrightarrow & \{j\} \end{array}$$

which proves the claim either way. □

7.4.3 Given an ∞-category C and a C-valued presheaf $F\colon I^{op} \to C$, for an object i of I, we define F/i as the presheaf on I/i obtained by composing with the canonical map $p\colon I/i \to I$ (i.e. $F/i = p^*(F)$). Similarly, we define $\partial F/i$ as the presheaf on $\partial I/i$ obtained as the restriction of F/i on $\partial I/i$. We shall say that F *is representable at the boundary of* i if the limit of $\partial F/i$ exists in C. If this is the case, we denote such a limit by

$$\partial F(i) = \varprojlim \partial F/i\,. \tag{7.4.3.1}$$

Since $(i, 1_i)$ is a final object of I/i there is a canonical identification $F(i) = \varprojlim F/i$, and thus a canonical map of the form

$$F(i) \to \partial F(i)\,. \tag{7.4.3.2}$$

Corollary 7.4.4 *Let I be a direct category of finite length. We let S be any subset of the set of objects i of I such that $\ell(i) = \ell(I)$. Let $F\colon N(I)^{op} \to C$ be a presheaf on I with values in an ∞-category C. We assume that the following conditions are verified.*

(a) The presheaf F is representable at the boundary of any object $i \in S$.
(b) The products $\prod_{i\in S} F(i)$ and $\prod_{i\in S} \partial F(i)$ exist in C.
(c) The limit of the restriction of F at $N(I \setminus S)$ exists in C.
(d) The pull-back along the induced map $\prod_{i\in S} F(i) \to \prod_{i\in S} \partial F(i)$ of the canonical map $\varprojlim F_{|N(I\setminus S)} \to \prod_{i\in S} \partial F(i)$ exists in C.

Then the limit of F exists, and there is a canonical Cartesian square in C of the form below.

$$\begin{array}{ccc} \varprojlim F & \longrightarrow & \varprojlim F_{|N(I\setminus S)} \\ \downarrow & & \downarrow \\ \prod_{i\in S} F(i) & \longrightarrow & \prod_{i\in S} \partial F(i) \end{array} \tag{7.4.4.1}$$

Proof Let us choose a universe $\mathbf{U}$ such that I is $\mathbf{U}$-small. If C has $\mathbf{U}$-small limits, then Proposition 7.4.2 allows us to use Proposition 7.3.5 (interpreted through Remark 7.3.6), which provides the Cartesian square (7.4.4.1). To reach the general case, we apply what precedes to $\underline{\mathrm{Hom}}(C^{op}, \mathcal{S})$, and get a Cartesian square of presheaves as in (7.4.4.1), but replacing F by its image by the Yoneda embedding $h_C(F)$. Since the Yoneda functor commutes with limits, our hypotheses imply right away that $\varprojlim h_C(F)$ is representable, hence that $\varprojlim F$ exists, and that (7.4.4.1) is Cartesian in C. □

Proposition 7.4.5 *Let I be a direct category, and $F\colon N(I)^{\mathrm{op}} \to C$ a presheaf with value in an ∞-category C. We assume that:*

(a) for each integer $n \geq 0$, the limit of the restriction of F at $N(I^{(n)})$ exists in C;
(b) the limit of the induced tower

$$(7.4.5.1)\qquad \cdots \to \varprojlim F_{|N(I^{(n+1)})} \to \varprojlim F_{|N(I^{(n)})} \to \cdots \to \varprojlim F_{|N(I^{(0)})}$$

exists in C.

Then the limit of F exists and there is a canonical invertible map

$$(7.4.5.2)\qquad \varprojlim F \simeq \varprojlim_{n\geq 0} \left(\varprojlim F_{|N(I^{(n)})} \right).$$

Proof Since the inclusions $I^{(n)} \subset I$ are sieves, this is again a direct application of Proposition 7.3.5 and of the fact that the Yoneda embedding preserves and detects limits. □

Definition 7.4.6 Let C be an ∞-category with a specified final object e. A *class of fibrations* of C is a subobject $\mathsf{F} \subset C$ which contains all identities and is closed under composition, such that, for any maps $v\colon y' \to y$ and $f\colon x \to y$ in C, such that $x \to e$, $y \to e$ and $y' \to e$ are in F, the pull-back $y' \times_y x$ exists, and, for any Cartesian square of the form

$$\begin{array}{ccc} x' & \xrightarrow{u} & x \\ {\scriptstyle f'}\downarrow & & \downarrow{\scriptstyle f} \\ y' & \xrightarrow{v} & y \end{array}$$

the map f' belongs to F.

7.4.7 We assume now that a class of fibrations F is fixed. Given a direct category I, we shall say that a presheaf $F\colon I^{\mathrm{op}} \to C$ is *Reedy fibrant* (or *Reedy fibrant with respect to* F, if there is any need for more precision) if, for any object i of I, F is representable at the boundary of i and if the canonical map $F(i) \to \partial F(i)$ belongs to F. An object x of C will be said to be *fibrant* (or *fibrant with respect to* F) if the morphism $x \to e$ belongs to F (or, equivalently, if x is Reedy fibrant when seen as a presheaf on $N([0]) = \Delta^0$).

A morphism $F \to G$ of Reedy fibrant C-valued presheaves over I is called a *Reedy fibration* (or a *Reedy fibration with respect to* F) if, for any object i of I, the fibre product $F(i) \times_{\partial G(i)} G(i)$ is representable in C, and the canonical map

$$F(i) \to \partial F(i) \times_{\partial G(i)} G(i)$$

induced by the canonical commutative square

$$\begin{array}{ccc} F(i) & \longrightarrow & G(i) \\ \downarrow & & \downarrow \\ \partial F(i) & \longrightarrow & \partial G(i) \end{array}$$

belongs to F. A morphism between fibrant objects $x \to y$ will be called a *fibration* (or a *fibration with respect to* F) if it belongs to F (or, equivalently, if it is a Reedy fibration when seen as a presheaf on $N([1]) = \Delta^1$).

Proposition 7.4.8 *Let $F: N(I)^{op} \to C$ be a Reedy fibrant presheaf on a finite direct category. Then the limit of F exists and is fibrant in C. Moreover, for any sieve $j: J \to I$, the presheaf $j^*(F)$ is Reedy fibrant and the induced map*

$$\varprojlim F \to \varprojlim j^*(F)$$

is a fibration.

Proof The first assertion follows right away from Corollary 7.4.4 by induction on the length of I. To prove the second assertion, we may consider the set S of objects i of I which are of length $\ell(I)$, but which are not in the image of j. We may assume that $S \neq \varnothing$, since, otherwise, $I = J$, and the assertion is trivial. The Cartesian square (7.4.4.1) implies that the map

$$\varprojlim F \to \varprojlim F_{|N(I \setminus S)}$$

is a fibration, since it is the pull-back of a product of fibrations between fibrant objects of the form $F(i) \to \partial F(i)$ (we use here the fact that finite products of elements of F are in F, which is a direct consequence of Corollary 6.6.10 and of the fact that F is closed under compositions and under pull-backs in the subcategory of fibrant objects). By induction on the number of objects of I, the comparison map

$$\varprojlim F_{|N(I \setminus S)} \to \varprojlim F_{|N(J)}$$

is a fibration as well. Therefore, the composition

$$\varprojlim F \to \varprojlim F_{|N(I \setminus S)} \to \varprojlim F_{|N(J)}$$

is a fibration. □

Proposition 7.4.9 *Let $F \to G$ be a Reedy fibration between Reedy fibrant C-valued presheaves on a finite direct category $N(I)$. Then the induced map $\varprojlim F \to \varprojlim G$ is a fibration.*

Proof We proceed by induction on the length of I. In the case where $\ell(I) \leq 0$, this assertion amounts to the stability of F by products. If $\ell(I) > 0$, since $\ell(\partial I) < \ell(I)$, the induced map

$$\varprojlim F_{|N(\partial I)} \to \varprojlim G_{|N(\partial I)} \tag{7.4.9.1}$$

is a fibration. Since, by virtue of Proposition 7.4.8, $\varprojlim G$ is fibrant, we can form the following Cartesian square.

$$\begin{array}{ccc} \varprojlim G \times_{\varprojlim G_{|N(\partial I)}} \varprojlim F_{|N(\partial I)} & \longrightarrow & \varprojlim F_{|N(\partial I)} \\ \downarrow & & \downarrow \\ \varprojlim G & \longrightarrow & \varprojlim G_{|N(\partial I)} \end{array} \tag{7.4.9.2}$$

If $S = \{i \in \mathrm{Ob}(I) \mid \ell(i) = \ell(I)\}$, since the functor $\prod_S$ commutes with limits, we have a canonical pull-back square of the following form.

$$\begin{array}{ccc} \prod_{i \in S} G(i) \times_{\partial G(i)} F(i) & \longrightarrow & \prod_{i \in S} \partial F(i) \\ \downarrow & & \downarrow \\ \prod_{i \in S} G(i) & \longrightarrow & \prod_{i \in S} \partial G(i) \end{array} \tag{7.4.9.3}$$

Since we also have Cartesian squares of the form (7.4.4.1) for F and for G, using Corollary 6.6.10, we deduce that there is a canonical Cartesian square of the form

$$\begin{array}{ccc} \varprojlim F & \longrightarrow & \varprojlim G \times_{\varprojlim G_{|N(\partial I)}} \varprojlim F_{|N(\partial I)} \\ \downarrow & & \downarrow \\ \prod_{i \in S} F(i) & \longrightarrow & \prod_{i \in S} G(i) \times_{\partial G(i)} F(i) \end{array} \tag{7.4.9.4}$$

in which the lower horizontal map is a finite product of fibrations, whence a fibration. □

Proposition 7.4.10 *Reedy fibrations between Reedy fibrant objects define a class of fibrations.*

Proof Any such fibration $F \to G$ induces a fibration between fibrant objects $F(i) \to G(i)$ for any object i (as a particular case of Proposition 7.4.9 applied to the direct category I/i). This proves the existence of pull-backs along Reedy fibrations of any map between Reedy fibrant presheaves. Let $F_0 \to F_1 \to F_2$ be two composable Reedy fibrations and i an object of I. We observe that there is a Cartesian square

$$\begin{array}{ccc} \partial F_0(i) \times_{\partial F_1(i)} F_1(i) & \longrightarrow & F_1(i) \\ \downarrow & & \downarrow \\ \partial F_0(i) \times_{\partial F_2(i)} F_2(i) & \longrightarrow & \partial F_1(i) \times_{\partial F_2(i)} F_2(i) \end{array}$$

whose left vertical map must be a fibration. Composing with the fibration

$$F_0(i) \to \partial F_0(i) \times_{\partial F_1(i)} F_1(i)$$

this shows that the composed map $F_0 \to F_2$ is a Reedy fibration.

Let $p \colon F \to F'$ be a Reedy fibration and $u' \colon G' \to F'$ be a morphism with Reedy fibrant domain. Since G' is fibrewise fibrant and since p is fibrewise a fibration, one may form the following Cartesian square.

$$\begin{array}{ccc} G & \xrightarrow{u} & F \\ {\scriptstyle q}\downarrow & & \downarrow{\scriptstyle p} \\ G' & \xrightarrow{u'} & F' \end{array}$$

Since p is a Reedy fibration, for any object i of I, $\partial F(i) \to \partial F'(i)$ is a fibration, by Proposition 7.4.9. Therefore, $\partial G(i) \simeq \partial G'(i) \times_{\partial F'(i)} \partial F(i)$ is representable in C. Finally, we see that there is a canonical Cartesian square

$$\begin{array}{ccc} G(i) & \longrightarrow & F(i) \\ \downarrow & & \downarrow \\ \partial G(i) \times_{\partial G'(i)} G'(i) & \longrightarrow & \partial F(i) \times_{\partial F'(i)} F'(i) \end{array}$$

which shows that q is a Reedy fibration. □

Proposition 7.4.11 *Let I be a finite direct category and $H' \colon N(\partial I)^{\mathrm{op}} \to C$ a Reedy fibrant presheaf. The data of fibrations*

$$\pi_i \colon H(i) \to \varprojlim H'_{|\partial I/i},$$

indexed by the set of objects i such that $\ell(i) = \ell(I)$, extend essentially uniquely to a Reedy fibrant presheaf $H \colon N(I)^{\mathrm{op}} \to C$ whose restriction to ∂I is equal to H', and such that the maps $H(i) \to \partial H(i)$ are equivalent to π_i for all i.

Furthermore, given any other presheaf $F \colon N(I)^{\mathrm{op}} \to C$, to extend a morphism

$$F_{|N(\partial I)} \to H'$$

it is sufficient to determine maps

$$F(i) \to H(i)$$

such that the diagram

$$\begin{array}{ccc} F(i) & \longrightarrow & H(i) \\ & \searrow & \downarrow \\ & & \partial H(i) \end{array}$$

commutes for all i of length $\ell(I)$.

Proof We observe that $N(\partial I/i) * \Delta^0 = N(I/i) = N(I)/i$ (since I is cycle-free). By virtue of Proposition 4.2.3, there is a canonical commutative triangle

$$\begin{array}{ccc} & N(\partial I/i) & \\ {}^{u_i}\swarrow & & \searrow^{v_i} \\ N(\partial I/i) \diamond \Delta^0 & \longrightarrow & N(\partial I/i) * \Delta^0 \end{array}$$

in which the horizontal map is a weak categorical equivalence, while the slanted maps are monomorphisms. This induces a commutative triangle of the form

$$\begin{array}{ccc} \underline{\mathrm{Hom}}(N(I/i)^{\mathrm{op}}, C) & \longrightarrow & \underline{\mathrm{Hom}}((N(\partial I/i) \diamond \Delta^0)^{\mathrm{op}}, C) \\ {}_{v_i^*}\searrow & & \swarrow_{u_i^*} \\ & \underline{\mathrm{Hom}}(N(\partial I)^{\mathrm{op}}, C) & \end{array}$$

in which the horizontal map is an equivalence of ∞-categories and the functors u_i^* and v_i^* are isofibrations. Therefore, the fibres of u_i^* and v_i^* at the object H' are equivalent. Let S be the set of objects of length $\ell(I)$. The data of maps π_i for $i \in S$ thus correspond to functors $H_i \colon N(I/i)^{\mathrm{op}} \to C$ such that $v_i^*(H_i) = H'_{|\partial I/i}$ for all $i \in S$. On the other hand, it follows from Propositions 7.3.11 and 7.4.2 that there is a canonical Cartesian square of the following form.

$$\begin{array}{ccc} \underline{\mathrm{Hom}}(N(I)^{\mathrm{op}}, C) & \longrightarrow & \underline{\mathrm{Hom}}(N(\partial I)^{\mathrm{op}}, C) \\ \downarrow & & \downarrow \\ \prod_{i \in S} \underline{\mathrm{Hom}}(N(I/i)^{\mathrm{op}}, C) & \xrightarrow{\prod_{i \in S} v_i^*} & \prod_{i \in S} \underline{\mathrm{Hom}}(N(\partial I/i)^{\mathrm{op}}, C) \end{array}$$

This square is even homotopy Cartesian because the horizontal maps are isofibrations. Therefore, the presheaves H_i, $i \in S$, and H' define a unique presheaf $H \colon N(I)^{\mathrm{op}} \to C$ which extends H'.

To prove the second assertion about the extension of morphisms, we observe that, since the restriction functor

$$\underline{\mathrm{Hom}}(\Delta^1 \times N(I)^{\mathrm{op}}, C) \to \underline{\mathrm{Hom}}(\Delta^1 \times N(\partial I)^{\mathrm{op}}, C)$$

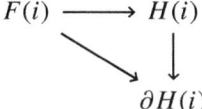

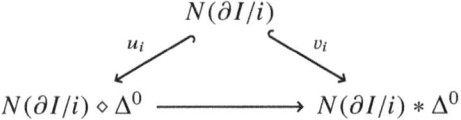

is an isofibration, we can work up to equivalence. Therefore, after having chosen a universe $\mathbf{U}$ such that C is $\mathbf{U}$-small, we may as well replace C by $\underline{\mathrm{Hom}}(C^{\mathrm{op}}, \mathcal{S})$, and the various protagonists by their images by the Yoneda embedding h_C. This means that we may assume that C has finite limits and the class of fibrations consists of all maps in C. In particular, any presheaf is Reedy fibrant. In this context, the problem of the extension of morphisms of presheaves on $N(\partial I)$ to presheaves on $N(I)$ is simply the problem of the extension of presheaves on $N(\partial J)$ to presheaves on $N(J)$ for $J = [1] \times I$. Applying the first part of the proof to the finite direct category J thus proves the proposition in full. □

Definition 7.4.12 An *∞-category with weak equivalences and fibrations* is a triple (C, W, Fib), where C is an ∞-category with a final object, $W \subset C$ is a subcategory which has the two-out-of-three property, and $Fib \subset C$ is a class of fibrations, such that the following properties are verified.

(i) For any Cartesian square of C

$$\begin{array}{ccc} x' & \xrightarrow{u} & x \\ {\scriptstyle p'}\downarrow & & \downarrow{\scriptstyle p} \\ y' & \xrightarrow{v} & y \end{array}$$

in which p is a fibration between fibrant objects, and y' is fibrant, if p belongs to W, so does p'.

(ii) For any map with fibrant codomain $f : x \to y$ in C, there exists a map $w : x \to x'$ in W and a fibration $p : x' \to y$ such that f is a composition of p and w.

We often will call *weak equivalences* the arrows in W. The maps of C which are both weak equivalences and fibrations will be called *trivial fibrations*.

Dually, one defines *∞-categories with weak equivalences and cofibrations* the triples (C, W, Cof) such that $(C^{\mathrm{op}}, W^{\mathrm{op}}, Cof^{\mathrm{op}})$ is an ∞-category with weak equivalences and fibrations. The morphisms in W are also called weak equivalences, while those in Cof are called *cofibrations*. The initial object of C will often be written $\varnothing$, and the objects x such that $\varnothing \to x$ is a cofibration will be said to be *cofibrant*.

From now on, we fix an ∞-category with weak equivalences and fibrations C.

Proposition 7.4.13 (Ken Brown's lemma) *For any map $f : x \to y$ between* *fibrant objects in C, there exist commutative triangles of the form*

$$
\begin{array}{ccccc}
 & & x & & \\
 & \swarrow\!\!\!= & \downarrow{\scriptstyle s} & \searrow{\scriptstyle f} & \\
x & \xleftarrow{p} & z & \xrightarrow{q} & y
\end{array}
$$

with p a trivial fibration, and q a fibration.

Proof Since x and y are fibrant, their product exists in C and is fibrant. The maps 1_x and f determine a map $g\colon x \to x \times y$. We consider a factorisation of g into a weak equivalence $s\colon x \to z$ followed by a fibration $\pi\colon z \to x \times y$. The maps p and q are defined by composing π with the projections $x \times y \to x$ and $x \times y \to y$, respectively. □

Corollary 7.4.14 *Let $F\colon C \to D$ be a functor and $V \subset D$ a subcategory which has the two-out-of-three property. If F sends trivial fibrations between fibrant objects into V, then F sends weak equivalences between fibrant objects into V.*

Proposition 7.4.15 *For any Cartesian square of C of the form*

$$
\begin{array}{ccccc}
x'' & \xrightarrow{p'} & x' & \xrightarrow{p} & x \\
{\scriptstyle u''}\downarrow & & {\scriptstyle u'}\downarrow & & \downarrow{\scriptstyle u} \\
y'' & \xrightarrow{q'} & y' & \xrightarrow{q} & y
\end{array}
$$

in which q as well as any composition of q and q' are fibrations, with both x and y fibrant, if q' is a weak equivalence, so is p'.

Proof For any fibrant object z of C, the slice category C/z has a natural structure of ∞-category with weak equivalences and fibrations: the weak equivalences (fibrations) are the maps whose image by the canonical projection $C/z \to C$ is a weak equivalence (a fibration, respectively). Let $C(z)$ be the subcategory of fibrant objects of C/z (i.e. the full subcategry of fibrations $f\colon z' \to z$). It follows easily from Theorem 6.6.9 that pulling back along u defines a left exact functor

$$u^*\colon C(y) \to C(x)\,.$$

The assertion of the proposition is that the functor u^* preserves weak equivalences between fibrant objects, which follows from Corollary 7.4.14. □

Proposition 7.4.16 *For any Cartesian square of C*

$$
\begin{array}{ccc}
x' & \xrightarrow{u} & x \\
{\scriptstyle p'}\downarrow & & \downarrow{\scriptstyle p} \\
y' & \xrightarrow{v} & y
\end{array}
$$

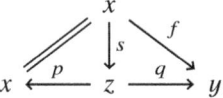

in which p is a fibration between fibrant objects, and y' is fibrant, if v is a weak equivalence, so is u.

Proof If v is a trivial fibration, so is u. Therefore, by virtue of Ken Brown's lemma (Proposition 7.4.13), we may assume, without loss of generality, that there is a trivial fibration $q\colon y \to y'$ such that the identity of y' is a composition of q and v. Let r be a composition of q and p. We form the following Cartesian square.

$$\begin{array}{ccc} x_0 & \xrightarrow{q_0} & x \\ {\scriptstyle r_0}\downarrow & & \downarrow{\scriptstyle r} \\ y & \xrightarrow{q} & y' \end{array}$$

We observe that q_0 is a trivial fibration, and that there is a map $i\colon x \to x_0$ such that the identity of x is a composition of q_0 and i, and such that p is a composition of r_0 and i. In particular, the map i is a weak equivalence. Using Corollary 6.6.10, we see that there are also Cartesian squares of the form

$$\begin{array}{ccccc} x' & \xrightarrow{i'} & x & \xrightarrow{r} & y' \\ {\scriptstyle u}\downarrow & & {\scriptstyle \lambda}\downarrow & & \downarrow{\scriptstyle v} \\ x & \xrightarrow{i} & x_0 & \xrightarrow{r_0} & y \end{array}$$

such that p' is a composition of r and i'. The previous proposition implies that the map i' is a weak equivalence. On the other hand, the identity of x is a composition of the map λ and of the trivial fibration q_0. Therefore, since i, i' and λ are weak equivalences, the map u is a weak equivalence as well. □

7.4.17 For a simplicial set X, we define $W(X, C)$ by forming the Cartesian square below.

(7.4.17.1)
$$\begin{array}{ccc} W(X,C) & \hookrightarrow & \underline{\mathrm{Hom}}(X,C) \\ \downarrow & & \downarrow{\scriptstyle ev} \\ \prod_{X_0} W & \hookrightarrow & \prod_{X_0} C \end{array}$$

The arrows of $W(X, C)$ are called the *fibrewise weak equivalences*.

Let I be a finite direct category. There is the notion of Reedy fibration in $\underline{\mathrm{Hom}}(N(I)^{\mathrm{op}}, C)$. Together with the fibrewise weak equivalences, we want to define a structure of ∞-category with weak equivalences and fibrations.

Proposition 7.4.18 *A Reedy fibration between Reedy fibrant presheaves $p\colon F \to G$ is a fibrewise weak equivalence if and only if, for any object i of I, the induced map $F(i) \to \partial F(i) \times_{\partial G(i)} G(i)$ is a trivial fibration.*

Proof If $F(i) \to \partial F(i) \times_{\partial G(i)} G(i)$ is a trivial fibration for all i, then applying Proposition 7.4.9 for the direct categories I/i and for the class of fibrations $\mathsf{F} = W \cap Fib$, we see that the maps

$$F(i) = \varprojlim F/i \to \varprojlim G/i = G(i)$$

all are trivial fibrations.

For the converse, we shall proceed by induction on the length of I. The case of $\ell(I) \leq 0$ is a consequence of the fact that trivial fibrations are closed under compositions and pull-backs. If $\ell(I) > 0$, then the induction hypothesis implies that we may apply Proposition 7.4.9 for the direct category $\partial I/i$ with respect to the class of fibrations $\mathsf{F} = W \cap Fib$. Therefore, the induced maps

$$\partial F(i) \to \partial G(i)$$

are trivial fibrations. Hence their pull-backs

$$\partial F(i) \times_{\partial G(i)} G(i) \to G(i)$$

are trivial fibrations. Since the map $F(i) \to G(i)$ is a weak equivalence for all i, this proves that p is a Reedy fibration with respect to the class $\mathsf{F} = W \cap Fib$. □

Proposition 7.4.19 *Let $u\colon F \to G$ be a morphism of C-valued presheaves on a finite direct category. There exists a Reedy fibration $p\colon H \to G$ and a fibrewise equivalence $w\colon F \to H$ such that u is a composition of w and p.*

Proof We proceed by induction on the length of I. For $\ell(I) < 0$, there is nothing to prove. If $\ell(I) \geq 0$, then we may choose a factorisation of

$$u_{|N(\partial I)}\colon F_{|N(\partial I)} \to G_{|N(\partial I)}$$

into a Reedy fibration $p'\colon H' \to G_{|N(\partial I)}$ and a fibrewise equivalence $w'\colon F_{|N(\partial I)} \to H'$. For each object i of length $\ell(I)$ in I, we choose a factorisation of the induced map

$$F(i) \to \varprojlim H'_{|N(\partial I/i)} \times_{\partial G(i)} G(i)$$

into a weak equivalence $w(i)\colon F(i) \to H(i)$ followed by a fibration q_i. We write $\pi_i\colon F(i) \to \varprojlim H'_{|N(\partial I/i)}$ for any choice of composition of q_i with the first projection. Then the first assertion of Proposition 7.4.11 shows that the maps π_i define a Reedy fibrant presheaf H on $N(I)$ which extends H'. The second part of Proposition 7.4.11 implies that the maps $w(i)$ determine a map $w\colon F \to H$ which extends w' and which is a fibrewise weak equivalence. Finally, any choices of compositions p_i of the maps q_i with the second projections towards $G(i)$ define a map $p\colon H \to G$ which extends p' and which is a Reedy fibration, such that u is a composition of w and p. □

Theorem 7.4.20 *For any finite direct category I, the ∞-category of diagrams $\underline{\mathrm{Hom}}(N(I)^{\mathrm{op}}, C)$, equipped with fibrewise weak equivalences and Reedy fibrations, is an ∞-category with weak equivalences and fibrations.*

Proof This is the conjunction of Proposition 7.4.18, of a double application of Proposition 7.4.10 (once for $\mathsf{F} = Fib$ and once for $\mathsf{F} = W \cap Fib$), and of Proposition 7.4.19. □

7.4.21 We define *fibrewise fibrations* to be the morphisms of presheaves $F \to G$ on a simplicial set X such that $F(x) \to G(x)$ is a fibration for any object x of X.

Corollary 7.4.22 *For any finite direct category I, the ∞-category of diagrams $\underline{\mathrm{Hom}}(N(I)^{\mathrm{op}}, C)$, equipped with fibrewise weak equivalences and fibrewise fibrations, is an ∞-category with weak equivalences and fibrations.*

Proof The only non-obvious aspect of this corollary is the existence of factorisations. Let $u\colon F \to G$ be a morphism of presheaves on $N(I)$, with G fibrewise fibrant. By virtue of Proposition 7.4.19, we may choose a commutative square

$$\begin{array}{ccc} F & \xrightarrow{f} & F' \\ {\scriptstyle u}\downarrow & & \downarrow{\scriptstyle q} \\ G & \xrightarrow{g} & G' \end{array}$$

in which F' and G' are Reedy fibrant, q is a Reedy fibration, and both f and g are weak equivalences. Since, in particular, q is a fibrewise fibration, we may form the pull-back $H = G \times_{G'} F'$. The two-out-of-three property implies that the induced map $w\colon F \to H$ is a fibrewise weak equivalence, and the projection $p\colon H \to G$ is a fibrewise fibration, since it is a pull-back of q. Hence u is the composition of a fibrewise weak equivalence and of a fibrewise fibration. □

7.5 Derived Functors

7.5.1 Let C be an ∞-category equipped with a subcategory of weak equivalences $W \subset C$. We shall often denote the localisation of C as $L(C) = W^{-1}C$. In the case where $C = N(\mathcal{C})$ is the nerve of a category, we shall also write $L(\mathcal{C}) = L(C)$.

Let C be an ∞-category with weak equivalences and fibrations. We denote by C_f the full subcategory of C whose objects are the fibrant objects. We define weak equivalences and fibrations in C_f in the obvious way: these are the maps which are weak equivalences or fibrations in C, respectively. It is then obvious that C_f is an ∞-category with weak equivalences and fibrations. We then have a canonical commutative square

$$(7.5.1.1)\qquad \begin{array}{ccc} C_f & \xrightarrow{\iota} & C \\ {\scriptstyle \gamma_f}\downarrow & & \downarrow{\scriptstyle \gamma} \\ L(C_f) & \xrightarrow{\bar{\iota}} & L(C) \end{array}$$

where γ and γ_f are the canonical functors, ι is the inclusion, and $\bar{\iota}$ is the functor induced from ι by the universal property of $L(C_f)$.

Definition 7.5.2 Let C and D be two ∞-categories with weak equivalences and fibrations. A functor $F\colon C \to D$ is *left exact* if it has the following properties.

(i) The functor F preserves final objects.
(ii) The functor F sends fibrations between fibrant objects to fibrations, and trivial fibrations between fibrant objects to trivial fibrations.
(iii) For any Cartesian square in C of the form

$$\begin{array}{ccc} x' & \xrightarrow{u} & x \\ {\scriptstyle p'}\downarrow & & \downarrow{\scriptstyle p} \\ y' & \xrightarrow{v} & y \end{array}$$

in which p is a fibration and y and y' are fibrant, the square

$$\begin{array}{ccc} F(x') & \xrightarrow{F(u)} & F(x) \\ {\scriptstyle F(p')}\downarrow & & \downarrow{\scriptstyle F(p)} \\ F(y') & \xrightarrow{F(v)} & F(y) \end{array}$$

is Cartesian in D.

Dually, if C and D are two ∞-categories with weak equivalences and cofibrations, a functor $F\colon C \to D$ is *right exact* if $F^{\mathrm{op}}\colon C^{\mathrm{op}} \to D^{\mathrm{op}}$ is left exact.

Remark 7.5.3 It follows right away from Ken Brown's lemma (Proposition 7.4.13) that any left exact functor preserves weak equivalences between fibrant objects.

Example 7.5.4 If C is an ∞-category with finite limits, it may be seen as an ∞-category with weak equivalences and fibrations by defining the weak equivalences to be the invertible maps and the fibrations to be all maps. For two ∞-categories with finite limits C and D, we see that a functor $f\colon C \to D$ is left exact if and only if it preserves final objects as well as Cartesian squares. By virtue of Proposition 7.3.28, a functor between ∞-categories with finite limits is left exact if and only if it commutes with finite limits.

Proposition 7.5.5 *Let $F\colon C \to D$ be a left exact functor between ∞-categories with weak equivalences and fibrations. For any finite direct category I, the induced functor*

$$F\colon \underline{\mathrm{Hom}}(N(I)^{\mathrm{op}}, C) \to \underline{\mathrm{Hom}}(N(I)^{\mathrm{op}}, D)$$

is left exact for the fibrewise weak equivalences and the Reedy fibrations. Furthermore, for any Reedy fibrant presheaf $\Phi\colon N(I)^{\mathrm{op}} \to C$, the canonical map

$$F(\varprojlim \Phi) \to \varprojlim F(\Phi)$$

is invertible.

Proof We proceed by induction on the length of I. If $\ell(I) \leq 0$, this is clear: F commutes with finite products of fibrant objects. Otherwise, since $\ell(\partial I/i) < \ell(I)$, the induction hypothesis shows that, for any Reedy fibrant presheaf Φ and any object i, there is a canonical isomorphism

$$F(\partial\Phi(i)) \simeq \partial(F(\Phi))(i)\,.$$

Since Φ is left exact, one deduces that it preserves Reedy fibrations between Reedy fibrant objects. The fact that F commutes with limits of Reedy fibrant presheaves comes by induction from the pull-back diagram (7.4.4.1). □

Proposition 7.5.6 *For any ∞-category with weak equivalences and fibrations C, the localisation $L(C_f)$ has finite limits, and the localisation functor $\gamma_f\colon C_f \to L(C_f)$ is left exact. Moreover, for any ∞-category D with finite limits, and for any left exact functor $F\colon C_f \to D$, the induced functor $\bar{F}\colon L(C_f) \to D$ is left exact.*

Proof It follows right away from Proposition 7.1.10 that $L(C_f)$ has final objects and that the functor γ_f preserves them. Ken Brown's lemma (Proposition 7.4.13) shows that the class of fibrations which are weak equivalences of C_f is indeed a class of trivial fibrations with respect to weak equivalences in the sense of Definition 7.2.20. Therefore, Theorem 7.2.25 applies here. Proposition 7.4.16 thus implies that, for any Cartesian square in C_f of the form

$$\begin{array}{ccc} x' & \xrightarrow{u} & x \\ {\scriptstyle p'}\downarrow & & \downarrow{\scriptstyle p} \\ y' & \xrightarrow{v} & y \end{array}$$

in which p is a fibration, the square

$$\begin{array}{ccc} \gamma_f(x') & \xrightarrow{\gamma_f(u)} & \gamma_f(x) \\ {\scriptstyle \gamma_f(p')}\downarrow & & \downarrow{\scriptstyle \gamma_f(p)} \\ \gamma_f(y') & \xrightarrow{\gamma_f(v)} & \gamma_f(y) \end{array}$$

is Cartesian in $L(C_f)$. On the other hand, the functor

$$\underline{\mathrm{Hom}}(\Lambda^2_2, \gamma_f) \colon \underline{\mathrm{Hom}}(\Lambda^2_2, C_f) \to \underline{\mathrm{Hom}}(\Lambda^2_2, L(C_f))$$

is essentially surjective: since trivial fibrations define a right calculus of fractions, by Corollary 7.2.18, formula (7.2.10.4) shows that any map in $L(C_f)$ is of the form $\gamma_f(p)\gamma_f(s)^{-1}$, where s is a trivial fibration. This shows that $L(C_f)$ has limits of type Λ^2_2. By virtue of Theorem 7.3.27, this proves that $L(C_f)$ has finite limits. The above also shows that all Cartesian squares in $L(C_f)$ are isomorphic to images of Cartesian squares of C_f in which all maps are fibrations. Therefore, for any ∞-category D with finite limits, and for any left exact functor $F \colon C_f \to D$, the induced functor $\bar{F} \colon L(C_f) \to D$ is left exact: it obviously preserves the final object and it preserves pull-back squares. □

Definition 7.5.7 An *∞-category of fibrant objects* is an ∞-category with weak equivalences and fibrations C in which all objects are fibrant. Dually, an *∞-category of cofibrant objects* is an ∞-category with weak equivalences and cofibrations C in which all objects are cofibrant.

Example 7.5.8 For any ∞-category with weak equivalences and fibrations C, the full subcategory of fibrant objects C_f is a category of fibrant objects.

Example 7.5.9 Any ∞-category with finite limits C, seen as an ∞-category with weak equivalences and fibrations (where the weak equivalences are the invertible maps, while all maps are fibrations), is an ∞-category of fibrant objects.

7.5.10 Let C be an ∞-category of fibrant objects, and $S \subset C$ a simplicial subset. We let C_S denote the ∞-category of fibrant objects whose underlying ∞-category is the same as C, as well as its subcategory of fibrations, while the weak equivalences of C_S are the maps in the smallest subcategory W_S of C which contains S as well as the weak equivalences of C, has the two-out-of-three property and satisfies condition (i) of Definition 7.4.12. Given an ∞-category D with finite limits, we denote by $\underline{\mathrm{Hom}}_{lex,S}(C, D)$ the full subcategory of $\underline{\mathrm{Hom}}(C, D)$ which consists of left exact functors which send the maps of S to invertible maps in D. We also write $\underline{\mathrm{Hom}}_{lex}(L(C_S), D)$ for the full subcategory of left exact functors from $L(C_S)$ to D. We have localisation functors

$$\ell\colon C \to L(C) \quad \text{and} \quad \ell_S\colon C_S \to L(C_S)$$

which both are left exact, by virtue of Proposition 7.5.6.

Proposition 7.5.11 *Under the assumptions above (paragraph 7.5.10), composing with the left exact localisation functor $\ell_S\colon C_S \to L(C_S)$ induces an equivalence of ∞-categories*

$$\underline{\mathrm{Hom}}_{lex,S}(C, D) \to \underline{\mathrm{Hom}}_{lex}(L(C_S), D)$$

for any ∞-category with finite limits D.

Proof It is clear that any left exact functor $F\colon C \to D$ which sends the maps of S to invertible maps in D sends the maps in W_S to invertible maps, and thus induces a unique left exact functor $C_S \to D$. Therefore, since ℓ_S is left exact, we have a commutative square

$$\begin{array}{ccc} \underline{\mathrm{Hom}}_{lex,S}(C, D) & \xrightarrow{\ell_S^*} & \underline{\mathrm{Hom}}_{lex}(L(C_S), D) \\ \downarrow & & \downarrow \\ \underline{\mathrm{Hom}}_{W_S}(C, D) & \xrightarrow{\ell_S^*} & \underline{\mathrm{Hom}}(L(C_S), D) \end{array}$$

in which, by definition, the lower horizontal arrow as well as both vertical ones are fully faithful functors. The upper horizontal arrow is thus fully faithful as well. The essential surjectivity of the latter follows from the last assertion of Proposition 7.5.6. □

Remark 7.5.12 The preceding proposition, which really is just a reformulation of Proposition 7.5.6, may be applied in the case where C is an ∞-category with finite limits (the weak equivalences being the invertible maps and the fibrations all maps): this explains how to invert maps in a way which is compatible with finite limits.

7.5.13 Let C be an ∞-category of fibrant objects. A *path object* of an object y of C is a diagram of the form

$$y \xrightarrow{s} y^I \xrightarrow{(d^0,d^1)} y \times y \tag{7.5.13.1}$$

such that the diagonal map $y \to y \times y$ is a composition of (d^0, d^1) and s, and such that (d^0, d^1) is a fibration and s is a weak equivalence; this implies that each the maps $d^\varepsilon\colon y^I \to y$ is a trivial fibration.

Two maps $f_0, f_1 \colon x \to y$ are *path-homotopic* if there exist a trivial fibration $p \colon x' \to x$ and a map $h \colon x' \to y^I$ such that $d^\varepsilon h = f_\varepsilon p$ in $ho(C)$ for $\varepsilon = 0, 1$.

We observe that d^0 and d^1 are both inverses of s in $ho(L(C))$, whence are equal. This implies that, if two maps f_0 and f_1 are path-homotopic in C, then they are equal in $ho(L(C))$.

Lemma 7.5.14 *Let C be an ∞-category with weak equivalences and fibrations. We consider two commutative diagrams of the following form in C*

$$\begin{array}{ccc} x & \xrightarrow{f} & t \\ \downarrow{\scriptstyle g} & & \downarrow{\scriptstyle p_i} \\ z & \xrightarrow{q_i} & y_i \end{array} \qquad i = 0, 1,$$

with y_i fibrant for $i = 0, 1$, g a weak equivalence, both p_1 and q_1 fibrations, and both p_0 and q_0 trivial fibrations. Then $p_1 p_0^{-1} = q_1 q_0^{-1}$ in $ho(L(C_f))$.

Proof We form a Cartesian square of the form

$$\begin{array}{ccc} x_0 & \xrightarrow{f_0} & t \\ \downarrow{\scriptstyle g_0} & & \downarrow{\scriptstyle p_0} \\ z & \xrightarrow{q_0} & y_0 \end{array}$$

in C. Then the pair (f, g) induces a map $h_0 \colon x \to x_0$ so that f and g are compositions of h_0 with f_0 and g_0, respectively. We observe that both f_0 and g_0 are trivial fibrations, so that h_0 is a weak equivalence. We choose a path object of y_1:

$$y_1 \xrightarrow{s} y_1^I \xrightarrow{(d^0, d^1)} y_1 \times y_1 \,.$$

There exists a commutative square of the form

$$\begin{array}{ccccc} x & \xrightarrow{v} & y_1 & \xrightarrow{s} & y_1^I \\ \downarrow{\scriptstyle h_0} & & & & \downarrow{\scriptstyle (d^0, d^1)} \\ x_0 & & \xrightarrow{(k,l)} & & y_1 \times y_1 \end{array}$$

where v is a composition of p_1 and f_1 as well as a composition of q_1 and g_1, while k and l are compositions of f_0 with p_1 and of g_0 with q_1, respectively. This defines a map

$$\xi \colon x \to x_1 = x_0 \times_{y_1 \times y_1} y_1^I$$

that is the composition of a weak equivalence $h \colon x \to x'$ and of a fibration

$r\colon x' \to x_1$. Let $a\colon x_1 \to x_0$ and $b\colon x_1 \to y_1^I$ be the first and second projections, respectively. In $ho(L(C_f))$, we have

$$d^0 br = kar \quad \text{and} \quad d^1 br = lar\,.$$

On the other hand, we see that any composition of a and r in C is a trivial fibration: it is a fibration because both a and r have this property, and it is a weak equivalence because the weak equivalence h_0 is a composition of the weak equivalence h with a and r. In other words k and l are path-homotopic, which implies that $k = l$ in $ho(L(C_f))$. In other words, if u denotes a composition of p_0 and of f_0, then we have the relations

$$p_1 p_0^{-1} = p_1 f_0 u^{-1} = k u^{-1} = l u^{-1} = q_1 g_0 u^{-1} = q_1 q_0^{-1}$$

in $ho(L(C_f))$. □

Lemma 7.5.15 *For any ∞-category with weak equivalences and fibrations C, the canonical functor $ho(L(C_f)) \to ho(L(C))$ is an equivalence of categories.*

Proof We first choose a *cleavage*, by which we mean the following data:

(a) for each object x of C, a choice of a weak equivalence $j_x\colon x \to R(x)$ with $R(x)$ fibrant (whenever x is fibrant, we choose $R(x) = x$ and $j_x = 1_x$);
(b) for each map $u\colon x \to y$, a choice of a diagram of the form below, in which $\sigma(u)$ is a weak equivalence, while $p(u)$ and $q(u)$ are fibrations, and whose image in $ho(C)$ is commutative

$$\begin{array}{ccccc} x & & \xrightarrow{\quad u \quad} & & y \\ {\scriptstyle j_x}\downarrow & \searrow^{\sigma(u)} & & & \downarrow{\scriptstyle j_y} \\ R(x) & \xleftarrow{q(u)} & S(u) & \xrightarrow{p(u)} & R(y) \end{array}$$

(whenever $u = 1_x$, we choose $S(1_x) = x$ and $p(u) = q(u) = 1_{R(x)}$).

For the existence of such a thing, part (a) is clear. For (b), given a map $u\colon x \to y$ which is not the identity, there is a canonical map $(j_x, j_y u)\colon x \to R(x) \times R(y)$, which we may factor as a weak equivalence $\sigma(u)\colon x \to S(u)$ followed by a fibration $\pi(u)\colon S(u) \to R(x) \times R(y)$. The maps $p(u)$ and $q(u)$ are obtained by composing $\pi(u)$ with the first and second projections of the product $R(x) \times R(y)$. We want to promote the assignment $x \mapsto R(x)$ to a functor

$$R\colon ho(C) \to ho(L(C_f))\,.$$

Given a map $u\colon x \to y$ in C, we put $R(u) = p(u)q(u)^{-1}$. This defines a functor indeed. For $u = 1_x$, we obviously have $R(1_x) = 1_{R(x)}$. It remains to check the compatibility with compositions. Let $u\colon x \to y$ and $v\colon y \to z$ be two maps in C. The commutative square

$$\begin{array}{ccccc} x & \xrightarrow{u} & y & \xrightarrow{\sigma(v)} & S(v) \\ \big\downarrow{\scriptstyle \sigma(u)} & & & & \big\downarrow{\scriptstyle q(v)} \\ S(u) & & \xrightarrow{\quad p(u)) \quad} & & R(y) \end{array}$$

defines a map $f\colon x \to T(v,u) = S(v)\times_{R(y)} S(u)$. We also choose a composition $w\colon x \to z$ of u and v. This fits into the following commutative diagram in C.

$$\begin{tikzcd} & S(u) \arrow[ld, "p(u)"'] \arrow[rrr, "q(u)"] & & & R(x) & \\ R(y) & & T(v,u) \arrow[lu, "pr_2"'] \arrow[ld, "pr_1"] & X \arrow[l, "f"'] \arrow[rr, "\sigma(w)"] & & S(w) \arrow[lu, "q(w)"'] \arrow[ld, "p(w)"] \\ & S(v) \arrow[lu, "q(v)"] \arrow[rrr, "p(v)"] & & & R(z) & \end{tikzcd}$$

Lemma 7.5.14 implies that

$$p(w)q(w)^{-1} = p(v)pr_1(q(u)pr_2)^{-1}$$

in $ho(L(C_f))$. Since we obviously have

$$p(v)pr_1(q(u)pr_2)^{-1} = p(v)q(v)^{-1}p(u)q(u)^{-1},$$

this proves that $R(v)R(u) = R(w)$. Keeping track of notation from diagram (7.5.1.1), there is a canonical functor

$$ho(\imath)\colon ho(C_f) \to ho(C)\,.$$

The composed functor $R \circ ho(\imath)$ is then the canonical functor $ho(\gamma_f)$ from $ho(C_f)$ to $ho(L(C_f))$. Similarly, by construction, for any map $u\colon x \to y$ in C, the square

$$\begin{array}{ccc} x & \xrightarrow{\quad u \quad} & y \\ {\scriptstyle j_x}\big\downarrow & & \big\downarrow{\scriptstyle j_y} \\ R(x) & \xrightarrow{p(u)q(u)^{-1}} & R(y) \end{array}$$

commutes in $ho(L(C))$. In other words, there is an invertible natural transformation $j\colon ho(\gamma) \to ho(\bar{\imath}) \circ R$, where γ is the localisation functor from C to $L(C)$. This readily implies that the functor induced by $\imath$ from $ho(L(C_f))$ to $ho(L(C))$ is an equivalence of categories. □

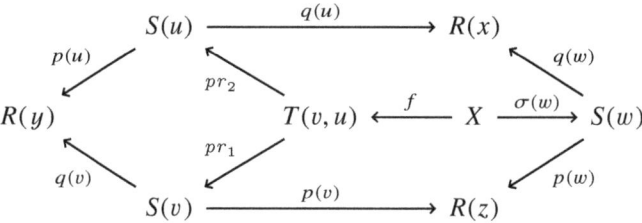

Proposition 7.5.16 *Let x be a fibrant object in an ∞-category with weak equivalences and fibrations C. The induced functor $C_f/\gamma_f(x) \to C/\gamma(x)$ is final.*

Proof Recall that

$$C_f/\gamma_f(x) = L(C_f)/\gamma_f(x) \times_{L(C_f)} C_f \quad \text{and} \quad C/\gamma(x) = L(C)/\gamma(x) \times_{L(C)} C$$

and the comparison functor is induced by the functor $\bar{\iota}\colon L(C_f) \to L(C)$. An object ξ of $L(C)/\gamma(x)$ is thus determined by a couple (c, u), where c is an object of C, and $u\colon \gamma(c) \to \gamma(x)$ is a map in $L(C)$. We have to prove that the coslice $\xi\backslash(C_f/\gamma_f(x))$ is weakly contractible. By Lemma 4.3.15, it is sufficient to consider given a functor of the form

$$F\colon E \to \xi\backslash(C_f/\gamma_f(x)),$$

where E is the nerve of a finite partially ordered set, and to prove that F is Δ^1-homotopic to a constant functor. In the case where E is empty, this means that we have to prove that $\xi\backslash(C_f/\gamma_f(x))$ is not empty. For this purpose, we choose a weak equivalence $p\colon c \to d$ with d fibrant. By virtue of Lemma 7.5.15 the map u defines a unique map $\delta\colon \gamma_f(d) \to \gamma_f(x)$ in $ho(L(C_f))$ so that $\bar{\iota}(\delta)\gamma(p) = u$. It remains to consider the case where E is not empty. In what follows, objects of C or of C_f will often be considered as E-indexed constant functors. A functor F as above is essentially determined by a functor $\Phi\colon E \to C_f$, equipped with a map $\varphi\colon \gamma_f(\Phi) \to \gamma_f(x)$, as well as with a map $\psi\colon c \to \Phi$, such that the diagram

$$\begin{array}{ccc} \gamma(c) & \xrightarrow{u} & \gamma(x) \\ {\scriptstyle\gamma(\psi)}\big\downarrow & \nearrow{\scriptstyle\bar{\iota}(\varphi)} & \\ \gamma(\Phi) & & \end{array}$$

commutes in $\underline{\mathrm{Hom}}(E, L(C))$. By virtue of Proposition 7.4.19, there exists a fibrewise weak equivalence $w\colon \Phi \to \Phi_0$ such that Φ_0 is Reedy fibrant. Since $\gamma_f(w)$ is invertible, one can find a map $\varphi_0\colon \gamma_f(\Phi_0) \to \gamma_f(x)$ such that φ is a composition of φ_0 and of $\gamma_f(w)$. Composing w with ψ defines a map $\psi_0\colon c \to \Phi_0$. The triple $(\Phi_0, \varphi_0, \psi_0)$ determines a functor

$$F_0\colon E \to \xi\backslash(C_f/\gamma_f(x)),$$

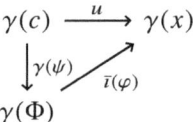

and w is a natural transformation $F \to F_0$. In other words, we may replace F by F_0, and assume, without loss of generality, that the underlying functor Φ is Reedy fibrant. Then, by virtue of Proposition 7.4.8, the limit of Φ exists and is fibrant in C. The map ψ being a cone, it determines a map

$$\pi\colon c \to \varprojlim \Phi\,.$$

We choose a commutative triangle in C of the form

$$\begin{array}{ccc} c & \xrightarrow{\pi} & \varprojlim \Phi \\ {\scriptstyle p}\big\downarrow & \nearrow{\scriptstyle q} & \\ d & & \end{array}$$

where p is a weak equivalence, while q is a fibration. The canonical morphism $\upsilon\colon \varprojlim \Phi \to \Phi$, together with a choice of composition with φ, defines a map $\lambda\colon \gamma_f(\varprojlim \Phi) \to \gamma_f(x)$. Composing λ with q defines a morphism between constant functors

$$\delta\colon \gamma_f(d) \to \gamma_f(x)\,.$$

Let e be an object of E, and $\delta_e\colon \gamma_f(d) \to \gamma_f(x)$ the map of $L(C_f)$ obtained by evaluating δ at e. Since u is a composition of $\bar{\imath}(\delta_e)$ with the inverse of $\gamma(p)$, the triple (d, p, δ_e) defines a constant functor from E to $\xi\backslash(C_f/\gamma_f(x))$. Finally, the commutative diagram

$$\begin{array}{ccccccc} \gamma(d) & \xrightarrow{\gamma(p)^{-1}} & \gamma(c) & \xrightarrow{\gamma(\psi)} & \gamma(\Phi) & \xrightarrow{\bar{\imath}(\varphi)} & \gamma(x) \\ & {\scriptstyle 1_{\gamma(d)}}\searrow & {\scriptstyle \gamma(p)}\big\downarrow & & \big\uparrow{\scriptstyle \gamma(\upsilon)} & \nearrow{\scriptstyle \bar{\imath}(\lambda)} & \\ & & \gamma(d) & \xrightarrow{\gamma(q)} & \gamma(\varprojlim \Phi) & & \end{array}$$

shows that any composition of υ and q defines a natural transformation from the constant functor (d, p, δ) to F. □

Corollary 7.5.17 *Let* **U** *be a universe and C a* **U***-small ∞-category with weak equivalences and fibrations. For any ∞-category D with* **U***-small colimits, and for any functor $F\colon C \to D$, the commutative square* (7.5.1.1) *induces an invertible map*

$$(\gamma_f)_!\iota^*(F) \simeq \bar{\imath}^*\gamma_!(F)\,.$$

Proof It is sufficient to prove that the evaluation of the canonical map

$$(\gamma_f)_!\iota^*(F) \to \bar{\imath}^*\gamma_!(F)$$

at any object x of C_f is invertible. By a double application of the dual version of Proposition 6.4.9, this evaluation is equivalent to the map

$$\varinjlim_{C_f/\gamma_f(x)} \iota^*(F)/\gamma_f(x) \to \varinjlim_{C/\gamma(x)} F/\gamma(x)\,,$$

where $F/\gamma(x)$ is the composition of F with the canonical map $C/\gamma(x) \to C$, and similarly for $\iota^*(F)/\gamma_f(x)$. By virtue of Theorem 6.4.5, the previous proposition implies that this map is invertible for all x. □

Theorem 7.5.18 *Let C be an ∞-category with weak equivalences and fibrations. The canonical functor $\bar{\iota}$: $L(C_f) \to L(C)$ is an equivalence of ∞-categories. In particular, the ∞-category $L(C)$ has finite limits and the functor $\gamma\colon C \to L(C)$ is left exact.*

Proof Since the functor $\bar{\iota}$ is essentially surjective, it is sufficient to prove that it is fully faithful. For this purpose, we may consider a universe **U** such that C is **U**-small, and, by Proposition 6.1.15, it is sufficient to prove that the induced functor

$$\bar{\iota}_!\colon \underline{\mathrm{Hom}}(L(C_f), \mathcal{S}) \to \underline{\mathrm{Hom}}(L(C), \mathcal{S})$$

is fully faithful. In other words, we want to prove that the unit map $1 \to \bar{\iota}^*\bar{\iota}_!$ is an invertible natural transformation. Since both functors $\bar{\iota}_!$ and $\bar{\iota}^*$ commute with colimits (because they have right adjoints), and since any $\mathcal{S}$-valued functor indexed by a **U**-small ∞-category is a colimit of representable ones, it is sufficient to prove that the map $F \to \bar{\iota}^*\bar{\iota}_!(F)$ is invertible for any representable functor F. Since the functor γ_f is essentially surjective, this implies that we only have to prove that the induced map

$$(\gamma_f)_! \to \bar{\iota}^*\bar{\iota}_!(\gamma_f)_!$$

is invertible. We have

$$\begin{aligned}(\gamma_f)_! &\simeq (\gamma_f)_!\iota^*\iota_! \\ &\simeq \bar{\iota}^*\gamma_!\iota_! \\ &\simeq \bar{\iota}^*\bar{\iota}_!(\gamma_f)_!,\end{aligned}$$

where the first isomorphism expresses the fact that ι is fully faithful, the second one is the base change formula of Corollary 7.5.17, while the last one comes from the commutativity of the square (7.5.1.1). The second part of the theorem follows straight away from the first and from Proposition 7.5.6. □

Corollary 7.5.19 *Let C be an ∞-category with weak equivalences and fibrations. For a morphism between fibrant objects $p\colon x \to y$, the following conditions are equivalent.*

(i) The morphism f has a section in $ho(L(C))$.
(ii) There exists a morphism $p'\colon x' \to x$ such that the composition of p' and p is a weak equivalence.

(iii) There exists a fibration $p' : x' \to x$ such that the composition of p' and p is a weak equivalence.

Proof The equivalence between conditions (ii) and (iii) follows right away from the fact that one can factorise any map with fibrant domain into a weak equivalence followed by a fibration. It is clear that (i) follows from (ii). To prove that condition (i) implies condition (ii), we deduce from the preceding theorem (in fact, from Lemma 7.5.15) that it is sufficient to consider the case where all the objects of C are fibrant (replacing C by C_f). We observe then that, in any commutative diagram of the form below,

$$\begin{array}{ccc} x & \xleftarrow{\;s\;} & z \\ {\scriptstyle s'}\uparrow & \overset{u}{\swarrow} & \downarrow{\scriptstyle \varphi} \\ z' & \xrightarrow{\;\varphi'\;} & y \end{array}$$

in which both s and s' are trivial fibrations, the map φ is a weak equivalence if and only if φ' has the same property. Therefore, using the right calculus of fractions under the precise form of formula (7.2.10.4) (with $W(x)$ the full subcategory of $C//x$ which consists of trivial fibrations $x' \to x$), we see that, if there is an equality of maps $s^{-1}\varphi$ and $t^{-1}\psi$ in $ho(L(C))$, with s and t two trivial fibrations in C, and φ and ψ two maps in C, then φ is a weak equivalence if and only if ψ has the same property. Contemplating the case where both t and ψ are identities, we observe that (i) implies (ii). □

Corollary 7.5.20 *Let C be an ∞-category with weak equivalences and fibrations. A commutative square in C of shape*

$$\begin{array}{ccc} x' & \xrightarrow{\;u\;} & x \\ {\scriptstyle p'}\downarrow & & \downarrow{\scriptstyle p} \\ y' & \xrightarrow{\;v\;} & y \end{array}$$

in which p is a fibration and both y and y' are fibrant becomes Cartesian in $L(C)$ if and only if the corresponding map $x' \to y' \times_y x$ becomes invertible in $L(C)$.

Proof This follows right away from the last assertion of Theorem 7.5.18. □

Corollary 7.5.21 *Let C be an ∞-category with weak equivalences and fibrations, and let*

$$\begin{array}{ccc} x' & \xrightarrow{u} & x \\ {\scriptstyle p'}\downarrow & & \downarrow{\scriptstyle p} \\ y' & \xrightarrow{v} & y \end{array}$$

be a Cartesian square of C, in which p is a fibration and y' is fibrant. If p (or v) becomes invertible in $L(C)$, then the map p' (or u, respectively) becomes invertible in $L(C)$.

Proof By virtue of the preceding corollary, we may replace C by $L(C)$. In other words, we may assume that the weak equivalences are the invertible maps and that the fibrations are all maps, in which case the assertion is a triviality. □

Remark 7.5.22 If C is an ∞-category with weak equivalences and fibrations, thanks to the preceding corollary, we get another ∞-category with weak equivalences and fibrations $\overline{C}$ as follows. The underlying ∞-category is the same as C, and similarly for the subcategory of fibrations. The weak equivalences are those which become invertible in $L(C)$ (i.e. the subcategory of weak equivalences is the pull-back of $k(L(C))$ in C). It is clear that $L(C) = L(\overline{C})$. This means that, in many situations, it is quite harmless to replace C by $\overline{C}$. A direct consequence of Corollary 7.5.19 is that the assignment $C \mapsto \overline{C}$ commutes with the formation of slices over fibrant objects: in other words, for any fibrant object x in C, a map in C/x induces an invertible map in $L(C/x)$ if and only if its image in C becomes invertible in $L(C)$.

7.5.23 Let C be an ∞-category with weak equivalences $W \subset C$, and let $F\colon C \to D$ be a functor. The pull-back functor

$$\gamma^*\colon \underline{\mathrm{Hom}}(L(C), D) \to \underline{\mathrm{Hom}}(C, D)$$

does not have any left adjoint in general, but one might still ask if the functor $\mathrm{Hom}(F, \gamma^*(-))$ is representable in $\underline{\mathrm{Hom}}(L(C), D)$. If this is the case, a representative is denoted by $\mathbf{R}F\colon L(C) \to D$, and is called the *right derived functor of F* (this is an abuse, of course: to speak of the right derived functor of F, one must also specify the natural transformation $F \to \mathbf{R}F \circ \gamma$ which exhibits $\mathbf{R}F$ as such).

Dually, if the functor $\mathrm{Hom}(\gamma^*(-), F)$ is representable, a representative is denoted by $\mathbf{L}F\colon L(C) \to D$ and is called the *left derived functor of F*. We remark that $(\mathbf{L}F)^{\mathrm{op}}\colon L(C)^{\mathrm{op}} = L(C^{\mathrm{op}}) \to D^{\mathrm{op}}$ is then the right derived functor of $F^{\mathrm{op}}\colon C^{\mathrm{op}} \to D^{\mathrm{op}}$.

Lemma 7.5.24 *If $F\colon C \to D$ sends weak equivalences to invertible maps, then the functor*

$$\bar{F}\colon L(C) \to D\,,$$

associated to F by the universal property of $L(C)$, is the right derived functor of F.

Proof We may assume that both C and D are $\mathbf{U}$-small for some universe $\mathbf{U}$. Let $G\colon L(C) \to D$ be any functor. Then the invertible map $\bar{F} \circ \gamma \simeq F$ and the equivalence of ∞-categories

$$\underline{\mathrm{Hom}}(L(C), D) \simeq \underline{\mathrm{Hom}}_W(C, D)$$

(where W is the subcategory of weak equivalences in C) produce invertible maps

$$\mathrm{Hom}(\bar{F}, G) \simeq \mathrm{Hom}(\bar{F} \circ \gamma, G \circ \gamma) \simeq \mathrm{Hom}(F, G \circ \gamma)$$

in $\mathcal{S}$, functorially in G. □

7.5.25 A consequence of the previous lemma is that, if C is an ∞-category with weak equivalences and fibrations, any functor $F\colon C \to D$ which sends weak equivalences between fibrant objects to invertible maps has a right derived functor $\mathbf{R}F$. More precisely, keeping the notation introduced in diagram (7.5.1.1), $\mathbf{R}F$ may be constructed as follows. One chooses a quasi-inverse $R\colon L(C) \to L(C_f)$ of the equivalence of ∞-categories $\bar{\iota}$ provided by Theorem 7.5.18. Let

$$\bar{F}\colon L(C_f) \to D \tag{7.5.25.1}$$

be a functor equipped with an invertible natural transformation $j\colon \bar{F} \circ \gamma_f \xrightarrow{\sim} F \circ \iota$. One defines

$$\mathbf{R}F = \bar{F} \circ R\,. \tag{7.5.25.2}$$

This functor $\mathbf{R}F$ is a right derived functor of F indeed. To see this, we observe first that the functor

$$\bar{\iota}^*\colon \underline{\mathrm{Hom}}(L(C), D) \to \underline{\mathrm{Hom}}(L(C_f), D)$$

is a right adjoint of R^*, by Proposition 6.1.6. Hence one may identify $\bar{F} \circ R$ with $\bar{\iota}_!(\bar{F})$. To prove that $\bar{\iota}_!(\bar{F})$ is the right derived functor of F, we may choose a universe $\mathbf{U}$ such that C and D are $\mathbf{U}$-small. Since the functor

$$(h_{D^{\mathrm{op}}})^{\mathrm{op}}\colon D \to \underline{\mathrm{Hom}}(D, \mathcal{S})^{\mathrm{op}} = D'$$

is fully faithful, we may replace D by D' (and F by $(h_{D^{\mathrm{op}}})^{\mathrm{op}}(F)$), and assume that D has **U**-small colimits. We have then to identify $\bar{\imath}_!(\bar{F})$ with $\gamma_!(F)$. But Lemma 7.5.24 identifies $\bar{F}$ with $\mathbf{R}(F \circ \imath) = (\gamma_f)_!(\imath^* F)$. Therefore, Corollary 7.5.17 gives a canonical isomorphism $\bar{F} \simeq \bar{\imath}^* \gamma_!(F)$. Since $\bar{\imath}$ is an equivalence of categories, this determines, by transposition, an invertible map $\bar{F} \circ R = \bar{\imath}_!(\bar{F}) \simeq \gamma_!(F)$.

This explicit construction shows furthermore that, for any other functor $F' : D \to D'$, we have

$$F' \circ \mathbf{R}F = \mathbf{R}(F' \circ F)\,. \tag{7.5.25.3}$$

Definition 7.5.26 Let C be an ∞-category with weak equivalences and fibrations, and D an ∞-category equipped with a subcategory of weak equivalences. Given a functor $F : C \to D$, which sends the weak equivalences between fibrant objects of C to weak equivalences of D, the *right derived functor of F* is defined as the right derived functor of the composition

$$C \xrightarrow{F} D \xrightarrow{\gamma_D} L(D)\,,$$

where γ_D is the localisation functor by the weak equivalences of D. This right derived functor of F is denoted by $\mathbf{R}F$. In other words, we put

$$\mathbf{R}F = \mathbf{R}(\gamma_D \circ F) : L(C) \to L(D)$$

(this makes sense, by applying the construction of paragraph 7.5.25).

Remark 7.5.27 Corollary 7.4.14 shows that the right derived functor of any left exact functor always exists. In particular, the right derived functor of the localisation functor $\gamma : C \to L(C)$ exists: it is the identity of $L(C)$. This fits well with the fact that the formation of derived functor behaves well with composition, as may be observed with Proposition 7.5.29.

Proposition 7.5.28 *For any left exact functor $F : C \to D$ between ∞-categories with weak equivalences and fibrations, the right derived functor $\mathbf{R}F : L(C) \to L(D)$ is left exact (hence commutes with finite limits).*

Proof Let C_f and D_f be the full subcategories of fibrant objects in C and D, respectively. There is an essentially commutative diagram of the form

$$\begin{array}{ccc} L(C_f) & \xrightarrow{\bar{F}} & L(D_f) \\ \downarrow & & \downarrow \\ L(C) & \xrightarrow{\mathbf{R}F} & L(D) \end{array}$$

where the vertical maps are equivalences of ∞-categories, and $\bar{F}$ is the functor induced by the restriction of F to C_f. It is thus sufficient to prove that $\bar{F}$ is left exact, which follows right away from the second assertion of Proposition 7.5.6. □

Proposition 7.5.29 *Let $F\colon C_0 \to C_1$ and $G\colon C_1 \to C_2$ be two left exact functors between ∞-categories with weak equivalences and fibrations. The canonical map $\mathbf{R}G \circ \mathbf{R}F \to \mathbf{R}(G \circ F)$ is invertible.*

Proof This follows straight away from formula (7.5.25.3). □

The following theorem, originally due to Quillen in the context of model categories, has been generalised by Maltsiniotis in [Mal07], with a proof which is robust enough to be promoted *mutatis mutandis* to ∞-categories. The particular case of model ∞-categories should arguably be attributed to Mazel-Gee [MG16b].

Theorem 7.5.30 (Quillen) *Let $F\colon C \rightleftarrows D\colon U$ be an adjunction with co-unit map and unit map $\varepsilon\colon FU \to 1_D$ and $\eta\colon 1_C \to UF$, respectively. We suppose that C is an ∞-category with weak equivalences and cofibrations, and that D is an ∞-category with weak equivalences and fibrations. If F sends weak equivalences between cofibrant objects to weak equivalences, and if G sends weak equivalences between fibrant objects to weak equivalences, then there is a canonical adjunction of the form*

$$\mathbf{L}F\colon L(C) \rightleftarrows L(D)\colon \mathbf{R}U\,.$$

Proof Let us denote by

$$\gamma_C\colon C \to L(C) \quad \text{and} \quad \gamma_D\colon D \to L(D)$$

the corresponding localisation functors. For any functor $G\colon L(D) \to E$, we have

$$G \circ \mathbf{L}F = \mathbf{L}(G \circ \gamma_D \circ F)\,.$$

Similarly, for any functor $G\colon L(C) \to E$, we have

$$G \circ \mathbf{R}U = \mathbf{R}(G \circ \gamma_C \circ U)\,.$$

Therefore, the canonical maps

$$\mathbf{L}F \circ \gamma_C \circ U \to \gamma_D \circ F \circ U \xrightarrow{\gamma_D * \varepsilon} \gamma_D$$

induce a map

$$\bar{\varepsilon}\colon \mathbf{L}F \circ \mathbf{R}U = \mathbf{R}(\mathbf{L}F \circ \gamma_C \circ U) \to \mathbf{R}(\gamma_D) = 1_{L(D)}\,.$$

Dually, the maps

$$\gamma_C \xrightarrow{\gamma_C * \eta} \gamma_C \circ U \circ F \to \mathbf{R}U \circ \gamma_D \circ F$$

induce a map

$$\bar{\eta}\colon 1_{L(C)} = \mathbf{L}(\gamma_C) \to \mathbf{L}(\mathbf{R}U \circ \gamma_D \circ F) = \mathbf{R}U \circ \mathbf{L}F\,.$$

We observe that, by construction, the canonical maps provide a commutative square of the following form.

$$\begin{array}{ccc} \mathbf{L}F \circ \gamma_C \circ U & \longrightarrow & \mathbf{L}F \circ \mathbf{R}U \circ \gamma_D \\ \downarrow & & \downarrow {\scriptstyle \bar{\varepsilon} * \gamma_D} \\ \gamma_D \circ F \circ U & \xrightarrow{\gamma_D * \varepsilon} & \gamma_D \end{array}$$

Composing with F on the right, this gives a commutative diagram

$$\begin{array}{cccc} \mathbf{L}F \circ \gamma_C \xrightarrow{\mathbf{L}F \circ \gamma_C * \eta} & \mathbf{L}F \circ \gamma_C \circ U \circ F & \longrightarrow & \mathbf{L}F \circ \mathbf{R}U \circ \gamma_D \circ F \\ & \downarrow & & \downarrow {\scriptstyle \bar{\varepsilon} * \gamma_D \circ F} \\ & \gamma_D \circ F \circ U \circ F & \xrightarrow{\gamma_D * \varepsilon * F} & \gamma_D \circ F \end{array}$$

such that the composed map $\mathbf{L}F \circ \gamma_C \to \gamma_D \circ F$ is the canonical one (corresponding to the identity of $\mathbf{L}F$, by transposition). This proves that the composition

$$\mathbf{L}F \xrightarrow{\mathbf{L}F * \bar{\eta}} \mathbf{L}F \circ \mathbf{R}U \circ \mathbf{L}F \xrightarrow{\bar{\varepsilon} * \mathbf{L}F} \mathbf{L}F$$

is the identity. Replacing C by D^{op} and D by C^{op} in the computation above gives the other expected identity, and thus achieves the proof. □

7.6 Equivalences of ∞-Categories with Finite Limits

In this section, given an ∞-category C, subcategories of weak equivalences $W \subset C$ always have the property that a simplex $x\colon \Delta^n \to C$ belongs to W if and only if its restrictions $x_{|\Delta^{\{i,i+1\}}}\colon \Delta^{\{i,i+1\}} \to C$ all belong to W for $0 \leq i < n$. In other words, the inclusion map $W \to C$ is required to be an inner fibration. Such a subcategory W contains all invertible maps of C if and only if the map $W \to C$ is an isofibration.

Definition 7.6.1 Let C and C' be ∞-categories equipped with subcategories of weak equivalences $W \subset C$ and $W' \subset C'$. A functor $f\colon C \to D$ has the *right approximation property* if the following two conditions are verified.

App 1. A morphism of C belongs to W if and only if its image by f is in W'.

App 2. For any objects y_0 and x_1 in D and C, respectively, and for any map

$\psi\colon y_0 \to f(x_1)$ in D, there is a map $\varphi\colon x_0 \to x_1$ in C and a weak equivalence $u\colon y_0 \to f(x_0)$ such that the following triangle commutes.

$$\begin{array}{ccc} y_0 & \xrightarrow{\psi} & f(x_1) \\ {\scriptstyle u}\downarrow{\scriptstyle \wr} & \nearrow{\scriptstyle f(\varphi)} & \\ f(x_0) & & \end{array}$$

Remark 7.6.2 Any functor $f\colon C \to D$ between ∞-categories such that the induced functor

$$ho(f)\colon ho(C) \to ho(D)$$

is an equivalence of categories has the right approximation property (where the weak equivalences are the invertible maps). Indeed, in this case, for any map $\psi\colon y_0 \to f(x_1)$ in D, the essential surjectivity of $ho(f)$ means that one may choose an invertible map of the form $u\colon y_0 \to f(x_0)$, and the fullness of $ho(f)$ implies that one may complete these data into a commutative triangle as above.

Example 7.6.3 For any ∞-category with weak equivalences and fibrations C, the inclusion $C_f \to C$ has the right approximation property.

Example 7.6.4 Let C be an ∞-category with weak equivalences and fibrations. If all objects are fibrant in C, and if the weak equivalences of C are precisely the maps which become invertible in $L(C)$, then the localisation functor $\gamma\colon C \to L(C)$ has the right approximation property. Indeed, this comes from the fact that, by virtue of Corollary 7.2.18 and of Proposition 7.4.13, the trivial fibrations define a right calculus of fractions, so that formula (7.2.10.4) holds.

Proposition 7.6.5 *If a functor $f\colon C \to D$ has the right approximation property, then so does the induced functor $C/x \to D/f(x)$ for any object x of C (where the weak equivalences in C/x are those maps whose image by the canonical functor $C/x \to C$ are weak equivalences, and similarly for D/y).*

Proof By virtue of Proposition 4.2.9, one may as well consider the functors $C/\!/x \to D/\!/f(x)$, in which case this is straightforward. □

Remark 7.6.6 In the case where the weak equivalences are the invertible maps, the preceding proposition does make sense: the functors $C/x \to C$ are conservative, since they are right fibrations.

Lemma 7.6.7 *Let $f\colon C \to C'$ be a left exact functor between ∞-categories with weak equivalences and fibrations. We write W and W' for the weak equivalences in C and in C', respectively. If f has the right approximation property, then the induced functor $W \to W'$ is final (whence a weak homotopy equivalence).*

Proof Let y be an object of W'. We want to prove that $y\backslash\backslash W$ is weakly contractible. For this purpose, we will use Lemma 4.3.15. Let E be the nerve of a finite partially ordered set, and $F\colon E \to y\backslash\backslash W$ be a functor. We want to prove that F is Δ^1-homotopic to a constant map. Such a functor F is determined by a functor $\Phi\colon E \to C$ which sends all maps of E to weak equivalences, equipped with a fibrewise weak equivalence from the constant functor with value y to $f\Phi$. By Proposition 7.4.19, one may choose a fibrewise weak equivalence $\Phi \to \Phi_0$ such that Φ_0 is Reedy fibrant. Replacing Φ by Φ_0, we may assume that Φ has a limit and that it is fibrant in C. Since f is left exact, by virtue of Proposition 7.5.5, the limit of $f\Phi$ also exists in C', and is nothing other than the image by f of the limit of Φ in C. Therefore, there is a canonical map $\psi\colon y \to f(\varprojlim \Phi)$ in C'. The right approximation property ensures that there exists a map $\varphi\colon x \to \varprojlim \Phi$ in C and a weak equivalence $u\colon y \to f(x)$ in C' such that ψ is a composition of $f(\varphi)$ and u. The map φ induces a natural transformation from the constant functor with value x to Φ. This natural transformation is a fibrewise weak equivalence because its image by f composed with the invertible map u is the given fibrewise weak equivalence from y to Φ, and because f detects weak equivalences. In other words, there is a natural transformation from the constant functor with value (x, u) to F. □

Proposition 7.6.8 *Let C be an ∞-category with weak equivalences and fibrations. We write W and W_f for the subcategories of weak equivalences in C and in the full subcategory of fibrant objects C_f, respectively. Then the functor $W_f \to W$ is final (hence a weak homotopy equivalence).*

Proof Since the inclusion $C_f \to C$ is left exact and has the right approximation property, we can apply the preceding lemma. □

Corollary 7.6.9 *Let C be an ∞-category with weak equivalences and fibrations. We assume that the subcategory of weak equivalences $W \subset C$ is saturated (i.e. that it has the property that a map in C is a weak equivalence if and only if it becomes invertible in $L(C)$; see Remark 7.1.5). Then the canonical functor $W \to k(L(C))$ is final (hence a weak homotopy equivalence).*

Proof We have a commutative diagram of the form

$$\begin{array}{ccc} W_f & \longrightarrow & W \\ \downarrow & & \downarrow \\ k(L(C_f)) & \longrightarrow & k(L(C)) \end{array}$$

in which the upper horizontal map is final, by the previous proposition, while, by virtue of Theorem 7.5.18, the lower horizontal map is an equivalence of

∞-groupoids (whence final). We have seen in Example 7.6.4 that the functor $C_f \to L(C_f)$ has the right approximation property. Since it is also left exact, by Proposition 7.5.6, we can apply Lemma 7.6.7 and see that the left vertical functor in the square above is final. We conclude the proof with Corollary 4.1.9. □

Theorem 7.6.10 *Let $f: C \to D$ be a functor between ∞-categories with finite limits. If f commutes with finite limits, then the following three conditions are equivalent.*

(i) The functor f is an equivalence of ∞-categories.
(ii) The functor $ho(f): ho(C) \to ho(D)$ is an equivalence of categories.
(iii) The functor f has the right approximation property.

Proof We already know that condition (i) implies condition (ii). The fact that condition (ii) implies condition (iii) has already been discussed in Remark 7.6.2. It is thus sufficient to prove that condition (iii) implies condition (i). Let us assume that f has the right approximation property. For any object y of D, if e is a final object of C, then $f(e)$ is a final object of D, hence there is a map $y \to f(e)$. The property of right approximation thus implies that there exists an invertible map from y to $f(x)$ for some object x of C. It remains to prove that f is fully faithful. Let y be an object of C. Lemma 7.6.7 implies that $k(f): k(C) \to k(D)$ is an equivalence of ∞-groupoids. But, by Proposition 7.6.5, the induced functor $C/y \to D/f(y)$ also has the right approximation property. By virtue of Corollary 7.3.29, the slice C/y has finite limits, and the functor from C/y to $D/f(y)$ commutes with finite limits. We may thus apply Lemma 7.6.7 once more and deduce that the induced map $k(C/y) \to k(D/f(y))$ is an equivalence of ∞-groupoids. On the other hand, since the functor $C/y \to C$ is conservative, for any object x of C, we have canonical homotopy pull-back squares of the following form.

$$\begin{array}{ccccc} C(x,y) & \longrightarrow & k(C/y) & \longrightarrow & C/y \\ \downarrow & & \downarrow & & \downarrow \\ \Delta^0 & \xrightarrow{x} & k(C) & \longrightarrow & C \end{array}$$

Similarly, $D(f(x), f(y))$ is the homotopy fibre of the map $k(D/f(y)) \to k(D)$. Since we have a commutative square

$$\begin{array}{ccc} k(C/y) & \longrightarrow & k(D/f(y)) \\ \downarrow & & \downarrow \\ k(C) & \longrightarrow & k(D) \end{array}$$

in which the two horizontal maps are weak homotopy equivalences, we see that the induced map $C(x, y) \to D(f(x), f(y))$ is an equivalence of ∞-groupoids. This proves that f is fully faithful and essentially surjective, whence that it is an equivalence of ∞-categories. □

Corollary 7.6.11 *Let $F\colon C \to D$ be a left exact functor between ∞-categories with weak equivalences and fibrations. Then the induced right derived functor $\mathbf{R}F\colon L(C) \to L(D)$ is an equivalence of ∞-categories if and only if the induced functor $ho(\mathbf{R}F)\colon ho(L(C)) \to ho(L(D))$ is an equivalence of categories.*

Proof This follows right away from Proposition 7.5.28 and from the previous theorem. □

7.6.12 If C is an ∞-category with weak equivalences and fibrations, for any fibrant object x, the slice C/x is an ∞-category with weak equivalences and fibrations: the weak equivalences (fibrations) of C/x are the maps whose image by the canonical projection $C/x \to C$ are weak equivalences (fibrations).

Corollary 7.6.13 *Let C be an ∞-category with weak equivalences and fibrations. We denote by $\gamma\colon C \to L(C)$ the localisation functor. For any fibrant object x of C, the canonical functor $C/x \to L(C)/\gamma(x)$ induces an equivalence of ∞-categories*

$$L(C/x) \simeq L(C)/\gamma(x)\,.$$

Proof It is harmless to assume that C is saturated (i.e. that a map in C is a weak equivalence if and only if its image in $L(C)$ is invertible); see Remark 7.5.22. We shall prove that the canonical functor $\varphi\colon L(C/x) \to L(C)/\gamma(x)$ has the approximation property. Since the functor $L(C)/\gamma(x) \to L(C)$ is conservative, to prove property App 1, it is sufficient to check that the functor $L(C/x) \to L(C)$ is conservative. Using the right calculus of fractions, this amounts to checking that a map of C/x becomes invertible in $L(C/x)$ if and only if it becomes invertible in $L(C)$. Since C is assumed to be saturated, this amounts to asserting that any map in C/x whose image in C is a weak equivalence is a weak equivalence, which is true by definition. To prove that φ satisfies axiom App 2, it is sufficient to check that the functor $C/x \to L(C)/\gamma(x)$ has this property. This would follow from the fact that the functor $\gamma\colon C \to L(C)$ satisfies property App 2. This, in turn, follows right away from the fact that, by Theorem 7.5.18, we have $L(C_f) \simeq L(C)$ (where C_f is the full subcategory of fibrant objects in C), and from Example 7.6.4. Now, since x is fibrant, as observed above, the slice C/x has a natural structure of ∞-category with weak equivalences and fibrations. Therefore, since the localisation functor γ is left exact, by virtue of

Lemma 7.5.24 and of Proposition 7.5.28, the functor φ is left exact. Finally, Theorem 7.6.10 shows that φ is an equivalence of ∞-categories. □

We can use the preceding corollary to understand when two maps become equivalent in the localisation in terms of path-homotopies (see paragraph 7.5.13).

Proposition 7.6.14 *Let C be an ∞-category of fibrant objects, and y an object of C. The relation of path-homotopy does not depend on the choice of a path object of y. Furthermore, two maps $f, g\colon x \to y$ are path-homotopic if and only if their images in $ho(L(C))$ are equal. In particular, it is an equivalence relation on the set* $\mathrm{Hom}_{ho(C)}(x, y)$, *and it is compatible with composition.*

Proof Given an object s in C, and maps $f\colon x \to s$ and $i\colon y \to s$, we choose a factorisation of i into a weak equivalence $w\colon y \to z'$ followed by a fibration $\pi\colon z' \to s$. Applying Theorem 7.5.18 to C/s, we may consider the right calculus of fractions defined by trivial fibrations between fibrant objects in C/s, and see that maps from (x, f) to (y, i) in $L(C/s)$ can all be described by commutative diagrams of the form

$$\begin{array}{ccccc} x & \xleftarrow{p} & z & \xrightarrow{q} & z' \\ & \searrow^{f} & \downarrow & \swarrow_{\pi} & \\ & & s & & \end{array}$$

in which p is a trivial fibration (*a priori*, we should have replaced f by a fibration, but we can reintroduce f itself using the fact that pull-backs of trivial fibrations along maps between fibrant objects exist in C, and are trivial fibrations). Applying what precedes for $s = y \times y$ and i the diagonal map, this shows the first assertion (because path objects of y are simply fibrant replacements of the diagonal map $y \to y \times y$, seen as an object of the slice category over $y \times y$). The canonical equivalences of ∞-categories provided by Corollary 7.6.13,

$$L(C/s) \simeq L(C)/\gamma(s),$$

and the fact that the localisation functor γ commutes with finite products show that, to prove the second assertion, we may assume, without loss of generality, that the weak equivalences of C are the invertible maps. The proposition is then obvious. □

Proposition 7.6.15 *Let $F\colon C \to D$ be a left exact functor between ∞-categories with weak equivalences and fibrations. The right derived functor $\mathbf{R}F\colon L(C) \to L(D)$ is an equivalence of ∞-categories if and only if the functor F has the following two properties.*

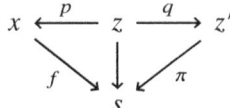

(a) *If a morphism between fibrant objects* $u\colon x \to y$ *in* C *becomes a weak equivalence in* D*, then there is a morphism with fibrant domain* $u'\colon x' \to x$ *and a composition* $uu'\colon x' \to y$ *which is a weak equivalence in* C*.*

(b) *For any fibrant objects* x_1 *in* C *and* y_0 *in* D*, and for any morphism* $\psi\colon y_0 \to F(x_1)$*, there exists a commutative square of the form*

$$\begin{array}{ccc} y_0 & \xrightarrow{\psi} & F(x_1) \\ \uparrow{\scriptstyle t} & & \uparrow{\scriptstyle F(\varphi)} \\ y_1 & \xrightarrow{s} & F(x_0) \end{array}$$

in D*, where both* s *and* t *are weak equivalences between fibrant objects, and* φ *is a morphism with fibrant domain in* C*.*

Proof By virtue of Theorem 7.5.18, it is sufficient to prove this proposition in the case where both C and D are ∞-categories of fibrant objects. On the other hand, we know that the derived functor $\mathbf{R}F$ is left exact. Therefore, by Theorem 7.6.10, it is an equivalence of ∞-categories if and only if it has the right approximation property of Definition 7.6.1. Moreover, by virtue of Ken Brown's lemma (Proposition 7.4.13), there is a right calculus of trivial fibrations in any ∞-category of fibrant objects (Corollary 7.2.18). It follows easily from the right calculus of fractions and from Corollary 7.5.19 that condition (a) is equivalent to the property that the derived functor $\mathbf{R}F$ is conservative. The right calculus of fractions and condition (b) also imply that condition App 2 of Definition 7.6.1 is satisfied by $\mathbf{R}F$. This shows that conditions (a) and (b) imply that $\mathbf{R}F$ is an equivalence of ∞-categories. To prove the converse, it remains to check that, if $\mathbf{R}F$ is an equivalence of ∞-categories, then condition (b) holds. Let $\psi\colon y_0 \to F(x_1)$ be a morphism in D. We see that condition (b) holds for ψ if and only if it holds for a composition of ψ with some trivial fibration. Since $\mathbf{R}F$ is full and essentially surjective, using the right calculus of fractions, we see that we may assume, without loss of generality, that $y_0 = F(x_0)$ for some object x_0, and that there is a map $\varphi\colon x_0 \to x_1$ such that ψ and $F(\varphi)$ are equal in $ho(L(D))$. By virtue of Proposition 7.6.14, this means that ψ and $F(\varphi)$ are path-homotopic in D, and we observe that this readily implies property (b). □

Corollary 7.6.13 may also be used to give sufficient conditions for a localised ∞-category to be locally Cartesian closed as follows, generalising a result of Kapulkin [Kap17].

Proposition 7.6.16 *Let* C *be an* ∞*-category with weak equivalences and fibrations. Given a fibrant object* x*, we write* $C(x)$ *for the full subcategory of*

C/x which consists of fibrations $x' \to x$. We see $C(x)$ as a category of fibrant objects, with weak equivalences (fibrations) the maps whose image by the standard functor $C(x) \to C$ is a weak equivalence (a fibration, respectively). We assume that, for any fibration between fibrant objects $p\colon x \to y$, the pull-back functor

$$p^*\colon C(y) \to C(x)\,, \quad (y' \to y) \mapsto (y' \times_y x \to x)$$

has a right adjoint $p_*\colon C(x) \to C(y)$ which preserves trivial fibrations. Then, for any map $p\colon x \to y$ in $L(C)$, the pull-back functor

$$p^*\colon L(C)/y \to L(C)/x\,, \quad (y' \to y) \mapsto (y' \times_y x \to x)$$

has a right adjoint.

Proof Let $\gamma\colon C \to L(C)$ be the localisation functor. By virtue of Theorem 7.5.18, using the right calculus of fractions as well as the existence of factorisation of maps between fibrant objects into a weak equivalence followed by a fibration, together with Propositions 6.1.6, 6.1.7 and 6.1.8, we see that it is sufficient to prove that, for any fibration between fibrant objects $p\colon x \to y$, the pull-back functor

$$\gamma(p)^*\colon L(C)/\gamma(y) \to L(C)/\gamma(x)$$

has a right adjoint. By virtue of Corollary 7.4.14, both functors of the adjunction

$$p^*\colon C(y) \rightleftarrows C(x) :p_*$$

preserve weak equivalences and thus induce an adjunction of localised ∞-categories, by Proposition 7.1.14:

$$\bar{p}^*\colon L(C(y)) \rightleftarrows L(C(x)) :\bar{p}_*\,.$$

But the functor p^* also has a left adjoint $p_!$ (induced by composition with p) which also preserves weak equivalences, whence also induces an adjunction

$$\bar{p}_!\colon L(C(x)) \rightleftarrows L(C(y)) :\bar{p}^*\,.$$

Theorem 7.5.18 and Corollary 7.6.13 imply that $L(C(z)) \simeq L(C)/\gamma(z)$ for all fibrant objects z, and that the functor $\bar{p}_!$ is equivalent to the functor $\gamma(p)_!$ in the adjunction

$$\gamma(p)_!\colon L(C)/\gamma(x) \rightleftarrows L(C)/\gamma(y) :\gamma(p)^*\,.$$

Therefore, the pull-back functor $\gamma(p)^*$ corresponds to the functor $\bar{p}^*$ above, which has a right adjoint. Hence, so does $\gamma(p)^*$. □

Theorem 7.6.17 *Let C be an ∞-category with weak equivalences and fibrations, and I a finite direct category. The canonical functor*

$$L(\underline{\mathrm{Hom}}(N(I)^{\mathrm{op}}, C)) \to \underline{\mathrm{Hom}}(N(I)^{\mathrm{op}}, L(C))$$

is an equivalence of ∞-categories.

Proof By virtue of Theorem 7.5.18, the localisation functor $C \to L(C)$ is left exact and thus induces a left exact functor

$$\underline{\mathrm{Hom}}(N(I)^{\mathrm{op}}, C) \to \underline{\mathrm{Hom}}(N(I)^{\mathrm{op}}, L(C)),$$

where the left-hand side is equipped with the fibrewise structure of ∞-category with weak equivalences and fibrations provided by Corollary 7.4.22. By Proposition 7.5.28, this induces a left exact functor

$$L(\underline{\mathrm{Hom}}(N(I)^{\mathrm{op}}, C)) \to \underline{\mathrm{Hom}}(N(I)^{\mathrm{op}}, L(C)).$$

We shall prove that this functor has the right approximation property and apply Theorem 7.6.10.

Let us prove the property of conservativity. Let $f: X \to Y$ be a morphism in $L(\underline{\mathrm{Hom}}(N(I)^{\mathrm{op}}, C))$. We may consider the latter ∞-category as a localisation of the Reedy structure of Theorem 7.4.20. Applying Theorem 7.5.18 to the latter, and using the right calculus of fractions by Reedy trivial fibrations, we may assume that X and Y are Reedy fibrant functors from $N(I)^{\mathrm{op}}$ to C and that f is a Reedy fibration in $\underline{\mathrm{Hom}}(N(I)^{\mathrm{op}}, C)$. The evaluation of f at any object i of I is invertible in $L(C)$. We shall prove by induction on the length of I that, under this assumption, f becomes invertible in $L(\underline{\mathrm{Hom}}(N(I)^{\mathrm{op}}, C))$. The case where $\ell(I) \leq 0$ is obvious: Proposition 7.1.13 implies that our functor is then an equivalence of categories. If $\ell(I) > 0$, the preceding corollary, applied to presheaves on $N(\partial I)^{\mathrm{op}}$, implies that there is a Reedy fibration $p': Z' \to X_{|N(\partial I)}$ whose composition with $f_{|N(\partial I)}$ is a fibrewise weak equivalence. We may form, for each maximal object i, the following diagram

$$\begin{array}{ccccc}
\partial Z'(i) \times_{\partial X(i)} X(i) & \xrightarrow{\pi_i} & X(i) & \xrightarrow{f(i)} & Y(i) \\
\downarrow & & \downarrow & & \downarrow \\
\partial Z'(i) & \xrightarrow{\partial p(i)} & \partial X(i) & \xrightarrow{\partial f(i)} & \partial Y(i)
\end{array}$$

in which $\partial Z'(i)$ is the limit of the restriction of Z' to $\partial I/i$. We observe that the maps $f(i)$ and $\partial f(i)$ are invertible in $L(C)$: the exactness of the localisation functor allows us to apply Proposition 7.5.5 to them. Therefore, the commutative square of the right-hand side is Cartesian in $L(C)$. The commutative square of the left-hand side is also Cartesian in $L(C)$: it is obtained by pulling back a fibration in the subcategory of fibrant objects of C. Therefore, the composed upper horizontal map is invertible in $L(C)$. Corollary 7.5.19 thus provides a fibration $q_i \colon Z(i) \to \partial Z'(i) \times_{\partial X(i)} X(i)$ whose composition with q_i and π_i gives a weak equivalence in C. By virtue of Proposition 7.4.11, we have proven that there is a Reedy fibration $p \colon Z \to X$ whose composition with f is a fibrewise weak equivalence. Applying what precedes to p, we see that there is a Reedy fibration $p' \colon Z' \to Z$ such that the composition of p' and p is a weak equivalence. This implies that f is invertible in $L(\underline{\mathrm{Hom}}(N(I)^{\mathrm{op}}, C))$.

Our functor is essentially surjective: using Proposition 7.4.11, this follows right away by induction on the length of I. Let X and Y be two Reedy fibrant presheaves on $N(I)$. It is now sufficient to show that any map f from X to Y in $L(\underline{\mathrm{Hom}}(N(I)^{\mathrm{op}}, C))$ can be realised as an inverse of a trivial fibration $s \colon Z \to X$ followed by a map $p \colon Z \to Y$ in $\underline{\mathrm{Hom}}(N(I)^{\mathrm{op}}, C)$. We proceed again by induction on $\ell(I)$. We may thus assume that $f_{|N(\partial I)}$ is realised as the inverse of a trivial fibration $s' \colon Z' \to X_{|N(\partial I)}$ and of a map $p' \colon Z' \to Y_{|N(\partial I)}$, and that, for each maximal object i of I, the map $f(i)$ is the inverse of a trivial fibration $u_i \colon Z_i \to X(i)$ and of a map $q_i \colon Z_i \to Y(i)$. We form the following pull-back square in C, in which the vertical map on the right-hand side is a trivial fibration.

$$\begin{array}{ccccc} Z(i) & & \longrightarrow & & \partial Z'(i) \\ \downarrow & & & & \downarrow \\ Z_i & \xrightarrow{u_i} & X(i) & \longrightarrow & \partial X(i) \end{array}$$

Hence the map $s(i) \colon Z(i) \to X(i)$ is a trivial fibration because the map u_i is a trivial fibration. Composing with q_i, we define a map $p(i) \colon Z(i) \to Y(i)$. Applying Proposition 7.4.11, we see that this defines a presheaf Z whose restriction to $N(\partial I)$ is Z' and a map $(s, p) \colon Z \to X \times Y$ such that s is a trivial fibration. One checks that the inverse of s composed with p is equivalent to f by induction, using Proposition 7.4.11. □

Corollary 7.6.18 *Let C be an ∞-category with weak equivalences and fibrations, and I a finite direct category. We denote by $W(N(I), C)$ the subcategory of fibrewise weak equivalences (see paragraph 7.4.17), and we write $W(N(I), C)_{Reedy}$ for its full subcategory of Reedy fibrant presheaves. If W is saturated, then the natural functors*

$$W(N(I), C)_{Reedy} \to W(N(I), C) \to k(\underline{\mathrm{Hom}}(N(I), L(C)))$$

are weak homotopy equivalences.

Proof Proposition 7.6.8 and Theorem 7.4.20 show together that the inclusion of $W(N(I), C)_{Reedy}$ into $W(N(I), C)$ is a weak homotopy equivalence. It follows from the first assertion of Corollary 3.5.12, applied to $L(C)$, that we may apply Corollary 7.6.9 to the fibrewise structure of ∞-category with weak equivalences and fibrations provided by Corollary 7.4.22, and get a weak homotopy equivalence of the form

$$W(N(I), C) \to k(L(\underline{\mathrm{Hom}}(N(I), C)))\,.$$

The preceding theorem implies that there is a canonical equivalence of ∞-groupoids

$$k(L(\underline{\mathrm{Hom}}(N(I), C))) \simeq k(\underline{\mathrm{Hom}}(N(I), L(C)))\,,$$

which gives our claim. □

7.7 Homotopy Completeness

Proposition 7.7.1 *Let I be a set, and C_i, $i \in I$, an I-indexed family of ∞-categories with weak equivalences and fibrations. Then the product $C = \prod_{i \in I} C_i$ has a natural structure of ∞-category with weak equivalences and fibrations, defined fibrewise over I. The natural functor*

$$L(C) \to \prod_{i \in I} L(C_i)$$

is an equivalence of ∞-categories.

Proof The first assertion is obvious. The product of the localisation functors $\gamma_i \colon C_i \to L(C_i)$ gives a left exact functor

$$\pi = \prod_{i \in I} \gamma_i \colon C \to \prod_{i \in I} L(C_i)$$

which induces a left exact functor

$$\mathbf{R}\pi \colon L(C) \to \prod_{i \in I} L(C_i)\,,$$

by Proposition 7.5.28. It follows from the right calculus of fractions defined by trivial fibrations and Theorem 7.5.18, and from Corollary 7.5.19, that the functor $\mathbf{R}\pi$ is conservative. As seen in Example 7.6.4, the right calculus of fractions also shows that each functor $\gamma_i \colon C_i \to L(C_i)$ has property App 2 of Definition 7.6.1. Therefore, so does the functor π, whence the functor $\mathbf{R}\pi$ has the approximation property. Theorem 7.6.10 thus implies that the functor $\mathbf{R}\pi$ is an equivalence of ∞-categories. □

Definition 7.7.2 Let $\mathbf{U}$ be a universe. An ∞-category with weak equivalences and fibrations C is *homotopy* $\mathbf{U}$*-complete* if the following conditions are verified.

(a) For any $\mathbf{U}$-small set I and any family $(x_i)_{i \in I}$ of fibrant objects in C, the product $\prod_{i \in I} x_i$ exists and is fibrant.
(b) For any $\mathbf{U}$-small set I and any I-indexed family $p_i \colon x_i \to y_i$ of fibrations (of trivial fibrations) between fibrant objects, the induced map $\prod_{i \in I} x_i \to \prod_{i \in I} y_i$ is a fibration (a trivial fibration, respectively).

A functor between homotopy $\mathbf{U}$-complete ∞-categories with weak equivalences and fibrations is *homotopy continuous* if it is left exact and if it commutes with $\mathbf{U}$-small products of fibrant objects.

Dually, an ∞-category with weak equivalences and cofibrations C is *homotopy* $\mathbf{U}$*-cocomplete* if C^{op} is homotopy $\mathbf{U}$-complete. A functor between homotopy $\mathbf{U}$-cocomplete ∞-categories with weak equivalences and cofibrations is *homotopy cocontinuous* if it is right exact and if it commutes with $\mathbf{U}$-small coproducts of fibrant objects.

Example 7.7.3 By virtue of Theorem 7.3.22, an ∞-category with finite limits is homotopy $\mathbf{U}$-complete if and only if it has $\mathbf{U}$-small limits, i.e. if it is $\mathbf{U}$-complete. Similarly, by Proposition 7.3.23, a functor between $\mathbf{U}$-complete ∞-categories is homotopy continuous if and only if it commutes with $\mathbf{U}$-small limits.

In the case of (nerves of) Quillen model categories, the following proposition is due to Barnea, Harpaz and Horel [BHH17].

Proposition 7.7.4 *Let C be a homotopy $\mathbf{U}$-complete ∞-category with weak equivalences and fibrations. The localisation $L(C)$ has $\mathbf{U}$-small limits, and the localisation functor $\gamma \colon C \to L(C)$ is homotopy continuous.*

Proof By virtue of Theorem 7.5.18, we may assume that all objects of C are fibrant. Corollary 7.4.14 then shows that any $\mathbf{U}$-small product of weak equivalences is a weak equivalence. Therefore, the existence of products in $L(C)$ comes from conjunction of Propositions 7.1.14 and 7.7.1. This also shows that γ is homotopy continuous. Theorem 7.3.22 then implies that $L(C)$ has $\mathbf{U}$-small limits. □

Corollary 7.7.5 *Let $F \colon C \to D$ be a homotopy continuous functor between homotopy $\mathbf{U}$-cocomplete ∞-categories with weak equivalences and fibrations. The associated right derived functor $\mathbf{R}F \colon L(C) \to L(D)$ commutes with $\mathbf{U}$-small limits.*

Proof If C_f is the full subcategory of fibrant objects in C, since the canonical functor $C_f \to L(C)$ commutes with $\mathbf{U}$-small products and is essentially surjective, it is sufficient to prove that the composed functor

$$C_f \to L(C) \xrightarrow{\mathbf{R}F} L(D)$$

commutes with $\mathbf{U}$-small products. But, by construction of $\mathbf{R}F$, this composed functor is equivalent to the functor

$$C_f \xrightarrow{F_{|C_f}} D \to L(D)\,.$$

The latter commutes with $\mathbf{U}$-small products by Proposition 7.7.4. Therefore, the functor $\mathbf{R}F$ commutes with $\mathbf{U}$-small products, and it is left exact, by Proposition 7.5.28. Hence it commutes with $\mathbf{U}$-small limits, by Proposition 7.3.23. □

7.7.6 Let C and D be two homotopy $\mathbf{U}$-complete ∞-categories of fibrant objects. A functor $F : C \to D$ is homotopy continuous if and only if it is left exact and commutes with $\mathbf{U}$-small products. We write $\underline{\mathrm{Hom}}_{h*}(C, D)$ for the full subcategory of $\underline{\mathrm{Hom}}(C, D)$ which consists of homotopy continuous functors. In the case where both C and D are ∞-categories with $\mathbf{U}$-small limits, we thus have the equality $\underline{\mathrm{Hom}}_{h*}(C, D) = \underline{\mathrm{Hom}}_{*}(C, D)$, where the right-hand side denotes the full subcategory of functors which commute with $\mathbf{U}$-small limits.

Theorem 7.7.7 *Let C be a homotopy $\mathbf{U}$-complete ∞-category of fibrant objects. The localisation functor $\gamma : C \to L(C)$ is the universal homotopy continuous functor whose codomain is an ∞-category with $\mathbf{U}$-small limits. In other words, for any ∞-category with $\mathbf{U}$-small limits D, the functor γ induces an equivalence of ∞-categories*

$$\gamma^* : \underline{\mathrm{Hom}}_{*}(L(C), D) \xrightarrow{\sim} \underline{\mathrm{Hom}}_{h*}(C, D)\,.$$

Proof If $W \subset C$ denotes the subcategory of weak equivalences in C, by definition of γ, we have a commutative diagram of the form

$$\begin{array}{ccc}
\underline{\mathrm{Hom}}_{*}(L(C), D) & \xrightarrow{\gamma^*} & \underline{\mathrm{Hom}}_{h*}(C, D) \\
\downarrow & & \downarrow \\
\underline{\mathrm{Hom}}(L(C), D) & \xrightarrow{\gamma^*} & \underline{\mathrm{Hom}}_{W}(C, D)
\end{array}$$

in which the vertical maps are fully faithful, and the lower horizontal map is an equivalence of ∞-categories. To finish the proof, it is sufficient to check that, if a functor $C \to D$ is homotopy continuous, then the induced functor $L(C) \to D$ commutes with $\mathbf{U}$-small limits. By virtue of Lemma 7.5.24, this is a particular case of Corollary 7.7.5. □

7.7.8 Since the sequel of this chapter will be written in terms of colimits, we will write the dual version of the previous theorem explicitly.

Let C and D be two homotopy **U**-cocomplete ∞-categories of fibrant objects. A functor $F: C \to D$ is homotopy cocontinuous if and only if it is right exact and commutes with **U**-small coproducts. We write $\underline{\mathrm{Hom}}_{h!}(C, D)$ for the full subcategory of $\underline{\mathrm{Hom}}(C, D)$ which consists of homotopy cocontinuous functors. In the case where both C and D are ∞-categories with **U**-small colimits, we thus have the equality $\underline{\mathrm{Hom}}_{h!}(C, D) = \underline{\mathrm{Hom}}_!(C, D)$, where the right-hand side denotes the full subcategory of functors which commute with **U**-small colimits.

Theorem 7.7.9 *Let C be a homotopy* **U***-cocomplete ∞-category of cofibrant objects. For any ∞-category with* **U***-small colimits D, the functor $\gamma: C \to L(C)$ induces an equivalence of ∞-categories*

$$\gamma^*: \underline{\mathrm{Hom}}_!(L(C), D) \xrightarrow{\sim} \underline{\mathrm{Hom}}_{h!}(C, D) .$$

Remark 7.7.10 In the theorem above, no assumption on local **U**-smallness is made. This is necessary because, in general, even if we assume that C is locally **U**-small, there is no reason for $L(C)$ to be locally **U**-small (there are reasonable sufficient conditions though, such as Corollary 7.10.5).

However, this theorem gives us a lot of freedom to construct *homotopy cocontinuous localisations*, i.e. to invert maps in a way which is compatible with homotopy colimits. Here is why and how. Let C be a homotopy **U**-cocomplete ∞-category with weak equivalences and cofibrations. Let S be a set of maps in C. We let C_S be the ∞-category with weak equivalences and cofibrations whose underlying ∞-category is C, whose subcategory of cofibrations *Cof* is the same as for C, and whose subcategory of weak equivalences is the intersection of all $W \subset C$ containing S and such that (C, W, Cof) is a homotopy **U**-cocomplete ∞-category with weak equivalences and cofibrations. Then the identity functor defines a homotopy continuous functor $C \to C_S$, and we put

$$L_S(C) = L(C_S) . \tag{7.7.10.1}$$

It is easy to see that the canonical functor $L(C) \to L_S(C)$ identifies $L_S(C)$ as the localisation of $L(C)$ the image of the subcategory of weak equivalences of C_S. Since the localisation functor $\gamma: C \to L(C)$ is homotopy continuous, for any functor which commutes with **U**-small colimits $f: L(C) \to D$ the subcategory $W_f = \gamma^{-1}(f^{-1}(k(D)))$ turns the triple (C, W_f, Cof) into a homotopy **U**-cocomplete ∞-category with weak equivalences and cofibrations. Therefore, $S \subset W_f$ if and only if f factors through the canonical functor $L(C) \to L_S(C)$. In other words, if we write

$$\underline{\mathrm{Hom}}_{h!,S}(C, D) = \underline{\mathrm{Hom}}_{h!}(C, D) \cap \underline{\mathrm{Hom}}_S(C, D) \tag{7.7.10.2}$$

and

$$\underline{\mathrm{Hom}}_{!,S}(L(C), D) = \underline{\mathrm{Hom}}_{!}(L(C), D) \cap \underline{\mathrm{Hom}}_{\gamma(S)}(L(C), D), \tag{7.7.10.3}$$

the functor $C \to C_S$ induces an equivalence of ∞-categories

$$\underline{\mathrm{Hom}}_{h!,S}(C, D) \xrightarrow{\sim} \underline{\mathrm{Hom}}_{h!}(C_S, D). \tag{7.7.10.4}$$

Therefore, it follows from Theorem 7.7.9 that we have an equivalence of ∞-categories

$$\underline{\mathrm{Hom}}_{!}(L_S(C), D) \xrightarrow{\sim} \underline{\mathrm{Hom}}_{!,S}(L(C), D). \tag{7.7.10.5}$$

We observe that all this may be applied for $C = L(C)$; in other words, what precedes is a way to construct localisations which are compatible with $\mathbf{U}$-small colimits; in this situation, we may speak of *cocontinuous localisations*.

7.8 The Homotopy Hypothesis

7.8.1 We fix a Grothendieck universe $\mathbf{U}$. We then have the ∞-category $\mathcal{S}$ of $\mathbf{U}$-small ∞-groupoids. We shall write *set* for the category of $\mathbf{U}$-small sets, by which we mean the sets which are elements of $\mathbf{U}$. This is thus a version of the category of sets which is small. Therefore, it makes sense to speak of its nerve $N(set)$. An object x of $\mathcal{S}$ is *discrete* if it classifies a Kan fibration of the form $X \to \Delta^0$ where X is a $\mathbf{U}$-small Kan complex such that, for any object x in X, and any positive integer n, the homotopy group $\pi_n(X, x)$ is trivial. A trivial consequence of Corollary 3.8.14 is the following lemma.

Lemma 7.8.2 *An object x of $\mathcal{S}$ is discrete if and only if it is equivalent to an object which classifies a Kan fibration of the form $\coprod_I \Delta^0 \to \Delta^0$, where I is a $\mathbf{U}$-small set.*

Proposition 7.8.3 *There is a canonical fully faithful functor $\jmath \colon N(set) \to \mathcal{S}$ whose essential image consists precisely of discrete objects. Furthermore, this functor has a left adjoint*

$$\pi_0 \colon \mathcal{S} \to N(set).$$

Proof Since the ∞-category $\mathcal{S}$ has $\mathbf{U}$-small coproducts, the category $ho(\mathcal{S})$ has $\mathbf{U}$-small coproducts. Therefore, there is an essentially unique functor $set \to ho(\mathcal{S})$ which commutes with $\mathbf{U}$-small coproducts and which preserves final objects. Using the full embedding of $ho(\mathcal{S})$ into the homotopy category $LFib(\Delta^0)$ of the Kan–Quillen model structure provided by Theorem 5.4.5, we see that this functor is fully faithful and that its essential image consists of discrete objects. We observe that, for any objects x and y in $\mathcal{S}$, if ever y is discrete, then the Kan complex $\mathcal{S}(x, y)$ is discrete: by virtue of Corollary 5.4.7, it is sufficient to prove that, if X and Y are Kan complexes such that Y is discrete, then the Kan complex $\mathrm{Map}(X, Y)$ is discrete, which is well known (and easy to prove). Let $\mathcal{S}_{(0)}$ be the full subcategory of $\mathcal{S}$ spanned by discrete objects. Then what precedes and paragraph 3.7.7 tell us that the canonical functor $\mathcal{S}_{(0)} \to N(ho(\mathcal{S}_{(0)}))$ is an equivalence of ∞-categories. Since there is an equivalence of categories $set \simeq ho(\mathcal{S}_{(0)})$, this provides a fully faithful functor

$$N(set) \simeq \mathcal{S}_{(0)} \subset \mathcal{S}.$$

To prove that it has a left adjoint, we consider an object x classifying the Kan fibration $X \to \Delta^0$, and we choose an object $\pi_0(x)$ classifying the Kan fibration $\pi_0(X) \to \Delta^0$. We choose a map $x \to \pi_0(x)$ corresponding to the canonical map $X \to \pi_0(X)$. Then, for any discrete object y of $\mathcal{S}$, the map $x \to \pi_0(x)$ induces an equivalence

$$\mathrm{Hom}_{\mathcal{S}}(\pi_0(x), y) \xrightarrow{\sim} \mathrm{Hom}_{\mathcal{S}}(x, y)\,.$$

To prove this, we observe that $\mathrm{Hom}_{\mathcal{S}}(x, y)$ is discrete: it corresponds to the Kan complex $\mathcal{S}(x, y)$. Therefore, it is sufficient to prove that the induced map

$$\mathrm{Hom}_{ho(\mathcal{S})}(\pi_0(x), y) \to \mathrm{Hom}_{ho(\mathcal{S})}(x, y)$$

is bijective, which follows from its analogue in the classical homotopy theory of Kan complexes, by Theorem 5.4.5. Proposition 6.1.11 thus shows that the assignment $x \mapsto \pi_0(x)$ can be promoted to a left adjoint of the inclusion functor $\mathcal{S}_{(0)} \subset \mathcal{S}$. □

Proposition 7.8.4 *Let $set_\bullet$ be the category of pointed* **U***-small sets. There is a homotopy pull-back square of the following form.*

$$\begin{array}{ccc} N(set_\bullet) & \longrightarrow & \mathcal{S}_\bullet \\ \downarrow & & \downarrow{\scriptstyle p_{univ}} \\ N(set) & \xrightarrow{J} & \mathcal{S} \end{array} \tag{7.8.4.1}$$

Proof Let us form the following pull-back.

$$\begin{array}{ccc} X & \longrightarrow & \mathcal{S}_\bullet \\ \downarrow & & \downarrow{\scriptstyle p_{univ}} \\ N(set) & \xrightarrow{J} & \mathcal{S} \end{array}$$

Since $\mathcal{S}_\bullet \simeq e\backslash\mathcal{S} \simeq e\backslash\backslash\mathcal{S}$, the final object e being discrete, one can see that X is canonically equivalent to $e\backslash\backslash\mathcal{S}_{(0)} \simeq e\backslash N(set) \simeq N(set_\bullet)$. □

7.8.5 Let A be a **U**-small category. The functor $\jmath\colon N(set) \to \mathcal{S}$ induces a functor

(7.8.5.1)
$$\jmath\colon N(\underline{\mathrm{Hom}}(A^{\mathrm{op}}, set)) = \underline{\mathrm{Hom}}(N(A)^{\mathrm{op}}, N(set)) \to \underline{\mathrm{Hom}}(N(A)^{\mathrm{op}}, \mathcal{S})\,.$$

The previous proposition means that we may choose $\jmath$ such that the square (7.8.4.1) is actually Cartesian in the category of simplicial sets (because we only care about the J-homotopy class of $\jmath$). This gives the following description of the functor (7.8.5.1). For any functor $F\colon A^{\mathrm{op}} \to set$, there is a discrete fibration $p_F\colon A/F \to A$, and we have a functor

$$\tilde{F}\colon A/F^{\mathrm{op}} \to set_{\bullet}$$

which sends a pair (a, s) to the pointed set $(F(a), s)$. This provides the Cartesian squares below.

(7.8.5.2)
$$\begin{array}{ccccc} N(A/F)^{\mathrm{op}} & \xrightarrow{N(\tilde{F})^{\mathrm{op}}} & N(set_{\bullet}) & \longrightarrow & \mathcal{S}_{\bullet} \\ {\scriptstyle N(p_F)^{\mathrm{op}}}\downarrow & & \downarrow & & \downarrow{\scriptstyle p_{univ}} \\ N(A)^{\mathrm{op}} & \xrightarrow{N(F)^{\mathrm{op}}} & N(set) & \xrightarrow{\jmath} & \mathcal{S} \end{array}$$

In other words, $\jmath(F)$ classifies the left fibration $N(p_F)^{\mathrm{op}}$. Therefore, there are an equality and a canonical invertible map

(7.8.5.3)
$$\jmath(F) = N(p_F)_!(e) \xleftarrow{\sim} \varinjlim_{(a,s)\in N(A/F)^{\mathrm{op}}} h_{N(A)}(a)$$

provided by Remark 6.1.24 and by Corollary 6.2.16, respectively.

Lemma 7.8.6 *The functor* (7.8.5.1) *has the following exactness property. Let I be a* **U***-small category, and $F\colon I \to \underline{\mathrm{Hom}}(A^{\mathrm{op}}, set)$ be a functor such that one of the following three conditions is satisfied.*

(i) The category I is discrete.
(ii) We have $N(I) = \Lambda^2_0$, and the map $F(0) \to F(1)$ is a monomorphism.
(iii) The category I is filtered, and, for any map $u\colon i \to j$ in I, the induced map $F(i) \to F(j)$ is a monomorphism.

Then the comparison map

$$\varinjlim \jmath(F) \to \jmath(\varinjlim F)$$

is invertible in $\underline{\mathrm{Hom}}(N(A)^{\mathrm{op}}, \mathcal{S})$.

Proof This follows right away from Proposition 7.3.5 and from formula (7.8.5.3). □

7.8.7 Let X be a $\mathbf{U}$-small simplicial set. We shall consider $N(\underline{\mathrm{Hom}}(\mathbf{\Delta}^{\mathrm{op}}/X, set))$ as an ∞-category of cofibrant objects whose weak equivalences are the weak equivalences of the contravariant model category structure over X, and whose cofibrations are the monomorphisms of simplicial sets over X. It is homotopy $\mathbf{U}$-cocomplete. Composing the embedding functor (7.8.5.1) for $A = \mathbf{\Delta}/X$ with the functor $(\tau_X)_!\colon \underline{\mathrm{Hom}}(N(\mathbf{\Delta}/X)^{\mathrm{op}}, \mathcal{S}) \to \underline{\mathrm{Hom}}(X^{\mathrm{op}}, \mathcal{S})$, this defines the functor

$$\rho\colon N(\underline{\mathrm{Hom}}((\mathbf{\Delta}/X)^{\mathrm{op}}, set)) \to \underline{\mathrm{Hom}}(X^{\mathrm{op}}, \mathcal{S})\,. \tag{7.8.7.1}$$

There is an essentially commutative diagram of the following form

$$\begin{array}{ccc} \underline{\mathrm{Hom}}((\mathbf{\Delta}/X)^{\mathrm{op}}, set) & \xrightarrow{ho(\rho)} & ho(\underline{\mathrm{Hom}}(X^{\mathrm{op}}, \mathcal{S})) \\ {\scriptstyle\gamma}\downarrow & & \downarrow \\ RFib(X) & \xrightarrow{\sim} & LFib(X^{\mathrm{op}}) \end{array} \tag{7.8.7.2}$$

in which the functor γ is (the restriction to $\mathbf{U}$-small objects of) the localisation functor, the equivalence $RFib(X) \simeq LFib(X^{\mathrm{op}})$ is induced by the functor $(-)^{\mathrm{op}}$, and the right vertical map is the fully faithful functor of Theorem 5.4.5. To see this, let us consider a presheaf $F\colon (\mathbf{\Delta}/X)^{\mathrm{op}} \to set$. It corresponds to a morphism of simplicial sets $q\colon Y \to X$. We denote by $\tilde{q}\colon \mathbf{\Delta}/Y \to \mathbf{\Delta}/X$ the corresponding functor. We then have a Cartesian square

$$\begin{array}{ccc} (\mathbf{\Delta}/Y)^{\mathrm{op}} & \longrightarrow & set_{\bullet} \\ \downarrow{\scriptstyle\tilde{q}} & & \downarrow \\ (\mathbf{\Delta}/X)^{\mathrm{op}} & \longrightarrow & set \end{array} \tag{7.8.7.3}$$

which induces a canonical isomorphism $\mathbf{\Delta}/Y \simeq (\mathbf{\Delta}/X)/F$. In other words, formula (7.8.5.3) for $A = \mathbf{\Delta}/Y$ gives us a canonical isomorphism

$$\jmath(F) \simeq N(\tilde{q})_!(e)\,. \tag{7.8.7.4}$$

On the other hand, we have a commutative diagram

$$\begin{array}{ccc} N(\mathbf{\Delta}/Y) & \xrightarrow{N(\tilde{q})} & N(\mathbf{\Delta}/X) \\ {\scriptstyle\tau_Y}\downarrow & & \downarrow{\scriptstyle\tau_X} \\ Y & \xrightarrow{q} & X \end{array} \tag{7.8.7.5}$$

and thus isomorphisms

$$\rho(F) = (\tau_X)_!\jmath(F) \simeq (\tau_X)_!N(\tilde{q})_!(e) \simeq q_!(\tau_Y)_!(e) \xrightarrow{\sim} q_!(e)\,, \tag{7.8.7.6}$$

where the last isomorphism is justified by the fact that τ_Y^* is fully faithful, since

τ_Y is a (weak) localisation, by Proposition 7.3.15. Since these identifications are functorial, Proposition 6.1.14 shows that diagram (7.8.7.2) is essentially commutative, as claimed above.

Lemma 7.8.8 *The functor* (7.8.7.1) *is homotopy cocontinuous.*

Proof By virtue of Lemma 7.8.6, we only have to check that the functor $\jmath$ sends weak equivalences to invertible maps. This follows right away from the essential commutativity of diagram (7.8.7.2). □

The following theorem was conjectured by Nichols-Barrer [NB07], at least up to a slight reformulation; see Remark 5.4.11.

Theorem 7.8.9 *The functor ρ exhibits $\underline{\mathrm{Hom}}(X^{\mathrm{op}}, \mathcal{S})$ as the localisation of the ∞-category $N(\underline{\mathrm{Hom}}((\boldsymbol{\Delta}/X)^{\mathrm{op}}, set))$ by the weak equivalences of the contravariant model category structure over X.*

Proof By virtue of Theorem 7.7.9, the preceding lemma implies that ρ factors through a $\mathbf{U}$-small colimit preserving functor

(7.8.9.1) $$\bar{\rho}\colon L(\underline{\mathrm{Hom}}((\boldsymbol{\Delta}/X)^{\mathrm{op}}, set)) \to \underline{\mathrm{Hom}}(X^{\mathrm{op}}, \mathcal{S})\,.$$

By Theorem 7.6.10, it is sufficient to prove that the induced functor

(7.8.9.2) $$ho(\bar{\rho})\colon ho(L(\underline{\mathrm{Hom}}((\boldsymbol{\Delta}/X)^{\mathrm{op}}, set))) \to ho(\underline{\mathrm{Hom}}(X^{\mathrm{op}}, \mathcal{S}))$$

is an equivalence of categories. Formula (7.8.7.6) shows that this functor is essentially surjective. On the other hand, the essential commutativity of diagram (7.8.7.2) and the identification of $ho(L(N(\underline{\mathrm{Hom}}((\boldsymbol{\Delta}/X)^{\mathrm{op}}, set))))$ with the full subcategory of $RFib(X)$ which consists of $\mathbf{U}$-small right fibrations of codomain X show that the functor (7.8.9.2) fits in an essentially commutative diagram of the form

(7.8.9.3) $$\begin{array}{ccc} ho(L(\underline{\mathrm{Hom}}((\boldsymbol{\Delta}/X)^{\mathrm{op}}, set))) & \xrightarrow{ho(\bar{\rho})} & ho(\underline{\mathrm{Hom}}(X^{\mathrm{op}}, \mathcal{S})) \\ \downarrow & & \downarrow \\ RFib(X) & \xrightarrow{\sim} & LFib(X^{\mathrm{op}}) \end{array}$$

in which the two vertical maps are fully faithful and the lower horizontal one is an equivalence of categories. Therefore, the functor (7.8.9.2) is fully faithful. □

Remark 7.8.10 In particular, in the case where $X = \Delta^0$, the preceding theorem provides an equivalence of ∞-categories from the localisation of the category of **U**-small simplicial sets by weak homotopy equivalences with the ∞-category $\mathcal{S}$ of **U**-small ∞-groupoids:

(7.8.10.1) $$\mathbf{L}\rho\colon L(\underline{\mathrm{Hom}}(\boldsymbol{\Delta}^{\mathrm{op}}, set)) \xrightarrow{\sim} \mathcal{S}.$$

This will be extended to simplicial presheaves in Corollary 7.9.9 below. We remark that the full subcategory of $\underline{\mathrm{Hom}}(\mathcal{S}, \mathcal{S})$ which consists of equivalences of ∞-categories from $\mathcal{S}$ to itself is equivalent to Δ^0: by virtue of Theorem 6.3.13, this ∞-category is equivalent to the full subcategory of final objects in $\mathcal{S}$. In other words, there is a unique automorphism of $\mathcal{S}$: the identity. This means that, given any ∞-category C, the ∞-category of equivalences of ∞-categories $C \to \mathcal{S}$ is either empty or equivalent to Δ^0.

Remark 7.8.11 There is a model category structure on the category of topological spaces, whose fibrations are the Serre fibrations, and whose weak equivalences are the weak homotopy equivalences: the continuous maps $f\colon X \to Y$ such that the induced map $\pi_0(X) \to \pi_0(Y)$ is bijective and the induced map $\pi_i(X, x) \to \pi_i(Y, f(x))$ is an isomorphism of groups for all $i > 0$ and any base point $x \in X$; see [Qui67, chapter II, 3.1, theorem 1]. The adjunction

$$|-|\colon sSet \rightleftarrows Top : Sing$$

introduced in Example 1.2.7 is then a Quillen adjunction. By virtue of a theorem of Milnor, the unit map $X \to Sing(|X|)$ is a homotopy equivalence for any Kan complex X; see [GZ67, chapter VII, 3.1]. Furthermore, Quillen showed that a morphism of simplicial sets $f\colon X \to Y$ is a weak homotopy equivalence if and only if the induced map $|X| \to |Y|$ is a homotopy equivalence; see [Qui67, chapter II, 3.19, proposition 4]. This implies that the adjunction above induces an equivalence of localised ∞-categories

(7.8.11.1) $$L(\underline{\mathrm{Hom}}(\boldsymbol{\Delta}^{\mathrm{op}}, set)) \simeq L(top),$$

where *top* denotes the category of **U**-small topological spaces; see (the proof of) Proposition 7.1.14.

On the other hand, Theorem 6.3.13 for $A = \Delta^0$ means that any final object $\Delta^0 \to \mathcal{S}$ exhibits the ∞-category $\mathcal{S}$ of **U**-small ∞-groupoids as the free completion of the point Δ^0 by **U**-small colimits. Therefore, equivalences (7.8.10.1) and (7.8.11.1) mean that, for any ∞-category C with **U**-small colimits, and any object X in C, there is a unique functor $F\colon L(top) \to C$ which commutes with colimits equipped with an invertible map $F(e) \simeq X$ (e being the one-point space). This should be compared with the axiomata of Eilenberg and Steenrod characterising ordinary homology theory: we may take for C the localisation $D(Ab)$ of (the nerve of) the category of chain complexes of **U**-small abelian groups by quasi-isomorphisms, and $X = \mathbf{Z}$, seen as a chain complex concentrated in degree zero (since, by virtue of [Hov99, theorem 2.3.11], this is the localisation of a bicomplete Quillen model category structure, the ∞-category $D(Ab)$ has $\mathbf{U}$-small colimits and limits, by Proposition 7.7.4).

7.9 Homotopy Limits as Limits

We fix a universe $\mathbf{U}$.

7.9.1 Let C be an ∞-category with $\mathbf{U}$-small colimits, equipped with a subcategory of weak equivalences $W \subset C$ which turns C into a $\mathbf{U}$-cocomplete ∞-category of cofibrant objects (the cofibrations being all maps). We denote, as usual, by $\gamma\colon C \to L(C)$ the localisation by W. Theorem 7.7.9 expresses the fact that γ is a localisation which is compatible with $\mathbf{U}$-small colimits. Given any $\mathbf{U}$-small simplicial set X, the ∞-category $\underline{\mathrm{Hom}}(X, C)$ also has $\mathbf{U}$-small colimits, and one can form the subcategory $W(X, C)$ of fibrewise weak equivalences (see the Cartesian square (7.4.17.1)). We thus have a localisation functor by the fibrewise weak equivalences

$$\gamma_X\colon \underline{\mathrm{Hom}}(X, C) \to L(\underline{\mathrm{Hom}}(X, C))\,. \tag{7.9.1.1}$$

As above, $\underline{\mathrm{Hom}}(X, C)$ can be seen as a $\mathbf{U}$-cocomplete ∞-category of cofibrant objects, so that the functor γ_X commutes with $\mathbf{U}$-small colimits.

Proposition 7.9.2 *For any $\mathbf{U}$-small simplicial set X, the canonical functor*

$$L(\underline{\mathrm{Hom}}(X, C)) \to \underline{\mathrm{Hom}}(X, L(C))$$

is an equivalence of ∞-categories.

Proof The functor γ induces a functor

$$\gamma_* = \underline{\mathrm{Hom}}(X, \gamma)\colon\ \underline{\mathrm{Hom}}(X, C) \to \underline{\mathrm{Hom}}(X, L(C))$$

which sends fibrewise weak equivalences to (fibrewise) invertible maps, whence induces a functor

$$\bar{\gamma}_*\colon L(\underline{\mathrm{Hom}}(X, C)) \to \underline{\mathrm{Hom}}(X, L(C))\,.$$

Since the functor γ commutes with $\mathbf{U}$-small colimits, so does γ_*. Henceforth, by virtue of Theorem 7.7.9, the functor $\bar{\gamma}_*$ commutes with $\mathbf{U}$-small colimits as well. Moreover, for any ∞-category D with $\mathbf{U}$-small colimits, there is a canonical equivalence of ∞-categories

$$\underline{\mathrm{Hom}}_!(L(\underline{\mathrm{Hom}}(X, C)), D) \xrightarrow{\sim} \underline{\mathrm{Hom}}_{h!}(\underline{\mathrm{Hom}}(X, C), D)$$

(where the subscript $h!$ means that we consider the subcategory of functors which commute with **U**-small colimits and which send fibrewise weak equivalences to invertible maps). On the other hand, by Theorem 6.7.2, there is a canonical equivalence of ∞-categories of the form

$$\underline{\mathrm{Hom}}_!(\underline{\mathrm{Hom}}(X, C), D) \xrightarrow{\sim} \underline{\mathrm{Hom}}_!(C, \underline{\mathrm{Hom}}(X^{\mathrm{op}}, D)) .$$

It restricts to an equivalence of ∞-categories

$$\underline{\mathrm{Hom}}_{h!}(\underline{\mathrm{Hom}}(X, C), D) \xrightarrow{\sim} \underline{\mathrm{Hom}}_{h!}(C, \underline{\mathrm{Hom}}(X^{\mathrm{op}}, D)) .$$

Another use of Theorem 7.7.9 gives an equivalence

$$\underline{\mathrm{Hom}}_!(L(C), \underline{\mathrm{Hom}}(X^{\mathrm{op}}, D)) \xrightarrow{\sim} \underline{\mathrm{Hom}}_{h!}(C, \underline{\mathrm{Hom}}(X^{\mathrm{op}}, D)) ,$$

and Theorem 6.7.2 gives an equivalence of ∞-categories:

$$\underline{\mathrm{Hom}}_!(\underline{\mathrm{Hom}}(X, L(C)), D) \xrightarrow{\sim} \underline{\mathrm{Hom}}_!(L(C), \underline{\mathrm{Hom}}(X^{\mathrm{op}}, D)) .$$

Finally, this proves that the functor $\bar{\gamma}_*$ induces a canonical equivalence of ∞-categories

$$\underline{\mathrm{Hom}}_!(\underline{\mathrm{Hom}}(X, L(C)), D) \xrightarrow{\sim} \underline{\mathrm{Hom}}_!(L(\underline{\mathrm{Hom}}(X, C)), D)$$

for any ∞-category D with **U**-small colimits. This readily implies the proposition. □

Remark 7.9.3 Under the hypothesises of the proposition above, if, furthermore, C is the nerve of a small category D, then $L(C)$ is also the nerve of a category. Indeed, the localisation $L(C)$ is equivalent to a filtered colimit of **U**-small ∞-categories X_i. For each index i, the canonical functor $X_i \to ho(X_i)$ induces an equivalence of ∞-categories

$$L(\mathrm{Hom}(ho(X_i), C)) \simeq L(\underline{\mathrm{Hom}}(X_i, C))$$

whence an equivalence

$$\underline{\mathrm{Hom}}(ho(X_i), L(C)) \simeq \underline{\mathrm{Hom}}(X_i, L(C)) .$$

Since the formation of $ho(X)$ commutes with filtered colimits, passing to the homotopy limit gives an equivalence of ∞-categories

$$\underline{\mathrm{Hom}}(ho(L(C)), L(C)) \simeq \underline{\mathrm{Hom}}(L(C), L(C)) .$$

In particular, the identity of $L(C)$ factors up to J-homotopy through $ho(L(C))$, which implies that the canonical functor

$$L(C) \to N(ho(L(C)))$$

is an equivalence of ∞-categories. In other words, to create genuine ∞-categories out of ordinary categories from homotopy cocontinuous localisations, we need to work with general categories with weak equivalences and cofibrations.

Definition 7.9.4   An ∞-category with weak equivalences and cofibrations C is U-*hereditary* if it has the following properties:

(i) it is homotopy U-cocomplete (see Definition 7.7.2);

(ii) a map in C is a weak equivalence if and only if it its image is invertible in $L(C)$;

(iii) for any U-small category I, the ∞-category of functors $\underline{\mathrm{Hom}}(N(I), C)$ has a structure of an ∞-category with weak equivalences and cofibrations where the weak equivalences and cofibrations are defined fibrewise.

An ∞-category with weak equivalences and fibrations C is U-*hereditary* if C^{op} is a U-*hereditary* ∞-category with weak equivalences and cofibrations.

Remark 7.9.5   One may always force property (ii) above to hold. See Remark 7.5.22.

Example 7.9.6   Assume that C is a homotopy U-cocomplete ∞-category with weak equivalences and cofibrations. Suppose that C has a functorial factorisation: for each map $f : x \to y$ in C such that x is cofibrant, there is a commutative diagram

(7.9.6.1)
$$
\begin{array}{ccc}
 & t(f) & \\
i(f) \nearrow & & \searrow p(f) \\
x & \xrightarrow{\ \ f\ \ } & y
\end{array}
$$

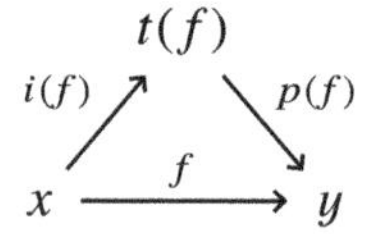

such that

(a) the map $i(f)$ is a cofibration, and the map $p(f)$ is a weak equivalence;

(b) the formation of diagram (7.9.6.1) is functorial in the sense that it comes from a functor from $\underline{\mathrm{Hom}}(\Delta^1, C)'$ to $\underline{\mathrm{Hom}}(\Delta^2, C)$, where $\underline{\mathrm{Hom}}(\Delta^1, C)'$ is the full subcategory of $\underline{\mathrm{Hom}}(\Delta^1, C)$ whose objects are the maps $f : x \to y$ with x cofibrant;

(c) a map of C is a weak equivalence if and only if it is invertible in $L(C)$.

Then C is a U-hereditary ∞-category with weak equivalences and cofibrations.

Remark 7.9.7   If C is a homotopy U-cocomplete ∞-category with weak equivalences and cofibrations, a sufficient condition for C to be U-hereditary is: for any sequence of maps

$$x_0 \xrightarrow{f_1} x_1 \xrightarrow{f_2} \cdots \to x_{n-1} \xrightarrow{f_n} x_n \to \cdots \tag{7.9.7.1}$$

such that each f_n is a cofibration (a trivial cofibration) between cofibrant objects, the colimit $x_\infty = \varinjlim_n x_n$ exists in C and the canonical map $x_0 \to x_\infty$ is a cofibration (a trivial cofibration, respectively).

In the case where C is the nerve of a category, the proof that this condition implies that C is $\mathbf{U}$-hereditary can be found in the work of Rădulescu-Banu [RB09, theorems 7.2.3 and 9.2.4]. The same methods apply for ∞-categories as well, using the first part of Section 7.4 (this is a good exercise we leave to the reader). We do not know if all homotopy $\mathbf{U}$-cocomplete ∞-categories with weak equivalences and cofibrations are $\mathbf{U}$-hereditary or not, though.

Theorem 7.9.8 *Let C be a $\mathbf{U}$-hereditary ∞-category with weak equivalences and cofibrations. Then, for any $\mathbf{U}$-small category I, the canonical functor*

$$L(\underline{\mathrm{Hom}}(N(I), C)) \to \underline{\mathrm{Hom}}(N(I), L(C))$$

is an equivalence of ∞-categories.

Proof The functor $\gamma\colon C \to L(C)$ induces a functor

$$\gamma_* = \underline{\mathrm{Hom}}(N(I), \gamma)\colon \underline{\mathrm{Hom}}(N(I), C) \to \underline{\mathrm{Hom}}(N(I), L(C)) \tag{7.9.8.1}$$

which sends fibrewise weak equivalences to (fibrewise) invertible maps, whence induces a functor

$$\bar{\gamma}_*\colon L(\underline{\mathrm{Hom}}(N(I), C)) \to \underline{\mathrm{Hom}}(N(I), L(C))\,. \tag{7.9.8.2}$$

Theorem 7.7.9 implies that the functor $\bar{\gamma}_*$ commutes with $\mathbf{U}$-small colimits.

By a multiple use of the dual version of Theorem 7.5.18, we see that we may assume that all objects of C are cofibrant. In particular, C has $\mathbf{U}$-small coproducts. There is a canonical functor

$$N(I^{\mathrm{op}}) \times C \to \underline{\mathrm{Hom}}(N(I), C) \tag{7.9.8.3}$$

defined by $(i, x) \mapsto i \otimes x = i_!(x)$. In other words, the left adjoint of the evaluation functor at i exists because we must have

$$i_!(x)(j) \simeq \coprod_{\mathrm{Hom}_I(i,j)} x \tag{7.9.8.4}$$

in C (to be more precise, we see that $i_!(x)$ is well defined in $C' = \underline{\mathrm{Hom}}(C, \mathcal{S})^{\mathrm{op}}$, where $\mathcal{S}'$ is the ∞-category of $\mathbf{V}$-small ∞-groupoids for a large universe $\mathbf{V}$, and the formula above shows that $i_!(x)$ actually belongs to C because the functor $(h_{C^{\mathrm{op}}})^{\mathrm{op}}\colon C \to C'$ is fully faithful and commutes with colimits). Using

Proposition 7.1.13, we see that inverting weak equivalences in the product $N(I^{\mathrm{op}}) \times C$ gives the product $N(I^{\mathrm{op}}) \times L(C)$. We thus get a canonical functor

$$\varphi\colon N(I) \times L(C) \to L(\underline{\mathrm{Hom}}(N(I), C))\,. \tag{7.9.8.5}$$

By virtue of Theorem 6.7.2, the latter extends uniquely into a functor

$$\varphi_!\colon \underline{\mathrm{Hom}}(N(I), L(C)) \to L(\underline{\mathrm{Hom}}(N(I), C)) \tag{7.9.8.6}$$

which commutes with **U**-small colimits. The composition $\bar{\gamma}_* \varphi_!$ is canonically isomorphic to the identity because $\bar{\gamma}_* \varphi$ is the canonical functor from $N(I) \times L(C)$ to $\underline{\mathrm{Hom}}(N(I), L(C))$. To prove the theorem, it is thus sufficient to prove that the functor $\varphi_!$ is an equivalence of ∞-categories. We claim that the functor $\bar{\gamma}_*$ is conservative. Indeed, using the calculus of fractions, it is sufficient to prove that any map $f\colon F \to G$ in $\underline{\mathrm{Hom}}(N(I), C)$ which is sent to an invertible map in $\underline{\mathrm{Hom}}(N(I), L(C))$ is a weak equivalence. But, under these assumptions, each map $F(i) \to G(i)$ is sent to an invertible map in $L(C)$, and thus f is a weak equivalence, by property (ii) of Definition 7.9.4. Furthermore, by virtue of Proposition 7.1.14, for any object i of I, the adjunction

$$i_!\colon C \rightleftarrows \underline{\mathrm{Hom}}(N(I), C) : i^* \tag{7.9.8.7}$$

induces an adjunction

$$i_!\colon L(C) \rightleftarrows L(\underline{\mathrm{Hom}}(N(I), C)) : i^*\,. \tag{7.9.8.8}$$

The explicit formula for $i_!$ also shows that, for any functor F from $N(I)$ to $L(C)$, there is a canonical invertible map

$$i_!(x) \simeq \varphi_!(i_!(x))\,. \tag{7.9.8.9}$$

This can be used to prove that the functor $\varphi_!$ is fully faithful as follows. Let F and G be two functors from $N(I)$ to $L(C)$. As seen in the proof of Lemma 6.7.7, there is a canonical invertible map of the form

$$\varinjlim_{x \to F(i)} i_!(x) \simeq F\,. \tag{7.9.8.10}$$

Here, the limit is indexed by $J = (N(I)^{\mathrm{op}} \times L(C))/F$ and Lemma 6.7.5 means that there exists a **U**-small simplicial set K as well as a final map $K \to J$, so that we can pretend that the indexing diagram of this limit is **U**-small, by Theorem

6.4.5. Therefore, we get

$$\begin{aligned}
\operatorname{Hom}(F,G) &\simeq \varprojlim_{x\to F(i)} \operatorname{Hom}(i_!(x),G)\\
&\simeq \varprojlim_{x\to F(i)} \operatorname{Hom}(x,G(i))\\
&\simeq \varprojlim_{x\to F(i)} \operatorname{Hom}(x,\varphi_!(G)(i))\\
&\simeq \varprojlim_{x\to F(i)} \operatorname{Hom}(i_!(x),\varphi_!(G))\\
&\simeq \varprojlim_{x\to F(i)} \operatorname{Hom}(\varphi_!(i_!(x)),\varphi_!(G))\\
&\simeq \operatorname{Hom}\Big(\varinjlim_{x\to F(i)} \varphi_!(i_!(x)),\varphi_!(G)\Big)\\
&\simeq \operatorname{Hom}(\varphi_!(F),\varphi_!(G))\,.
\end{aligned}$$

Finally, it remains to prove that the functor $\varphi_!$ is essentially surjective. For this purpose, we observe that, for any object i of I and any object x of $L(C)$, we have

(7.9.8.11) $$\operatorname{Hom}(i_!(x),F)\simeq \operatorname{Hom}(x,F(i))\simeq \operatorname{Hom}(i_!(x),\bar{\gamma}_*(F))\,.$$

This may be interpreted as follows. Let F be an object of $L(\underline{\operatorname{Hom}}(N(I),C))$. We define the relative slice

$$(N(I)^{\mathrm{op}}\times L(C))/F = (N(I)^{\mathrm{op}}\times L(C))\times_{L(\underline{\operatorname{Hom}}(N(I),C))} L(\underline{\operatorname{Hom}}(N(I),C))/F$$

and similarly, replacing $L(\underline{\operatorname{Hom}}(N(I),C))$ by $\underline{\operatorname{Hom}}(N(I),L(C))$, we define the relative slice $(N(I)^{\mathrm{op}}\times L(C))/\bar{\gamma}_*(F)$. The identification (7.9.8.11) means that the canonical comparison functor

(7.9.8.12) $$(N(I)^{\mathrm{op}}\times L(C))/F\to (N(I)^{\mathrm{op}}\times L(C))/\bar{\gamma}_*(F)$$

is a fibrewise equivalence between right fibrations over $N(I)^{\mathrm{op}}\times L(C)$. Whence the map (7.9.8.12) is an equivalence of ∞-categories, by Theorem 4.1.16. Lemma 6.7.5 means that there is final functor from a $\mathbf{U}$-small simplicial set to the relative slice $(N(I)^{\mathrm{op}}\times L(C))/\bar{\gamma}_*(F)$. Therefore, colimits of type $(N(I)^{\mathrm{op}}\times L(C))/F$ exist in any ∞-category with $\mathbf{U}$-small colimits. In other words, the following definition of $G\colon N(I)\to L(C)$ makes sense

(7.9.8.13) $$G=\varinjlim_{x\to F(i)} i_!(x)$$

and the maps $i_!(x)\to F$ induce a map $G\to F$. The fibrewise equivalence

(7.9.8.12) and formula (7.9.8.10) imply that

$$\varinjlim_{x \to F(i)} i_!(x) \simeq \bar{\gamma}_*(F)\,. \tag{7.9.8.14}$$

Since $\bar{\gamma}_*$ commutes with **U**-small colimits, this proves that the image of the map $G \to F$ by $\bar{\gamma}_*$ is invertible. Since the functor $\bar{\gamma}_*$ is conservative, this proves that $G \simeq F$. But we already know that the functor $\varphi_!$ is fully faithful and commutes with **U**-small colimits. Since the objects of type $i_!(x)$ are in the essential image of $\varphi_!$ (7.9.8.9), this proves that F is in the essential image of $\varphi_!$. □

Corollary 7.9.9 *For any* **U***-small category A, there is a canonical equivalence of* ∞*-categories from the localisation of the nerve of the category of* **U***-small simplicial presheaves on A by the class of fibrewise weak homotopy equivalences and the* ∞*-category* $\underline{\mathrm{Hom}}(N(A)^{\mathrm{op}}, \mathcal{S})$ *of presheaves of* **U***-small* ∞*-groupoids over* $N(A)$.

Proof This is a direct consequence of Theorem 7.8.9 for $X = \Delta^0$ and of Theorem 7.9.8 for $I = A^{\mathrm{op}}$. □

Remark 7.9.10 In the classical literature on homotopical algebra (including the second chapter of this very book), given a complete model category $\mathcal{C}$ one defines homotopy limits as the right adjoint of the functor

$$ho(\mathcal{C}) \to ho(\underline{\mathrm{Hom}}(I, \mathcal{C}))$$

induced by the constant diagram functor. But Theorem 7.9.8 means that the latter functor is obtained by applying the Boardman–Vogt construction to the constant functor

$$L(\mathcal{C}) \to \underline{\mathrm{Hom}}(N(I), L(\mathcal{C}))\,.$$

Since, as observed in Remark 6.1.5, the Boardman–Vogt construction is compatible with adjunctions, this means that homotopy limits are induced by the limit functor in $L(\mathcal{C})$.

Remark 7.9.11 Although it is stated and proved using a different language (using a particular explicit description of the localisation $L(C)$, as opposed to using its universal property only), the particular case of Theorem 7.9.8 where C is the nerve of a category is a result of Lenz [Len18, theorem 3.33]. Yet another approach to prove Theorem 7.9.8 consists in adapting the proof of Theorem 7.6.17 in order to deal with the nerve of any small direct category (as opposed to finite ones only), and then using the second assertion of Proposition 7.3.15 (and a little bit of work) to reduce the general case to the case of direct categories.

7.10 Mapping Spaces in Locally Small Localisations

We fix a universe $\mathbf{U}$.

Proposition 7.10.1 *Let C be an ∞-category with finite limits. If, for any object x of C, the category $ho(C/x)$ is locally $\mathbf{U}$-small, then C is locally $\mathbf{U}$-small.*

Proof Let us assume that each category $ho(C/x)$ is locally small. By virtue of Corollary 5.7.9, it is sufficient to prove that, for any map $f\colon x \to y$ in C, the homotopy groups $\pi_n(C(x, y), f)$ are $\mathbf{U}$-small for any integer $n > 0$.

We observe that there is a canonical invertible map

$$\mathrm{Hom}_{C/x}(x, x \times y) \simeq \mathrm{Hom}_C(x, y)$$

where $x \times y$ is considered as an object of C/x via the first projection $x \times y \to x$. Replacing C by C/x, we may thus assume that $x = e$ is the final object. One can form the loop space of y at f by forming the pull-back below.

$$\begin{array}{ccc} \Omega(y, f) & \longrightarrow & e \\ \downarrow & & \downarrow f \\ e & \xrightarrow{f} & y \end{array}$$

If we still denote by f the map $e \to \Omega(y, f)$ induced by the commutative square

$$\begin{array}{ccc} e & \longrightarrow & e \\ \downarrow & & \downarrow f \\ e & \xrightarrow{f} & y \end{array}$$

we may define the iterated loop spaces at the point f as follows:

$$\Omega^0(y, f) = y\,, \quad \Omega^{n+1}(y, f) = \Omega(\Omega^n(y, f), f) \quad \text{for } n \geq 0.$$

Since the functor $\mathrm{Hom}_C(e, -)$ commutes with limits, we have

$$\mathrm{Hom}_C(e, \Omega^n(y, f)) \simeq \Omega^n(\mathrm{Hom}_C(e, y), f)\,.$$

Hence, by Remark 6.6.11,

$$\pi_n(C(e, y), f) \simeq \mathrm{Hom}_{ho(C)}(e, \Omega^n(y, f))\,.$$

This proves that C is locally $\mathbf{U}$-small. □

Definition 7.10.2 Let C be an ∞-category with weak equivalences and fibrations. A *class of cofibrant objects* is a set Q of objects of C such that:

(i) for any object x in C, there exists a weak equivalence $x' \to x$ with x' in Q;

(ii) for any trivial fibration between fibrant objects $p\colon x \to y$ and any map $v\colon a \to y$ in C with $a \in Q$, there exists a map $u\colon a \to x$ such that v is a composition of u and p.

If such a class Q is fixed, the elements of Q will often be called *cofibrant objects*.

Remark 7.10.3 For any fibrant object x, the slice category C/x has a natural structure of ∞-category with weak equivalences and fibrations: the weak equivalences (fibrations) of C/x are the maps whose image in C is a weak equivalence (fibration). If Q is a class of cofibrant objects, then the objects of the form (a, u) in C/x, where $u\colon a \to x$ is a map with domain in Q, form a class of cofibrant objects of C/x.

Proposition 7.10.4 *Let C be an ∞-category with weak equivalences and fibrations equipped with a class of cofibrant objects. For any cofibrant objects a and any fibrant object y, the set $\mathrm{Hom}_{ho(L(C))}(a, y)$ is a quotient of $\mathrm{Hom}_{ho(C)}(a, y)$.*

Proof Let $w\colon a \to x$ be a weak equivalence with fibrant codomain. By virtue of Theorem 7.5.18, we may describe

$$\mathrm{Hom}_{ho(L(C))}(x, y) \simeq \mathrm{Hom}_{ho(L(C))}(a, y)$$

using the right calculus of fractions by trivial fibrations provided by Corollary 7.2.18. Therefore, any map f from a to y in $L(C)$ is obtained from a trivial fibration $s\colon z \to x$ and a map $p\colon z \to y$ such that $f = ps^{-1}w$ in $ho(L(C))$. But there is a map $v\colon a \to z$ such that w is the composition of s and v. Hence $s^{-1}w = v$ in $ho(L(C))$. Therefore, $f = pv$ in $ho(L(C))$. □

Corollary 7.10.5 *Let C be a locally $\mathbf{U}$-small ∞-category with weak equivalences and fibrations. If C has a class of cofibrant objects in the sense of Definition 7.10.2, then $L(C)$ is locally $\mathbf{U}$-small.*

Proof By virtue of Corollaries 5.7.8 and 7.6.13, of Remark 7.10.3 and of Proposition 7.10.1, it is sufficient to prove that the category $ho(L(C))$ is locally $\mathbf{U}$-small. By virtue of the preceding proposition, this follows from the fact that $ho(C)$ is locally $\mathbf{U}$-small. □

Proposition 7.10.6 *Let C be a locally $\mathbf{U}$-small ∞-category with $\mathbf{U}$-small limits. We fix an object x of C and consider a functor*

$$\mathrm{Map}(x, -)\colon C \to \mathcal{S}$$

which commutes with limits and which is equipped with a functorial bijection

$$\mathrm{Hom}_{ho(C)}(x, y) \simeq \pi_0(\mathrm{Map}(x, y))$$

for all objects y. Then there is a canonical equivalence

$$\mathrm{Hom}_C(x, y) \simeq \mathrm{Map}(x, y)$$

functorially in y.

Proof The identity of x defines an element of $\pi_0(\mathrm{Map}(x, x))$. Let us consider a representative of the latter $u\colon e \to \mathrm{Map}(x, x)$. By the Yoneda lemma, there is a unique functorial map

$$\mathrm{Hom}_C(x, y) \to \mathrm{Map}(x, y)$$

which sends 1_x to u. This map induces an isomorphism after we apply the functor π_0 for all y. Given an object a of $\mathcal{S}$ classifying a Kan complex A, we denote by y^a the limit of the A^{op}-indexed constant diagram with value y. We then have

$$\begin{aligned}
\mathrm{Hom}_{ho(\mathcal{S})}(a, \mathrm{Hom}_C(x, y)) &\simeq \mathrm{Hom}_{ho(\mathcal{S})}(e, \mathrm{Hom}_C(x, y)^a)\\
&\simeq \pi_0(\mathrm{Hom}_C(x, y^a))\\
&\simeq \pi_0(\mathrm{Map}(x, y^a))\\
&\simeq \mathrm{Hom}_{ho(\mathcal{S})}(e, \mathrm{Map}(x, y^a))\\
&\simeq \mathrm{Hom}_{ho(\mathcal{S})}(e, \mathrm{Map}(x, y)^a)\\
&\simeq \mathrm{Hom}_{ho(\mathcal{S})}(a, \mathrm{Map}(x, y))\,.
\end{aligned}$$

Therefore, the Yoneda lemma applied to $ho(\mathcal{S})$ implies the proposition. □

7.10.7 Given an object X in a small and locally **U**-small Quillen model category $\mathcal{C}$ with **U**-small limits whose category of fibrant objects is denoted by $\mathcal{C}_f$, a *mapping space* of X is a functor

$$\mathrm{Map}(X, -)\colon \mathcal{C}_f \to \underline{\mathrm{Hom}}(\mathbf{\Delta}^{\mathrm{op}}, set)_f$$

with the following property and structure.

(a) Its nerve is a homotopy continuous functor when we consider the structure of a **U**-complete ∞-category of fibrant objects on $N(\mathcal{C}_f)$ obtained in the obvious way from the weak equivalences and fibrations of $\mathcal{C}$, and similarly for $N(\underline{\mathrm{Hom}}(\mathbf{\Delta}^{\mathrm{op}}, set)_f)$ with respect to the Kan–Quillen model category structure.
(b) There is a given functorial bijection

$$\pi_0(\mathrm{Map}(X, Y)) \simeq \mathrm{Hom}_{ho(\mathcal{C})}(X, Y)$$

for any fibrant object Y of $\mathcal{C}$.

Remark 7.10.8 In practice, we have $\mathrm{Map}(X,Y)_0 = \mathrm{Hom}_{\mathcal{C}}(X,Y)$ (functorially in Y), in which case, if ever X is cofibrant, condition (b) is a consequence of condition (a).

Remark 7.10.9 There are systematic ways to construct mapping spaces as above. See [Hov99, corollary 5.4.4], for instance.

Example 7.10.10 In the case where $\mathcal{C} = \underline{\mathrm{Hom}}(\mathbf{\Delta}^{\mathrm{op}}, set)$, with the Joyal model category structure, such a mapping space is provided by $k(\underline{\mathrm{Hom}}(X,-))$ (this is an easy consequence of Theorem 3.5.11, of Corollary 3.6.4 and of Remark 3.6.5).

Example 7.10.11 In the case where $\mathcal{C} = \underline{\mathrm{Hom}}((\mathbf{\Delta}/C)^{\mathrm{op}}, set)$, with the contravariant model category structure on a $\mathbf{U}$-small simplicial set C, such a mapping space is provided by construction (4.1.12.1): this follows right away from Propositions 4.1.13 and 4.1.14 (see the proof of Theorem 4.1.16).

7.10.12 Let us assume that a mapping space of an object X is given, as above. We denote by

(7.10.12.1) $$\mathbf{R}\,\mathrm{Map}(X,-)\colon L(\mathcal{C}) \to \mathcal{S}$$

the functor obtained by composing the induced functor

$$L(\mathcal{C}_f) \to L(\underline{\mathrm{Hom}}(\mathbf{\Delta}^{\mathrm{op}}, set))$$

with a quasi-inverse of the equivalence of ∞-categories provided by Theorem 7.5.18 and with the equivalence of ∞-categories (7.8.10.1). Corollary 7.7.5 ensures that the functor $\mathbf{R}\,\mathrm{Map}(X,-)$ commutes with $\mathbf{U}$-small limits. Given an object Y of $\mathcal{C}$, we shall still denote by Y its image in $L(\mathcal{C})$. Since $\pi_0(\mathbf{R}\,\mathrm{Map}(X,X))$ is the set of endomorphisms of X in $ho(\mathcal{C})$, there is a map $u\colon e \to \mathbf{R}\,\mathrm{Map}(X,X)$ corresponding to the identity of X. By the Yoneda lemma applied to the ∞-category $L(\mathcal{C})^{\mathrm{op}}$, u determines a canonical map

(7.10.12.2) $$\mathrm{Hom}_{L(\mathcal{C})}(X,Y) \to \mathbf{R}\,\mathrm{Map}(X,Y)$$

which is functorial in Y, seen as an object of $L(\mathcal{C})$.

Proposition 7.10.13 *The map* (7.10.12.2) *is invertible.*

Proof We know that $L(\mathcal{C})$ is locally $\mathbf{U}$-small, by Corollary 7.10.5. Therefore, we may apply Proposition 7.10.6. □

Example 7.10.14 Given a Grothendieck universe $\mathbf{U}$, there is an ∞-category ∞-*Cat* of $\mathbf{U}$-small ∞-categories: we put $\infty\text{-}Cat = L(\mathcal{C})$ where $\mathcal{C}$ denotes the category of $\mathbf{U}$-small simplicial sets equipped with the restriction of the Joyal model category structure. Applying the preceding proposition to Example

7.10.10 we see that, for any **U**-small ∞-categories A and B, the ∞-groupoid $\mathrm{Hom}_{\infty\text{-}Cat}(A, B)$ canonically classifies the Kan fibration $k(\underline{\mathrm{Hom}}(A, B)) \to \Delta^0$ up to homotopy. Theorem 7.8.9 (possibly with the help of a little bit of calculus of fractions) implies that the ∞-category $\mathcal{S}$ can be canonically identified with the full subcategory of ∞-*Cat* which consists of **U**-small ∞-groupoids. The inclusion $\mathcal{S} \subset \infty\text{-}Cat$ has a left adjoint: it sends an ∞-category A to the ∞-groupoid $A^{-1}A$ (this may be deduced, by definition of localisation, from the identification of $\mathrm{Hom}_{\infty\text{-}Cat}$ as above, using Proposition 6.1.11). There is also a right adjoint, which sends an ∞-category A to the ∞-groupoid $k(A)$ (this follows from Corollary 3.5.3 and from Proposition 7.1.14, for instance).

7.11 Presentable ∞-Categories

Definition 7.11.1 A *left Bousfield localisation* of an ∞-category C is a functor $\gamma\colon C \to L$ which has a fully faithful right adjoint.

A *right Bousfield localisation* of an ∞-category C is a functor $\gamma\colon C \to L$ which has a fully faithful left adjoint.

Proposition 7.11.2 (Gabriel and Zisman) *For a functor $\gamma\colon C \to L$ (between ∞-categories), the following conditions are equivalent.*

(a) The functor γ has a right adjoint, and there is a subcategory W of C such that γ exhibits L as the localisation of C by W.
(b) The functor γ has a right adjoint, and exhibits L as the localisation of C by $W = \gamma^{-1}(k(L))$.
(c) The functor γ is a left Bousfield localisation.

Furthermore, if γ is a left Bousfield localisation, there is a canonical left calculus of fractions of $\gamma^{-1}(k(L))$ in C.

Proof Proposition 7.1.18 tells us that condition (b) follows from (c). Conversely, it follows from Proposition 7.1.17 that (a) implies (c). Since (b) clearly implies (a), this shows that the three conditions are equivalent. Let us assume that γ is a left Bousfield localisation. There is a right adjoint r of γ. Let y be an object of C. One checks that the full subcategory of $y\backslash C$ whose objects are maps $y \to y'$ such that $\gamma(y) \to \gamma(y')$ is invertible has a final object, namely the canonical map $y \to r(\gamma(y))$ (using Proposition 4.3.10, for instance). Therefore, by Theorem 6.4.5, for any object x of C, there is a canonical invertible map of the form

$$\varinjlim_{y \to y'} \mathrm{Hom}_C(x, y') \xrightarrow{\sim} \mathrm{Hom}_C(x, r(\gamma(y)))$$

where the colimit is indexed by the category of maps $y \to y'$ in $\gamma^{-1}(k(L))$. Since the functor $r\gamma$ obviously takes the elements of $\gamma^{-1}(k(L))$ to invertible maps in L, this proves the last assertion. □

7.11.3 We fix a universe $\mathbf{U}$. Let A be a $\mathbf{U}$-small simplicial set and V a $\mathbf{U}$-small subcategory of $\widehat{A} = \underline{\mathrm{Hom}}(A^{\mathrm{op}}, \mathcal{S})$. We denote by $L_V(\widehat{A})$ the cocontinuous localisation of $\widehat{A}$ by V (obtained by applying the construction of Remark 7.7.10 for $C = \widehat{A}$ and $S = V$, the weak equivalences of $\widehat{A}$ being the invertible maps, and the cofibrations being all maps).

A presheaf $F\colon A^{\mathrm{op}} \to \mathcal{S}$ will be called *V-local* if, for any map $X \to Y$ in V, the induced map

$$\mathrm{Hom}_{\widehat{A}}(Y, F) \to \mathrm{Hom}_{\widehat{A}}(X, F)$$

is invertible. We shall write $\widehat{A}_V$ for the full subcategory of V-local objects in $\widehat{A}$.

Proposition 7.11.4 *The localisation functor $\gamma\colon \widehat{A} \to L_V(\widehat{A})$ has a right adjoint, and thus exhibits $L_V(\widehat{A})$ as a left Bousfield localisation of $\widehat{A}$. Furthermore, the essential image of the right adjoint of γ consists of the V-local objects, so that γ induces an equivalence of ∞-categories*

$$\widehat{A}_V \simeq L_V(\widehat{A}) .$$

Proof We denote by $\mathcal{C}$ the category of $\mathbf{U}$-small simplicial sets over A, equipped with the contravariant model structure, which we regard as a category of cofibrant objects. We have a canonical equivalence of ∞-categories $L(\mathcal{C}) \simeq \widehat{A}$, by Theorem 7.8.9. Using the calculus of fractions, we see that we may assume that V is the image of a subcategory V_0 of $\mathcal{C}$. We can then apply the construction of Remark 7.7.10 for $C = \mathcal{C}$ and $S = V_0$ and get a new category of cofibrant objects $\mathcal{C}_0 = C_S$. Theorem 7.7.9 implies that $L(\mathcal{C}_0)$ and $L_V\widehat{A}$ satisfy the same universal property and are thus canonically equivalent. Using the factorisation of maps into a monomorphism followed by a map with the right lifting property with respect to monomorphisms, we may assume that all the maps in V_0 are monomorphisms. There is then a model category structure on the category of $\mathbf{U}$-small simplicial sets over A, which we call $\mathcal{D}$: it is obtained by applying Theorem 2.4.19 to the homotopical structure generated by the cylinder object associated to the interval J and to the smallest class of J-anodyne extensions containing right anodyne extensions over A as well as maps of the form

$$\begin{array}{ccc} R & \xrightarrow{\;i\;} & S \\ & \searrow \quad \swarrow & \\ & A & \end{array}$$

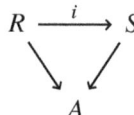

with i in V_0. By inspection of the construction of this class (Example 2.4.13), we see that all its elements are weak equivalences of $\mathcal{C}_0$ (because the latter is homotopy cocomplete, by definition). Therefore, we deduce from Remark 2.4.41 that $L(\mathcal{C}_0) = L(\mathcal{D})$. By virtue of Propositions 2.4.40 and 5.3.1, we see that we may apply Theorem 7.5.30 for $C = \mathcal{C}$, $D = \mathcal{D}$ (seen as a category with weak equivalences and fibrations), and $F = U$ the identity functor, which proves that the localisation functor $\gamma = \mathbf{L}F : L(\mathcal{C}) \to L(\mathcal{C}_0) = L(\mathcal{D})$ has a right adjoint.

Let r be a right adjoint of γ. Since r sends any map in V to an invertible map, it is clear that $r(F)$ is V-local for any object F of $L_V(\widehat{A})$. Conversely, let F be a V-local object. Then the functor

$$\mathrm{Hom}_{\widehat{A}}(-, F)^{\mathrm{op}} : \widehat{A} \to \mathcal{S}^{\mathrm{op}}$$

is (homotopy) cocontinuous and sends maps in V to invertible ones, so that, by virtue of Theorem 7.7.9, it factors through $L_V(\widehat{A})$. In particular, it sends any map of $\widehat{A}$ which becomes invertible in $L_V(\widehat{A})$ to an invertible map. This means that the putative left calculus of fractions defined by the identity of F is actually a left calculus of fractions (i.e. that the identity of F is a right calculus of fractions on the opposite of $\widehat{A}$). Therefore, the left calculus of fractions formula (i.e. the dual version of Theorem 7.2.8) gives us an identification of the form

$$\mathrm{Hom}(X, F) \simeq \mathrm{Hom}(X, r(\gamma(F)))$$

functorially in X. This implies that the map $\eta : F \to r(\gamma(F))$ is invertible, by the Yoneda lemma. Henceforth the functor r induces an equivalence of categories from $L_V(\widehat{A})$ to $\widehat{A}_V$, by Theorem 3.9.7. The restriction of γ to $\widehat{A}_V$ remains a left adjoint of this equivalence of ∞-categories, which implies that it induces an equivalence of ∞-categories as well. □

Definition 7.11.5 An ∞-category $\mathcal{C}$ is *presentable* if there exists a **U**-small ∞-category A and a **U**-small subcategory V of $\widehat{A}$, as well as an equivalence of ∞-categories

$$L_V(\widehat{A}) \simeq \mathcal{C}$$

(with the notation of paragraph 7.11.3).

Remark 7.11.6 This notion strongly depends on a (possibly implicit) choice of a universe **U**. Since the formation of $\widehat{A}$ is compatible with weak categorical equivalences, we may as well have taken an arbitrary **U**-small simplicial set A in this definition.

Proposition 7.11.7 *Any presentable ∞-category is locally* $\mathbf{U}$*-small and has* $\mathbf{U}$*-small limits as well as* $\mathbf{U}$*-small colimits.*

Proof Given a simplicial set A, we already know that $\widehat{A}$ is locally $\mathbf{U}$-small (Proposition 5.7.3) and has $\mathbf{U}$-small limits (Corollary 6.3.8) as well as $\mathbf{U}$-small colimits (Proposition 6.2.12). Therefore, Proposition 6.2.17 tells us that any left Bousfield localisation of $\widehat{A}$ has $\mathbf{U}$-small limits and colimits. Proposition 7.11.4 thus tells us that any presentable ∞-category also has this property. □

Proposition 7.11.8 *Let $F\colon \mathcal{C} \to \mathcal{D}$ be a functor. We assume that $\mathcal{C}$ is presentable and that $\mathcal{D}$ is locally small as well as cocomplete. Then F is cocontinuous if and only if it has a right adjoint.*

Proof We choose a small ∞-category A, a small subcategory V of $\widehat{A}$, and an equivalence of ∞-categories

$$\varphi\colon L_V(\widehat{A}) \to \mathcal{C}\,.$$

Composing with the localisation functor $\gamma\colon \widehat{A} \to L_V(\widehat{A})$, we thus have a functor

$$G = F\varphi\gamma\colon \widehat{A} \to \mathcal{D}\,.$$

By virtue of Theorem 7.7.9, the functor F is cocontinuous if and only if G has the same property, and by Proposition 6.3.9, the functor G is cocontinuous if and only if it has a right adjoint. Moreover, Proposition 7.11.4 provides the existence of a right adjoint of γ. By virtue of Proposition 6.1.8, it is thus sufficient to prove that, if G has a right adjoint U, then F has a right adjoint. But it is clear that the image of any object of $\mathcal{D}$ by U is V-local. Therefore, G restricts to an adjunction between the full subcategory $\widehat{A}_V$ of V-local presheaves and $\mathcal{D}$. Using the equivalence $\widehat{A}_V \simeq \mathcal{C}$ induced by φ and by Proposition 7.11.4, this provides a right adjoint to F. □

Proposition 7.11.9 *Let $\mathcal{C}$ be a presentable ∞-category. Then, for any* $\mathbf{U}$*-small simplicial set X, the ∞-category $\underline{\mathrm{Hom}}(X, \mathcal{C})$ is presentable.*

Proof Let A be a $\mathbf{U}$-small ∞-category and V a $\mathbf{U}$-small subcategory of $\widehat{A}$. In $\widehat{X^{\mathrm{op}} \times A} = \underline{\mathrm{Hom}}(X, \widehat{A})$, we define the set of maps V_A as the one whose elements are of the form

$$x_!(i)\colon x_!(R) \to x_!(S)$$

with $i\colon R \to S$ in V and x an object of X. The formula

$$\mathrm{Hom}(x_!(S), F) \simeq \mathrm{Hom}(S, F(x))$$

implies that the V_X-local presheaves on $X^{\mathrm{op}} \times A$ are precisely the functors $F \colon X \to \widehat{A}$ such that $F(x)$ is V-local for any object x of X. In other words,

$$\widehat{X^{\mathrm{op}} \times A}_{V_X} = \underline{\mathrm{Hom}}(X, \widehat{A}_V)\,.$$

We conclude with Proposition 7.11.4. □

Proposition 7.11.10 *Any cocontinuous localisation of any presentable ∞-category by a* **U***-small subcategory is presentable.*

Proof Let $\mathcal{C}$ be a presentable ∞-category, and S a **U**-small subcategory of $\mathcal{C}$. We choose an equivalence of $\mathcal{C}$ with $L_V(\widehat{A})$, the cocontinuous localisation of presheaves on a **U**-small ∞-category A by a small subcategory V of $\widehat{A}$. Let S be a **U**-small subcategory of $\mathcal{C}$. We want to prove that the cocontinuous localisation $L_S(\mathcal{C})$ (as constructed in Remark 7.7.10) is presentable. Using the calculus of fractions, we see that we may assume that S is the image of a small subcategory T of $\widehat{A}$. We then have a canonical equivalence of ∞-categories

$$L_{V \cup T}(\widehat{A}) \simeq L_S(\mathcal{C})\,,$$

simply because both ∞-categories have the same universal property, by Theorem 7.7.9. Therefore, $L_S(\mathcal{C})$ is presentable. □

Proposition 7.11.11 *The (nerve of the) category of* **U***-small sets is presentable.*

Proof We define $S^{-1} = \varnothing$ to be the initial object of $\mathcal{S}$, and then, by induction, we define S^n by forming the following coCartesian square for $n \geq 0$ (where, as usual, e is the final object).

$$\begin{array}{ccc} S^{n-1} & \longrightarrow & e \\ \downarrow & & \downarrow \\ e & \longrightarrow & S^n \end{array}$$

Let $V = \{S^1 \to e\}$. By Proposition 7.11.4, the cocontinuous localisation of $\mathcal{S}$ by V is equivalent to the full subcategory of $\mathcal{S}$ which consists of V-local objects, i.e. of ∞-groupoids x such that the map $S^1 \to e$ induces an equivalence:

$$x \simeq \mathrm{Hom}(e, x) \xrightarrow{\sim} \mathrm{Hom}(S^1, x)\,.$$

By virtue of Proposition 7.8.3, it is sufficient to prove that an object x of $\mathcal{S}$ is discrete if and only if it is V-local. The Cartesian squares

$$\begin{array}{ccc} \mathrm{Hom}(S^n, x) & \longrightarrow & x \\ \downarrow & & \downarrow \\ x & \longrightarrow & \mathrm{Hom}(S^{n-1}, x) \end{array}$$

imply that x is V-local if and only if the maps $S^n \to e$ induce invertible maps

$$x \simeq \mathrm{Hom}(e, x) \xrightarrow{\sim} \mathrm{Hom}(S^n, x), \quad \text{for } n > 0.$$

Using Proposition 3.8.11, we see that this implies that the canonical map $x \to \pi_0(x)$ is invertible, hence that x is discrete. On the other hand, the coCartesian squares of sets (due to the fact that π_0 is a left adjoint)

$$\begin{array}{ccc} \pi_0(S^{n-1}) & \longrightarrow & e \\ \downarrow & & \downarrow \\ e & \longrightarrow & \pi_0(S^n) \end{array}$$

provide an inductive proof that $\pi_0(S^n) \simeq e$ for all $n > 0$. This implies that any discrete object is V-local. □

Corollary 7.11.12 *For any* **U***-small category* C*, the (nerve of the) category of presheaves of* **U***-small sets on* C *is presentable.*

Combining the proofs of Propositions 7.11.9 and 7.11.11, we also obtain the following.

Proposition 7.11.13 *A presentable* ∞*-category is equivalent to the nerve of a category if and only if it is the cocontinuous localisation of the nerve of a category of presheaves of* **U***-small sets on a* **U***-small category by a* **U***-small set of maps.*

Definition 7.11.14 A *presentable* category is a category whose nerve is presentable. A *combinatorial* model category is a cofibrantly generated model category whose underlying category is presentable.

Remark 7.11.15 Any reasonable algebraic structure defines a presentable category. For instance, the models of any Lawvere theory define a presentable category. Therefore, the category of groups, and the category of abelian groups, the category of rings all are presentable. Similarly, the category of complexes of left modules over a ring is presentable. This means that most of the classical model structures on categories of simplicial groups, simplicial abelian groups, simplicial rings or of complexes of left modules over a ring, are combinatorial. For instance, the model category of complexes of left modules over a ring, whose weak equivalences are the quasi-isomorphisms and whose fibrations are the degreewise surjections, is combinatorial. Obviously, all the model structures provided by Theorem 2.4.19 are combinatorial. A systematic treatment of combinatorial model structures is given in Lurie's book [Lur09, section A.2.6]. For the record, we mention the following important characterisation of presentable ∞-categories.

Theorem 7.11.16 (Dugger) *For any combinatorial model category $\mathcal{C}$, the localisation $L(\mathcal{C})$ is a presentable ∞-category.*

Proof Given a **U**-small category C and a **U**-small set S of morphisms of **U**-small simplicial presheaves on C, Dugger defines in [Dug01b] a model structure UC/S, the universal homotopy theory in which C embeds and in which the elements of S are weak equivalences. More precisely, taking into account Corollary 7.9.9 and Proposition 7.10.13, one can describe $L(UC/S)$ as the full subcategory of $\widehat{C}$ which consists of S-local objects (see [Dug01b, definition 5.4]). Hence Proposition 7.11.4 for $A = C$ and $V = S$ shows that Dugger's ∞-category $L(UC/S)$ is presentable. Now, [Dug01a, theorem 1.1] provides a Quillen equivalence between a model category of the form UC/S and $\mathcal{C}$. Combined with Corollary 7.6.11, this shows that $L(\mathcal{C})$ is equivalent to $L(UC/S) \simeq \widehat{C}_S$, and thus that $L(\mathcal{C})$ is presentable. □

Remark 7.11.17 The proof of Proposition 7.11.4 is also a proof that, up to equivalence, any presentable ∞-category is of the form $L(\mathcal{C})$ for some combinatorial model category $\mathcal{C}$. Therefore, the preceding theorem means that an ∞-category is presentable if and only if it is the localisation of (the nerve of) a combinatorial model category. The fundamental role of presentable ∞-categories goes back to early work of Simpson [Sim99], where one can find various intrinsic characterisations of presentable ∞-categories; for a modern treament, see [Lur09, theorem 5.5.1.1]. Lurie took the theory to another level in his famous monograph [Lur09], where a whole chapter is devoted to them. In particular, Lurie developed powerful tools to manipulate presentable ∞-categories, which are at the heart of his theory of higher topoi (and of quite a few other things).

In particular, the ∞-category ∞-*Cat* of **U**-small ∞-categories of Example 7.10.14 is presentable. However, such a property does not only come from an abstract result such as Dugger's theorem above: Rezk's homotopy theory of complete Segal spaces provides an explicit description of ∞-*Cat* as a presentable ∞-category: the proof of this fact is essentially the content of Joyal and Tierney's paper [JT07], which is in fact a particular case of a general process, explained later by Ara [Ara14]. Such an explicit presentation of ∞-*Cat* may be used to determine it axiomatically; see [Toë05].

Bibliography

[AFR17] David Ayala, John Francis, and Nick Rozenblyum, *A stratified homotopy hypothesis*, preprint, arXiv: 1502.01713v4, 2017.

[Ara14] Dimitri Ara, *Higher quasi-categories vs higher Rezk spaces*, J. K-Theory **14** (2014), no. 3, 701–749. MR 3350089

[Bar16] Clark Barwick, *On the algebraic K-theory of higher categories*, J. Topol. **9** (2016), no. 1, 245–347. MR 3465850

[Ber07] Julia E. Bergner, *Three models for the homotopy theory of homotopy theories*, Topology **46** (2007), no. 4, 397–436. MR 2321038

[Ber12] ———, *Homotopy limits of model categories and more general homotopy theories*, Bull. Lond. Math. Soc. **44** (2012), no. 2, 311–322. MR 2914609

[Ber18] ———, *The homotopy theory of $(\infty, 1)$-categories*, London Mathematical Society Student Texts, vol. 90, Cambridge University Press, Cambridge, 2018.

[BHH17] Ilan Barnea, Yonatan Harpaz, and Geoffroy Horel, *Pro-categories in homotopy theory*, Algebr. Geom. Topol. **17** (2017), no. 1, 567–643. MR 3604386

[BK12a] Clark Barwick and Daniel M. Kan, *A characterization of simplicial localization functors and a discussion of DK equivalences*, Indag. Math. (N.S.) **23** (2012), no. 1–2, 69–79. MR 2877402

[BK12b] ———, *Relative categories: another model for the homotopy theory of homotopy theories*, Indag. Math. (N.S.) **23** (2012), no. 1–2, 42–68. MR 2877401

[BM11] Andrew J. Blumberg and Michael A. Mandell, *Algebraic K-theory and abstract homotopy theory*, Adv. Math. **226** (2011), no. 4, 3760–3812. MR 2764905

[BR13] Julia E. Bergner and Charles Rezk, *Reedy categories and the Θ-construction*, Math. Z. **274** (2013), no. 1–2, 499–514. MR 3054341

[Bro73] Kenneth S. Brown, *Abstract homotopy theory and generalized sheaf cohomology*, Trans. Amer. Math. Soc. **186** (1973), 419–458. MR 0341469

[BS10] John C. Baez and Michael Shulman, *Lectures on n-categories and cohomology*, Towards higher categories, IMA Volumes in Mathematics and its Applications, vol. 152, Springer, New York, 2010, pp. 1–68. MR 2664619

[BV73] J. Michael Boardman and Rainer M. Vogt, *Homotopy invariant algebraic structures on topological spaces*, Lecture Notes in Mathematics, vol. 347, Springer, Berlin, 1973. MR 0420609

[Cis06] Denis-Charles Cisinski, *Les préfaisceaux comme modèles des types d'homotopie*, Astérisque (2006), no. 308, xxiv+392. MR 2294028

[Cis08] *Propriétés universelles et extensions de Kan dérivées*, Theory Appl. Categ. **20** (2008), no. 17, 605–649. MR 2534209

[Cis09] *Locally constant functors*, Math. Proc. Cambridge Philos. Soc. **147** (2009), no. 3, 593–614. MR 2557145

[Cis10] *Catégories dérivables*, Bull. Soc. Math. France **138** (2010), no. 3, 317–393. MR 2729017

[Cis16] *Catégories supérieures et théorie des topos*, Astérisque (2016), no. 380, Séminaire Bourbaki, vol. 2014/2015, Exp. No. 1097, 263–324. MR 3522177

[DHKS04] William G. Dwyer, Philip S. Hirschhorn, Daniel M. Kan, and Jeffrey H. Smith, *Homotopy limit functors on model categories and homotopical categories*, Mathematical Surveys and Monographs, vol. 113, American Mathematical Society, Providence, RI, 2004. MR 2102294

[DK80a] William G. Dwyer and Daniel M. Kan, *Calculating simplicial localizations*, J. Pure Appl. Algebra **18** (1980), no. 1, 17–35. MR 578563

[DK80b] *Function complexes in homotopical algebra*, Topology **19** (1980), no. 4, 427–440. MR 584566

[DK80c] *Simplicial localizations of categories*, J. Pure Appl. Algebra **17** (1980), no. 3, 267–284. MR 579087

[DK87] *Equivalences between homotopy theories of diagrams*, Algebraic topology and algebraic K-theory (Princeton, NJ, 1983), Annals of Mathematics Studies, vol. 113, Princeton University Press, Princeton, NJ, 1987, pp. 180–205. MR 921478

[DS11] Daniel Dugger and David I. Spivak, *Mapping spaces in quasi-categories*, Algebr. Geom. Topol. **11** (2011), no. 1, 263–325. MR 2764043

[Dug01a] Daniel Dugger, *Combinatorial model categories have presentations*, Adv. Math. **164** (2001), no. 1, 177–201. MR 1870516

[Dug01b] *Universal homotopy theories*, Adv. Math. **164** (2001), no. 1, 144–176. MR 1870515

[Fre70] Peter Freyd, *Homotopy is not concrete*, The Steenrod algebra and its applications (Proc. Conf. to celebrate N. E. Steenrod's sixtieth birthday, Battelle Memorial Inst., Columbus, Ohio, 1970), Lecture Notes in Mathematics, vol. 168, Springer, Berlin, 1970, pp. 25–34. MR 0276961

[GJ99] Paul G. Goerss and John F. Jardine, *Simplicial homotopy theory*, Progress in Mathematics, vol. 174, Birkhäuser, Basel, 1999. MR 1711612

[GZ67] Pierre Gabriel and Michel Zisman, *Calculus of fractions and homotopy theory*, Ergebnisse der Mathematik und ihrer Grenzgebiete, Band 35, Springer, New York, 1967. MR 0210125

[HM15] Gijs Heuts and Ieke Moerdijk, *Left fibrations and homotopy colimits*, Math. Z. **279** (2015), no. 3–4, 723–744. MR 3318247

[Hor16] Geoffroy Horel, *Brown categories and bicategories*, Homology Homotopy Appl. **18** (2016), no. 2, 217–232. MR 3575996

[Hov99] Mark Hovey, *Model categories*, Mathematical Surveys and Monographs, vol. 63, American Mathematical Society, Providence, RI, 1999. MR 1650134

[Hoy17] Marc Hoyois, *The six operations in equivariant motivic homotopy theory*, Adv. Math. **305** (2017), 197–279. MR 3570135

[Joy02] André Joyal, *Quasi-categories and Kan complexes*, J. Pure Appl. Algebra **175** (2002), no. 1–3, 207–222, Special volume celebrating the 70th birthday of Professor Max Kelly. MR 1935979

[Joy08a] André Joyal, *Notes on quasi-categories*, preprint, 2008.

[Joy08b] *The theory of quasi-categories and its applications*, preprint, 2008.

[JT07] André Joyal and Myles Tierney, *Quasi-categories vs Segal spaces*, Categories in algebra, geometry and mathematical physics, Contemporary Mathematics, vol. 431, American Mathematical Society, Providence, RI, 2007, pp. 277–326. MR 2342834

[Kan58] Daniel M. Kan, *Adjoint functors*, Trans. Amer. Math. Soc. **87** (1958), 294–329. MR 0131451

[Kap17] Krzysztof Kapulkin, *Locally cartesian closed quasi-categories from type theory*, J. Topol. **10** (2017), no. 4, 1029–1049. MR 3743067

[KL16] Krzysztof Kapulkin and Peter Lefanu Lumsdaine, *The simplicial model of univalent foundations (after Voevodsky)*, preprint, arXiv: 1211.2851, 2016.

[KM08] Bruno Kahn and Georges Maltsiniotis, *Structures de dérivabilité*, Adv. Math. **218** (2008), no. 4, 1286–1318. MR 2419385

[Lei14] Tom Leinster, *Basic category theory*, Cambridge Studies in Advanced Mathematics, vol. 143, Cambridge University Press, Cambridge, 2014. MR 3307165

[Len18] Tobias Lenz, *Homotopy (pre-)derivators of cofibration categories and quasi-categories*, preprint, arXiv: 1712.07845; to appear in Algebr. Geom. Topol., 2018.

[LMG15] Zhen Lin Low and Aaron Mazel-Gee, *From fractions to complete Segal spaces*, Homology Homotopy Appl. **17** (2015), no. 1, 321–338. MR 3350085

[Lur09] Jacob Lurie, *Higher topos theory*, Annals of Mathematics Studies, vol. 170, Princeton University Press, Princeton, NJ, 2009. MR 2522659

[Lur17] *Higher algebra*, Harvard University, Cambridge, MA, 2017.

[Mal05a] Georges Maltsiniotis, *La théorie de l'homotopie de Grothendieck*, Astérisque (2005), no. 301, vi+140. MR 2200690

[Mal05b] *Structures d'asphéricité, foncteurs lisses, et fibrations*, Ann. Math. Blaise Pascal **12** (2005), no. 1, 1–39, English translation available at arXiv: 0912.2432. MR 2126440

[Mal07] *Le théorème de Quillen, d'adjonction des foncteurs dérivés, revisité*, C. R. Math. Acad. Sci. Paris **344** (2007), no. 9, 549–552. MR 2323740

[Mal12] *Carrés exacts homotopiques et dérivateurs*, Cah. Topol. Géom. Différ. Catég. **53** (2012), no. 1, 3–63. MR 2951712

[Mei16] Lennart Meier, *Fibration categories are fibrant relative categories*, Algebr. Geom. Topol. **16** (2016), no. 6, 3271–3300. MR 3584258

[MG16a] Aaron Mazel-Gee, *Goerss–Hopkins obstruction theory via model ∞-categories*, Ph.D. thesis, University of California, Berkeley, 2016.

[MG16b] ———, *Quillen adjunctions induce adjunctions of quasicategories*, New York J. Math. **22** (2016), 57–93. MR 3484677

[Mor06] Fabien Morel, *Homotopy theory of schemes*, SMF/AMS Texts and Monographs, vol. 12, American Mathematical Society, Providence, RI; Société Mathématique de France, Paris, 2006, Translated from the 1999 French original by James D. Lewis. MR 2257774

[NB07] Joshua Paul Nichols-Barrer, *On quasi-categories as a foundation for higher algebraic stacks*, Ph.D. thesis, Massachusetts Institute of Technology, 2007.

[Nik11] Thomas Nikolaus, *Algebraic models for higher categories*, Indag. Math. (N.S.) **21** (2011), no. 1–2, 52–75. MR 2832482

[NRS18] Hoang Kim Nguyen, George Raptis, and Christoph Schrade, *Adjoint functor theorems for ∞-categories*, preprint, arXiv: 1803.01664, 2018.

[Nui16] Joost Nuiten, *Localizing ∞-categories with hypercovers*, preprint, arXiv: 1612.03800, 2016.

[Qui67] Daniel G. Quillen, *Homotopical algebra*, Lecture Notes in Mathematics, vol. 43, Springer, Berlin, 1967. MR 0223432

[Qui73] Daniel Quillen, *Higher algebraic K-theory. I*, Algebraic K-theory I, Lecture Notes in Mathematics, vol. 341, Springer, Berlin, 1973, pp. 85–147. MR 0338129

[RB09] Andrei Rădulescu-Banu, *Cofibrations in homotopy theory*, preprint, arXiv: math/0610009v4, 2009.

[Rez01] Charles Rezk, *A model for the homotopy theory of homotopy theory*, Trans. Amer. Math. Soc. **353** (2001), no. 3, 973–1007. MR 1804411

[Rez10a] ———, *A Cartesian presentation of weak n-categories*, Geom. Topol. **14** (2010), no. 1, 521–571. MR 2578310

[Rez10b] ———, *Correction to "A Cartesian presentation of weak n-categories"*, Geom. Topol. **14** (2010), no. 4, 2301–2304. MR 2740648

[Rie14] Emily Riehl, *Categorical homotopy theory*, New Mathematical Monographs, vol. 24, Cambridge University Press, Cambridge, 2014. MR 3221774

[Rie17] ———, *Category theory in context*, Aurora: Dover Modern Math Originals, Dover Publications, Mineola, NY, 2017.

[RV16] Emily Riehl and Dominic Verity, *Homotopy coherent adjunctions and the formal theory of monads*, Adv. Math. **286** (2016), 802–888. MR 3415698

[RV17a] ———, *Fibrations and Yoneda's lemma in an ∞-cosmos*, J. Pure Appl. Algebra **221** (2017), no. 3, 499–564. MR 3556697

[RV17b] ———, *Kan extensions and the calculus of modules for ∞-categories*, Algebr. Geom. Topol. **17** (2017), no. 1, 189–271. MR 3604377

[SGA72] Séminaire de Géométrie Algébrique du Bois-Marie 1963–1964 (SGA 4), Dirigé par M. Artin, A. Grothendieck, et J. L. Verdier. Avec la collaboration de N. Bourbaki, P. Deligne et B. Saint-Donat. *Théorie des topos et cohomologie étale des schémas. Tome 1: Théorie des topos*, Lecture Notes in Mathematics, vol. 269, Springer, Berlin, 1972. MR 0354652

[Shu08] Michael Shulman, *Set theory for category theory*, preprint, arXiv: 0810.1279v2, 2008.

[Sim99] Carlos Simpson, *A Giraud-type characterization of the simplicial categories associated to closed model categories as ∞-pretopoi*, preprint, arXiv: math/9903167, 1999.

[Sim12] ———, *Homotopy theory of higher categories*, New Mathematical Monographs, vol. 19, Cambridge University Press, Cambridge, 2012. MR 2883823

[Spi10] Markus Spitzweck, *Homotopy limits of model categories over inverse index categories*, J. Pure Appl. Algebra **214** (2010), no. 6, 769–777. MR 2580656

[Szu16] Karol Szumiło, *Homotopy theory of cofibration categories*, Homology Homotopy Appl. **18** (2016), no. 2, 345–357. MR 3576003

[Szu17] ———, *Homotopy theory of cocomplete quasicategories*, Algebr. Geom. Topol. **17** (2017), no. 2, 765–791. MR 3623671

[Tho79] Robert W. Thomason, *Homotopy colimits in the category of small categories*, Math. Proc. Cambridge Philos. Soc. **85** (1979), no. 1, 91–109. MR 510404

[Toë05] Bertrand Toën, *Vers une axiomatisation de la théorie des catégories supérieures*, K-Theory **34** (2005), no. 3, 233–263. MR 2182378

[TV05] Bertrand Toën and Gabriele Vezzosi, *Homotopical algebraic geometry. I. Topos theory*, Adv. Math. **193** (2005), no. 2, 257–372. MR 2137288

[TV08] ———, *Homotopical algebraic geometry. II. Geometric stacks and applications*, Mem. Amer. Math. Soc. **193** (2008), no. 902, x+224. MR 2394633

[Uni13] The Univalent Foundations Program, *Homotopy type theory – univalent foundations of mathematics*, Institute for Advanced Study (IAS), Princeton, NJ, 2013. MR 3204653

[VK14] Ya. Varshavskiĭ and D. Kazhdan, *The Yoneda lemma for complete Segal spaces*, Funktsional. Anal. i Prilozhen. **48** (2014), no. 2, 3–38, translation in Funct. Anal. Appl. **48** (2014), no. 2, 81–106. MR 3288174

[Wei99] Michael Weiss, *Hammock localization in Waldhausen categories*, J. Pure Appl. Algebra **138** (1999), no. 2, 185–195. MR 1689629

Notation

Index

For EU product safety concerns, contact us at Calle de José Abascal, 56–1°, 28003 Madrid, Spain or eugpsr@cambridge.org.

www.ingramcontent.com/pod-product-compliance
Ingram Content Group UK Ltd.
Pitfield, Milton Keynes, MK11 3LW, UK
UKHW040004250426
470322UK00005BA/15

* 9 7 8 1 1 0 8 4 7 3 2 0 0 *